A Complete O Level Mathematics

2nd Edition

Also from Stanley Thornes (Publishers) Ltd.

A Greer C.S.E. MATHEMATICS—The Core Course (*1978*)

A Greer C.S.E. MATHEMATICS—The Special Topics (*1978*)

A Greer ARITHMETIC FOR COMMERCE (*1976*)

Bostock and Chandler APPLIED MATHEMATICS I (*1975*) (*for "A" G.C.E. Courses*)

Bostock and Chandler APPLIED MATHEMATICS II (*1976*) (*for "A" G.C.E. Courses*)

Bostock and Chandler PURE MATHEMATICS I (*1978*) (*for "A" G.C.E. Courses*)

A Complete O Level Mathematics

2nd Edition

A. Greer

Senior Lecturer, Gloucester Technical College

Stanley Thornes (Publishers) Ltd.

First published in 1973 by Stanley Thornes (Publishers) Ltd
EDUCA House, Liddington Estate,
Leckhampton Road, CHELTENHAM GL53 0DN

Reprinted with corrections and additional chapters 1974
Reprinted 1975
Second edition (with additional material and further corrections) 1976
Reprinted 1977
Reprinted 1978 (Twice)
Reprinted 1979

ISBN 0 85950 001 2

Printed in Great Britain by The Pitman Press, Bath

Preface

This is a revision course in mathematics intended for those students seeking an 'O' level GCE qualification either in an organised educational institution situation, or through the medium of a correspondence or similar guided tuition course arrangement. The contents include all the major topics required by the syllabuses of the various examining boards.

A basic knowledge of fractions and decimals has been assumed, but otherwise references to all the elementary aspects of mathematics have been included whenever this seemed to be desirable. Because this is a revision book the sections on Arithmetic, Algebra, Geometry, and Trigonometry have been dealt with separately, The emphasis is on a simple approach and a teacher and his class may work, if desired, through the book chapter by chapter.

A very large number of graded examples of the type commonly found in examination papers has been provided in each section of the work, together with the answers, and students will find it possible to work from relatively easy problems to a situation in which confidence in dealing with harder questions is acquired. At the end of most chapters there are sets of objective type questions which have been called 'self tests'. This type of question is growing in popularity and many of the examining boards are now including them in their examination papers. By working through these self tests the student will be prepared for any objective type question which occurs in his examination paper.

Finally I would like to thank Dr. A. S. Hicks for his invaluable work in correcting the manuscript, and checking the answers to problems, and also for his suggestions for the improvement of the text.

A. Greer

Gloucester 1973

Note to the Second Impression

The production of the first reprint of the book has allowed an opportunity to add two further chapters, viz. Chapter 38 on Arithmetical and Geometrical Series, and Chapter 39 on Compound interest. The section of Proofs of Theorems has been modified and a number of small corrections inserted.

A. Greer

Gloucester 1974

Note to the Second Edition

In response to several requests I have included the topics of radian measure and stock, shares and bankruptcy in this edition. Also, because of the effects of inflation, some of the topics on money such as exchange rates etc. have become dated and these topics have been altered to bring them up to date.

I would like to thank all those who have written with constructive suggestions for improving the book and to those who have taken the trouble to report errors in the script and in the answers.

A. Greer

Gloucester 1976

Contents

Chapter 1 Arithmetical Operations

SEQUENCE OF OPERATIONS

Numbers are often combined in a series of arithmetical operations. When this happens a definite sequence must be observed.

1) Brackets are used if there is any danger of ambiguity. The contents of the bracket must be evaluated before performing any other operation. Thus

$2 \times (7+4) = 2 \times 11 = 22$
$15-(8-3) = 15-5 = 10$

2) Multiplication and division must be done before addition and subtraction. Thus

$5 \times 8 + 7 = 40 + 7 = 47$ (*not* 5×15)
$8 \div 4 + 9 = 2 + 9 = 11$ (*not* $8 \div 13$)
$5 \times 4 - 12 \div 3 + 7 = 20 - 4 + 7 = 27 - 4 = 23$

Exercise 1

Find values for the following:

1) $3+5\times 2$

2) $3\times 6-8$

3) $7\times 5-2+4\times 6$

4) $8\div 2+3$

5) $7\times 5-12\div 4+3$

6) $11-9\div 3+7$

7) $3\times(8+7)$

8) $2+8\times(3+6)$

9) $17-2\times(5-3)$

10) $11-12\div 4+3\times(6-2)$

OPERATIONS WITH FRACTIONS

The sequence of operations when dealing with fractions is the same as that used with whole numbers. It is, in order:

1) Work out brackets;
2) Multiply;
3) Divide;
4) Add;
5) Subtract.

Examples

1) Simplify $\frac{1}{5} \div \left(\frac{1}{3} \div \frac{1}{2}\right)$

$$\frac{1}{5} \div \left(\frac{1}{3} \div \frac{1}{2}\right) = \frac{1}{5} \div \left(\frac{1}{3} \times \frac{2}{1}\right) = \frac{1}{5} \div \frac{2}{3}$$

$$= \frac{1}{5} \times \frac{3}{2} = \frac{3}{10}$$

2) Simplify $(3\frac{3}{4} \times 1\frac{1}{5}) - 2\frac{5}{9}$

$$(3\tfrac{3}{4} \times 1\tfrac{1}{5}) - 2\tfrac{5}{9} = \left(\frac{15}{4} \times \frac{6}{5}\right) - \frac{23}{9}$$

$$= \frac{9}{2} - \frac{23}{9} = \frac{81-46}{18} = \frac{35}{18} = 1\tfrac{17}{18}$$

3) Simplify $\dfrac{2\frac{4}{5}+1\frac{1}{4}}{3\frac{3}{5}} - \dfrac{5}{16}$

$$2\tfrac{4}{5} + 1\tfrac{1}{4} = 3\,\frac{16+5}{20} = 3 + \frac{21}{20} = 4\tfrac{1}{20}$$

$$\frac{4\frac{1}{20}}{3\frac{3}{5}} - \frac{5}{16} = (4\tfrac{1}{20} \div 3\tfrac{3}{5}) - \frac{5}{16}$$

$$= \left(\frac{81}{20} \div \frac{18}{5}\right) - \frac{5}{16} = \left(\frac{81}{20} \times \frac{5}{18}\right) - \frac{5}{16}$$

$$= \frac{9}{8} - \frac{5}{16} = \frac{13}{16}$$

Exercise 2

Simplify the following:

1) $3\frac{3}{14}+\left(1\frac{1}{49}\times\frac{7}{10}\right)$

2) $\frac{1}{4}\div\left(\frac{1}{8}\times\frac{2}{5}\right)$

3) $1\frac{2}{3}\div\left(\frac{3}{5}\div\frac{9}{10}\right)$

4) $(1\frac{7}{8}\times 2\frac{2}{5})-3\frac{2}{3}$

5) $\dfrac{2\frac{2}{3}+1\frac{1}{5}}{5\frac{4}{5}}$

6) $3\frac{2}{3}\div\left(\frac{2}{3}+\frac{4}{5}\right)$

7) $\dfrac{5\frac{3}{5}-(3\frac{1}{2}\times\frac{2}{3})}{2\frac{1}{3}}$

8) $\frac{2}{5}\times\left(\frac{2}{3}-\frac{1}{4}\right)+\frac{1}{2}$

9) $\dfrac{3\frac{9}{16}\times\frac{4}{9}}{2+(6\frac{1}{4}\times 1\frac{1}{5})}$

10) $\dfrac{\frac{5}{9}-\frac{7}{15}}{1-(\frac{5}{9}\times\frac{7}{15})}$

APPROXIMATE NUMBERS

One way of obtaining an approximate number is to state the number to so many *significant figures*. The rules are as follows:

1) *If the first figure to be discarded is 5 or more the previous figure is increased by 1.* Thus,

7·392 5 = 7·393 correct to 4 significant figures
= 7·39 correct to 3 significant figures
= 7·4 correct to 2 significant figures

2) *Zeros must be kept to show the position of the decimal point, or to indicate that the zero is a significant figure.*

Thus

14 384 = 14 380 correct to 4 significant figures
= 14 000 correct to 2 significant figures
0·0887 = 0·089 correct to 2 significant figures
= 0·09 correct to 1 significant figure
29·8604 = 29·860 correct to 5 significant figures
= 29·9 correct to 3 significant figures

A second way of obtaining an approximate number is to state the number to so many *decimal places*. The number of figures after the decimal point determines the number of decimal places. Thus

0·0493 = 0·049 correct to 3 decimal places
= 0·05 correct to 2 decimal places
11·273 = 11·27 correct to 2 decimal places
= 11·3 correct to 1 decimal place

When the answer to a division is required correct to, say, three decimal places the division should be continued to four decimal places and the answer then expressed correct to three decimal places. Similar considerations apply to answers required to a certain number of significant figures.

Exercise 3

1) Write down the following numbers correct to the number of significant figures stated.

a) 24·865 82
 i) to 6 ii) to 4 iii) to 2
b) 0·008 357 1
 i) to 4 ii) to 3 iii) to 2
c) 4·978 48
 i) to 5 ii) to 3 iii) to 1
d) 21·987 to 2
e) 35·603 to 4
f) 28 387 617
 i) to 5 ii) to 2
g) 4·149 76
 i) to 5 ii) to 4 iii) to 3
h) 9·204 8 to 3

2) Write down the following numbers correct to the number of decimal places stated.

a) 5·149 87
 i) to 4 ii) to 3 iii) to 2
b) 35·285
 i) to 2 ii) to 1
c) 0·004 977
 i) to 5 ii) to 4 iii) to 3
d) 8·407 6
 i) to 3 ii) to 2 iii) to 1
e) 0·852 9
 i) to 3 ii) to 2

3) Find the value of:

a) 18·76 ÷ 14·3 correct to 2 decimal places
b) 0·0396 ÷ 2·51 correct to 2 significant figures

c) $7{\cdot}21 \div 0{\cdot}038$ correct to 3 significant figures
d) $5{\cdot}13 \times 7{\cdot}34$ correct to 2 decimal places
e) $27 \times 13 \div 17$ correct to 3 significant figures

ROUGH CHECKS FOR CALCULATIONS

The worst mistake that can be made in a calculation is that of misplacing the decimal point. To place it wrongly, even by one place, makes the answer ten times too large or ten times too small. To prevent this occurring it is always worth while doing a rough check by using approximate numbers. When doing rough checks always try to select numbers so that they multiply out easily or so that they cancel.

Examples

1) $0{\cdot}23 \times 0{\cdot}56$
For a rough estimate we will take $0{\cdot}2 \times 0{\cdot}6$

Product roughly $= 0{\cdot}2 \times 0{\cdot}6 = 0{\cdot}12$
Correct product $= 0{\cdot}128\,8$

(The rough check shows that the answer is 0·1288 and *not* 1·288 or 0·012 88)

2) $173{\cdot}3 \div 27{\cdot}8$
For a rough estimate we will take $180 \div 30$

answer roughly $= 180 \div 30 = 6$
Correct answer $= 173{\cdot}3 \div 27{\cdot}8 = 6{\cdot}23$

(Note that the rough answer and the correct answer are of the same order)

3) $\dfrac{8{\cdot}198 \times 19{\cdot}56 \times 30{\cdot}82 \times 0{\cdot}198}{6{\cdot}52 \times 3{\cdot}58 \times 0{\cdot}823}$

answer roughly $= \dfrac{8 \times 20 \times 30 \times 0{\cdot}2}{6 \times 4 \times 1} = 40$

Correct answer $= 50{\cdot}94$

(Although there is a big difference between the approximate and correct answers, the rough check shows that the answer is 50·94 and not 509·4 or 5·094)

Exercise 4

Find rough checks for the following:

1) $223{\cdot}6 \times 0{\cdot}0048$

2) $32{\cdot}7 \times 0{\cdot}259$

3) $0{\cdot}682 \times 0{\cdot}097 \times 2{\cdot}38$

4) $78{\cdot}41 \div 23{\cdot}78$

5) $0{\cdot}059 \div 0{\cdot}00268$

6) $33{\cdot}2 \times 29{\cdot}6 \times 0{\cdot}031$

7) $\dfrac{0{\cdot}728 \times 0{\cdot}00625}{0{\cdot}0281}$

8) $\dfrac{27{\cdot}5 \times 30{\cdot}52}{11{\cdot}3 \times 2{\cdot}73}$

SELF TEST 1

In questions 1 to 15 state the letter or letters corresponding to the correct answer (or answers).

1) $3 + 7 \times 4$ is equal to
a 40 b 31 c 84

2) $6 \times 5 - 2 + 4 \times 6$ is equal to
a 52 b 42 c 18

3) $7 \times 6 - 12 \div 3 + 1$ is equal to
a 40 b 39 c 21

4) $17 - 2 \times (6 - 4)$ is equal to
a 30 b 1 c 13

5) $3 \times \left(\frac{1}{2} - \frac{1}{3}\right)$ is equal to

a $1\frac{1}{2} - \frac{1}{3}$ b $\frac{1}{2}$ c $3 \times \frac{1}{2} - 3 \times \frac{1}{3}$

6) $\frac{5}{8} + \frac{1}{2} \times \frac{1}{4}$ is equal to

a $\frac{9}{32}$ b $\frac{3}{4}$ c $\frac{9}{16}$

7) The answer to $13{\cdot}0063 \times 1000$ is
a 13·063 b 1300·63 c 130·063
d 13 006·3

8) The answer to $1{\cdot}5003 \div 100$ is

a 0·015003 b 0·15003 c 0·153
d 1·53

9) The correct answer to $18{\cdot}2 \times 0{\cdot}013 \times 5{\cdot}21$ is

a 12·32686 b 123·2686
c 1·232686 d 0·1232686

10) The exact value of $\frac{121 \times 125 \times 81}{7425}$ is

a 166 b 165 c 1650 d 1660

11) The exact value of $\frac{312 \times 11 \times 19}{39}$ is

a 16172 b 16171 c 1672
d 1671

12) The number 158861 correct to 2 significant figures is

a 15 b 150000 c 16
d 160000

13) The number 0·081778 correct to 3 significant figures is

a 0·082 b 0·081 c 0·0818
d 0·0817

14) The number 0·075538 correct to 2 decimal places is

a 0·076 b 0·075 c 0·07
d 0·08

15) The number $0{\cdot}1\dot{6}$ correct to 4 significant figures is

a 0·1616 b 0·1617 c 0·1667
d 0·1666

In questions 16 to 30 decide whether the answer should be "true" or "false".
Is it true or false that

16) $7+5\times3=22$ -----

17) $7-2\times3=15$ -----

18) $6\times5-3+2\times7=26$ -----

19) $10\div2+3=8$ -----

20) $7\times5-12\div4+2=10$ -----

21) $3\times5-12\div(3+1)=12$ -----

22) $18-10\div2+3\times(5-2)=22$ -----

23) $\frac{27}{9}$ is the same as $27\div9$ -----

24) $\frac{28}{35}\div\frac{7}{16}$ is the same as $\frac{28}{35}\times\frac{16}{7}$ -----

25) $3\times\left(\frac{1}{2}+\frac{1}{4}\right)$ is the same as $3\times\frac{3}{4}$ -----

26) $\frac{5}{8}+\frac{1}{4}\times\frac{1}{2}$ is the same as $\frac{7}{8}\times\frac{1}{2}$ -----

27) 20963 correct to 2 significant figures is 21000 -----

28) 0·09983 correct to 3 significant figures is 0.099 -----

29) 0·007891 correct to 3 decimal places is 0·008 -----

30) $0{\cdot}1\dot{8}$ is equal to 0·189 correct to 3 significant figures -----

Chapter 2 **Decimal Currency**

THE BRITISH SYSTEM

The British system of decimal currency uses the pound as the basic unit. The only sub-unit used is the penny such that

100 pence = 1 pound

The abbreviation p is used for pence and the abbreviation £ is used for pounds.
A decimal point is used to separate the pounds from the pence, for example

£3·58 meaning 3 pounds and 58 pence.

There are two ways of expressing amounts less than £1. For example, 74 pence may be written as £0·74 or as 74p; 5 pence may be written as £0·05 or as 5p.
The smallest unit used is the half-penny which is always written as the fraction $\frac{1}{2}$. Thus £5·17$\frac{1}{2}$ means 5 pounds and 17$\frac{1}{2}$ pence. 53$\frac{1}{2}$ pence is written as either £0·53$\frac{1}{2}$ or 53$\frac{1}{2}$p.
Note that $\frac{1}{2}$p = £0·005.
The addition of sums of money is done in almost the same way as the addition of decimals. The exception occurs with the half-penny.

Examples

1) add together £3·78, £5·23 and £8·19

£ 3·78
£ 5·23
£ 8·19
£17·20 = total

2) add together £2.58$\frac{1}{2}$, £3·27$\frac{1}{2}$ and £5·73$\frac{1}{2}$

£ 2·58$\frac{1}{2}$
£ 3·27$\frac{1}{2}$
£ 5·73$\frac{1}{2}$
£11·59$\frac{1}{2}$ = total

3) add together 39p, 84$\frac{1}{2}$p and £1.73

39p = £0·39
84$\frac{1}{2}$p = £0·84$\frac{1}{2}$
£1·73
Total = £2.96$\frac{1}{2}$

Multiplication and division are similar to the multiplication and division of decimals.

Examples

1) Find the cost of 23 articles if each costs 27p.

Now 27p = £0·27
23 × £0.27 = £6.21

Hence the cost of 23 articles at 27p each is £6.21

2) Find the total cost of 15 articles costing 12p each, 20 articles costing 7$\frac{1}{2}$p each and 9 articles costing 18$\frac{1}{2}$p each.

15 at 12p each = 15 × £0.12 = £1·80
20 at 7$\frac{1}{2}$p each = 20 × £0·07$\frac{1}{2}$ = £1·50
9 at 18$\frac{1}{2}$p each = 9 × £0.18$\frac{1}{2}$ = £1·66$\frac{1}{2}$
Total = £4·96$\frac{1}{2}$

Hence the total cost of the articles is £4·96$\frac{1}{2}$

3) If 127 identical articles cost £14·60$\frac{1}{2}$, find the cost of each article.
The cost of each article = £14·60$\frac{1}{2}$ ÷ 127

Now £14·60$\frac{1}{2}$ is a mixture of decimals and fractions. We may make the amount wholly decimal if we remember that the $\frac{1}{2}$ represents $\frac{1}{2}$ of £0·01 = £0·005. Thus £14·60$\frac{1}{2}$ becomes £14·605 and the calculation becomes £14·605 ÷ 127

```
      0·115
127)14·605
     12 7
     ----
      1 90
      1 27
      ----
        635
        635
        ---
        ...
```

Hence the cost of each article is £0·115 or £0·11$\frac{1}{2}$ or 11$\frac{1}{2}$p.

Exercise 5

1) Express the following as pence:
a) £0·71 b) £0·63 c) £0·58$\frac{1}{2}$

2) Express the following as pounds:
a) 6p b) 72$\frac{1}{2}$p c) 97$\frac{1}{2}$p

3) Add the following sums of money together.
a) £2·15, £3·28, £4·63
b) £8·28, £109·17, £27·98, £70·15
c) £0·17$\frac{1}{2}$, £1·63$\frac{1}{2}$, £1·71, £1·90$\frac{1}{2}$
d) 82p, 71p, 82p
e) 17$\frac{1}{2}$p, 27p, 81$\frac{1}{2}$p, 74$\frac{1}{2}$p

4) Find the cost of 10 articles costing 11p each.

5) Find the cost of 100 articles costing 9$\frac{1}{2}$p each.

6) Find the cost of 72 articles costing 12$\frac{1}{2}$p each.

7) Find the total cost of each of the following:
a) 18 articles costing 7$\frac{1}{2}$p each, 20 articles costing 9$\frac{1}{2}$p each and 7 articles costing 11$\frac{1}{2}$p each.
b) 16 articles costing 5p each, 17 articles costing 7$\frac{1}{2}$p each and 3 articles costing 18$\frac{1}{2}$p each.
c) 5 articles costing 27$\frac{1}{2}$p each, 8 articles costing 60$\frac{1}{2}$p each, 9 articles costing 40$\frac{1}{2}$p each and 11 articles costing 19$\frac{1}{2}$p each.

8) Find the cost of the following:
a) 10 articles costing £18·11 each.
b) 10 articles costing £7·03$\frac{1}{2}$ each.
c) 100 articles costing £0·91$\frac{1}{2}$ each.
d) 100 articles costing £2·17$\frac{1}{2}$ each.

9) Find the cost of the following:
a) 21 articles costing £3·16 each.
b) 53 articles costing £0·23$\frac{1}{2}$ each.
c) 79 articles costing £1·87$\frac{1}{2}$ each.

10) Find the total cost of each of the following:
a) 12 articles costing £1·70 each, 8 articles costing £2.75 each and 9 articles costing £3·71 each.
b) 17 articles costing £0·30$\frac{1}{2}$ each, 2 articles costing £17·16 each and 7 articles costing £1·29$\frac{1}{2}$ each.
c) 5 articles costing £2·16$\frac{1}{2}$ each, 6 articles costing £4·28$\frac{1}{2}$ each, 9 articles costing £3·18$\frac{1}{2}$ and 4 articles costing £11·98 each.

11) If 12 identical articles cost £1·56, find the cost of each article.

12) If 241 identical articles cost £51·81$\frac{1}{2}$, find the cost of each article.

13) If 5000 articles cost £6525, find the cost of each article.

14) If 125 articles cost £270·62$\frac{1}{2}$ find the cost of each article.

FOREIGN EXCHANGE

Every country has its own monetary system. If there is to be trade and travel between any two countries there must be a rate at which money of one country can be converted into money of the other country.

Examples

1) If £1 = 120 Spanish pesetas, find to the nearest half-penny the value in British money of 1000 pesetas.

Since 120 pesetas = £1

$$1000 \text{ pesetas} = £\,\frac{1000}{120} = £8{\cdot}33$$

FOREIGN MONETARY SYSTEMS & EXCHANGE RATES 1975

COUNTRY	MONETARY UNIT	EXCHANGE RATE
Belgium	100 centimes = 1 franc	BF 83·25 = £1
France	100 centimes = 1 franc	F 9·20 = £1
Germany	100 pfennig = 1 mark	DM 5·38 = £1
Greece	100 lepta = 1 drachma	DR 68 = £1
Italy	100 centesimi = 1 lira	Lit 1490 = £1
Spain	100 céntimos = 1 peseta	Ptas 120 = £1
Sweden	100 öre = 1 krona	Skr 9·10 = £1
Switzerland	100 centimes = 1 franc	SWF 5·54 = £1
United States	100 cents = 1 dollar	$2·03 = £1

2) A tourist changes travellers cheques for £40·00 into French currency at 11·85 francs to the £1. He spends 350 francs and changes the remainder back into sterling at 11·80 francs to the £1. Find to the nearest penny how much he receives.

£40·00 = 40 × 11·85 francs = 474 francs
money spent = 350 francs
money remaining = 474 francs − 350 francs
= 124 francs

$$= £\frac{124}{11{\cdot}80} = £10{\cdot}51$$

Exercise 6

(Where necessary give the answers to 3 places of decimals.)
Using the exchange rates given above find:

1) The number of German marks equivalent to £15·00.

2) The number of Spanish pesetas equivalent to £25·00.

3) The number of U.S. dollars equivalent to £32·00.

4) The number of £ equivalent to 223 U.S. dollars.

5) The number of £ equivalent to 8960 Italian lire.

6) The number of £ equivalent to 560 Swedish Krona.

7) A transistor radio set costs £26·50 in the United Kingdom. An American visitor wishes to purchase a set but wishes to pay in U.S. dollars. Taking £1 = $2·38 find the equivalent price in dollars.

8) A tourist changes travellers cheques for £50·00 into Greek currency at 72·00 drachma to the £1. He spends 3220 drachma and changes the remainder back into sterling at 72·80 drachma to the £1. How much did the tourist receive?

9) Calculate the rate of exchange if a bank exchanges
a) 20 Canadian dollars for £7·71
b) 50 Swiss francs for £4·38
c) 1000 Swedish Krona for £90

10) In 1969 German marks were quoted as DM 9·60 = £1, whilst French francs were quoted as F 11·85 = £1. How many German marks could be obtained for 760 francs?

11) In 1969 the value of the £1 was changed from $2·80 to $2·40. An American visitor had to change £40 into dollars at the new rate. What was the tourist's loss in dollars and cents.

12) A person on holiday in France changes £40·00 into francs at a rate of 11·80 francs to the £1. His hotel expenses were 36 francs per day for eight days and his other expenses were 152 francs. On returning to England he changed what francs he had left into sterling at 11·60 francs to the £1. How much did he receive to the nearest penny.

SELF TEST 2

In questions 1 to 15 decide whether the answer is "true" or "false".

1) In the new British system of decimal currency, 100p = £1. - - - - -

2) The abbreviation NP is used for pence. - - - - -

3) A decimal point is used to separate the pounds from the pence. - - - - -

4) £1·78 means one pound and 78 pence. - - - - -

5) 36 pence may be written as £·36. - - - - -

6) One half penny is equal to £0·005 - - - - -

7) 58½ pence plus 27 pence is equal to £0·85½. - - - - -

8) 73½ pence may be written £0·735. - - - - -

9) The cost of 5 articles each costing 7½ pence is £0·37½. - - - - -

10) If 10 articles cost £3.75 the cost of each article is 37½p. - - - - -

11) If 120 Belgian francs are worth £1 then 1020 Belgian francs are worth £8·50. - - - - -

12) The exchange rate for Canadian dollars is $2·60 = £1. 1 Canadian dollar is therefore worth about 38p. - - - - -

13) The exchange rate for French francs is F12 = £1. F20 is therefore worth £1·50. - - - - -

14) In Italy 100 centisimi = 1 lira. If Lit 1500 = £1 then 1 centisimi is equal to about 0.007 pence. - - - - -

15) In Sweden 100 öre = 1 krona. If Skr 12·5 = £1 then 1 öre is worth about $\frac{1}{80}$th of a penny. - - - - -

Chapter 3 **Ratio and Proportion**

RATIO

A ratio is a comparison between two similar quantities. If the length of a certain ship is 120 metres long and the model of it is 1 metre long, then the length of the model is $\frac{1}{120}$th of the length of the ship. In making the model all the dimensions of the ship are reduced in the ratio of 1 to 120.

The ratio 1 to 120 is usually written 1 : 120.

A ratio can also be written as a fraction, as indicated above, and a ratio of 1 : 120 means the same as the fraction $\frac{1}{120}$.
The units must be the same before a ratio can be stated. We can state the ratio between 7 cm and 2 metres provided both lengths are brought to the same units. Thus if we convert 2 metres to 200 cm the ratio between the two lengths is 7 : 200.

Examples

1) Express the ratio 5 cm to 2 metres in its simplest form.

2 metres $= 2 \times 100$ cm $= 200$ cm.

$$5 : 200 = \frac{5}{200} = \frac{1}{40}$$

2) Express the ratio $4 : \frac{1}{4}$ in its lowest terms.

$$4 : \frac{1}{4} = 4 \div \frac{1}{4} = 4 \times \frac{4}{1} = \frac{16}{1}$$

$$\therefore 4 : \frac{1}{4} = 16 : 1$$

3) Two lengths are in the ratio 8 : 5. If the first length is 120 metres what is the second?

$$\text{The second length} = \frac{5}{8} \text{ of the first length}$$

$$= \frac{5}{8} \times 120 = 75 \text{ metres}$$

4) Two speeds are in the ratio 12 : 7. If the second is 21 kilometres per hour what is the first?

$$\text{The first speed} = \frac{12}{7} \times \text{the second speed}$$

$$= \frac{12}{7} \times 21 = 36 \text{ kilometres per hour}$$

Exercise 7

Express the following ratios as fractions in their lowest terms:

1) 8 : 3   2) 4 : 6   3) 12 : 4

4) 10 : 3   5) 8 : 12   6) 9 : 15

Express the following ratios as fractions in their lowest terms:

7) 1·5 kg to 200 g   8) 5 mm to 3 cm

9) 8 cm to 3 m   10) 30p to £2

11) 1·5 m to 120 mm   12) £5 to 80p

13) Two lengths are in the ratio 7 : 5. If the first length is 210 m, what is the second?

14) Two speeds are in the ratio 8 : 5. If the second speed is 120 kilometre per hour what is the first speed?

15) Express each of the following ratios as fractions in their lowest terms:
a) a speed of 20 km/h to 50 km/h
b) a cost of 80p per kg to a cost of 120p kg
c) 1 km to 20 m
d) a speed of 30 km/h to 20 m/s.

PROPORTIONAL PARTS

The diagram (Fig. 3.1) shows the line AB whose length is 160 mm divided into two parts in the ratio 3 : 5. As can be seen from the diagram the line has been divided into a total of 8 parts. The length AC contains 3 parts and the length BC contains 5 parts. Each part is 20 mm long; hence AC is $3 \times 20 = 60$ mm long and $BC = 5 \times 20 = 100$ mm long. We could tackle this problem as follows:

Total number of parts $= 3 + 5 = 8$
Length of each part $= 160 \div 8 = 20$ mm
Length of AC $= 3 \times 20 = 60$ mm
Length of BC $= 5 \times 20 = 100$ mm

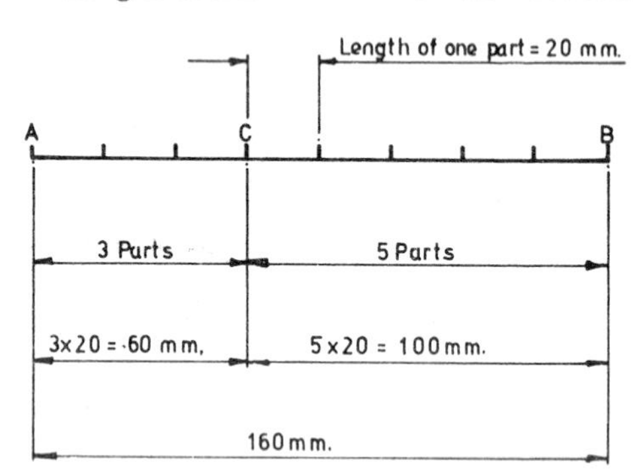

Fig. 3.1

Examples

1) Divide £1100 into two parts in the ratio 7 : 3

Total number of parts $= 7 + 3 = 10$

Amount of each part $= \dfrac{1100}{10} = £110$

Amount of first part $= 7 \times 110 = £770$

Amount of second part $= 3 \times 110 = £330$

2) An aircraft carries 2880 litres of fuel distributed in three tanks in the ratio 3 : 5 : 4. Find the quantity in each tank.

Total number of parts $= 3 + 5 + 4 = 12$

Amount of each part $= \dfrac{2880}{12} = 240$ litres

Amount of 3 parts $= 3 \times 240 = 720$ litres

Amount of 5 parts $= 5 \times 240 = 1200$ litres

Amount of 4 parts $= 4 \times 240 = 960$ litres

Exercise 8

1) Divide £800 in the ratio 5 : 3.

2) Divide £80 in the ratio 4 : 1.

3) A sum of money is divided into two parts in the ratio of 5 : 7. If the smaller amount is £200, find the larger amount.

4) Divide £120 in the ratio 5 : 4 : 3.

5) An alloy consists of copper, zinc and tin in the ratios of 3 : 4 : 1·5. Find the amount of each metal in 75 kg of the alloy.

6) A line is to be divided into 3 parts in the ratios of 2 : 7 : 11. If the line is 840 mm long find the length of each part.

7) A, B and C share a sum of money in the ratio 5 : 8 : 12. If C receives £12 more than B, find the sum of money that was shared.

8) 60 kg of alloy A are mixed with 100 kg of alloy B. If alloy A has lead and tin in the ratio 3 : 2 and alloy B has tin and copper in the ratio 1 : 4 find how much lead, tin and copper there is in the new alloy.

9) A quantity of alloy having a mass of 336 kg contains copper, lead and tin in the ratios 16 : 4 : 1 by mass. To this alloy are added 4 kg of copper, 36 kg of lead and 4 kg of tin. Calculate the ratios, by mass, of the copper, lead and tin in the new alloy.

10) Two villages have populations of 336 and 240 respectively. The two villages are to share a grant of £10 728 in proportion to their populations. Calculate how much each village receives.

SELF TEST 3

In questions 1 to 12 decide whether the answer is "true" or "false".

1) The ratio 6 : 3 is the same as the ratio 2 : 1 - - - - -

2) The ratio 5 : 10 is the same as the ratio 2 : 1 - - - - -

3) The ratio 20 : 100 : 300 is the same as the ratio 1 : 5 : 15 - - - - -

4) The ratio 9 : 2 may be written as 4·5 : 1 - - - - -

5) The fraction $\frac{2}{3}$ means the same as the ratio 2 : 3 - - - - -

6) The ratio 18 : 24 is the same as $\frac{3}{5}$ - - - - -

7) The ratio 18 : 30 is the same as $\frac{3}{5}$ - - - - -

8) The ratio 70p: £1·40 is the same as $\frac{1}{2}$ - - - - -

9) The ratio 8 cm : 4 m is the same as $\frac{2}{1}$ - - - - -

10) The ratio 5 km/min: 20 km/h is the same as $\frac{15}{1}$ - - - - -

11) When £600 is divided in the ratio 3 : 2 the two amounts are £360 and £240. - - - - -

12) When £900 is divided in the ratio 2 : 3 : 5 the three amounts are £200, £300 and £400. - - - - -

In questions 13 to 18 put a ring round the letter (or letters) corresponding to the correct answer (or answers).

13) When £1200 is divided in the ratio 7 : 5 the smallest amount is

a £700 b £500 c £240 d £480

14) When £360 is divided in the ratio 5 : 4 : 3 the smallest amount is

a £30 b £150 c £120 d £90

15) A, B and C share a sum of money in the ratio 7 : 5 : 14. If B receives £18 less than C then C receives

a £28 b £14 c £10 d £52

16) A quantity of alloy has a mass of 400 kg. It contains copper, lead and tin in the ratios by mass of 15 : 3 : 2. The mass of lead in the alloy is

a 300 kg b 60 kg c 40 kg d 200 kg

17) A line 920 cm long is divided into four parts in the ratios 15 : 13 : 10 : 8. The longest part is

a 260 cm b 200 cm c 300 cm d 160 cm

18) X, Y and Z share a sum of money in the ratios 7 : 8 : 16. Z receives £27 more than X. The sum of money that was shared is

a £279 b £558 c £93 d £48

Chapter 4 Percentages

When comparing fractions it is often convenient to express them with a denominator of 100. Fractions expressed with a denominator of 100 are called percentages.

$$\frac{1}{2}=\frac{50}{100}=50 \text{ per cent}$$

$$\frac{2}{5}=\frac{40}{100}=40 \text{ per cent}$$

The sign % is often used instead of the words per cent.

To convert a fraction into a percentage we multiply the fraction by 100.

Examples

1) $\frac{3}{4}=\frac{3}{4}\times 100\%=75\%$

2) $0{\cdot}3=0{\cdot}3\times 100\%=30\%$

3) $0{\cdot}245=0{\cdot}245\times 100\%=24{\cdot}5\%$

To convert a percentage into a fraction we divide the percentage by 100.

Examples

1) $45\%=\frac{45}{100}=0{\cdot}45$

2) $3{\cdot}9\%=\frac{3{\cdot}9}{100}=0{\cdot}039$

Exercise 9

Convert the following to percentages:

1) $\frac{3}{10}$ 2) $\frac{11}{20}$ 3) $\frac{9}{25}$ 4) $\frac{4}{5}$ 5) $\frac{31}{50}$

6) 0·63 7) 0·813 8) 0·667 9) 0·723

10) 0·027

Convert the following into vulgar fractions:

11) 32% 12) 78% 13) 6%

14) 24% 15) 31·5% 16) 48·2%

17) 2·5% 18) 1·25% 19) 3·95%

20) 20·1%

PERCENTAGE OF A QUANTITY

It is easy to find the percentage of a quantity if we express the percentage as a vulgar fraction.

Examples

1) What is 10% of 40?

Expressing 10% as a fraction it is $\frac{10}{100}$ and the problem then becomes: what is $\frac{10}{100}$ of 40?

$$10\% \text{ of } 40=\frac{10}{100} \text{ of } 40=\frac{10}{100}\times 40=4$$

2) What is 25% of £50?

$$25\% \text{ of } £50=\frac{25}{100} \text{ of } £50$$

$$=\frac{25}{100}\times £50=£12{\cdot}50$$

3) 22% of a certain length is 55 cm. what is the complete length?

22% of the length = 55 cm

$$1\% \text{ of the length}=\frac{55}{22}\text{ cm}=\frac{5}{2}\text{ cm.}$$

Now the complete length will be 100%

$$\therefore 100\% \text{ of the length} = \frac{5}{2} \times 100 \text{ cm} = 250 \text{ cm}$$

Alternatively

22% of the length = 55 cm

$$\text{Complete length} = \frac{100}{22} \times 55 \text{ cm} = 250 \text{ cm}$$

4) In an election the winning candidate received 16000 votes which represented 40% of the electorate. The only other candidate received 25% of the votes. How many of the electorate voted?
The total electorate is represented by 100%

Hence total electorate

$$= \frac{100}{40} \times 16000 = 40000$$

% of the electorate who voted
$=40+25=65\%$

Number of electorate who voted
$=65\%$ of 40000

$$= \frac{65}{100} \times 40000$$

$=26000$

Exercise 10

1) What is
a) 20% of 50 b) 30% of 80
c) 5% of 120 d) 12% of 20
e) 20·3% of 105 f) 3·7% of 68?

2) What percentage is
a) 25 of 200 b) 30 of 150
c) 24 of 150 d) 29 of 178
e) 15 of 33?
Where necessary give the answer correct to three significant figures.

3) A boy scores 36 marks out of 60 in an examination. What percentage is this? If the percentage needed to pass the examination is 45% how many marks are needed to pass?

4) If 20% of a length is 23 cm what is the complete length?

5) Given that 13·3 cm is 15% of a certain length, what is the complete length?

6) Express the following statements in the form of a percentage:
a) 3 eggs are bad in a box containing 144 eggs.
b) In a school of 650 pupils, 20 are absent.
c) In a school of 980 pupils, 860 eat lunches.

7) During 1964, in a certain country, the average number of children eating lunches at school each day was 29 336 which represents 74% of the total number of children attending school.
Calculate the total number of children attending school in that country.

8) In an election the winning candidate received 15 720 votes which represented 48% of the electorate. The only other candidate received the support of (or votes of) 22% of the electorate. Calculate the number of people who did not use their votes.

PERCENTAGE PROFIT AND LOSS

When a dealer buys or sells goods the cost price is the price at which he buys the goods and the selling price is the price at which he sells the goods. If the selling price is greater than the cost price a profit is made. The profit is the difference between the selling price and the cost price, that is

Profit = selling price − cost price

The profit per cent is always calculated on the cost price, that is

$$\text{Profit } \% = \frac{\text{selling price} - \text{cost price}}{\text{cost price}} \times 100$$

If a loss is made the cost price is greater than the selling price. The loss is the difference between the cost price and the selling price.

That is,

$$\text{Loss} = \text{cost price} - \text{selling price}$$

$$\text{Loss } \% = \frac{\text{cost price} - \text{selling price}}{\text{cost price}} \times 100$$

Examples

1) A shopkeeper buys an article for £5.00 and sells it for £6·00. What is his percentage profit?

Cost price
= £5·00 and selling price = £6·00

% profit

$$= \frac{6{\cdot}00 - 5{\cdot}00}{5{\cdot}00} \times 100 = 20\%$$

2) A dealer buys 20 articles at a total cost of £5·00. He sells them for 30p each. Find his percentage profit.

Since £5·00 = 500p

$$\text{Cost price per article} = \frac{500}{20} = 25\text{p}$$

Profit per article = 30p − 25p = 5p

$$\text{Profit } \% = \frac{5}{25} \times 100 = 20\%$$

3) A man buys a car for £700 and sells it for £560. What is his percentage loss?

Loss = cost price − selling price

= £700 − £560 = £140

$$\% \text{ loss} = \frac{140}{700} \times 100 = 20\%$$

Exercise 11

1) A shopkeeper buys an article for 80p and sells it for £1·00. Calculate his percentage profit.

2) Calculate the profit per cent when:
   a) Cost price is £1·50 and selling price is £1·80
   b) Cost price is 30p and selling price is 35p

3) Calculate the loss per cent when:
   a) Cost price is £0·75 and selling price is £0·65
   b) Cost price is £6·53 and selling price is £5·88

4) A greengrocer buys a box of 200 oranges for £5·00. He sells them for 3p each. Calculate his profit per cent.

5) A dealer buys 1000 articles for £600 and sells them at 80p each. Find his profit %.

6) A dealer buys 30 articles at 8p each. Three are damaged and unsaleable but he sells the others at 10p each. What is the profit %.

7) A car is bought for £850 and sold for £700. What is the loss %.

8) The price of coal has increased from £20·00 to £22·00 per 1000 kilogrammes. What is the percentage increase in the price of coal?

PERCENTAGE CHANGE

An increase of 5% in a number means that the number has been increased by $\frac{5}{100}$ of itself. Thus if the number is represented by 100, the increase is 5 and the new number is 105. The ratio of the new number to the old number is 105 : 100.

Examples

1) An increase of 10% in salaries makes the wage bill for a factory £5500.
   a) What was the wage bill before the increase?
   b) What is the amount of the increase?

   a) If 100% represents the wage bill before the increase, then 110% represents the wage bill after the increase.
   ∴ Wage bill before the increase

$$= \frac{100}{110} \times £5500 = £5000$$

b) The amount of the increase

$$= 10\% \text{ of } 5000$$

$$= \frac{10}{100} \times 5000 = £500$$

2) By selling an article for £4·23 a dealer makes a profit of $12\frac{1}{2}\%$ on his cost price. What is his profit.

If 100% represents the cost price then $112\frac{1}{2}\%$ represents the selling price.

$$\therefore \text{ Cost price} = \frac{100}{112\frac{1}{2}} \times £4{\cdot}23 = \frac{200}{225} \times £4{\cdot}23$$

$$= £3{\cdot}76$$

Profit = selling price − cost price

$$= £4{\cdot}23 - £3{\cdot}76 = £0{\cdot}47$$

A decrease of 5% in a number means that if the original number is represented by 100 then the decrease is 5 and the new number is 95. The ratio of the new number to the old number is 95 : 100.

Example

1) An article was sold for £30 which was a loss on the cost price of 10%. What was the cost price?

If 100% represents the cost price then 90% represents the selling price.

$$\therefore \text{ Cost price} = \frac{100}{90} \times 30 = £33{\cdot}33$$

Exercise 12

1) Calculate the selling price when
   a) Cost price is £5·00 and profit per cent is 20%
   b) Cost price is £3·75 and profit per cent is 16%

2) Calculate the cost price when
   a) selling price is £20·00 and profit is 25%
   b) selling price is 63p and profit is $12\frac{1}{2}\%$

3) By selling an article for £10·80 a shopkeeper makes a profit of 8%. What should be the selling price for a profit of 20%?

4) a) An article can be bought from a shopkeeper for a single cash payment of £74 or by 18 monthly instalments of £4·30. Calculate the extra cost of paying by instalments.
   b) By selling for £74 the shopkeeper made a profit of 25%. Find how much he paid for the article.

5) A shopkeeper marks an article to allow himself 25% profit on the cost price. If he sells it for £80 how much was the cost price?

6) If 8% of a sum of money is £2·40 find $9\frac{1}{2}\%$ of the sum.

7) The duty on an article is 20% of its value. If the duty is 60p, find the value of the article.

8) When a sum of money is decreased by 10% it becomes £18. What was the original sum?

9) A man sells a car for £425 thus losing 15% of what he paid for the car. How much did the car cost him?

10) Equipment belonging to a firm is valued at £15 000. Each year 10% of the value of the equipment is written off for depreciation. Find the value of the equipment at the end of two years.

DISCOUNT

When a customer buys an article from a dealer he often asks for a *discount*. This discount is the amount which the dealer will take off his *selling price* thus reducing his profit.

Examples

1) A radiogram is offered for sale at £60. A customer is offered a 10% discount for cash. How much does the customer actually pay?

$$\text{Discount} = 10\% \text{ of } £60 = \frac{10}{100} \times 60 = £6{\cdot}00$$

Amount actually paid = £60 − £6 = £54·00
(Alternatively: since only 90% of selling

price is paid amount actually paid = 90% of £60 = £54·00)

2) A suit is marked at a price of £19·50, but during a sale a discount of 10p in the £1 is allowed. When sold at the sale price a profit of $12\frac{1}{2}$% is made on the cost price. What would have been the percentage profit on the cost price if the suit had been sold at the marked price of £19·50?

Discount = 19·50 × 0·10 = £1·95
Sale price = £19·50 − £1·95 = £17·55
If the cost price is represented by 100%
then the sale price is represented by $112\frac{1}{2}$%.

$$\therefore \text{ Cost price} = \frac{100}{112\frac{1}{2}} \times £17{\cdot}55 = £15{\cdot}60$$

Profit at the marked price

= £19·50 − £15·60

= £3·90

Profit % at the marked price

$$= \frac{3{\cdot}90}{15{\cdot}60} \times 100 = 25\%$$

Exercise 13

1) A chair marked at £7·00 is sold for cash at a discount of 10%. What is the cash price?

2) A shopkeeper who allows 10% discount for cash states that the cash price of a suite of furniture is £180. What is the credit price?

3) A tailor charges £20 for a suit of clothes but allows a discount of 5% for cash. What is the cash price?

4) A housewife's grocery bill, after a deduction of $2\frac{1}{2}$% discount, amounts to £5·85. What is the gross amount of the bill?

5) A manufacturer offers a discount of 25% on an article for which the retailer pays £72. What is the manufacturer's price?

6) A car is listed by the manufacturer at £850. A garage owner buys it from the manufacturer for £595. What is the trade discount per cent?

7) When a shopkeeper sells articles for £18·90 each he makes a profit of 26% on the cost price. During a sale the articles are marked at £15·60 each. Calculate the percentage profit on an article sold during the sale.

8) A radiogram is offered for sale at £58·00 which represents a profit to the dealer of 16% on the cost price. If the radiogram was sold at a discount of 8% off the marked price of £58·00, calculate the percentage profit the dealer made.

9) A suit is marked at a price of £22·00, but during a sale a discount of 15 pence in the pound is allowed. When sold at the sale price a profit of 10% is made on the cost price. What is the cost price of the suit?

10) A wholesaler sells an article to a retailer for £260 which represents a profit to the wholesaler of 30%. The retailer then sells the article to a customer at a profit of 25%. Calculate the total percentage profit based on the price the wholesaler paid.

SELF TEST 4

In questions 1 to 14 decide whether the answer is 'true' or 'false'.

1) A fraction expressed with a denominator of 100 is called a percentage. _ _ _ _ _

2) $\frac{13}{25}$ is the same as 42%. _ _ _ _ _

3) 0·725 is the same as 72·5%. _ _ _ _ _

4) 3·5% is the same as $\frac{7}{20}$. _ _ _ _ _

5) 20·45% is the same as 2·045. _ _ _ _ _

6) 20% of 80 is 16. _ _ _ _ _

7) If 15% of a certain length is 45 mm the complete length is 300 mm. _ _ _ _ _

8) The total electorate for a certain constituency is 53 000. If 30% did not vote in an election then 37 100 did vote. - - - - -

9) When a shopkeeper buys an article for £4·00 and sells it for £5·00 his percentage profit is 20%. - - - - -

10) A dealer buys an article for £8·00 and sells it for £5·00. His loss is 37·5% - - - - -

11) A man's salary is increased by 20% and he now receives £36 per week. Hence his salary before the increase was £30 - - - - -

12) By selling an article for £9·00 a dealer made a profit of $33\frac{1}{3}$%. He therefore paid £6·00 for the article. - - - - -

13) A shopkeeper marks an article at £20 and by selling it at this price he makes a profit of 30%. On a cash sale he allows a discount of 20%. His profit on a cash sale is therefore 10%. - - - - -

14) A wholesaler sells goods to a retailer at a profit of 20%. The retailer sells them to a customer at a profit of 10%. The overall profit is therefore 32%. - - - - -

In questions 15 to 25 put a ring round the letter (or letters) corresponding to the correct answer or answers.

15) 35% is the same as

a $\frac{35}{100}$ b $\frac{7}{20}$ c $\frac{35}{10}$ d 0·35

16) $\frac{11}{25}$ is the same as

a 4·4% b 44% c 440%

17) 30% of a certain length is 600 mm. The complete length is

a 20 mm b 200 mm c 2000 mm
d 2 m

18) When a dealer sells an article for £18 he makes a profit of £3. His percentage is therefore

a 20% b $16\frac{2}{3}$% c $14\frac{2}{7}$%
d 25%

19) When a shopkeeper buys an article for £20 and sells it for £25 his percentage profit is

a 20% b 30% c 25%
d 80%

20) When a dealer sells an article for £50 he makes a profit of 25%. The price he paid for the article is

a £37·50 b £40·00 c £20
d £40·50

21) A dealer buys 40 articles at a total cost of £10·00. He sells them at 30p each. His percentage profit is

a $16\frac{2}{3}$% b 20% c 30%
d 25%

22) An article was sold for £60 which was a loss on the cost price of 10%. The cost price was therefore

a £54·00 b £66·00 c £66·67
d £70·50

23) The duty on an article is 25% of its value. If the duty paid is 80p, the value of the article is

a 800p b 320p c 180p
d 100p

24) An article is offered for sale at £120 which represents a profit of 20% to the dealer. On a cash sale he allows a discount of 10%. His profit on a cash sale is therefore

a 10% b 8% c 30%
d 18%

25) When a shopkeeper sells articles for £37·80 each he makes a profit of 26% on the cost price. During a sale the articles are marked at £31·20 each. He therefore makes a profit of

a $3\frac{1}{3}$% b 3·8% c 5%
d 4%

Chapter 5 **Simple Interest**

Interest is the profit return on investment. The money which is invested is called the *principal*. The percentage return per annum is called the *rate per cent*. Thus interest at a rate of 7% means that the interest on a principal of £100 is £7 per annum. The total formed by adding the principal and the interest for any length of time is called the *amount* after that time. The amount is therefore the total sum of money which remains invested after the period of time.

When a bank offers an interest rate of 5% per annum this means that the bank will, at the end of the year, add 5% of the money invested (i.e. the principal) to the money invested. If this interest is cashed the money invested will remain intact. If this is done every year for four years, then the total interest cashed would be 5% for four years. An arrangement of this kind is called *simple interest*.

Example

Find the simple interest on £500 for 4 years at 5% per annum.

The interest on £100 for 1 year is £5

The interest on £500 for 1 year is

$$\frac{500}{100} \times 5 = £25$$

The interest on £500 for 4 years is

£25 × 4 = £100

If P stands for the principal

T stands for the time in years

R stands for the rate per cent per annum

I stands for the interest

then $I = \frac{PRT}{100}$

Note that P and I must be in the same monetary units but any convenient unit may be used.

This formula can be transposed to give P, R and T in terms of the other letters. Thus

$$T = \frac{100\,I}{PR}$$

$$R = \frac{100\,I}{PT}$$

$$P = \frac{100\,I}{RT}$$

Examples

1) £800 is invested for 5 years at 7% per annum. Calculate the amount of simple interest earned.

We are given P = 800, R = 7 and T = 5.

$$I = \frac{PRT}{100} = \frac{800 \times 7 \times 5}{100} = £280$$

2) £700 is invested at 4% per annum. How long will it take for the simple interest to amount to £84.

We are given I = 84, P = 700 and R = 4.

$$T = \frac{100\,I}{PR} = \frac{100 \times 84}{700 \times 4} = 3$$

The length of time needed is 3 years.

3) The simple interest on £400 invested for 3 years is £63. What is the rate per cent?

We are given P = 400, I = 63 and T = 3

$$R = \frac{100\,I}{PT} = \frac{100 \times 63}{400 \times 3} = 5\tfrac{1}{4}$$

Hence the rate is $5\frac{1}{4}$% per annum.

Exercise 14

1) Find the simple interest on £700 invested for 3 years at 6% per annum.

2) Find the simple interest on £500 invested for 6 months at 8% per annum.

3) Find the simple interest on £1200 invested for 4 months at $7\frac{5}{8}$% per annum.

4) In what length of time will £500 be the simple interest on £2500 which is invested at 5% per annum?

5) In what length of time will £16 be the simple interest on £480 invested at 8% per annum?

6) In what length of time will £75 be the simple interest on £500 invested at 6% per annum?

7) The interest on £600 invested for 5 years is £210. What is the rate per cent?

8) The interest on £200 invested for 4 months is £6. What is the rate per cent?

9) What principal is needed so that the interest will be £48 if it is invested at 3% per annum for 5 years?

10) Which receives the more interest per annum:

£150 invested at 4% or £180 invested at $3\frac{1}{2}$%.

What is the annual difference?

11) A man invests £700 at 6% per annum and £300 at 8% per annum. What is his total annual interest on these investments?

12) A man deposited £350 in a bank and £14 interest was added at the end of the first year. The whole amount was left in the bank for a second year at the same rate of interest. Find the amount in the bank at the end of the second year.

13) A man invests £2000 at 5% per annum simple interest in order to pay school fees for his son. He pays the fees of £400 partly from the interest and partly from the capital invested. How much of the capital is left after 3 years?

14) How long will it take for a sum of money invested at 5% per annum simple interest to increase in value by 30%?

15) A man buys a house for £10000. He pays 20% of its value on purchase and pays the remainder in 2 years. If the simple interest is charged at 10% per annum, how much did the man pay altogether?

SELF TEST 5

In questions 1 to 10 state the letter (or letters) corresponding to the correct answer (or answers).

1) The simple interest on £500 for 4 years at 7% per annum is
a £1400 b £14000 c £20
d £140

2) The simple interest on £800 invested at 8% per annum for 6 months is
a £48 b £32 c £64 d £60

3) The simple interest on £800 invested for 5 years was £240. The rate of interest per annum was therefore
a 6% b 5% c 3% d 7%

4) The simple interest on £800 invested at 5% per annum over a number of years amounted to £160. The cash was therefore invested for
a 7 years b 5 years c 4 years
d 6 years

5) The simple interest on £400 invested for 4 months was £12. The rate of interest per annum is
a 36% b 4·8% c 7% d 9%

6) A man invests £700 at 5% per annum and £300 at 6% per annum. The entire investment has an interest rate of
a $5\frac{1}{2}$% b 5·3% c 11%
d 53%

7) A man invests £9000 at 10% per annum and £1000 at 8% per annum. His return on the complete investment is
a 18% b 2% c 10%
d 9·8%

8) What sum of money must be invested to give £30 simple interest if the rate is 6% per annum and the time is 2 years?

a £250   b £300   c £400

d £360

9) A sum of money is invested at 8% per annum for 4 years and the simple interest is £160. The amount of money invested is

a £500   b £640   c £320

d £1280

10) A sum of money was invested at 5% per annum for 4 years. The total amount lying to the credit of the investor at the end of the 4 years was £600. The amount originally invested was

a £800   b £600   c £500

d £400

Chapter 6 Salaries, Household Bills, Rates and Taxes

SALARIES

Everyone who works for an employer receives a wage or a salary in return for his labours. Wages may be paid at so much per hour, so much per week or so much per month. Most employees work a basic week of so many hours and it is this basic week which fixes the wages of the employee. Overtime is time worked over and above the basic week and it is often paid for at enhanced rates of pay which is paid extra to the basic wage.

Examples

1) A man's basic wage for a 40 hour week is £28·80. By working overtime he increased his wages to £34·32. If the overtime rate is 20p per hour more than the basic rate per hour find how many hours of overtime he worked.

Basic rate per hour

$$= \frac{£28{\cdot}80}{40 \text{ hours}} = £0{\cdot}72 \text{ or } 72\text{p}$$

Overtime rate per hour

$$= 72\text{p} + 20\text{p} = 92\text{p}$$

Amount earned in overtime

$$= £34{\cdot}32 - 28{\cdot}80 = £5{\cdot}52$$

Amount of overtime

$$= \frac{£5{\cdot}52}{92\text{p}} = \frac{552\text{p}}{92\text{p}} = 6 \text{ hours}$$

2. Workmen asking for a rise of 10% of their wages were granted only a rise of 4%, which brought their weekly wages up to £20·80 for each man. Calculate how much the weekly wage would have been if the 10% rise had been granted.

Let 100 represent the weekly wage before any rise was granted.
Then the weekly wage after the 4% rise is 104 and the weekly wage after a 10% rise would be 110.
∴ Weekly wage based on a 10% rise

$$= \frac{110}{104} \times £20{\cdot}80 = £22{\cdot}00$$

Exercise 15

1) A man's basic wage for a 38 hour week is £30·40. What is his basic hourly rate? In a certain week he earned £35·10 by working overtime. If he worked 5 hours overtime, what is the overtime rate?

2) A man's basic hourly rate is 72p. Overtime is paid for at $1\frac{1}{4}$ times the basic rate. If the basic week is 40 hours, how many hours of overtime must be worked in order for the man to earn £34·20 for the week?

3) A man's basic pay is £20·00 for a 40 hour week. Overtime from Monday to Friday is paid for at the basic rate, but on Saturday it is paid for at $1\frac{1}{2}$ times the basic rate whilst on Sunday it is paid for at twice the basic rate. If the man worked 5 hours overtime between Monday and Friday, 4 hours on Saturday and 7 hours on Sunday, how much was his total wage for the week?

4) A man's basic pay for a 40 hour week is £32·00. In a certain week he worked overtime and his total wage for the week was £39·00. If overtime is paid for at 20p per hour more than the basic rate find how many hours of overtime he worked.

5) After obtaining a rise of 5%, the weekly wage for an office worker became £21·00. How much was his wage before the rise?

6) After spending one-seventh of my income on rent and two-sevenths of the remainder on household expenses I have £750 left. Calculate my income.

7) Workmen asking for a rise of 8% of their weekly wage were awarded a rise of $3\frac{1}{2}$%. This brought the weekly wage of each man up to £20·70. Calculate how much the weekly wage would have been if the 8% rise had been granted.

8) A man is paid a basic rate of 80p per hour for a 40 hour week. On weekdays he is paid overtime at $1\frac{1}{4}$ times the basic rate. If his wage for a certain week was £38·00. Calculate how many hours of overtime he worked.

HOUSEHOLD BILLS

The most important household bills are gas and electric bills. There are various ways by which the total bill for these items may be calculated. Some of the ways are as follows:

1) A standard rate of so much per unit for all the units of gas or electricity used.
2) A fixed charge plus so much per unit for all the units used.

3) A rate of so much per unit for the first so many units used and a different rate for the remainder of the units used.

The following examples illustrate the use of each of these methods.

Examples
1) A householder uses 300 units of electricity in a certain quarter. If the electricity is charged for at 1·5p per unit calculate the amount of the householder's bill.

Cost of 300 units @ 1·5p unit
$= 300 \times 1{\cdot}5 = 450\text{p} = £4{\cdot}50$

The householder's bill will be £4·50.

2) In a certain area a householder is charged for electricity as follows: a fixed charge of £4·00 plus 0·75p per unit of electricity used. If his bill for electricity is £5·80, how many units did he use?

Cost of units used $= £5{\cdot}80 - £4{\cdot}00 = £1{\cdot}80$

$$\text{Number of units used} = \frac{£1{\cdot}80}{0{\cdot}75\text{p}} = \frac{180}{0{\cdot}75} = 240$$

3) A householder is charged for electricity as follows: the first 80 units used are charged for at 4p per unit and the remainder of the units used are charged for at 1·2p per unit. If the householder uses 800 units in a quarter find the amount of his bill.

Cost of 80 units @ 4p per unit
$= 80 \times 4 = £3{\cdot}20$
Cost of 720 units @ 1·2p per unit
$= 720 \times 1{\cdot}2 = £8{\cdot}64$
Total $= £11{\cdot}84$

Exercise 16

1) A householder pays for all the electricity he uses at 2·5p per unit. If he uses 520 units in a quarter, what will be his quarterly bill?

2) In a certain area electricity is charged for at a fixed rate of 2·3p for every unit used. If a householder receives a bill for £16·79, how many units has he used?

3) A householder receives an electricity bill for £5·50 and he used 250 units. If the electricity is charged for at so much per unit find the cost per unit.

4) In a certain area a householder pays for his electricity as follows: a fixed charge of £7·00 plus 1·3p per unit for each of the units used. Find his bill if he uses 370 units.

5) A householder pays for his electricity by means of a fixed charge of £5·50 plus 1·25p for each unit of electricity he uses. If his bill is £11·00, find how many units he used.

6) A householder has the choice of paying for his electricity as follows:

1) At a fixed rate of 2·5p per unit for each unit he uses;

2) A fixed charge of £8·00 plus 1·25p per unit for each unit he uses.

If the householder estimates that he will use 650 units of electricity which method of payment should he choose?

7 A man is charged for his electricity as follows: 4p per unit for the first 120 units used and 1·5p per unit for the remainder. If he uses 800 units how much will he pay for his electricity?

8) If a householder is charged 3·5p for the first 80 units of electricity used and 1·3p per unit for the remainder, find how many units the householder used if the bill is £5·66

9) A householder has the choice of paying for his electricity as follows:

1) A fixed charge of £5·00 plus 2p per unit for each unit used;
2) A charge of 4p per unit for the first 150 units plus a charge of 1·8p for each subsequent unit.

If the householder estimates that he will use 600 units, which method of payment should he use and how much will he save?

RATES

Every property in a town or city is given a rateable value which is fixed by the Local District Valuer. This rateable value depends upon the size, condition and position of the property. The rates of a town or city are levied by the Local Government Authority at so much in the £1 of rateable value, for instance £0·70 in the £1. The money brought in by the rates is used to pay for such things as education, sanitation, libraries, etc.

Examples

1) The rateable value of a house is £120. If the rates are £0·75 in the £1, how much must the owner pay in rates per annum?

Rates payable per annum
$= £120 \times 0{\cdot}75 = £90$

2) The estimated product of a one pence rate in a certain city is £41 550. Find the total rateable value of all the property in the city.

£0·01 is the amount of rates collected per £1 of rateable value.

∴ £41 550 is the amount of rates collected on $\frac{41\,550}{0{\cdot}01} = £4\,155\,000$ of rateable value.

3) The cost of highways and bridges in a city is equivalent to a rate of 5·10 pence in the £1. If the rateable value of all the property in the city is £5 225 000, find how much money is available for expenditure on highways and bridges in the financial year.

$$\text{Amount available} = 5{\cdot}10 \times 5\,225\,000 \text{ pence}$$

$$= £\frac{5{\cdot}10 \times 5\,225\,000}{100}$$

$$= £266\,475$$

Exercise 17

1) The rateable value of a house is £90. Calculate the rates payable by a householder per year when the rates are £0·70 in the £1.

2) A householder pays £90 in rates when the rate levied is £0·75 in the £1. What is the rateable value of the house?

3) A house is assessed at a rateable value of £45. The owner pays £40·50 in rates for the year. At what rate were the rates levied for that year?

4) The rateable value of all the property in a city is £850 000. What is the product of a penny rate? How much must the rates be if the total expenses for the city for a year are £790 500?

5) The total rateable value for all the property in a city is £8 796 000. Calculate the cost of the public libraries if a rate of 2·3 pence in the £1 must be levied for the purpose.

6) The expenditure of a town is £450 000 and its rates are 87 pence in the £1. The cost of

the library is £15 000. What rate in the £1 is needed for the upkeep of the library?

7) The total rateable value of a town is £1 560 000. Calculate the value of a penny rate for the town. The finance committee estimates that expenditure next year will be increased by £46 800 and they decide to obtain the extra money by raising the rate.
Calculate,

a) the necessary increase in the rate per £1;
b) the additional payment which a householder will have to make if his house has a rateable value of £126.

8) The rateable value for a house used to be £63 and the annual amount paid in rates was £64·89. The new rateable value is £174 and the rate is £0·41 in the £1. Find the increase or decrease in the amount paid in rates. The water rate was 8% of the old rateable value and it is $2\frac{3}{4}$% of the new one, both reckoned to the nearest penny. Find the increase or decrease in the amount paid for water.

INCOME TAX

Taxes are levied by the Chancellor of the Exchequer in order to produce income to pay for the armed services, the Civil Service, the National Health Services and other expenditures. The largest producer of revenue is Income Tax.
Every person who has an income above a certain minimum amount has to pay part of that income to the Government in income tax. Tax is not paid on the whole income—certain allowances are made as follows:

1) an allowance the amount of which varies according whether the taxpayer is a single person or a married man.

2) Allowances for children, dependent relatives etc.

3) Allowances for superannuation contributions, mortgage payments etc.

The residue left after the allowances have been deducted is called the taxable income.
The following example shows the method used in calculating income tax.

Example
A man's salary is £2700 per year. His taxable income is found by deducting the following from his salary:

1 A married man's allowance of £775

2 A children's allowance of £430

3) Superannuation payments of £135

4) Interest on building society mortgage £360

He then pays tax at the basic rate of 30%.
Calculate

a) His taxable income
b) The total amount paid in income tax.

a) To find the taxable income deduct the following from the salary of £2700.

Married man's allowance	=£775
Children's allowance	=£430
Superannuation payments	=£135
Interest to building society	=£360
Total allowances	=1700

Taxable income = £2700 − £1700 = £1000

b) Total tax paid = 30% of £1000 = £300

Exercise 18

Use the following allowances for the following questions:

Single person's allowance	£595
Married man's allowance	£775
Child under 11 years old	£200
Child 11–16 years old	£235
Child over 16 years old	£265
Dependent relative	£100

1) A person's taxable income is £1200. If tax is paid at 30% find the amount paid in income tax.

2) When income tax is levied at 30% a man pays £60 in income tax. What is his taxable income.

3) Calculate the income tax to be paid by a single man earning £3000 per annum when tax is levied at 30%.

4) A married man with two children under 11 years old earns £2500 per annum. If these are his only allowances find how much he pays in income tax when this is charged at 30%

5) A married man with one child aged 15 years and a second aged 17 years earns £4000 per annum. He has a dependent relative that he helps to support and also pays building society interest of £300 per annum. If his superannuation payments are 5% of his gross salary calculate the amount he pays in income tax per annum, when this is charged at 30%.

SELF TEST 6

In questions 1 to 11 state the letter (or letters) corresponding to the correct answer (or answers).

1) A man's basic wage for a 35 hour week is £22·75. His overtime rate is 10 pence per hour more than his basic rate. If he works 6 hours of overtime in a certain week his wages for that week will be

a £23·35 b £27·25 c £28·00
d £26.35

2) Office workers asked for a rise of 12% but were granted only 5% which brought their weekly wage up to £33·60. If the 12% rise had been granted their weekly wage would have been

a £32 b £36 c £35·84
d £35·96

3) A man's basic pay for a 40 hour week is £20·00. Overtime is paid for at 25% above the basic rate. In a certain week he worked overtime and his total wage was £25·00. He therefore worked a total of

a 48 hours b 45 hours c 50 hours
d 47 hours

4) A householder is charged for electricity as follows: the first 80 units are charged for at 3p per unit and each subsequent unit used at 1·3p per unit. If the electricity bill amounted to £8·12 the number of units used was

a 600 b 550 c 440 d 520

5) A householder pays for his electricity by means of a fixed charge of £3·60 plus 1·2p for each unit of electricity used. If he used 330 units of electricity his bill was

a £3·96 b £7·56 c £8·05
d £4·36

6) The rateable value of a house is £150. If the rates are 80p in the £ the rates payable are

a £150 b £12 c £120
d £140

7) In a certain city the total rateable value of all the property in the city is £8 000 000. The product of a penny rate is therefore

a £80 000 b £8000 c £16 000
d £20 000

8) The cost of highways in a town is equivalent to a rate of 6·2 pence in the £. If the rateable value of all the property in the town is £3 500 000 then the cost of highways is

a £21 700 b £22 000 c £220 000
d £217 000

9) The expenditure of a town is £300 000 and its rates are 75 pence in the £. The cost of the library is £20 000. The rate needed for the upkeep of the library is

a 4p in the £ b 5p in the £1
c 8p in the £ d 7·5p in the £

10) When income tax was levied at 30% a man paid £90 in income tax. His taxable income was therefore

a £30 b £27 c £270 d £300

Chapter 7 Averages

AVERAGES

To find the average of a set of quantities, add the quantities together and divide by the number of quantities in the set. Thus,

$$\text{average} = \frac{\text{sum of the quantities}}{\text{number of quantities}}$$

Examples

1) A boy makes the following scores at cricket: 8, 20, 3, 0, 5, 9, 15 and 12. What is his average score?

$$\text{average score} = \frac{8+20+3+0+5+9+15+12}{8}$$

$$= \frac{72}{8} = 9$$

2) The oranges in a box have a mass of 4680 gm. If the average mass of an orange is 97·5 gm find the number of oranges in the box.

Total mass = average mass of an orange × number of oranges in the box

$$\therefore \text{Number of oranges in the box} = \frac{4680}{97{\cdot}5} = 48$$

3) Find the average age of a team of boys given that four of them are each 15 years 4 months old and the other three boys are each 14 years 9 months old.

Total age of 4 boys at 15 years 4 months = 61 years 4 months
Total age of 3 boys at 14 years 9 months = 44 years 3 months
Total age of 7 boys = 105 years 7 months

$$\text{Average age} = \frac{\text{105 years 7 months}}{7}$$

$$= \text{15 years 1 month}$$

4) The average age of the teachers in a school is 39 years and their total age is 1170 years, whereas the pupils whose average age is 14 years have a total age of 6580 years. Find the average age of all the people in the school.

The first step is to find the number of teachers:

Number of teachers

$$= \frac{\text{Total age of the teachers}}{\text{average age of the teachers}}$$

$$= \frac{1170}{39} = 30$$

We now find the number of pupils:

$$\text{Number of pupils} = \frac{6580}{14} = 470$$

We can now find the average age of all the people in the school:

Total age of all the people in the school = 1170 + 6580 = 7750 years

Total number of people in the school = 30 + 470 = 500

Average age of all the people in the school

$$= \frac{7750}{500} = 15{\cdot}5 \text{ years}$$

Exercise 19

1) Find the average of the following readings: 22·3 mm, 22·5 mm, 22·6 mm, 21·8 mm and 22·0 mm.

2) Find the average mass of 22 boxes if 9 have a mass of 12 kg, 8 have a mass of $12\frac{1}{2}$ kg and 5 have a mass of $11\frac{3}{4}$ kg.

3) 4 kg of apples costing 20p per kg are mixed with 8 kg costing 14p per kg. What is the average price per kg?

4) 30 litres of petrol costing 8p per litre is mixed with 40 litres costing 9p per litre. Find the average price of the mixture.

5) The average of nine numbers is 72 and the average of four of them is 40. What is the average of the other five?

6) The apples in a box have a mass of 240 kg. If the average mass of an apple is 120 g find the number of apples in the box.

7) Find the average of a team of boys if 5 of them are each 15 years old and 6 of them are 14 years 1 month old.

8) A grocer sells 40 tins of soup at 8p per tin, 50 at 9p per tin and 60 tins at 10p per tin. Find the average price per tin.

9) The average mark of 24 candidates taking an examination is 42. Find what the average mark would have been if one candidate, who scored 88, had been absent.

10) The average of three numbers is 58. The average of two of them is 49. Find the third number.

11) In a school, three classes took the same examination. Class A contained 30 pupils and the average mark for the class was 66. Class B contained 22 pupils and their average mark was 54. Class C contained 20 pupils. The average obtained by all the pupils together was 61·5. Calculate the average mark of Class C.

12) A farmer sent 50 sheep fleeces yielding a total of 168 kg of wool to a merchant. For one-third of the wool the farmer recieved 25p per kg for a quarter of the wool he received 20p per kg and for the rest he received 15p per kg. Calculate the total amount received by the farmer and the average price received for *each* fleece.

AVERAGE SPEED

The average speed is defined as total distance travelled divided by the total time taken. The unit of speed depends on the unit of distance and the unit of time. For instance, if the distance travelled is in kilometres (km) and the time taken is in hours (h) then the speed will be stated in kilometres per hour (km/h). If the distance is given in metres (m) and the time in seconds (s) then the speed is in metres per second (m/s).

Examples

1) A car travels a total distance of 200 km in 4 hours. What is its average speed?

$$\text{Average speed} = \frac{\text{distance travelled}}{\text{time taken}} = \frac{200}{4}$$

$$= 50 \text{ km/h}$$

2) A car travels 30 km at 30 km/h and 30 km at 40 km/h. Find its average speed.

Time taken to travel 30 km at 30 km/h

$$= \frac{30}{30} = 1 \text{ hour}$$

Time taken to travel 30 km at 40 km/h

$$= \frac{30}{40} = 0{\cdot}75 \text{ hour}$$

Total distance travelled $= 30 + 30 = 60$ km
Total time taken $= 1 + 0{\cdot}75 = 1{\cdot}75$ hour

$$\therefore \text{ Average speed} = \frac{60}{1{\cdot}75} = 34{\cdot}3 \text{ km/h}$$

3) A train travels for 4 hours at an average speed of 64 km/h. For the first 2 hours its average speed is 50 km/h. What is its average speed for the last 2 hours.

Total distance travelled in 4 hours

$= \text{average speed} \times \text{time taken} = 64 \times 4$

$= 256$ km

Distance travelled in first two hours

$= 50 \times 2 = 100$ km

$\therefore$ Distance travelled in last two hours

$$=256-100=156 \text{ km}$$

Average speed for the last two hours

$$=\frac{\text{distance travelled}}{\text{time taken}}=\frac{156}{2}=78 \text{ km/h}$$

Exercise 20

1) A train travels 300 km in 4 hours. What is its average speed?

2) A car travels 200 km at an average speed of 50 km/h. How long does it take?

3) If a car travels for 5 hours at an average speed of 70 km/h how far has it gone?

4) For the first $1\frac{1}{2}$ hours of a 91 km journey the average speed was 30 km/h. If the average speed for the remainder of the journey was 23 km/h, calculate the average speed for the entire journey.

5) A motorist travelling at a steady speed of 90 km/h covers a section of motorway in 25 minutes. After a speed limit is imposed he finds that, when travelling at the maximum speed allowed he takes 5 minutes longer than before to cover the same section. Calculate the speed limit.

6) In winter a train travels between two towns 264 km apart at an average speed of 72 km/h. In summer the journey takes 22 minutes less than in the winter. Find the average speed in summer.

7) A train travels between two towns 135 km apart in $4\frac{1}{2}$ hours. If on the return journey the average speed is reduced by 3 km/h, calculate the time taken for the return journey.

8) A car travels 272 km at an average speed of 32 km/h. On the return journey the average speed is increased to 48 km/h. Calculate the average speed over the whole journey.

SELF TEST 7

In the following questions state the letter (or letters) corresponding to the correct answer (or answers).

1) The average of 11·2, 11·3, 11·5, 11·1 and 11·2 is

a 11·3 b 11·4 c 11·22 d 11·26

2) The average weight of two adults in a family is 81 kg and the average weight of the three children in the family is 23 kg. The average weight of the whole family is

a 45·3 kg b 46·2 kg c 52·0 kg d 20·8 kg

3) 50 litres of oil costing 8p per litre is mixed with 70 litres of oil costing 9p per litre. The average price of the mixture is about

a 8·6 p b 8·5p c 9·6p d 10·8p

4) A grocer sells 20 tins of soup at 5p per tin, 30 at 8p per tin and 40 at 7p per tin. The average price of the soup per tin is

a $6\frac{2}{3}$p b $6\frac{8}{9}$ c 6p d $7\frac{1}{2}$p

5) The average of three numbers is 116. The average of two of them is 98. The third number is

a 18 b 107 c 110 d 152

6) An aeroplane flies non-stop for $2\frac{1}{4}$ hours and travels 1620 km. Its average speed in km/h is

a 720 km/h b 800 km/h c 3645 d 364·5

7) A car travels 50 km at 50 km/h and 70 km at 70 km/h. Its average speed is about

a 60 km/h b 65 km/h c 58 km/h d 62 km/h

8) A car travels for 3 hours at a speed of 45 km/h and for 4 hours at a speed of 50 km/h. It has therefore travelled a distance of

a 27·5 km b 95 km c 335 km d 353 km

9) A car travels 540 km at an average speed of 30 km/h. On the return journey the average speed is doubled to 60 km/h. The average speed over the entire journey is

a 45 km/h b 42 km/h c 40 km/h d 35 km/h.

10) A car travels between two towns 270 km apart in 9 hours. On the return journey the speed is increased by 10 km/h. The time taken for the return journey is

a $6\frac{1}{2}$ hours b $6\frac{3}{4}$ hours c 2·7 hours d $4\frac{1}{2}$ hours

Miscellaneous Exercise

Exercise 21

This exercise is divided into two sections A and B. The questions in Section A are intended to be done very quickly, but those in Section B should take about 20 minutes each to complete. All the questions are of the type found in O Level examination papers.

Section A

1) Find the cost of 12·7 kg of butter at 4·50 francs per kg.

2) Simplify $(2\frac{1}{2}-1\frac{1}{3})\div 1\frac{5}{9}$

3) Powder of total weight 168 kg is packed in 1344 packets all containing equal amounts. Find the number of grammes of powder in each packet.

4) Simplify $\dfrac{2{\cdot}7\times 0{\cdot}6}{0{\cdot}75}$

5) Taking £1 as 13·60 francs express 8·50 francs as pence correct to the nearest penny.

6) The average amount of oil in 3 cans is 16 litres per can. Two of the cans contain amounts which have an average of 13 litres. Find the number of litres contained in the third can.

7) Find the number of years in which a sum of £225 will yield £27 simple interest at the rate of 4% per annum.

8) A cask contains 400 litres of ginger beer. From it 300 glasses are filled each containing $\frac{3}{5}$ of a litre. Calculate the number of cups, each holding $\frac{2}{5}$ of a litre which may then be filled from the remaining contents of the cask.

9) For all goods sold to the value of £1, a salesman receives $4\frac{1}{2}$p commission. Find the amount of his commission after he has sold goods to the total value of £350.

10) A greengrocer bought 200 kg of potatoes at £2·70 per 100 kg and 300 kg of potatoes at £2·20 per 100 kg. He sold them all for a price of 5 kg for 15p. Calculate the total profit he made on the potatoes. Find also the price per 5 kg at which he should have sold these potatoes to make a profit of 75% of his total outlay.

11) Simplify $2\frac{8}{9}\div(1\frac{2}{3}+\frac{1}{2})$

12) Find the simple interest on a sum of £350 for 10 months at $4\frac{1}{2}$% per annum.

13) Find the exact value of 46·002 ÷ 374

14) One machine cuts 1000 kg of coal in 8 minutes and a second machine cuts 1000 kg of coal in 4 minutes. Find the number of minutes taken by both machines working together to cut 1000 kg of coal.

15) A car which travels 10 km on a litre of petrol requires $22\frac{1}{2}$ litres for a journey. Find the number of litres which would be required for the same journey by a car which travels 4 km on a litre of petrol.

16) After 6% of a man's total wage has been deducted he receives a net amount of £32·90. Calculate the amount deducted.

17) The rent of a house was £1·25 per week and the rates in a certain year were £0·81 in the £1 on a rateable value of £80. Find the total of rent and rates paid in a year of 52 weeks.

18) Simplify $(2\frac{1}{2}-1\frac{3}{8})\times 1\frac{1}{3}$

19) Find in francs, the price of 100 kg of coal when 850 kg of this coal costs 199·75 francs.

20) Find the simple interest paid on a loan of £315 for 10 months at 4% per annum.

21) A man's salary of £1575 is increased by 6%. Find his new salary.

22) The annual rent of a field amounts to £93·50 and is shared by two farmers in the ratio of 15 : 7. Find the difference between their shares.

23) The total cost of making 46 km of road was 14·49 million marks. Find the average cost per metre.

24) Simplify $(2\frac{2}{9} \times 1\frac{1}{5}) + 1\frac{5}{6}$

25) The total yield of apples from 144 trees was 23 400 kg. Find in kg the average yield per tree.

26) Simplify $\dfrac{0{\cdot}26 \times 14}{6{\cdot}5}$

27) When a petrol tank is $\frac{7}{8}$ full it contains 31·5 litres of petrol. Find its total capacity.

28) Find the sum which was lent for 4 years at $4\frac{1}{2}$% per annum simple interest if it yielded £81 in interest in that time.

29) If £1 is equivalent to 2·76 dollars find how much 46 cents is worth correct to the nearest half-penny.

30) Find the rateable value of a house on which the amount paid in rates was £96·00 when the rate was 80p in the £1.

31) An airline allows passengers 20 kg of luggage free but charges 10p per kg on excess, a fraction of 1 kg being treated as 1 kg. Find the charge to a passenger who has $39\frac{3}{4}$ kg of luggage.

32) When a number is divided by 73 the quotient is 11 and the remainder is 27. Find the number.

33) Simplify $\dfrac{(3\frac{1}{2} \times 1\frac{1}{2}) - 3}{9}$

34) When the rate of exchange is 13·76 francs to the £1, how many francs does a man receive for £5·75.

35) A sum of £144·50 is divided between two men in the ratio 10 : 7. How much does each receive.

36) The price of a coat was £68·25 and this is reduced by 5%. Calculate the new price of the coat correct to the nearest penny.

37) Twelve bottles of claret cost £8·16. How many bottles can be bought for £10·20?

38) A car runs 4 km on a litre of petrol which costs 8p. What would be the cost of petrol for a journey of 392 km.

39) Calculate correct to the nearest penny how much English money a man would receive for 75 francs when the rate of exchange was 13·6 francs to the £1.

40) For how many months would the simple interest on £60 be £1·40 at 4% per annum?

41) A man buys eggs at £0·18 for 10 and sells them at 2p each. Calculate his percentage profit.

42) A sum of £40·3 is divided between two persons in the ratio 8 : 5. What is the difference in the sums received?

43) Simplify $\dfrac{17{\cdot}6 \times 1{\cdot}5}{0{\cdot}33}$

44) In a class of 24 children the average height was 138 cm. After 4 children left the average height of those remaining was 135 cm. Calculate the average height of the 4 children who left.

Section B

45) A married man is allowed £775 and a single man £595 as a personal allowance. In addition a married man is allowed £200 for every child under 11 years old and £235 for children over 11 years old. Tax is levied at 35% on the taxable income. On the basis of these allowances calculate:

a) the amount paid in income tax by a married man with children aged 9, 12 and 15 if his earnings are £3200 per annum,

b) the amount paid in tax by a single man earning £2800 per annum.

46) A man earns £4800 per year. His taxable income is found by deducting the following allowances:

A married man's allowance of £775
A childrens' allowance of £635
Life assurance relief of £75
Allowance for a mortgage £180.

Calculate his taxable income. Tax is then paid on his taxable income at 35%. Calculate the amount of tax paid per annum.

47) The statistics published by an examining board show that the number of entries for a certain subject was 7962 in the year 1963 and 9519 in the following year. Find the increase per cent, correct to one decimal place. If the increase for the year 1965 was 24% of the entry in 1964, find the entry in 1965. The pass rate fell from 52·4% in 1963 to 49·3% in 1964. Find to the nearest ten, how many more candidates passed in 1964 than in 1963.

48) a) The population of a city increases each year by 2% of its value at the beginning of that year. On 1st Jan., 1965 the population was 1 875 000. Calculate the population of the city on 1st Jan., 1968.

b) A household gas meter was read on Monday, 8th Jan., 1971 and again on Monday 8th April, 1971 the readings being 13 000 cubic metres and 13 200 cubic metres respectively. If the householder is charged 15p per therm for the first 12 therms used and 10p per therm for the remainder, calculate the average weekly cost of gas given that 1000 $m^3 = 160$ therms.

49) The total rateable value of the property in a locality is £1 800 000. Calculate the income from a penny rate. The rateable value of a house on which the owner pays £52·50 in rates per half year is £150. Calculate the rate in the pound which is levied. Find, to the nearest one-tenth of a penny, the rate which must be levied to pay for the collection of refuse in the area if this costs £70 765 annually.

50) A car uses petrol at the rate of 10 km per litre when being driven in rural areas and at the rate of 5 km per litre in towns. If one litre of petrol costs 8p calculate the cost of petrol consumed on a 450 km journey five-sixths of which is through rural areas and the remainder through towns. If the other expenses for the journey amounted to 3p per km calculate the total cost of the journey. If the car sets out on the journey at 5 a.m. and travels at an average speed of 60 km per hour through rural areas and 25 km per hour through towns, calculate the time of completing the journey allowing 3 hour 10 minutes for stops on the way.

51) On 1st Jan. a man deposits £250 in a new account on which 4% simple interest is paid. Calculate the total amount in his account at the end of 3 years. Income tax at 40p in the £ is now paid on the total interest over the three years. Calculate the amount of tax paid. After allowing for this tax payment, express his total net return from the deposit over the whole period of 3 years as a percentage of his original deposit.

52) A man finds in one year that the total cost of running his car is £180 which includes £17·50 road tax and £50·50 petrol tax. If on every pound earned the man has to pay 40p income tax how much does the man have to earn to pay for the running of his car? Find the percentage of this amount he paid in taxes.

53) A television set can be bought for £67·10, maintained at £8·65 per year for 7 years, then sold for £10·50. Renting the set, maintenance

included, costs £2 a calendar month for 3 years, £1·90 a month for the next 2 years and £1·75 a month thereafter. Find how much is saved in 7 years by buying the set.

54) The rateable value of a house used to be £63 and the annual amount paid was £75·60. Find how much in the £ this was. The new rateable value is £174 and the rate is 48p in the £. Find the increase or decrease in the amount paid. The water rate was 8% of the old rateable value and is 3% of the new one, both reckoned to the nearest penny. Find the increase or the decrease in the amount paid for water.

55) A certain mixture of glycerine and water contains 35% of glycerine by weight. To every 100 grammes of the mixture are added 25 grammes of water. Find what percentage of the weight of the diluted mixture is glycerine.

Chapter 8 Basic Algebra

FINDING THE VALUES OF ALGEBRAIC EXPRESSIONS

The value of an algebraic expression is found by substituting the given values for the symbols in the expression.

Examples

1) Find the value of $3x+7$ when $x=5$

When $x=5$, $3x+7=(3\times5)+7=15+7=22$ (note that the multiplication sign appears when x is given its numerical value).

2) If $x=3$, $y=4$ and $z=5$, find the value of $8yz+5xz$

$$8yz+5xz=(8\times4\times5)+(5\times3\times5)$$
$$=160+75=235$$

POWERS AND INDICES

The quantity $2\times2\times2\times2$ is written 2^4 and is called the **fourth power** of 2. The small figure 4, which gives the number of 2's to be multiplied together, is called the **index** (plural: **indices**). Also,

$$a\times a\times a=a^3$$
$$y\times y\times y\times y\times y=y^5$$

The expression $3x^2$ means $3\times x\times x$ whereas the expression $(3x)^2=3x\times3x=9x^2$. Similarly

$$a^2b=a\times a\times b \quad \text{and} \quad ab^2=a\times b\times b$$

Examples

1) Find the value of a) $7a^2bc^3$ b) $(3a^2bc)^2$ if $a=2$, $b=4$ and $c=5$

a) $7a^2bc^3=7\times2^2\times4\times5^3=7\times4\times4\times125$
$=14\,000$

b) $(3a^2bc)^2=(3\times2^2\times4\times5)^2=240^2$
$=57\,600$

2) Find the value of $\dfrac{b^2c^3}{a^4}$ when $b=5$, $c=3$ and $a=2$.

$$\frac{b^2c^3}{a^4}=\frac{5^2\times3^3}{2^4}=\frac{25\times27}{16}=\frac{675}{16}=42\tfrac{3}{16}$$

Exercise 22

1) If $a=1$, $b=2$ and $c=3$ find the values of the following:

a) $2a+3$ b) $b-2$
c) $2b$ d) $4c+4$
e) $3abc$ f) $4b-4$
g) $\dfrac{c}{3}$ h) $a+2b+c$
i) $3a+b-c$ j) $a-b+c$
k) $c+b-a$ l) $\dfrac{5bc}{a}$

2) If $x=5$, $y=2$ and $z=3$ find the values of the following:

a) $7xy+3z$ b) $\dfrac{x}{5}+y+\dfrac{2z}{3}$
c) $\dfrac{x+y}{z+x}$ d) $\dfrac{2xy}{3z+8}$
e) $\dfrac{3z-x}{2x-3y}$ f) $\dfrac{3xy+7y}{8z}$

3) If $p=3$, $q=4$ and $r=2$ find the values of the following:

a) p^2 b) q^3 c) $4q^2$
d) p^2q^3r e) q^2-p^2 f) $6p^2q^3$
g) $(p^2q)^3$ h) $(3r^2q^2)^2$ i) $4p+3r^3$
j) $\dfrac{p^3q^2}{r^2}$ k) $\sqrt{3p^3q}$ l) $\sqrt{\dfrac{r^4}{p^2q^2}}$

DIRECTED NUMBERS

Figure 8.1 shows part of a centigrade thermometer. The freezing point of water is 0°C. Temperatures above freezing point may be read off the scale directly and so may those below freezing. We now have to decide on a method of showing whether a temperature is above or below zero. We may say that a temperature is 6°C above zero or a temperature is 5°C below zero. These statements are not compact enough for calculations. Therefore we say that a temperature of $+6$°C is a temperature 6°C above freezing point. A temperature of 5°C below freezing point is then written as -5°C. Thus we have used the signs $+$ and $-$ to indicate a *change of direction.*

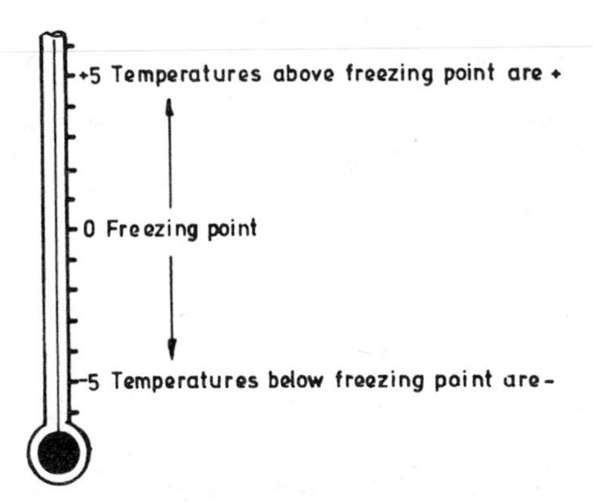

Fig. 8.1

Again if starting from a given point distances measured to the *right* are regarded as *positive* then distances measured to the *left* are regarded as *negative*.
Numbers which have a sign attached to them are called **directed numbers.** Thus $+7$ is a positive number and -7 is a negative number.

RULES FOR THE USE OF DIRECTED NUMBERS

The addition of positive and negative numbers

In Fig. 8.2 below, a movement from left to right (that is, in the direction OA) is regarded as positive, whilst a movement from right to left (that is, in the direction OB) is regarded as negative.

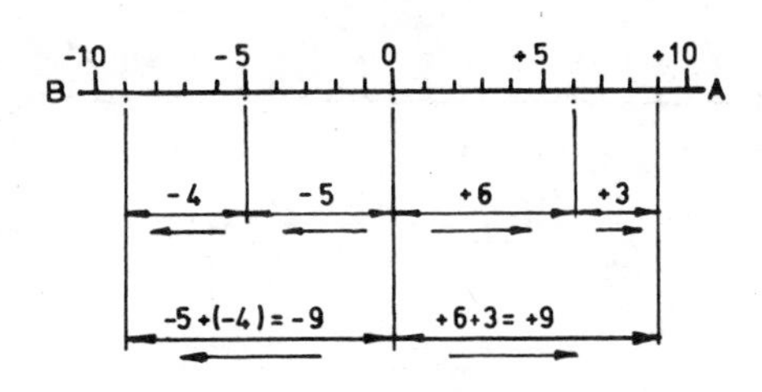

Fig. 8.2

To find the value of $+6+3$ Measure 6 units to the right of O and then a further 3 units to the right. The final position is 9 units to the right of O.

$$\therefore +6+3=+9$$

To find the value of $-5+(-4)$. Measure 5 units to the left of O and then a further 4 units to the left. The final position is 9 units to the left of O.

$$\therefore -5+(-4)=-9$$

From these results we obtain the rule:
To add several numbers together whose signs are the same add the numbers together. The sign of the result is the same as the sign of each of the numbers.

Positive signs are frequently omitted, as shown in the following examples:

Examples

1) $+5+9=+14$

More often this is written

$$5+9=14$$

2) $-7+(-9)=-16$

More often this is written

$$-7-9=-16$$

3) $-7-6-4=-17$

The addition of numbers having different signs

To find the value of $-4+11$. Measure 4 units to the left of O and from this point measure

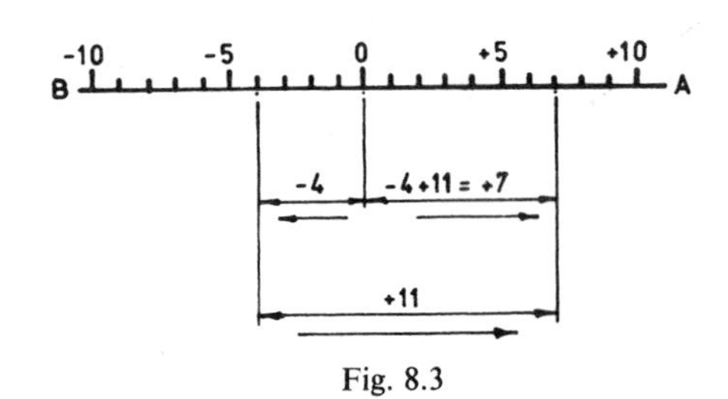

Fig. 8.3

11 units to the right, as shown in Fig. 8.3. Thus,

$$-4+11=7$$

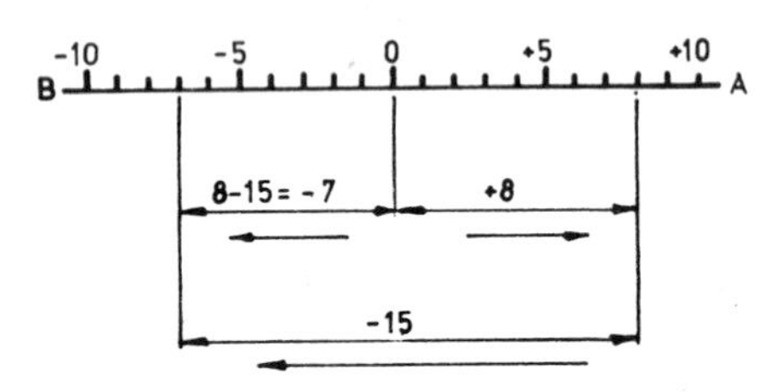

Fig. 8.4

To find the value of $8-15$. Measure 8 units to the right of O and from this point measure 15 units to the left, as shown in Fig. 8.4. Thus,

$$8-15=-7$$

From these results we obtain the rule:
To add two numbers together whose signs are different, subtract the numerically smaller from the larger. The sign of the result will be the same as the sign of the larger number.

Examples
1) $-12+6=-6$

2) $11-16=-5$

When dealing with several numbers having mixed signs add the positive and negative numbers together separately. The set of numbers is then reduced to two numbers, one positive and one negative, which are added in the way just shown. For example:

$$-16+11-7+3+8=-23+22=-1$$

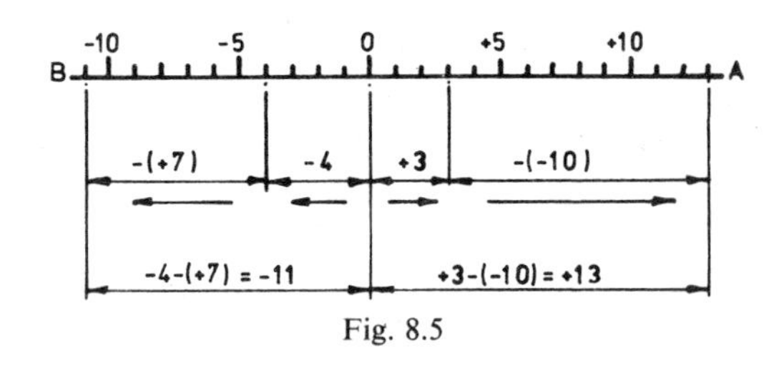

Fig. 8.5

The subtraction of directed numbers

To find the value of $-4-(+7)$ From Fig. 8.5 it will be seen that $-(+7)$ is the same as -7 and, therefore,

$$-4-(+7)=-4-7=-11$$

To find the value of $+3-(-10)$. From Fig. 8.5 it will be seen that $-(-10)$ is the same as $+10$ and, therefore,

$$+3-(-10)=3+10=13$$

The rule is:
To subtract a directed number change its sign and add the resulting number.

Examples
1) $-10-(-6)=-10+6=-4$

2) $7-(+8)=7-8=-1$

3) $8-(-3)=8+3=11$

Multiplication of directed numbers

Now

$$5+5+5=15$$

that is,

$$3\times 5=15$$

Thus two positive numbers multiplied together give a positive result. Now

$$(-5)+(-5)+(-5)=-15$$

that is

$$3\times(-5)=-15$$

Thus a positive number multiplied by a negative number gives a negative result. Suppose we wish to find the value of $(-3)\times(-5)$. We can write (-3) as $-(+3)$, and hence

$$(-3)\times(-5)=-(+3)\times(-5)$$
$$=-(-15)=+15$$

Thus a negative number multiplied by a negative number gives a positive result. We may summarize the results as follows:

$(+)\times(+)=+$ $\quad(-)\times(+)=-$

$(+)\times(-)=-$ $\quad(-)\times(-)=+$

and the rule is:
The product of two numbers with like signs is positive whilst the product of two numbers with unlike signs is negative.

Examples
1) $7\times 4=28$

2) $7\times(-4)=-28$

3) $(-4)\times 7=-28$

4) $(-7)\times(-4)=28$

Division of directed numbers

The rules must be similar to those used for multiplication since, if $3\times(-5)=-15$,

$$\frac{-15}{3}=-5$$

Also

$$\frac{-15}{-5}=3$$

The rule is:
When dividing, numbers with like signs give a positive answer and numbers with unlike signs give a negative answer.

Examples

1) $\frac{20}{4}=5$ $\quad$ 2) $\frac{20}{-4}=-5$

3) $\frac{-20}{4}=-5$ $\quad$ 4) $\frac{-20}{-4}=5$

5) $\frac{(-9)\times(-4)\times 5}{3\times(-2)}=\frac{36\times 5}{-6}=\frac{180}{-6}=-30$

Exercise 23

Find the values of the following:

1) $+8+7$ $\quad$ 2) $-7-5$

3) $-15-17$ $\quad$ 4) $8+6$

5) $+6-11$ $\quad$ 6) $7-16$

7) $-8-6$ $\quad$ 8) $-5+10$

9) $12-7$ $\quad$ 10) $-5-15$

11) $-7-6-3$ $\quad$ 12) $-8+9-2$

13) $15-7-8$ $\quad$ 14) $23-21-8-2$

15) $-7-11-8+19-13$

16) $8-(+6)$ $\quad$ 17) $-5-(-8)$

18) $8-(-6)$ $\quad$ 19) $-3-(-7)$

20) $-4-(-5)$ $\quad$ 21) $-2-(+3)$

22) $-4-(-7)-(-4)$

23) $-11-(+4)-(-3)$

24) $7\times(-6)$ $\quad$ 25) $(-6)\times 7$

26) 7×6 $\quad$ 27) $(-7)\times(-6)$

28) $(-2)\times(-4)\times(-6)$ $\quad$ 29) $(-2)^2$

30) $3\times(-4)\times(-2)\times 5$ $\quad$ 31) $(-3)^3$

32) $6\div(-2)$ $\quad$ 33) $(-6)\div 2$

34) $(-6)\div(-2)$ $\quad$ 35) $6\div 2$

36) $(-10)\div 5$ $\quad$ 37) $1\div(-1)$

38) $(-2)\div(-4)$ $\quad$ 39) $(-3\div 3)$

40) $8\div(-4)$ $\quad$ 41) $\frac{(-6)\times(-4)}{(-2)}$

42) $\frac{(-8)}{(-4)\times(-2)}$

43) $\frac{-3\times(-4)\times(-2)}{(-2)\times(-6)}$

44) $\frac{4\times(-6)\times(-8)}{(-3)\times(-2)\times(-4)}$

45) $(-1)^3\times(-6)\div(-2)$

46) $2^3\times(-3)^2\div(-1)^3$

ADDITION OF ALGEBRAIC TERMS

Like terms are numerical multiples of the same algebraic quantity. Thus,

$7x$, $5x$ and $-3x$

are three like terms.
An expression consisting of like terms can be reduced to a single term by adding the numerical coefficients together. Thus

$$7x-5x+3x=(7-5+3)x=5x$$

$$3b^2+7b^2=(3+7)b^2=10b^2$$

$$-3y-5y=(-3-5)y=-8y$$

$$q-3q=(1-3)q=-2q$$

Only like terms can be added or subtracted. Thus $7a+3b-2c$ is an expression containing three unlike terms and it cannot be simplified any further. Similarly with $8a^2b+7ab^3-6a^2b^2$ which are all unlike terms.
It is possible to have several sets of like terms in an expression and each set can then be simplified:

$$8x+3y-4z-5x+7z-2y+2z$$
$$=(8-5)x+(3-2)y+(-4+7+2)z$$
$$=3x+y+5z$$

MULTIPLICATION AND DIVISION OF ALGEBRAIC QUANTITIES

The rules are exactly the same as those used with directed numbers:

$$(+x)(+y)=+(xy)=+xy=xy$$

$$5x\times 3y=5\times 3\times x\times y=15xy$$

$$(x)(-y)=-(xy)=-xy$$

$$(2x)(-3y)=-(2x)(3y)=-6xy$$

$$(-4x)(2y)=-(4x)(2y)=-8xy$$

$$(-3x)(-2y)=+(3x)(2y)=6xy$$

$$\frac{+x}{+y}=+\frac{x}{y}=\frac{x}{y}$$

$$\frac{-3x}{2y}=-\frac{3x}{2y}$$

$$\frac{-5x}{-6y}=+\frac{5x}{6y}=\frac{5x}{6y}$$

$$\frac{4x}{-3y}=-\frac{4x}{3y}$$

When *multiplying* expressions containing the same symbols, indices are used:

$$m\times m=m^2$$

$$3m\times 5m=3\times m\times 5\times m=15m^2$$

$$(-m)\times m^2=(-m)\times m\times m=-m^3$$

$$5m^2n\times 3mn^3$$
$$=5\times m\times m\times n\times 3\times m\times n\times n\times n$$
$$=15m^3n^4$$

$$3mn\times(-2n^2)$$
$$=3\times m\times n\times(-2)\times n\times n=-6mn^3$$

When *dividing* algebraic expressions, cancellation between numerator and denominator is often possible. Cancelling is equivalent to dividing both numerator and denominator by the same quantity:

$$\frac{pq}{p}=\frac{\not p\times q}{\not p}=q$$

$$\frac{3p^2q}{6pq^2}=\frac{3\times\not p\times p\times\not q}{6\times\not p\times\not q\times q}=\frac{3p}{6q}=\frac{p}{2q}$$

$$\frac{18x^2y^2z}{6xyz}=\frac{18\times\not x\times x\times\not y\times y\times\not z}{6\times\not x\times\not y\times\not z}=3xy$$

Exercise 24

Simplify the following:

1) $7x+11x$
2) $7x-5x$
3) $3x-6x$
4) $-2x-4x$
5) $-8x+3x$
6) $-2x+7x$
7) $8a-6a+7a$
8) $5m+13m-6m$
9) $6b^2-4b^2+3b^2$
10) $6ab-3ab-2ab$
11) $14xy+5xy-7xy+2xy$
12) $-5x+7x-3x-2x$

13) $-4x^2-3x^2+2x^2-x^2$

14) $3x-2y+4z-2x-3y+5z+6x+2y-3z$

15) $3a^2b+2ab^3+4a^2b^2-5ab^3+11b^4+6a^2b$

16) $1{\cdot}2x^3-3{\cdot}4x^2+2{\cdot}6x+3{\cdot}7x^2+3{\cdot}6x-2{\cdot}8$

17) $pq+2{\cdot}1qr-2{\cdot}2rq+8qp$

18) $2{\cdot}6a^2b^2-3{\cdot}4b^3-2{\cdot}7a^3-3a^2b^2-2{\cdot}1b^3+1{\cdot}5a^3$

19) $2x\times 5y$

20) $3a\times 4b$

21) $3\times 4m$

22) $\frac{1}{4}q\times 16p$

23) $x\times(-y)$

24) $(-3a)\times(-2b)$

25) $8m\times(-3n)$

26) $(-4a)\times 3b$

27) $8p\times(-q)\times(-3r)$

28) $3a\times(-4b)\times(-c)\times 5d$

29) $12x\div 6$

30) $4a\div(-7b)$

31) $(-5a)\div 8b$

32) $(-3a)\div(-3b)$

33) $4a\div 2b$

34) $4ab\div 2a$

35) $12x^2yz^2\div 4xz^2$

36) $(-12a^2b)\div 6a$

37) $8a^2bc^2\div 4ac^2$

38) $7a^2b^2\div 3ab$

39) $a\times a$

40) $b\times(-b)$

41) $(-m)\times m$

42) $(-p)\times(-p)$

43) $3a\times 2a$

44) $5X\times X$

45) $5q\times(-3q)$

46) $3m\times(-3m)$

47) $(-3pq)\times(-3q)$

48) $8mn\times(-3m^2n^3)$

49) $7ab\times(-3a^2)$

50) $2q^3r^4\times 5qr^2$

51) $(-3m)\times 2n\times(-5p)$

52) $5a^2\times(-3b)\times 5ab$

53) $m^2n\times(-mn)\times 5m^2n^2$

BRACKETS

Brackets are used for convenience in grouping terms together. When removing brackets *each term* within the bracket is multiplied by the quantity outside the bracket:

$3(x+y)=3x+3y$

$5(2x+3y)=5\times 2x+5\times 3y=10x+15y$

$4(a-2b)=4\times a-4\times 2b=4a-8b$

$m(a+b)=ma+mb$

$3x(2p+3q)=3x\times 2p+3x\times 3q=6px+9qx$

$4a(2a+b)=4a\times 2a+4a\times b=8a^2+4ab$

When a bracket has a minus sign in front of it, the signs of all the terms inside the bracket are changed when the bracket is removed. The reason for this rule may be seen from the following examples:

$$-3(2x-5y)=(-3)\times 2x+(-3)\times(-5y)$$
$$=-6x+15y$$

$-(m+n)=-m-n$

$-(p-q)=-p+q$

$-2(p+3q)=-2p-6q$

When simplifying expressions containing brackets first remove the brackets and then add the like terms together:

$$(3x+7y)-(4x+3y)=3x+7y-4x-3y$$
$$=-x+4y$$

$$3(2x+3y)-(x+5y)=6x+9y-x-5y$$
$$=5x+4y$$

$$x(a+b)-x(a+3b)=ax+bx-ax-3bx$$
$$=-2bx$$

$$2(5a+3b)+3(a-2b)=10a+6b+3a-6b$$
$$=13a$$

Exercise 25

Remove the brackets in the following:

1) $3(x+4)$

2) $2(a+b)$

3) $3(3x+2y)$

4) $\frac{1}{2}(x-1)$

5) $5(2p-3q)$

6) $7(a-3m)$

7) $-(a+b)$

8) $-(a-2b)$

9) $-(3p-3q)$

10) $-(7m-6)$

11) $-4(x+3)$

12) $-2(2x-5)$

13) $-5(4-3x)$

14) $2k(k-5)$

15) $-3y(3x+4)$

16) $a(p-q-r)$

17) $4xy(ab-ac+d)$

18) $3x^2(x^2-2xy+y^2)$

19) $-7P(2P^2-P+1)$

20) $-2m(-1+3m-2n)$

Remove the brackets and simplify:

21) $3(x+1)+2(x+4)$

22) $5(2a+4)-3(4a+2)$

23) $3(x+4)-(2x+5)$

24) $4(1-2x)-3(3x-4)$

25) $5(2x-y)-3(x+2y)$

26) $\frac{1}{2}(y-1)+\frac{1}{3}(2y-3)$

27) $-(4a+5b-3c)-2(2a+3b-4c)$

28) $2x(x-5)-x(x-2)-3x(x-5)$

29) $3(a-b)-2(2a-3b)+4(a-3b)$

30) $3x(x^2+7x-1)-2x(2x^2+3)-3(x^2+5)$

SELF TEST 8

In questions 1 to 50 the answer is either 'true' of 'false'.

1) The sum of two numbers can be represented by the expression $a+b$ _____

2) The expression $a-b$ represents the difference of two numbers a and b _____

3) The product of 8 and x is $8+x$ _____

4) 3 times a number minus 7 can be written as $3x-7$ _____

5) Two numbers added together minus a third number and the result divided by a fourth number may be written as $(a+b-c)\div d$ _____

6) The value of $3a+7$ when $a=5$ is 36. _____

7) The value of $8x-3$ when $x=3$ is 21. _____

8) The value of $3b-2c$ when $b=4$ and $c=3$ is 6. _____

9) The value of $8ab\div 3c$ when $a=6$, $b=4$ and $c=2$ is 32. _____

10) The quantity $a\times a\times a\times a$ is written a^3 _____

11) The quantity $y\times y\times y$ is written y^3 _____

12) a^3b^2 is equal to $a\times a\times a\times b\times b$ _____

13) The value of a^4 when $a=3$ is 81 _____

14) When $x=2$, $y=3$ and $z=4$ the value of $2x^2y^3z$ is 258 _____

15) $13-8-9$ is equal to 4 _____

16) $-8+11-20$ is equal to -17 _____

17) $-11+3-(-8)$ is equal to -16 _____

18) $5-(-7)+8$ is equal to 20 _____

19) $(-5)\times(-7)$ is equal to -35 _____

20) $(-8)\times(-3)$ is equal to 24 _____

21) $(-3)\times 7\times(-8)\times(-2)$ is equal to -336 _____

22) $(-5)^3$ is equal to 125 _____

23) $(-8)^2$ is equal to -64 _____

24) $(-2)^5$ is equal to -32 _____

25) $(-8)\div(-4)$ is equal to 2 _____

26) $(-6)\times(-4)\times 3\div(-8)$ is equal to 9 _____

27) $5x+8x$ is equal to $13x^2$ _____

28) $3x+6x$ is equal to $9x$ _____

29) $8x-5x$ is equal to 3 _____

30) $7x-2x$ is equal to $5x$ _____

31) $15xy+7xy-3xy-2xy$ is equal to $17xy$ _____

32) $8a \times 5a$ is equal to $40a$ - - - - -

33) $9x \times 5x$ is equal to $45x^2$ - - - - -

34) $(-5x) \times (-8x) \times 3x$ is equal to $120x^3$ - - - - -

35) a^2b is the same as ba^2 - - - - -

36) $5x^3y^2z$ is the same as $5y^2zx^3$ - - - - -

37) $8a^3b^2c^4$ and $16a^3b^3c^3$ are like terms - - - - -

38) $6x^2 \div (-3x)$ is equal to $3x$ - - - - -

39) $(-5pq^2) \times (-8p^2q)$ is equal to $40p^3q^3$ - - - - -

40) $a^2b^2 \times (-a^2b^2) \times 5a^2b^2$ is equal to $5a^2b^2$ - - - - -

41) $3(2x+7)$ is equal to $6x+7$ - - - - -

42) $5(3x+4)$ is equal to $15x+20$ - - - - -

43) $4(x+8)$ is equal to $4x+32$ - - - - -

44) $-(3x+5y)$ is equal to $-3x+5y$ - - - - -

45) $-(2a+3b)$ is equal to $-2a-3b$ - - - - -

46) $4x(3x-2xy)$ is equal to $12x^2-8x^2y$ - - - - -

47) $-8a(a-3b)$ is equal to $-8a^2-24ab$ - - - - -

48) $3(x-y)-5\,(2x-3y)$ is equal to $12y-7x$ - - - - -

49) $2x(x-2)-3x(x^2-5)$ is equal to $-3x^3+2x^2-19x$ - - - - -

50) $3a(2a^2+3a-1)-2a(3a^2+3)$ is equal to $9a^2+3a$ - - - - -

Chapter 9 Factorisation

HIGHEST COMMON FACTOR (H.C.F.)

The H.C.F. of a set of algebraic expressions is the highest expression which is a factor of each of the given expressions. To find the H.C.F. we therefore select the lowest power of each of the quantities which occur in *all* of the expressions and multiply them together. The method is shown in the following examples.

Examples

1) Find the H.C.F. of ab^2c^2, $a^2b^3c^3$, $a^2b^4c^4$.
Each expression contains the quantities a, b and c. To find the H.C.F. choose the *lowest* power of each of the quantities which occur in the three expressions and multiply them together. The lowest power of a is a, the lowest power of b is b^2 and the lowest power of c is c^2. Thus,

$$\text{H.C.F.} = ab^2c^2$$

2) Find the H.C.F. of x^2y^3, $x^3y^2z^2$, xy^2z^3. We notice that only x and y appear in *all* three expressions. The quantity z appears in only two of the expressions and cannot therefore appear in the H.C.F. To find the H.C.F. choose the lowest powers of x and y which occur in the three expressions and multiply them together. Thus,

$$\text{H.C.F.} = xy^2$$

3) Find the H.C.F. of $3m^2np^3$, $6m^3n^2p^2$, $24m^3p^4$. Dealing with the numerical coefficients 3, 6 and 24 we note that 3 is a factor of each of them. The quantities m and p occur in all three expressions, their lowest powers being m^2 and p^2. Hence,

$$\text{H.C.F.} = 3m^2p^2$$

FACTORISING

A factor is a common part of two or more terms which make up an algebraic expression. Thus the expression $3x+3y$ has two terms which have the number 3 common to both of them. Thus $3x+3y=3(x+y)$. We say that 3 and $(x+y)$ are the factors of $3x+3y$.
To factorise algebraic expressions of this kind, we first find the H.C.F. of all the terms making up the expression. The H.C.F. then appears outside the bracket. To find the terms inside the bracket divide each of the terms making up the expression by the H.C.F.

Examples

1) Find the factors of $ax+bx$.
The H.C.F. of ax and bx is x

$$\therefore ax+bx = x(a+b)$$

$$\left(\text{since } \frac{ax}{x} = a \text{ and } \frac{bx}{x} = b\right)$$

2) Find the factors of m^2n-2mn^2.
The H.C.F. of m^2n and $2mn^2$ is mn

$$\therefore m^2n-2mn^2 = mn(m-2n)$$

$$\left(\text{since } \frac{m^2n}{mn} = m \text{ and } \frac{2mn^2}{mn} = 2n\right)$$

3) Find the factors of $3x^4y+9x^3y^2-6x^2y^3$.
The H.C.F. of $3x^4y$, $9x^3y^2$ and $6x^2y^3$ is $3x^2y$

$$\therefore 3x^4y+9x^3y^2-6x^2y^3 = 3x^2y(x^2+3xy-2y^2)$$

$$\left(\text{since } \frac{3x^4y}{3x^2y} = x^2,\ \frac{9x^3y^2}{3x^2y} = 3xy \text{ and } \frac{6x^2y^3}{3x^2y} = 2y^2\right)$$

4) Find the factors of $\frac{ac}{x}+\frac{bc}{x^2}-\frac{cd}{x^3}$

The H.C.F. of $\frac{ac}{x}$, $\frac{bc}{x^2}$ and $\frac{cd}{x^3}$ is $\frac{c}{x}$

$$\therefore \frac{ac}{x}+\frac{bc}{x^2}-\frac{cd}{x^3}=\frac{c}{x}\left(a+\frac{b}{x}-\frac{d}{x^2}\right)$$

$$\left(\text{since } \frac{ac}{x}\div\frac{c}{x}=a,\ \frac{bc}{x^2}\div\frac{c}{x}=\frac{b}{x} \text{ and } \frac{cd}{x^3}\div\frac{c}{x}=\frac{d}{x^2}\right)$$

Exercise 26

Find the H.C.F. of the following:

1) p^3q^2, p^2q^3, p^2q

2) $a^2b^3c^3$, a^3b^3, ab^2c^2

3) $3mn^2$, $6mnp$, $12m^2np^2$

4) $2ab$, $5b$, $7ab^2$

5) $3x^2yz$, $12x^2yz$, $6xy^2z^3$, $3xyz^2$

Factorise the following:

6) $2x+6$

7) $4x-4y$

8) $5x-5$

9) $4x-8xy$

10) $mx-my$

11) $ax+bx+cx$

12) $\frac{x}{2}-\frac{y}{8}$

13) $5a-10b+15c$

14) ax^2+ax

15) $2\pi r^2+\pi rh$

16) $3y-9y^2$

17) ab^3-a^2b

18) $x^2y^2-axy+bxy^2$

19) $5x^3-10x^2y+15xy^2$

20) $9x^3y-6x^2y^2+3xy^5$

21) $I_0+I_0\alpha t$

22) $\frac{x}{3}-\frac{y}{6}+\frac{z}{9}$

23) $2a^2-3ab+b^2$

24) x^3-x^2+7x

25) $\frac{m^2}{pn}-\frac{m^3}{pn^2}+\frac{m^4}{p^2n^2}$

26) $\frac{x^2y}{2a}-\frac{2xy^2}{5a^2}+\frac{xy^3}{a^3}$

27) $\frac{l^2m^2}{15}-\frac{l^2m}{20}+\frac{l^3m^2}{10}$

28) $\frac{a^3}{2x^3}-\frac{a^2b}{4x^4}-\frac{a^2c}{6x^3}$

THE PRODUCT OF TWO BINOMIAL EXPRESSIONS

A binomial expression consists of two terms. Thus $3x+5$, $a+b$, $2x+3y$ and $4p-q$ are all binomial expressions.
To find the product of $(a+b)(c+d)$ consider the diagram (Fig. 9.1).

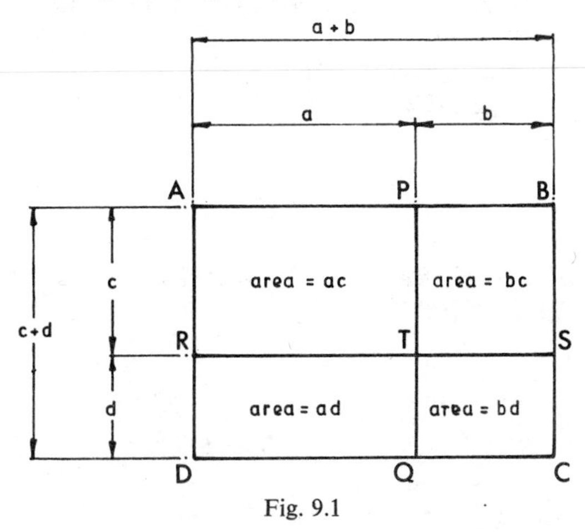

Fig. 9.1

In Fig 9.1 the rectangular area ABCD is made up as follows:

$$ABCD = APTR + TQDR + PBST + STQC$$

i.e. $(a+b)(c+d)=ac+ad+bc+bd$

It will be noticed that the expression on the right hand side is obtained by multiplying each term in the one bracket by each term in the other bracket. The process is illustrated below where each pair of terms connected by a line are multiplied together.

$$(a+b)(c+d)=ac+ad+bc+bd$$

Examples

1) $(3x+2)(4x+5)$
$$=3x\times 4x+3x\times 5+2\times 4x+2\times 5$$
$$=12x^2+15x+8x+10$$
$$=12x^2+23x+10$$

2) $(2p-3)(4p+7)$
$$=2p\times 4p+2p\times 7-3\times 4p-3\times 7$$
$$=8p^2+14p-12p-21$$
$$=8p^2+2p-21$$

3) $(z-5)(3z-2)$
$$=z\times 3z+z\times(-2)-5\times 3z-5\times(-2)$$
$$=3z^2-2z-15z+10$$
$$=3z^2-17z+10$$

4) $(2x+3y)(3x-2y)$
$$=2x\times 3x+2x\times(-2y)+3y\times 3x+3y\times(-2y)$$
$$=6x^2-4xy+9xy-6y^2$$
$$=6x^2+5xy-6y^2$$

THE SQUARE OF A BINOMIAL EXPRESSION

$$(a+b)^2=(a+b)(a+b)=a^2+ab+ba+b^2$$
$$=a^2+2ab+b^2$$

$$(a-b)^2=(a-b)(a-b)=a^2-ab-ba+b^2$$
$$=a^2-2ab+b^2$$

The square of a binomial expression is the sum of the squares of the two terms and twice their product.

Examples

1) $(2x+5)^2=(2x)^2+2\times 2x\times 5+5^2$
$$=4x^2+20x+25$$

2) $(3x-2)^2=(3x)^2+2\times 3x\times(-2)+(-2)^2$
$$=9x^2-12x+4$$

3) $(2x+3y)^2=(2x)^2+2\times 2x\times 3y+(3y)^2$
$$=4x^2+12xy+9y^2$$

THE SUM AND DIFFERENCE OF TWO TERMS

$$(a+b)(a-b)=a^2-ab+ba-b^2=a^2-b^2$$

This result is the difference of the squares of the two terms.

Examples

1) $(8x+3)(8x-3)=(8x)^2-3^2=64x^2-9$

2) $(2x+5y)(2x-5y)$
$$=(2x)^2-(5y)^2=4x^2-25y^2$$

Exercise 27

Find the products of the following:

1) $(x+1)(x+2)$
2) $(x+3)(x+1)$
3) $(x+4)(x+5)$
4) $(2x+5)(x+3)$
5) $(3x+7)(x+6)$
6) $(5x+1)(x+4)$
7) $(2x+4)(3x+2)$
8) $(5x+1)(2x+3)$
9) $(7x+2)(3x+5)$
10) $(x-1)(x-3)$
11) $(x-4)(x-2)$
12) $(x-6)(x-3)$
13) $(2x-1)(x-4)$
14) $(x-2)(3x-5)$
15) $(x-8)(4x-1)$
16) $(2x-4)(3x-2)$
17) $(3x-1)(2x-5)$
18) $(7x-5)(3x-2)$
19) $(x+3)(x-1)$
20) $(x-2)(x+7)$
21) $(x-5)(x+3)$
22) $(2x+5)(x-2)$
23) $(3x-5)(x+6)$
24) $(3x+5)(x+6)$
25) $(3x+5)(2x-3)$
26) $(6x-7)(2x+3)$
27) $(3x-5)(2x+3)$
28) $(3x+2y)(x+y)$
29) $(2p-q)(p-3q)$
30) $(3v+2u)(2v-3u)$
31) $(2a+b)(3a-b)$
32) $(5a-7)(a-6)$
33) $(3x+4y)(2x-3y)$
34) $(x+1)^2$
35) $(2x+3)^2$
36) $(3x+7)^2$
37) $(x-1)^2$
38) $(3x-5)^2$
39) $(2x-3)^2$
40) $(2a+3b)^2$
41) $(x+y)^2$
42) $(P+3Q)^2$
43) $(a-b)^2$
44) $(3x-4y)^2$

45) $(M-2N)^2$ 46) $(x-1)(x+1)$

47) $(x-3)(x+3)$ 48) $(x+7)(x-7)$

49) $(2x+5)(2x-5)$ 50) $(3x-7)(3x+7)$

51) $(2x-1)(2x+1)$ 52) $(a+b)(a-b)$

53) $(3x-2y)(3x+2y)$ 54) $(5a+2b)(5a-2b)$

FACTORISING BY GROUPING

To factorise the expression $ax+ay+bx+by$ first group the terms in pairs so that each pair of terms has a common factor. Thus,

$$\begin{aligned} ax+ay+bx+by &= (ax+ay)+(bx+by) \\ &= a(x+y)+b(x+y) \end{aligned}$$

Now notice that in the two terms $a(x+y)$ and $b(x+y)$, $(x+y)$ is a common factor. Hence,

$$a(x+y)+b(x+y)=(x+y)(a+b)$$
$$\therefore ax+ay+bx+by=(x+y)(a+b)$$

Similarly,

$$\begin{aligned} np+mp-qn-qm &= (np+mp)-(qn+qm) \\ &= p(n+m)-q(n+m) \\ &= (n+m)(p-q) \end{aligned}$$

FACTORS OF QUADRATIC EXPRESSIONS

A quadratic expression is one in which the highest power of the symbol used is the square. For instance, x^2-5x+3 and $3x^2-9$ are both quadratic expressions.

Case 1. Where the coefficient of the squared term is unity

$$(x+4)(x+3)=x^2+7x+12$$

Note that in the quadratic expression $x^2+7x+12$ the last term 12 has the factors 4×3. Also the coefficient of x is 7 which is the sum of the factors 4 and 3. This example gives the clue whereby quadratic expressions may be factorised.

Examples

1) Factorise x^2+6x+5
We note that $5=5\times 1$ and $5+1=6$

$$\therefore x^2+6x+5=(x+5)(x+1)$$

2) Factorise $x^2+10x+16$
Now $16=16\times 1$ or 8×2 or 4×4
But the sum of the factors must be 10. The only factors which add up to 10 are 8 and 2.

$$\therefore x^2+10x+16=(x+8)(x+2)$$

3) Factorise x^2+4x-5
Now $-5=(-5)\times 1$ or $5\times(-1)$
But the sum of the factors must be 4. Since $5+(-1)=4$ then,

$$x^2+4x-5=(x+5)(x-1)$$

4) Factorise $x^2+3x-10$
Now $-10=(-10)\times 1$ or $10\times(-1)$ or $(-5)\times 2$ or $5\times(-2)$.
Since the sum of the factors must be 3, the only pair that meets this requirement is 5 and -2. Hence

$$x^2+3x-10=(x+5)(x-2)$$

5) Factorise $x^2-8x+15$
Now $15=15\times 1$ or 5×3 or $(-5)\times(-3)$ or $(-1)\times(-15)$. But the sum of the factors must be -8. Since $-5+(-3)=-8$ then,

$$x^2-8x+15=(x-5)(x-3)$$

Case 2. Where the coefficient of the squared term is not unity

In this case we find all the possible factors of the first and last terms of the quadratic expression. Then, by trying the various combinations the combination which gives the correct middle term may be found.

Examples

1) Factorise $2x^2+5x-3$

Factors of $2x^2$	Factors of -3
$2x$ x	-3 $+1$
	$+3$ -1

The combinations of these factors are:

$(2x-3)(x+1)$
$=2x^2-x-3$ which is incorrect
$(2x+1)(x-3)$
$=2x^2-5x-3$ which is incorrect
$(2x+3)(x-1)$
$=2x^2+x-3$ which is incorrect
$(2x-1)(x+3)=2x^2+5x-3$ which is correct

Hence $2x^2+5x-3=(2x-1)(x+3)$

2) Factorise $12x^2-35x+8$

Factors of $12x^2$		Factors of 8	
$12x$	x	1	8
$6x$	$2x$	-1	-8
$4x$	$3x$	2	4
		-2	-4

By trying each combination in turn, it is found that the only one that will produce the correct middle term of $-35x$ is $(3x-8)(4x-1)$.
Hence $12x^2-35x+8=(3x-8)(4x-1)$

Case 3. Where the factors form a perfect square

It has been shown that:

$(a+b)^2=a^2+2ab+b^2$

and $(a-b)^2=a^2-2ab+b^2$

The square of a binomial expression therefore consists of: (square of 1st term)+(twice the product of the two terms)+(square of 2nd term).
Thus, to factorise $9a^2+12ab+4b^2$ we note that:

$9a^2=(3a)^2$ and $4b^2=(2b)^2$

also $12ab=2\times 3a\times 2b$.
Hence $9a^2+12ab+4b^2=(3a+2b)^2$

Examples
1) Factorise $16x^2-40x+25$

$16x^2=(4x)^2$; $25=5^2$, $40x=2\times 4x\times 5$.
$\therefore 16x^2-40x+25=(4x-5)^2$

2) Factorise $25x^2+20x+4$

$25x^2=(5x)^2$; $4=2^2$; $20x=2\times 5x\times 2$
$\therefore 25x^2+20x+4=(5x+2)^2$

Case 4. The factors of the difference of two squares

It has been previously shown that:

$(a+b)(a-b)=a^2-b^2$

The factors of the difference of two squares are therefore the sum and difference of the square roots of each of the given terms

Examples
1) Factorise $9m^2-4n^2$

Now $9m^2=(3m)^2$ and $4n^2=(2n)^2$
$\therefore 9m^2-4n^2=(3m+2n)(3m-2n)$

2) Factorise $4x^2-9$

Now $4x^2=(2x)^2$ and $9=3^2$
$\therefore 4x^2-9=(2x+3)(2x-3)$

Exercise 28

Factorise the following:

1) $ax+by+bx+ay$

2) $mp+np-mq-nq$

3) $a^2c^2+acd+acd+d^2$

4) $2pr-4ps+qr-2qs$

5) $4ax+6ay-4bx-6by$

6) $ab(x^2+y^2)-cd(x^2+y^2)$

7) $mn(3x-1)-pq(3x-1)$

8) $k^2l^2-mnl-k^2l+mn$

9) x^2+4x+3
10) x^2+6x+8

11) $x^2+9x+20$
12) x^2-3x+2

13) x^2-6x+8
14) $x^2-7x+12$

15) $x^2+2x-15$
16) $x^2+3x-28$

17) x^2+6x-7
18) x^2-x-12

19) $x^2-5x-14$
20) $x^2-2xy+y^2$

21) $a^2+4ab+3b^2$

22) $p^2-9pq+8q^2$

23) $m^2-5mn-24n^2$

24) $3p^2+p-2$

25) $2x^2+13x+15$

26) $3m^2-8m-28$

27) $4x^2-10x-6$

28) $10a^2+19a-15$

29) $21x^2+37x+10$

30) $26p^2+33p-9$

31) $6x^2+x-35$

32) $6p^2+7pq-3q^2$

33) $2a^2+7ab+6b^2$

34) $30a^2-43ab+15b^2$

35) $12x^2-5xy-2y^2$

36) $4x^2+12x+9$

37) $x^2+2xy+y^2$

38) $9x^2+6x+1$

39) $p^2+4pq+4q^2$

40) $\frac{1}{x^2}+\frac{2}{xy}+\frac{1}{y^2}$

41) $\frac{m^2}{4}+\frac{m}{3}+\frac{1}{9}$

42) $25x^2-20x+4$

43) $m^2-2mn+n^2$

44) $a^2-a+\frac{1}{4}$

45) x^2-4x+4

46) $25-\frac{20}{R}+\frac{4}{R^2}$

47) $\frac{1}{x^2}-\frac{2}{x}+1$

48) $4x^2-y^2$

49) m^2-n^2

50) $x^2-\frac{1}{9}$

51) $9p^2-4q^2$

52) $\frac{1}{x^2}-\frac{1}{y^2}$

53) $121p^2-64q^2$

54) $1-b^2$

55) $y^2-\frac{9}{16}x^2$

56) $(x+y)^2-q^2$

57) $a^2-(p+q)^2$

58) $4x^2-(x+3)^2$

MORE DIFFICULT FACTORISATION

Many algebraic expressions need a combination of several of the methods of factorising discussed previously. The following examples show the methods that should be adopted.

Examples

1) Factorise $(3p-q)^2+2r(3p-q)$
The H.C.F. is $(3p-q)$

$$\therefore (3p-q)^2+2r(3p-q)=(3p-q)(3p-q+2r)$$

2) Factorise $a^2-b^2+2a+2b$

$$\begin{aligned} a^2-b^2+2a+2b &= (a^2-b^2)+(2a+2b) \\ &= (a+b)(a-b)+2(a+b) \\ &= (a+b)(a-b+2) \end{aligned}$$

[since the H.C.F. is $(a+b)$].

3) Factorise $2x^3-50xy^2$
The H.C.F. is $2x$

$$\begin{aligned} \therefore 2x^3-50xy^2 &= 2x(x^2-25y^2) \\ &= 2x(x+5y)(x-5y). \end{aligned}$$

4) Factorise $4(a-b)-c(b-a)$
The term $-c(b-a)$ may be written

$$\begin{aligned} (-1)\times(-c)\times(-1)(b-a) &= +c(-b+a) \\ &= +c(a-b). \end{aligned}$$

The expression is unaltered because $(-1)\times(-1)=1$.

$$\begin{aligned} \therefore 4(a-b)-c(b-a) &= 4(a-b)+c(a-b) \\ &= (a-b)(4+c) \end{aligned}$$

5) Factorise a^2-b^2-a-b

$$\begin{aligned} a^2-b^2-a-b &= (a^2-b^2)-(a+b) \\ &= (a+b)(a-b)-(a+b) \\ &= (a+b)(a-b-1) \end{aligned}$$

[since the H.C.F. is $(a+b)$].

Exercise 29

Factorise the following:

1) $(a-b)^2-2x(a-b)$

2) $3(x+y)+(x+y)^2$

3) $5(3m-n)^2-5a(3m-n)$

4) x^2-y^2+x-y

5) $a^2-b^2-3a-3b$

6) $3x^2-3y^2-2x-2y$

7) $2m^2-2n^2+4m+4n$

8) $\pi R^2+2\pi Rh$

9) $4y^3-9a^2y$

10) $\pi lR^2-\pi lr^2$

11) $20x^3-45xy^2$

12) $\frac{2}{3}\pi r^3+\frac{1}{3}\pi r^2h$

13) $a^2-2ab-2bc+ac$

14) $(x-1)^2-4y^2$

15) $xy(2p-3)-y(2p-3)$

16) $a^2-b^2-(a+b)$

17) $18p^3-2p$

18) $3(x-y)^2-2(y-x)$

19) $8(x-y)^2-4(y-x)$

20) $a^2-b^2-3(b-a)$

SELF TEST 9

In questions 1 to 30 the answer is either 'true' or 'false'.

1) The H.C.F. of a^2bc^3 and ab^2 is ab. ------

2) The H.C.F. of x^2y^3z and x^2yz^2 is $x^2y^3z^2$ ------

3) The H.C.F. of $a^3b^2c^2$, $a^2b^3c^3$ and ab^4c^4 is ab^2c^2. ------

4) The factors of $a^2x^3+bx^2$ are $x^2(a^2x+b)$ ------

5) The factors of $3a^3y+6a^2x+9a^4z$ are $3a(a^2y+2ax+3a^3z)$ ------

6) The factors of $\frac{bz}{y}-\frac{cz}{y^2}+\frac{dz}{y^3}$ are $\frac{z}{y}\left(b-\frac{c}{y}+\frac{d}{y^2}\right)$ ------

7) One factor of $a^2b^3-a^3b^4+ab^2$ is $(ab-a^2b^2+1)$. The other factor is ab^2. ------

8) $(2p+5)(3p-7)$ is equal to $6p^2+29p-35$ ------

9) $(3x+7)(2x-7)$ is equal to $6x^2-7x+49$ ------

10) $(5x+2)(3x-8)$ is equal to $15x^2-34x-16$ ------

11) $(2x+3)^2$ is equal to $4x^2+6x+9$ ------

12) $(3x+5)^2$ is equal to $9x^2+30x+25$ ------

13) $(7x-2)^2$ is equal to $49^2-28x-4$ ------

14) $(3x-4)^2$ is equal to $9x^2-24x+16$ ------

15) $(2x-3)(2x+3)$ is equal to $4x^2-9$ ------

16) $(5x-2)(5x+2)$ is equal to $25x^2+4$ ------

17) x^2+5x+6 is equal to $(x+3)(x+2)$ ------

18) $x^2-8x+15$ is equal to $(x+3)(x+5)$ ------

19) $x^2-9x+14$ is equal to $(x-2)(x-7)$ ------

20) $x^2-2x-15$ is equal to $(x+3)(x-5)$ ------

21) $x^2+2x-35$ is equal to $(x-5)(x+7)$ ------

22) $9x^2-3x-12$ is equal to $(3x+4)(3x-3)$ ------

23) $8x^2-10x-25$ is equal to $(2x-5)(4x+5)$ ------

24) $25x^2+10x-3$ is equal to $(5x-1)(5x+3)$ ------

25) $(3x+2)^2$ is equal to $9x^2+12x+4$ ------

26) $4x^2-12x+9$ is equal to $(2x-3)^2$ ------

27) $25x^2-10x-1$ is equal to $(5x-1)^2$ ------

28) $49x^2+70x+25$ is equal to $(7x+5)^2$ ------

29) $4x^2-25$ is equal to $(2x+5)(2x-5)$ ------

30) $9x^2-49$ is equal to $(3x-7)(3x+7)$ ------

In questions 31 to 45 complete the bracket which has been left blank.

31) $x^2-x-6=(x+2)(\qquad)$

32) $x^2-12x+35=(x-5)(\qquad)$

33) $6x^2+31x+40=(3x+8)(\qquad)$

34) $10x^2-31x+15=(2x-5)(\qquad)$

35) $3a^2-7ab-6b^2=(a-3b)(\qquad)$

36) $8x^2-26xy+15y^2=(4x-3y)(\qquad)$

37) $x^2+2x+1=(\qquad)^2$

38) $x^2-2x+1=(\qquad)^2$

39) $9p^2-25=(3p+5)(\qquad)$

40) $4x^2-16y^2=4(x-2y)(\qquad)$

41) $(p-q)^2-r(p-q)=(p-q)(\qquad)$

42) $x^2-y^2+3x+3y=(x+y)(\qquad)$

43) $3(a-b)-x(b-a)=(3+x)(\qquad)$

44) $x^2-y^2-x-y=(x+y)(\qquad)$

45) $18x^3-2x=2x(\qquad)(\qquad)$

Chapter 10 Algebraic Fractions

MULTIPLICATION AND DIVISION OF FRACTIONS

As with ordinary arithmetic fractions, numerators can be multiplied together as can denominators, in order to form a single fraction. Thus,

$$\frac{a}{b} \times \frac{c}{d} = \frac{a \times c}{b \times d} = \frac{ac}{bd}$$

and

$$\frac{3x}{2y} \times \frac{p}{4q} \times \frac{r^2}{s} = \frac{3x \times p \times r^2}{2y \times 4q \times s} = \frac{3xpr^2}{8yqs}$$

Factors which are common to both numerator and denominator may be cancelled. It is important to realise that this cancelling means dividing the numerator and denominator by the same quantity. For instance,

$$\frac{8ab}{3mn} \times \frac{9n^2m}{4ab^2} = \frac{\overset{2}{\cancel{8}} \times \cancel{a} \times \cancel{b} \times \overset{3}{\cancel{9}} \times \cancel{n} \times n \times \cancel{m}}{\cancel{3} \times \cancel{m} \times \cancel{n} \times \cancel{4} \times \cancel{a} \times \cancel{b} \times \cancel{b}} = \frac{6n}{b}$$

and

$$\frac{7ab}{8mn^2} \times \frac{3m^2n^3}{2ab^3} \times \frac{16an}{63bm}$$

$$= \frac{\cancel{7} \times \cancel{a} \times \cancel{b} \times \cancel{3} \times \cancel{m} \times \cancel{m} \times \cancel{n} \times \cancel{n} \times n \times \cancel{16} \times a \times n}{\cancel{8} \times \cancel{m} \times \cancel{n} \times \cancel{n} \times \cancel{2} \times \cancel{a} \times b \times b \times b \times \underset{3}{\cancel{63}} \times \cancel{b} \times \cancel{m}}$$

$$= \frac{an^2}{3b^3}$$

Before attempting to simplify, factorise where this is possible and then cancel factors which are common to both numerator and denominator. Remember that the contents of a bracket may be regarded as a single term.

The expressions $(x-y)$, $a(b+c)$ and $(x-3)(x-5)$ may be regarded as single terms.

Examples

1) Simplify $\dfrac{x^2-x}{x-1}$

$$x^2-x=x(x-1)$$

Hence,

$$\frac{x^2-x}{x-1} = \frac{x(x-1)}{(x-1)} = x$$

2) Simplify $\dfrac{3x-12}{4x^2-8} \times \dfrac{2x^2-4}{9x-36}$

Factorising where this is possible:

$$\frac{3x-12}{4x^2-8} \times \frac{2x^2-4}{9x-36} = \frac{3(x-4)}{4(x^2-2)} \times \frac{2(x^2-2)}{9(x-4)} = \frac{1}{6}$$

3) Simplify $\dfrac{3xy-6x+y-2}{y^2-4}$

$$\begin{aligned} \text{now } 3xy-6x+y-2 &= (3xy-6x)+(y-2) \\ &= 3x(y-2)+(y-2) \\ &= (y-2)(3x+1) \end{aligned}$$

also $y^2-4=(y+2)(y-2)$

$$\therefore \frac{3xy-6x+y-2}{y^2-4} = \frac{(y-2)(3x+1)}{(y+2)(y-2)}$$

$$= \frac{3x+1}{y+2}$$

4) Simplify $\dfrac{4x^2-9}{4x^2+12x+9}$

now $4x^2-9=(2x+3)(2x-3)$
and $4x^2+12x+9=(2x+3)^2$

$$\therefore \frac{4x^2-9}{4x^2+12x+9} = \frac{(2x+3)(2x-3)}{(2x+3)^2} = \frac{2x-3}{2x+3}$$

To divide by a fraction invert it and then multiply:

Examples

1) Simplify $\dfrac{ax^2}{by} \div \dfrac{a^2}{b^2y^2}$

$$\frac{ax^2}{by} \div \frac{a^2}{b^2y^2} = \frac{ax^2}{by} \times \frac{b^2y^2}{a^2} = \frac{bx^2y}{a}$$

2) Simplify $\dfrac{3a^2+3am}{4a+6m} \div \dfrac{a^2+am}{4a+8m}$

$$\frac{3a^2+3am}{4a+6m} \times \frac{4a+8m}{a^2+am}$$

$$= \frac{3a(a+m)}{2(2a+3m)} \times \frac{4(a+2m)}{a(a+m)}$$

$$= \frac{6(a+2m)}{2a+3m}$$

Exercise 30

Simplify the following:

1) $\dfrac{a}{bc^2} \times \dfrac{b^2c}{a}$

2) $\dfrac{3pq}{r} \times \dfrac{qs}{2t} \times \dfrac{3rs}{pq^2}$

3) $\dfrac{2z^2y}{3ac^2} \times \dfrac{6a^2}{5zy^2} \times \dfrac{10c^3}{3y^2}$

4) $\dfrac{3x-6}{5x-10}$

5) $\dfrac{4m-2}{3m^2-15} \times \dfrac{5m^2-25}{8m-4}$

6) $\dfrac{2az+6bz}{6az+3bz} \times \dfrac{8a+4b}{az+bz} \times \dfrac{2az+4bz}{3a+9b}$

7) $\dfrac{3pq}{5rs} \div \dfrac{p^2}{15s^2}$

8) $\dfrac{6ab}{5cd} \div \dfrac{4a^2}{7bd}$

9) $\left(\dfrac{a}{bc^2} \times abc^3\right) \div \dfrac{a}{cb^2}$

10) $\dfrac{2m-5}{3m+2} \div \dfrac{4m^2-10m}{9m^2+6m}$

11) $\dfrac{3x+3y}{2x^2+4xy} \div \dfrac{6x+6y}{4x}$

12) $\left(\dfrac{3x+6}{2x+6} \times \dfrac{5x}{3x^2+6x}\right) \div \dfrac{2x+8}{x^2+3x}$

13) $\dfrac{a^2-b^2}{a+b}$

14) $\dfrac{3a-2b}{9a^2-12ab+4b^2}$

15) $\dfrac{3x+2}{12x^2+23x+10}$

16) $\dfrac{4x+7}{8x^2+2x-21}$

17) $\dfrac{2x-1}{2x^2+5x-3}$

18) $\dfrac{a^2-b^2+2a+2b}{a-b+2}$

19) $\dfrac{2a^3-18ax^2}{2a+6x}$

20) $\dfrac{(2x-y)^2+3a(2x-y)}{3a-y+2x}$

LOWEST COMMON MULTIPLE (L.C.M.)

In arithmetic the L.C.M. of two or more given numbers is the smallest number into which the given numbers will divide.

Example. Find the L.C.M. of 12, 40 and 45.

$12=2^2\times 3$; $40=2^3\times 5$; $45=3^2\times 5$

L.C.M. $=2^3\times 3^2\times 5=360$

(Note that in finding the L.C.M. we have selected the highest power of each of the prime factors which occur in *any* of the given numbers.)

To find the L.C.M. of a set of algebraic expressions we select the *highest power* of each factor which occurs in *any* of the expressions.

Examples

1) Find the L.C.M. of a^3b^2, abc^3, ab^3c
The highest powers of a, b and c which occur in any of the given expressions are a^3, b^3 and c^3.

$$\therefore \text{L.C.M.} = a^3b^3c^3$$

2) Find the L.C.M. of $5a^4b^4$, $10a^2b^3$, $6a^4b$
The L.C.M. of the numerical coefficients 5, 10 and 6 is 30. The highest powers of a and b which occur are a^4 and b^4. Hence,

$$\text{L.C.M.} = 30a^4b^4$$

3) Find the L.C.M. of $(x+4)^2$, $(x+4)(x+1)$.
Since the contents of a bracket may be regarded as a single symbol.

$$\text{L.C.M.} = (x+4)^2(x+1)$$

4) Find the L.C.M. of $(a+b)$, (a^2-b^2).
We note that a^2-b^2 factorises to give $(a+b)(a-b)$

$$\therefore \text{L.C.M.} = (a+b)(a-b)$$

ADDITION AND SUBTRACTION OF FRACTIONS

The method for algebraic fractions is the same as for arithmetical fractions, that is:

1) Find the L.C.M. of the denominators.
2) Express each fraction with the common denominators.

3) Add or subtract the fractions.

Examples

1) Simplify $\frac{a}{2}+\frac{b}{3}-\frac{c}{4}$

The L.C.M. of 2, 3, and 4 is 12

$$\frac{a}{2}+\frac{b}{3}-\frac{c}{4} = \frac{6a}{12}+\frac{4b}{12}-\frac{3c}{12} = \frac{6a+4b-3c}{12}$$

2) Simplify $\frac{2}{x}+\frac{3}{2x}+\frac{4}{3x}$

The L.C.M. of x, $2x$ and $3x$ is $6x$.

$$\frac{2}{x}+\frac{3}{2x}+\frac{4}{3x} = \frac{12+9+8}{6x} = \frac{29}{6x}$$

The sign in front of a fraction applies to the fraction as a whole. The line which separates the numerator and denominator acts as a bracket.

Examples.

1) Simplify $\frac{m}{12}+\frac{2m+n}{4}-\frac{m-2n}{3}$

The L.C.M. of 12, 4 and 3 is 12.

$$\therefore \frac{m}{12}+\frac{2m+n}{4}-\frac{m-2n}{3}$$

$$= \frac{m+3(2m+n)-4(m-2n)}{12}$$

$$= \frac{m+6m+3n-4m+8n}{12}$$

$$= \frac{3m+11n}{12}$$

2) Express as a single fraction in its lowest terms

$$\frac{1}{x-1}+\frac{2x}{1-x^2}$$

now $1-x^2 = (1+x)(1-x)$

We can rewrite $\frac{1}{x-1}$ as

$$\frac{(-1)\times 1}{(-1)\times(x-1)} = -\frac{1}{1-x}$$

$$\therefore \frac{1}{x-1}+\frac{2x}{1-x^2} = -\frac{1}{1-x}+\frac{2x}{(1+x)(1-x)}$$

The L.C.M. of $(1-x)$ and $(1+x)(1-x)$ is $(1+x)(1-x)$

$$\therefore \frac{-1}{1-x}+\frac{2x}{(1+x)(1-x)}=\frac{-1(1+x)+2x}{(1+x)(1-x)}$$

$$=\frac{-1-x+2x}{(1+x)(1-x)}$$

$$=\frac{x-1}{(1+x)(1-x)}$$

3) Simplify $\dfrac{2x}{x^2+x-6}+\dfrac{1}{x-2}$

$$\frac{2x}{x^2+x-6}+\frac{1}{x-2}=\frac{2x}{(x+3)(x-2)}+\frac{1}{x-2}$$

$$=\frac{2x+(x+3)}{(x+3)(x-2)}$$

$$=\frac{3x+3}{(x+3)(x-2)}$$

$$=\frac{3(x+1)}{(x+3)(x-2)}$$

Exercise 31

Find the L.C.M. of the following:

1) $4,\ 12x$

2) $3x,\ 6y$

3) $2ab,\ 4a,\ 6b$

4) $ab,\ bc,\ ac$

5) $3m^2pq,\ 9mp^2q,\ 12mpq,\ 3mn^2$

6) $5a^2b^3,\ 10ab^4,\ 2a^2b^3$

7) $(m-n)^2,\ (m-n)$

8) $(x+3)^2,\ (x+3),\ (x+1)$

9) $(x-1),\ (x^2-1)$

10) $(9a^2-b^2),\ (3a+b),\ (3a-b)$

Simplify the following:

11) $\dfrac{x}{3}+\dfrac{x}{4}+\dfrac{x}{5}$

12) $\dfrac{5a}{12}-\dfrac{7a}{18}$

13) $\dfrac{2}{q}-\dfrac{3}{2q}$

14) $\dfrac{3}{y}-\dfrac{5}{3y}+\dfrac{4}{5y}$

15) $\dfrac{3}{5p}-\dfrac{2}{3q}$

16) $\dfrac{3x}{2y}-\dfrac{5y}{6x}$

17) $3x-\dfrac{4y}{5z}$

18) $1-\dfrac{2x}{5}+\dfrac{x}{8}$

19) $3m-\dfrac{2m+n}{7}$

20) $\dfrac{3a+5b}{4}-\dfrac{a-3b}{2}$

21) $\dfrac{3m-5n}{6}-\dfrac{3m-7n}{2}$

22) $\dfrac{x-2}{4}+\dfrac{2}{5}$

23) $\dfrac{x-5}{3}-\dfrac{x-2}{4}$

24) $\dfrac{3x-5}{10}+\dfrac{2x-3}{15}$

25) $\dfrac{4}{x-5}-\dfrac{15}{x(x-5)}-\dfrac{3}{x}$

26) $\dfrac{3}{2x-1}-\dfrac{2x}{4x^2-1}$

27) $\dfrac{5x}{x^2-x-6}-\dfrac{2}{x+2}$

28) $\dfrac{7}{x^2+3x-10}-\dfrac{2}{x^2+5x}-\dfrac{2}{x^2-2x}$

29) $\dfrac{x+2}{x+3}-\dfrac{x-2}{x-3}$

30) $\dfrac{3x}{x^2-y^2}-\dfrac{x+3}{(x+y)^2}$

SELF TEST 10

In the questions below state the letter (or letters) corresponding to the correct answer (or answers).

1) $\dfrac{a^2-a}{a-1}$ is equal to

a a b a^2-1 c a^2 d $1-a^2$

2) $\dfrac{2x-6}{4x^2-8x}\times\dfrac{3x^2-6x}{4x-16}$ is equal to

a $\frac{3x-9}{8x-32}$ b $\frac{1}{4}$ c $\frac{3(x-3)}{8(x-4)}$ d 0

3) $\frac{9x^2-25}{9x^2-9x-10}$ is equal to

a $\frac{1-5x}{1-9x}$ b $\frac{3x+5}{3x+2}$ c $\frac{5}{2}$ d 2

4) $\left(\frac{6y+12}{4y+12}\times\frac{5y}{3y^2+6y}\right)\div\frac{4y+16}{2y^2+6y}$ is equal to

a $\frac{5y}{4y+16}$ b $\frac{1}{4}$ c $\frac{75}{(3y+2)(2y+3)}$

d $\frac{5y}{4(y+4)}$

5) The L.C.M. of a^4b^3, ab^2c^4 and ab^3c^3 is

a abc b ab^2 c $a^4b^3c^4$

d $a^6b^8c^7$

6) The L.C.M. of $(x+3)^2$, $(x+2)(x+3)$ and $(x+2)^2$ is

a $(x+3)^3(x+2)^3$ b $(x+3)^2(x+2)^2$

c $(x+2)(x+3)$ d $x+5$

7) The L.C.M. of $a^2+2ab+b^2$ and a^2-b^2 is

a $2a^2b^2$ b $(a^2-b^2)(a^2+2ab+b^2)$

c $2a^3b^3$ d $(a-b)(a+b)^2$

8) $\frac{6m-2n}{2}$ is equal to

a $\frac{3m-2n}{2}$ b $6m-n$ c $3m-2n$

d $3m-n$

9) $t-\frac{3-t}{2}+\frac{3+t}{2}$ is equal to

a t b $2t$ c $3+2t$ d $4t$

10) $\frac{x}{x^2-1}-\frac{2x}{x-1}$ is equal to

a $\frac{1-2x}{x-1}$ b $\frac{x(3-2x)}{x^2-1}$

c $\frac{x(2x+1)}{1-x^2}$ d $\frac{1}{x-1}+2$

11) $\frac{3x-7}{3}-\frac{2x-5}{2}$ is equal to

a -12 b -2 c $\frac{1}{6}$ d $-\frac{29}{6}$

12) $\frac{5x-10}{5}-\frac{3x-6}{3}$ is equal to

a 0 b -4 c $2x-4$ d $2x$

Chapter 11 Simple Equations

EQUATIONS

An equation is a statement that two quantities are equal, for instance, 1000 mm = 1m. More often an equation contains an *unknown quantity* which is represented by a symbol. It is the value of this unknown quantity which we desire to find.
In the equation $3x-4=23$, x is the unknown quantity. There is only one value of x such that the left hand side of the equation is equal to the right hand side (this value is $x=9$). When we have calculated this value of x we have solved the equation and the value of x so obtained is called the *solution* (i.e. the solution is $x=9$).
In the process of solving an equation the appearance of the equation may be considerably altered but the values on both sides must remain the same. We must maintain this equality and hence whatever we do to one side of the equation we must do exactly the same to the other side.
There are many types of equations which occur in mathematics and these are classified according to the *highest power* of the unknown quantity. Thus, the equation $3x-4=23$ contains only the first power of x; the equation $5x^2-3x+5=0$ contains x^2 as the highest power of x, that is the second power of x.

SIMPLE EQUATIONS

Simple equations contain only the first power of the unknown quantity. Thus

$$7t-5=4t+7$$

$$\frac{5x}{3}=\frac{2x+5}{2}$$

are both examples of simple equations.
After an equation is solved, the solution should be checked by substituting the result in each side of the equation separately. If each side of the equation then has the same value the solution is correct.

SOLVING SIMPLE EQUATIONS

Equations requiring multiplication and division.

1) Solve the equation $\frac{x}{6}=3$

Multiplying each side by 6, we get

$$\frac{x}{6}\times 6=3\times 6$$

$$x=18$$

Check: when $x=18$, L.H.S. $=\frac{18}{6}=3$

R.H.S. $=3$

Hence the solution is correct.

2) Solve the equation $5x=10$
Dividing each side by 5, we get

$$\frac{5x}{5}=\frac{10}{5}$$

$$x=2$$

Check: when $x=2$, L.H.S. $=5\times 2=10$
R.H.S. $=10$

Hence the solution is correct

Equations requiring addition and subtraction

1) Solve $x-4=8$

If we add 4 to each side, we get

$$x-4+4=8+4$$
$$x=12$$

The operation of adding 4 to each side is the same as transferring -4 to the R.H.S. but in so doing the sign is changed from a minus to a plus. Thus,

$$x-4=8$$
$$x=8+4$$
$$x=12$$

Check: when $x=12$, L.H.S. $=12-4=8$
R.H.S. $=8$

Hence the solution is correct.

2) Solve $x+5=20$
If we subtract 5 from each side, we get

$$x+5-5=20-5$$
$$x=15$$

Alternatively moving $+5$ to the R.H.S.

$$x=20-5$$
$$x=15$$

Check: when $x=15$, L.H.S. $=15+5=20$
R.H.S. $=20$

$\therefore$ The solution is correct.

Equations containing the unknown quantity on both sides.

In equations of this kind group all the terms containing the unknown quantity on one side of the equation and the remaining terms on the other side.

Examples
1) Solve $7x+3=5x+17$
Transferring $5x$ to the L.H.S. and $+3$ to the R.H.S.

$$7x-5x=17-3$$
$$2x=14$$
$$x=\frac{14}{2}$$
$$x=7$$

Check: when $x=7$, L.H.S. $=7\times7+3=52$
R.H.S. $=5\times7+17=52$

Hence the solution is correct.

2) Solve $3x-2=5x+6$

$$3x-5x=6+2$$
$$-2x=8$$
$$x=\frac{8}{-2}$$
$$x=-4$$

Check: when $x=-4$,

L.H.S. $=3\times(-4)-2=-14$
R.H.S. $=5\times(-4)+6=-14$

Hence the solution is correct.

Equations containing brackets

When an equation contains brackets remove these first and then solve as shown previously.

Examples
1) Solve $2(3x+7)=16$
Removing the bracket,

$$6x+14=16$$
$$6x=16-14$$
$$6x=2$$
$$x=\frac{2}{6}$$
$$x=\frac{1}{3}$$

Check: when $x=\frac{1}{3}$,

L.H.S. $=2\times\left(3\times\frac{1}{3}+7\right)=2\times(1+7)$
$=2\times8=16$

R.H.S. $=16$

Hence the solution is correct.

2) Solve $3(x+4)-5(x-1)=19$

Removing the brackets

$$3x+12-5x+5 = 19$$
$$-2x+17 = 19$$
$$-2x = 19-17$$
$$-2x = 2$$
$$x = \frac{2}{-2}$$
$$x = -1$$

Check: when $x=-1$,

$$\text{L.H.S.} = 3\times(-1+4)-5\times(-1-1)$$
$$= 3\times 3-5\times(-2)$$
$$= 9+10 = 19$$

$$\text{R.H.S.} = 19$$

Hence the solution is correct.

Equations containing fractions

When an equation contains fractions, *multiply each term of the equation* by the L.C.M. of the denominators.

Examples

1) Solve $\frac{x}{4}+\frac{3}{5}=\frac{3x}{2}-2$

The L.C.M. of the denominators 2, 4 and 5 is 20.
Multiplying each term by 20 gives,

$$\frac{x}{4}\times 20+\frac{3}{5}\times 20=\frac{3x}{2}\times 20-2\times 20$$
$$5x+12=30x-40$$
$$5x-30x=-40-12$$
$$-25x=-52$$
$$x=\frac{-52}{-25}$$
$$\therefore\ x=\frac{52}{25}$$

The solution may be verified by the check method shown in the previous examples.

2) Solve the equation $\frac{x-4}{3}-\frac{2x-1}{2}=4$

In solving equations of this type remember that the line separating the numerator and denominator acts as a bracket. The L.C.M. of of the denominators 3 and 2 is 6. Multiplying *each term* of the equation by 6,

$$\frac{x-4}{3}\times 6-\frac{2x-1}{2}\times 6=4\times 6$$
$$2(x-4)-3(2x-1)=24$$
$$2x-8-6x+3=24$$
$$-4x-5=24$$
$$-4x=24+5$$
$$-4x=29$$
$$x=\frac{29}{-4}$$
$$x=-\frac{29}{4}$$

3) Solve the equation $\frac{5}{2x+5}=\frac{4}{x+2}$

The L.C.M. of the denominators is $(2x+5)(x+2)$. Multiplying each term of the equation by this gives:

$$\frac{5}{2x+5}\times(2x+5)(x+2)=\frac{4}{x+2}\times(2x+5)(x+2)$$
$$\therefore\ 5(x+2)=4(2x+5)$$
$$5x+10=8x+20$$
$$5x-8x=20-10$$
$$-3x=10$$
$$x=\frac{10}{-3}$$
$$x=-\frac{10}{3}$$

Exercise 32

Solve the equations

1) $x+2=7$

2) $t-4=3$

3) $2q=4$

4) $x-8=12$

5) $q+5=2$

6) $3x=9$

7) $\frac{y}{2}=3$

8) $\frac{m}{3}=4$

9) $2x+5=9$

10) $5x-3=12$

11) $6p-7=17$

12) $3x+4=-2$

13) $7x+12=5$

14) $6x-3x+2x=20$

15) $14-3x=8$

16) $5x-10=3x+2$

17) $6m+11=25-m$

18) $3x-22=8x+18$

19) $0{\cdot}3d=1{\cdot}8$

20) $1{\cdot}2x-0{\cdot}8=0{\cdot}8x+1{\cdot}2$

21) $2(x+1)=8$

22) $5(m-2)=15$

23) $3(x-1)-4(2x+3)=14$

24) $5(x+2)-3(x-5)=29$

25) $3x=5(9-x)$

26) $4(x-5)=7-5(3-2x)$

27) $\frac{x}{5}-\frac{x}{3}=2$

28) $\frac{x}{3}+\frac{x}{4}+\frac{x}{5}=\frac{5}{6}$

29) $\frac{m}{2}+\frac{m}{3}+3=2+\frac{m}{6}$

30) $3x+\frac{3}{4}=2+\frac{2x}{3}$

31) $\frac{3}{m}=3$

32) $\frac{5}{x}=2$

33) $\frac{4}{t}=\frac{2}{3}$

34) $\frac{7}{x}=\frac{5}{3}$

35) $\frac{4}{7}y-\frac{3}{5}y=2$

36) $\frac{1}{3x}+\frac{1}{4x}=\frac{7}{20}$

37) $\frac{x+3}{4}-\frac{x-3}{5}=2$

38) $\frac{2x}{15}-\frac{x-6}{12}-\frac{3x}{20}=\frac{3}{2}$

39) $\frac{2m-3}{4}=\frac{4-5m}{3}$

40) $\frac{3-y}{4}=\frac{y}{3}$

41) $x-5=\frac{3x-5}{6}$

42) $\frac{x-2}{x-3}=3$

43) $\frac{3}{x-2}=\frac{4}{x+4}$

44) $\frac{3}{x-1}=\frac{2}{x-5}$

45) $\frac{3}{2x+7}=\frac{5}{3(x-2)}$

46) $\frac{x}{3}-\frac{3x-7}{5}=\frac{x-2}{6}$

47) $\frac{4p-1}{3}-\frac{3p-1}{2}=\frac{5-2p}{4}$

48) $\frac{3m-5}{4}-\frac{9-2m}{3}=0$

49) $\frac{x}{3}-\frac{2x-5}{2}=0$

50) $\frac{4x-5}{2}-\frac{2x-1}{6}=x$

MAKING EXPRESSIONS

It is important to be able to translate information into symbols thus making up algebraic expressions. The following examples will illustrate how this is done.

Examples

1) Find an expression which will give the total mass of a box containing x articles if the box has a mass of 7 kg and each article has a mass of 1·5 kg.

The total mass of x articles is $1{\cdot}5\,x$

$\therefore$ Total mass of the box of articles is $1{\cdot}5x+7$

2) If the price of an article is reduced from x**p** to y**p** make an expression giving the number of extra articles that can be bought for 80**p**.

At x**p** each the number of articles that can be bought for 80**p** is $\frac{80}{x}$

At y**p** each the number of articles that can be bought for 80**p** is $\frac{80}{y}$

The extra articles that can be bought is $\frac{80}{y} - \frac{80}{x}$

3) If x apples can be bought for 6**p** write down the cost of y apples.

If x apples cost 6**p**

Then 1 apple costs $\frac{6}{x}$**p**

Hence y apples cost $\frac{6}{x} \times y = \frac{6y}{x}$**p**

Exercise 33

1) A boy is x years old now. How old was he 5 years ago?

2) Find the total cost of 3 pencils at a pence each and 8 pens at b pence each.

3) A man works x hours per weekday except Saturday when he works y hours. If he works z hours on Sunday how many hours does he work per week?

4) What is the perimeter of a rectangle l mm long and b mm wide?

5) A man A has £a and a man B has £b. If A gives B £x how much will each have?

6) How many minutes are there between x minutes to 10 o'clock and 12 o'clock.

7) Find in pounds the total cost of a gramophone costing £Y and n records costing X pence each.

8) m articles are bought for x pence. Find the cost in pounds of buying n articles at the same rate.

9) A householder buys two daily papers at a pence each and 3 sunday papers at b pence each. What is his yearly expenditure (in pounds) on newspapers.

10) A reel of wire has n metres wound on it. If the reel itself has a mass of P kg and the wire has a mass of Q kg per metre find the total mass.

11) In one innings a batsman hit a sixes, b fours and c singles. How many runs did he score?

12) A factory employs M men, N boys and P women. If a man earns £x per week, a boy £y per week and a woman £z per week what is the total wage bill per week?

13) A man earns £u per week when he is working and he is paid £v per week when he is on holiday. If he is on holiday for 3 weeks per year find his total annual salary.

14) The price of m articles was £M but the price of each article is increased by n pence. How many articles can be bought for £N?

15) A man starts a job at a salary of £u per week. His salary is increased by y pence per week at the end of each year's service. What will be his salary after x years?

16) During a sale a shop gives a reduction of g pence in the pound on the marked price of articles. If a customer buys articles marked at £X, £Y and £Z how much will he actually pay?

17) A number m is divided into two parts. If a is one part what is the product of the two parts?

18) A man pays income tax at the rate of x pence in the pound. If his income is £M of which £Q is tax free how much tax (in pounds) does he pay.

19) After spending one-seventh of my income on rent and two-sevenths of the remainder on household expenses I have £X left. What is my income?

20) The cost of a supply of electricity is as follows. There is a fixed charge of £a, for the rent of the meter the charge is £b and the electricity is charged for at c pence per unit. If n units of electricity are used find an expression for the total cost (in £).

CONSTRUCTION OF SIMPLE EQUATIONS

If often happens that we are confronted with mathematical problems that are difficult or impossible to solve by arithmetical methods. We then represent the quantity that has to be found by a symbol. Then by constructing an equation which conforms to the data of the problem we can solve it to give us the value of the unknown quantity. It is stressed that both sides of the equation must be in the same units.

Examples

1) The perimeter of a rectangle is 56 cm. If one of the two adjacent sides is 4 cm longer than the other, find the dimensions of the rectangle.

Let x cm = length of the shorter side
Then $(x+4)$cm = length of the longer side

$$\text{Total perimeter} = x+x+(x+4)+(x+4) = (4x+8)\text{cm}$$

But the total perimeter = 56 cm

$$\text{Hence } 4x+8=56$$
$$4x=56-8$$
$$4x=48$$
$$x=12$$

Hence the shorter side is 12 cm long and the longer side is $12+4=16$ cm long.

2) £43·00 is spent on buying 16 articles, some at £3·00 each and the remainder at £2·00 each. How many articles of each kind are bought.
Let x be the number of articles bought at £3·00.
The $16-x$ is the number bought at £2·00.
The cost of x articles at £3·00 each is £$3x$
The cost of $16-x$ articles at £2·00 each is £$2(16-x)$
The total cost is £$[3x+2(16-x)]$.
But we are given that the total cost is £43·00.

$$\text{Hence, } 3x+2(16-x)=43$$
$$3x+32-2x=43$$
$$x+32=43$$
$$x=43-32$$
$$x=11$$

Hence 11 articles are bought at £3·00 each and 5 articles are bought at £2·00 each.

3) A 60 page book has n lines to a page. If the number of lines were reduced by 3 the number of pages would have to be increased by 10 to give the same writing space. Find the value of n.
At n lines to the page the total number of lines in the book is $60n$.
At $(n-3)$ lines to the page the total number of lines in the book is $70\ (n-3)$, since the number of pages is now 70.
Because the number of lines in the book has to remain the same:

$$60n=70(n-3)$$
$$60n=70n-210$$
$$60n-70n=-310$$
$$-10n=-210$$
$$n=\frac{-210}{-10}$$
$$n=21.$$

Exercise 34

1) £24·00 is divided between three men A, B and C. If B has three times as much as A and C has four times as much as A how much does each receive?

2) Divide the number 12 up into two parts so that 5 times the first part added to 3 times the second part equals 40.

3) Two men A and B have £120 in cash between them. If A gives B £10, B would have twice as much as A. How much has A?

4) Find two numbers whose sum is 9, which are such that 5 times the first number exceeds four times the second number by 9.

5) 15 articles are bought. Some cost 5 pence each and the others cost 8 pence each. If the total amount paid for them is 90 pence how many of each are bought?

6) 18 books are bought by a library. Some cost £1·00 and the remainder cost £1·25. How many of each are bought if the total cost is £20·00.

7) Find three *consecutive* whole numbers so that their sum is 48.

8) A room is 1·5 metres longer than it is wide. If its perimeter is 63 metres, find the dimensions of the room.

9) A lift A can carry 4 more people than lift B. When both lifts are full, when B makes three journeys it carries as many people as A does in two journeys. Find how many people each of the lifts can carry.

10) Two tanks contain equal amounts of liquid. They are connected by a pipe and 2500 litres flow from one to the other. One tank then contains 5 times as much liquid as the other. How many litres did each tank contain originally?

11) A man buys an article costing £688 on the hire purchase system. He pays a deposit and twelve monthly installments. The deposit is 4 times as large as the installment. How much is the deposit?

12) In a club share out £1720 is to be shared between 200 members. Male members are to receive £10 each, female members £8 each and juvenile members £4 each. If there are 5 times as many male members as juvenile members, how many female members are there?

13) Find the number which when added to the numerator and denominator of the fraction $\frac{5}{7}$ makes a new fraction which is equal to $\frac{4}{5}$.

14) Two taps are used to fill a tank which has a capacity of 600 litres. If it takes 16 minutes to fill the tank and one tap delivers water at twice the rate of the other find how many litres per minute each tap delivers.

15) The perimeter of a triangle ABC is 260 mm. The side BC is two-thirds of the length of the side AB and also 20 mm longer than the side AC. Find the lengths of the three sides of the triangle.

SELF TEST 11

In questions 1 to 25 the answer is either 'true' or 'false.'

1) If $\frac{x}{7}=3$ then $x=21$ - - - - -

2) If $\frac{x}{5}=10$ then $x=2$ - - - - -

3) If $\frac{x}{4}=16$ then $x=64$ - - - - -

4) If $5x=20$ then $x=4$ - - - - -

5) If $3x=6$ then $x=18$ - - - - -

6) If $x-5=10$ then $x=5$ - - - - -

7) If $x+8=16$ then $x=2$ - - - - -

8) If $x+7=14$ then $x=21$ - - - - -

9) If $x+3=6$ then $x=3$ - - - - -

10) If $x-7=14$ then $x=21$ - - - - -

11) If $3x+5=2x+10$ then $x=3$ - - - - -

12) If $2x+4=x+8$ then $x=4$ - - - - -

13) If $5x-2=3x-8$ then $x=-6$ - - - - -

14) If $3x-8=2-2x$ then $x=10$ - - - - -

15) If $2(3x+5)=18$ then $x=1$ - - - - -

16) If $2(x+4)-5(x-7)=7$ then $x=12$ ------

17) If $6y=10(8-y)$ then $y=15$ ------

18) If $\frac{5}{y}=10$ then $y=2$ ------

19) If $\frac{8}{y}=4$ then $y=2$ ------

20) $\frac{x}{3}+\frac{x}{4}=\frac{2x}{5}-11$ then $x=60$ ------

21) $\frac{x}{2}-1=\frac{x}{3}-\frac{1}{2}$ then $x=3$ ------

22) If $\frac{3}{x+5}=\frac{4}{x-2}$ then $x=26$ ------

23) If $\frac{3}{x-6}=\frac{2}{x-4}$ then $x=-2$ ------

24) If $\frac{x-4}{2}-\frac{x-3}{3}=4$ then $x=10$ ------

25) If $\frac{2x-3}{2}-\frac{x-6}{5}=3$ then $x=3$ ------

In questions 26 to 40 state the letter (or letters) corresponding to the correct answer (or answers).

26) If $3(2x-5)-2(x-3)=3$ then x is equal to

a 3 b 6 c $\frac{5}{4}$ d $\frac{11}{4}$

27) If $2(x+6)-3(x-4)=1$ then x is equal to

a 25 b 17 c 23 d -7

28) If $\frac{x-5}{3}=\frac{x+2}{2}$ then x is equal to

a 16 b -16 c 7 d -7

29) If $3(x-2)-5(x-7)=12$ then x is equal to

a $-8\frac{1}{2}$ b $8\frac{1}{2}$ c -7 d 0

30) If $\frac{3-2y}{4}=\frac{2y}{6}$ then y is equal to

a $\frac{18}{20}$ b $\frac{9}{10}$ c 3 d -3

31) The cost of electricity is obtained as follows: A fixed charge of £a, rent of a meter £b and a charge of c pence for each unit of electricity supplied. The total cost of using n units of electricity is therefore

a £$(a+b+nc)$ b £$(100(a+b)+nc)$
c $[100(a+b)+nc]$ pence

d £$\left(a+b+\frac{nc}{100}\right)$

32) At a factory p men earn an average wage of £a, q women earn an average wage of £b and r apprentices earn an average wage of £c. The average wage for all these employers is

a £$(a+b+c)$ b £$\left(\frac{a}{p}+\frac{b}{q}+\frac{c}{r}\right)$

c £$\frac{(ap+bq+cr)}{a+b+c}$ d £$\left(\frac{ap+bq+cr}{p+q+r}\right)$

33) A dealer ordered N tools from a manufacturer. The manufacturer can produce p tools per day but $x\%$ of these are faulty and unfit for sale. The number of days it takes the manufacturer to complete the order is

a $\frac{N}{p(100-x)}$ b $\frac{100\,N}{p(100-x)}$

c $\frac{N}{p(1-x)}$ d $\frac{p(100-x)}{N}$

34) A shopkeeper pays £c for x kg of apples. He sells them for b pence per kg. His percentage profit is therefore

a $\frac{bx-c}{c}$ b $\frac{bx-100c}{100c}$

c $\frac{100(bx-c)}{c}$ d $\frac{bx-100c}{c}$

35) A man walked for t hours at v km per hour and then cycled a distance of x km at c km per hour. His average speed for the whole journey was

a $\frac{vt+x}{t+c}$ b $\frac{vt+x}{t}$

c $\dfrac{c(vt+x)}{ct+x}$ d $\dfrac{vt+x}{t+\frac{x}{c}}$

36) The cost of hiring a bus is £60 . If nine of the seats are unoccupied the cost per person is £1 more than each person would have to pay if all the seats were full. If n is the number of seats in the bus then

a $\dfrac{60}{n}-\dfrac{60}{n-9}=1$ b $\dfrac{60}{n-9}-\dfrac{60}{n}=1$

c $\dfrac{60}{n}=n$ d $\dfrac{60}{n-9}=n$

37) At the beginning of term a student bought x books at a total cost of £22. A few days later he bought three more books for a further expenditure of £4. He found that this purchase had reduced the average cost per book by 20 pence. An equation from which x can be found is

a $\dfrac{26}{x}=20$ b $\dfrac{26}{x+3}=0{\cdot}20$

c $\dfrac{22}{x}-\dfrac{26}{x+3}=0{\cdot}20$ c $\dfrac{26}{x+3}-\dfrac{22}{x}=0{\cdot}20$

38) The smallest of three consecutive even numbers is m. Twice the square of the largest is greater than the sum of the squares of the other two numbers by 244. Hence

a $2(m+2)^2=(m+1)^2+m^2+244$
b $2(m+2)^2-(m+1)^2+m^2=244$
c $2(m+4)^2=(m+2)^2+m^2+244$
d $2(m+4)^2-(m+2)^2+m^2=244$

39) A bill for £74 was paid with £5 and £1 notes, a total of 50 notes being used. If x is the number of £5 notes used then

a $x+5(50-x)=74$
b $5x+(50-x)=74$
c $5x+(74-x)=50$
d $x+5(74-x)=50$

40) A householder can choose to pay for his electricity by one of the two following methods:

1) a basic charge of £1·60 together with a charge of 0·5 pence of each unit of electricity used;

2) a basic charge of £2·60 together with a charge of 0·4 pence for each unit of electricity used.

The number of units N for which the bill would be the same by either method may be found from the equation

a $1{\cdot}60+0{\cdot}5N=2{\cdot}60+0{\cdot}4N$
b $1{\cdot}60N+50=2{\cdot}60N+40$
c $160+0{\cdot}5N=260+0{\cdot}4N$
d $160N+50=260N+40$

Chapter 12 Formulae

EVALUATING FORMULAE

The statement $F=ma$ is described as a formula for F in terms of m and a. The value of F may be found by arithmetic after substituting the given values of m and a.

Examples

1) The formula $E=IR$ is used in electrical calculations. Find the value of E if $I=6$ and $R=4$

$E=IR=6\times 4=24$

2) If $P=\dfrac{RT}{V}$ find the value of P when $R=50$, $T=200$ and $V=5$

When $R=50$, $T=200$ and $V=5$

$P=\dfrac{50\times 200}{5}=2000$

Exercise 35

Find the values of the following:

1) $C=\pi D$ when $\pi=3{\cdot}14$ and $D=7$

2) $A=\pi rl$ when $\pi=3{\cdot}14$, $r=1$ and $l=9$

3) $I=PRT$ when $P=30$, $R=54$ and $T=100$

4) $P=\dfrac{1}{n}$ when $n=25$

5) $I=\dfrac{E}{R}$ when $E=240$ and $R=30$

6) $A=\frac{1}{2}BH$ when $B=8$ and $H=5$

7) $K=\dfrac{Wv^2}{2g}$ when $W=100$, $v=25$ and $g=32$

8) $S=90(2n-4)$ when $n=5$

9) $y=\dfrac{3t}{\sqrt{c}}$ when $t=7$ and $c=16$

10) $k=\dfrac{3n+2}{n+1}$ when $n=5$

FORMULAE GIVING RISE TO SIMPLE EQUATIONS

In the formula $H=wS(T-t)$, H is called the **subject** of the formula. It may be that we are given values of H, w, T and t and we have to find the value of S. The method is shown in the following examples.

Examples

1) Find S from the formula $H=wS(T-t)$ when $H=3$, $w=2$, $T=20$ and $t=12$

Substituting the given values in the equation,

$3=2S(20-12)$

$3=2\times S\times 8$

$3=16S$

$S=\dfrac{3}{16}$

2) Find H from the formula $A=2\pi R(R+H)$ when $A=1760$, $\pi=\dfrac{22}{7}$ and $R=14$.

Substituting the given values in the equation

$1760=2\times\dfrac{22}{7}\times 14\times(14+H)$

$1760=88\times(14+H)$

$\dfrac{1760}{88}=14+H$

$20 = 14 + H$

$20 - 14 = H$

$\therefore H = 6$

If the unknown letter occurs twice in an equation then terms containing this unknown letter should be brought to one side of the equation (see example 3).

3) If $C = \dfrac{nE}{R + nr}$ find the value of n

when $C = 1{\cdot}5$, $E = 3$, $R = 6{\cdot}3$ and $r = 1{\cdot}1$

Substituting the given values,

$$1{\cdot}5 = \frac{3n}{6{\cdot}3 + 1{\cdot}1n}$$

Multiplying both sides of the equation by $(6{\cdot}3 + 1{\cdot}1n)$,

$$1{\cdot}5(6{\cdot}3 + 1{\cdot}1n) = 3n$$

$$1{\cdot}5 \times 6{\cdot}3 + 1{\cdot}5 \times 1{\cdot}1n = 3n$$

$$9{\cdot}45 + 1{\cdot}65n = 3n$$

$$9{\cdot}45 = 3n - 1{\cdot}65n$$

$$9{\cdot}45 = 1{\cdot}35n$$

$$n = \frac{9{\cdot}45}{1{\cdot}35}$$

$$n = 7$$

4) If $D = 1{\cdot}2\sqrt{dL}$ find L when $D = 3{\cdot}6$ and $d = 2$.
Substituting the given values,

$$3{\cdot}6 = 1{\cdot}2\sqrt{2L}$$

Dividing both sides by $1{\cdot}2$

$$\frac{3{\cdot}6}{1{\cdot}2} = \sqrt{2L}$$

$$3 = \sqrt{2L}$$

Squaring both sides of the equation

$$3^2 = 2L$$

$$9 = 2L$$

$$L = \frac{9}{2}$$

$$L = 4{\cdot}5$$

Exercise 36

1) Find T from the formula $D = \dfrac{T+2}{P}$ when $D = 7{\cdot}5$ and $P = 12$.

2) Find d from the formula $E = \dfrac{p-d}{p}$ when $p = 3$ and $E = \dfrac{7}{8}$.

3) Find y from the formula $A = \frac{1}{2}h(x+y)$ when $A = 44$, $h = 8$ and $x = 4{\cdot}5$.

4) Find n from the formula $A = P\left(1 + \dfrac{nr}{12}\right)$ when $A = 510$, $P = 500$ and $r = 0{\cdot}04$.

5) If $P - mg = \dfrac{mv^2}{r}$ find the value of r when $P = 145$, $m = 2$, $g = 32$ and $v = 9$.

6) If $y = \dfrac{ab}{a-b}$ find the value of a when $y = 5$ and $b = 3$.

7) If $k = \dfrac{3n+2}{n+1}$ find the value of n when $k = 7$.

8) If $d = \sqrt{2hr}$ find the value of h when $d = 36$ and $r = 3$.

9) Find V from the formula $d = \sqrt{\dfrac{D^2S}{V}}$ when $d = 2$, $D = 3$ and $S = 6$.

10) If $H = h + \dfrac{v^2}{2g}$ find the value of g when $v = 16$, $H = 9$ and $h = 5$.

11) If $V = \sqrt{2gh}$ find the value of h when $V = 16$ and $g = 32$.

12) If $a = \sqrt{\dfrac{b}{b+c}}$ find the value of b when $a = 3$ and $c = 2$.

13) If $y=\dfrac{3a}{\sqrt{b}}$ find the value of b when $y=4$ and $a=8$.

14) If $x=\sqrt{\dfrac{a^2-b^2}{ay}}$ find the value of y when $x=2$, $a=5$ and $b=3$.

15) If $c=2\sqrt{2hr-h^2}$ find the value of r when $h=3$ and $c=4$.

TRANSPOSITION OF FORMULAE

Considering again, the formula $H=wS(T-t)$. We may be given several corresponding values of H, w, T and t and we want to find the corresponding values of S. We can, of course, find these values by the method shown in the previous examples but considerable time and effort will be spent in solving the resulting equations. Much of this effort would be saved if we could express the formula with S as the subject, because then we need only substitute the given values of H, w, T and t in the rearranged formula.

The process of rearranging a formula so that one of the other symbols becomes the subject is called **transposing the formula**. The rules used in transposition are the same as those used in solving equations. The methods used are as follows:

Symbols connected as a product

Examples

1) Transpose the formula $F=ma$ to make a the subject.
Divide both sides by m, then

$$\frac{F}{m}=\frac{ma}{m}$$

$$\frac{F}{m}=a$$

$$\text{or } a=\frac{F}{m}$$

2) make h the subject of the formula $V=\pi r^2h$
Divide both sides by πr^2, then

$$\frac{V}{\pi r^2}=\frac{\pi r^2h}{\pi r^2}$$

$$\frac{V}{\pi r^2}=h$$

$$\text{or } h=\frac{V}{\pi r^2}$$

Symbols connected as a quotient

Examples

1) Transpose $x=\dfrac{y}{b}$ for y.

Multiply both sides by b, then

$$x\times b=\frac{y}{b}\times b$$

$$bx=y$$

or $y=bx$

2) Transpose $M^3=\dfrac{3x^2w}{p}$ for p.

Multiply both sides by p, then

$$M^3p=3x^2w$$

Divide both sides by M^3, then

$$\frac{M^3p}{M^3}=\frac{3x^2w}{M^3}$$

$$p=\frac{3x^2w}{M^3}$$

Symbols connected by a plus or minus sign

Remember that when a term is transferred from one side of a formula to the other its sign is changed.

Examples

1) Transpose $x=3y+5$ for y.
Subtract 5 from both sides of the equation,

$$x-5=3y$$

Divide both sides by 3,

$$\frac{x-5}{3}=y$$

or $y=\dfrac{x-5}{3}$

2) Transpose $w=H+Cr$ for r.
Subtract H from both sides then,

$$w-H=Cr$$

Divide both sides by C,

$$\frac{w-H}{C}=r$$

or $r=\dfrac{w-H}{C}$

Formulae containing brackets

Examples

1) Transpose $y=a+\dfrac{x}{b}$ for x.

Subtract a from both sides then,

$$y-a=\frac{x}{b}$$

Multiply both sides by b,

$$b(y-a)=x$$

or $x=b(y-a)$

2) Transpose $l=a+(n-1)d$ for n.
Subtract a from both sides then,

$$l-a=(n-1)d$$

Divide both sides by d,

$$\frac{l-a}{d}=n-1$$

add 1 to each side,

$$\frac{l-a}{d}+1=n$$

or $n=\dfrac{l-a}{d}+1$

3) Transpose $y=\dfrac{ab}{a-b}$ for a.
Multiply both sides by $a-b$,

$$y(a-b)=ab$$

Remove the brackets as a is on both sides,

$$ay-yb=ab$$

Group the terms containing a on the L.H.S. and other terms on the R.H.S.

$$ay-ab=yb$$

Factorising the L.H.S.

$$a(y-b)=yb$$

Divide both sides by $y-b$,

$$a=\frac{yb}{y-b}$$

4) Transpose $Q=\dfrac{w(H-h)}{T-t}$ for t.
Multiply both sides by $(T-t)$,

$$Q(T-t)=w(H-h)$$

Divide both sides by Q,

$$T-t=\frac{w(H-h)}{Q}$$

Take t to the R.H.S. so that it becomes positive and take $\dfrac{w(H-h)}{Q}$ to the L.H.S.

$$T-\frac{w(H-h)}{Q}=t$$

or $t=T-\dfrac{w(H-h)}{Q}$

Formulae containing roots.

In tackling formulae containing square roots it must be remembered that when a term containing a root is squared, all that happens is that the root sign disappears. Thus,

$$(\sqrt{H})^2=H$$

$$(\sqrt{gh})^2=gh$$

Examples

1) Transpose $d=\sqrt{2hr}$ for r.
Squaring both sides,

$$d^2=(\sqrt{2hr})^2$$

$$d^2=2hr$$

Dividing both sides by $2h$,

$$\frac{d^2}{2h}=r$$

$$\text{or } r=\frac{d^2}{2h}$$

2) Transpose $d=\sqrt{\frac{b(x-b)}{c}}$ for x.

Squaring both sides,

$$d^2=\frac{b(x-b)}{c}$$

Multiplying both sides by c,

$$cd^2=b(x-b)$$

Dividing both sides by b,

$$\frac{cd^2}{b}=x-b$$

Adding b to both sides,

$$\frac{cd^2}{b}+b=x$$

$$\text{or } x=\frac{cd^2}{b}+b$$

3) Transpose $y=\frac{3t}{\sqrt{c}}$ for c.

Multiplying both sides by $\sqrt{c}$,

$$y\sqrt{c}=3t$$

Dividing both sides by y,

$$\sqrt{c}=\frac{3t}{y}$$

Squaring both sides

$$c=\left(\frac{3t}{y}\right)^2$$

$$\text{or } c=\frac{3t}{y}\times\frac{3t}{y}=\frac{9t^2}{y^2}$$

Exercise 37

Transpose the following:

1) $C=\pi d$ for d
2) $S=\pi dn$ for d
3) $PV=c$ for V
4) $A=\pi rl$ for l
5) $v^2=2gh$ for h
6) $I=PRT$ for R
7) $x=\frac{a}{y}$ for y
8) $I=\frac{E}{R}$ for R
9) $x=\frac{u}{a}$ for u
10) $P=\frac{RT}{V}$ for T
11) $d=\frac{0{\cdot}866}{N}$ for N
12) $S=\frac{ts}{T}$ for t
13) $H=\frac{PLAN}{33\,000}$ for L
14) $V=\frac{\pi d^2h}{4}$ for h
15) $p=P-14{\cdot}7$ for P
16) $v=u+at$ for t
17) $n=p+cr$ for r
18) $y=ax+b$ for x
19) $y=\frac{x}{5}+17$ for x
20) $H=S+qL$ for q
21) $a=b-cx$ for x
22) $D=B-1{\cdot}28d$ for d
23) $V=\frac{2R}{R-r}$ for r
24) $C=\frac{E}{R+r}$ for E
25) $S=\pi r(r+h)$ for h
26) $H=wS(T-t)$ for T
27) $C=\frac{N-n}{2p}$ for N
28) $T=\frac{12(D-d)}{L}$ for d
29) $V=\frac{2R}{R-r}$ for R

30) $P=\dfrac{S(C-F)}{C}$ for C

31) $V=\sqrt{2gh}$ for h

32) $w=k\sqrt{d}$ for d

33) $t=2\pi\sqrt{\dfrac{l}{g}}$ for l

34) $t=2\pi\sqrt{\dfrac{W}{gf}}$ for f

35) $P-mg=\dfrac{mv^2}{r}$ for m

36) $Z=\sqrt{\dfrac{x}{x+y}}$ for x

37) $k=\dfrac{3n+2}{n+1}$ for n

38) $a=\dfrac{3}{4t+5}$ for t

39) $v^2=2k\left(\dfrac{1}{x}-\dfrac{1}{a}\right)$ for x

40) $d=\dfrac{2(S-an)}{n(n-l)}$ for a

41) $c=2\sqrt{2hr-h^2}$ for r

42) $x=\dfrac{dh}{D-d}$ for d

43) $\dfrac{D}{d}=\sqrt{\dfrac{f+p}{f-p}}$ for f

SELF TEST 12

In the following questions state the letter (or letters) corresponding to the correct answer (or answers).

1) The value of $\dfrac{ab-c^2}{a^2-bc}$ when $a=2$, $b=-2$ and $c=-3$ is

a $-6{\cdot}5$ b $6{\cdot}5$ c $-3{\cdot}5$ d $3{\cdot}5$

2) The value of $x^2+y^2+z^2-3yz$ when $x=-2$, $y=3$ and $z=-4$ is

a 25 b -47 c -7 d 65

3) The value of $(2a-5b)^2+8ab$ when $a=3$ and $b=-2$ is

a 208 b -32 c 304 d 16

4) The value of $\dfrac{x}{y^2}-\dfrac{y}{z^2}-\dfrac{z}{x^2}$ when $x=-2$, $y=3$ and $z=-4$ is

a $-\dfrac{149}{144}$ b $\dfrac{85}{144}$ c $-\dfrac{203}{144}$ d $\dfrac{139}{144}$

5) The value of $ab(b-2c)-3abc$ when $a=3$, $b=-4$ and $c=1$ is

a -36 b 60 c -60 d 108

6) If $v=u-at$ then t is equal to

a $\dfrac{v-u}{a}$ b $\dfrac{u-v}{a}$ c $\dfrac{-v}{au}$ d $\dfrac{u}{av}$

7) If $a=b+t\sqrt{x}$ then x is equal to

a $\dfrac{(a-b)^2}{t}$ b $\dfrac{t}{(a-b)^2}$ c $\dfrac{t^2}{(a-b)^2}$ d $\dfrac{(a-b)^2}{t^2}$

8) If $x=2y-\dfrac{w}{v}$ then v is equal to

a $\dfrac{w}{2y-x}$ b $\dfrac{2y-x}{w}$ c $\dfrac{-w}{x-2y}$ d $\dfrac{-w}{2y-x}$

9) If $y=\dfrac{1-t^2}{1+t^2}$ then t is equal to

a $\left(\dfrac{1-y}{y+1}\right)^2$ b $\sqrt{\dfrac{1-y}{y+1}}$ c $\sqrt{\dfrac{1-y}{2}}$ d $\dfrac{(1-y)^2}{4}$

10) If $F=\dfrac{W(v-u)}{gt}$ then u is equal to

a $v-\dfrac{WF}{gt}$ b $\dfrac{Fgt}{W}-v$ c $v-\dfrac{Fgt}{W}$ d $Fgt-Wv$

11) If $y=\dfrac{1-3x}{1+5x}$ then x is equal to

a $\dfrac{1-y}{8}$ b $\dfrac{1-y}{2}$ c $\dfrac{1-y}{5y+3}$

d $\dfrac{1-y}{5y-3}$

12) If $A=2\pi R(R+H)$ then H is equal to

a $A-2\pi R^2$ b $\dfrac{A}{2\pi R}-R$

c $R-\dfrac{A}{2\pi R}$ d $\dfrac{A-2\pi R^2}{2\pi R}$

13) If $S=90(2n-4)$ then n is equal to

a $\dfrac{S}{180}+4$ b $\dfrac{S}{180}-2$ c $\dfrac{S}{180}+2$

d $\dfrac{S+360}{180}$

14) If $k=\dfrac{3n+2}{n+1}$ then n is equal to

a $\dfrac{2-k}{k-3}$ b $\dfrac{k-2}{3-k}$ c $\dfrac{1}{k-3}$

d $\dfrac{1}{3-k}$

15) If $T=2\pi\sqrt{\dfrac{R-H}{g}}$ then R is equal to

a $\dfrac{T^2}{2\pi}+\dfrac{H}{g}$ b $\dfrac{gT^2}{2\pi}+H$

c $\dfrac{gT^2+2\pi H}{2\pi}$ d $\dfrac{gT^2}{4\pi^2}+H$

16) If $K=\dfrac{Wv^2}{2g}$ then v is equal to

a $\sqrt{\dfrac{K}{2g}-W}$ b $\sqrt{2Kg-W}$

c $\sqrt{\dfrac{2Kg}{W}}$ d $\sqrt{2KgW}$

17) If $H=wS(T-t)$ then t is equal to

a $\dfrac{H}{wS}-T$ b $\dfrac{H-wST}{wS}$ c $\dfrac{T-H}{wS}$

d $\dfrac{wST-H}{wS}$

18) If $y=\dfrac{ab}{a-b}$ then b is equal to

a $\dfrac{ya}{a+1}$ b $\dfrac{ya}{a-1}$ c $\dfrac{ya}{a+y}$

d $\dfrac{ya}{a-y}$

19) If $a=\sqrt{\dfrac{b}{b+c}}$ then b is equal to

a $\dfrac{a^2c}{1+a^2}$ b $\dfrac{a^2c}{1-a^2}$ c $\dfrac{c}{1+a^2}$

d $\dfrac{c}{1-a^2}$

20) If $x=\sqrt{\dfrac{a^2-b^2}{ay}}$ then b is equal to

a $a-x^2ay$ b $\sqrt{x^2ay-a^2}$

c $\sqrt{a^2-x^2ay}$ d $\sqrt{a(a-x^2y)}$

Chapter 13 Simultaneous Equations

Consider the two equations:

$$2x+3y=13 \qquad \ldots(1)$$

$$3x+2y=12 \qquad \ldots(2)$$

Each equation contains the unknown quantities x and y. The solutions of the equations are the values of x and y which satisfy both equations. Equations such as these are called *simultaneous equations.*

ELIMINATION METHOD IN SOLVING SIMULTANEOUS EQUATIONS

The method will be shown by considering the following examples.

Examples

1) Solve the equations

$$3x+4y=11 \qquad \ldots(1)$$

$$x+7y=15 \qquad \ldots(2)$$

If we multiply equation (2) by 3 we shall have the same coefficient of x in both equations:

$$3x+21y=45 \qquad \ldots(3)$$

We can now eliminate x by subtracting equation (1) from equation (3)

$$3x+21y=45 \qquad \ldots(3)$$

$$3x+\ \ 4y=11 \qquad \ldots(1)$$

$$17y=34$$

$$y=2$$

To find x we substitute for $y=2$ in either of the original equations. Thus, substituting for $y=2$ in equation (1),

$$3x+4\times 2=11$$

$$3x+8=11$$

$$3x=11-8$$

$$3x=3$$

$$x=1$$

Hence the solutions are

$$x=1 \quad \text{and} \quad y=2$$

To check these values substitute them in equation (2). There is no point in substituting them in equation (1) because this was used in finding the value of x. Thus,

$$\text{L.H.S.}=1+7\times 2=15=\text{R.H.S.}$$

Hence the solutions are correct since the L.H.S. and R.H.S. are equal.

2) Solve the equations

$$5x+3y=29 \qquad \ldots(1)$$

$$4x+7y=37 \qquad \ldots(2)$$

The same coefficient of x can be obtained in both equations if equation (1) is multiplied by 4 (the coefficient of x in equation (2)) and equation (2) is multiplied by 5 (the coefficient of x in equation (1)).

Multiplying equation (1) by 4,

$$20x+12y=116 \qquad \ldots(3)$$

Multiplying equation (2) by 5,

$$20x+35y=185 \qquad \ldots(4)$$

Subtracting equation (3) from equation (4),

$$23y=69$$

$$y=3$$

Substituting for $y=3$ in equation (1),

$$5x+3\times 3=29$$

$$5x+9=29$$

$$5x=20$$

$$x=4$$

Hence the solutions are:

$y=3$ and $x=4$

Check in equation (2),

L.H.S. $=4\times 4+7\times 3=16+21=37=$ R.H.S.

3) Solve the equations

$$7x+4y=41 \quad \ldots(1)$$

$$4x-2y=2 \quad \ldots(2)$$

In these equations it is easier to eliminate y because the same coefficient of y can be obtained in both equations by multiplying equation (2) by 2.
Multiplying equation (2) by 2,

$$8x-4y=4 \quad \ldots(3)$$

adding equations (1) and (3),

$$15x=45$$

$$x=3$$

Substituting for $x=3$ in equation (1),

$$7\times 3+4y=41$$

$$21+4y=41$$

$$4y=20$$

$$y=5$$

Hence the solutions are

$x=3$ and $y=5$

Check in equation 2,

L.H.S. $=4\times 3-2\times 5=12-10=2=$ R.H.S.

4) Solve the equations

$$\frac{2x}{3}-\frac{y}{4}=\frac{7}{12} \quad \ldots(1)$$

$$\frac{3x}{4}-\frac{2y}{5}=\frac{3}{10} \quad \ldots(2)$$

It is best to clear each equation of fractions before attempting to solve.
In equation (1) the L.C.M. of the denominators is 12. Hence by multiplying equation (1) by 12,

$$8x-3y=7 \quad \ldots(3)$$

In equation (2) the L.C.M. of the denominators is 20. Hence by multiplying equation (2) by 20,

$$15x-8y=6 \quad \ldots(4)$$

We now proceed in the usual way.
Multiplying equation (3) by 8,

$$64x-24y=56 \quad \ldots(5)$$

Multiplying equation (4) by 3,

$$45x-24y=18 \quad \ldots(6)$$

Subtracting equation (6) from equation (5),

$$19x=38$$

$$x=2$$

Substituting for $x=2$ in equation (3),

$$8\times 2-3y=7$$

$$16-3y=7$$

$$-3y=-9$$

$$y=3$$

Hence the solutions are

$x=2$ and $y=3$

Since equation (3) came from equation (1) we must do the check in equation (2).

$$\text{L.H.S.}=\frac{3\times 2}{4}-\frac{2\times 3}{4}=\frac{6}{4}-\frac{6}{5}$$

$$=\frac{30-24}{20}=\frac{6}{20}=\frac{3}{10}=\text{R.H.S.}$$

Exercise 38

Solve the following equations for x and y and check the solutions:

1) $3x+2y=7$
$x+y=3$

2) $4x-3y=1$
$x+3y=19$

3) $x+3y=7$
$2x-2y=6$

4) $7x-4y=37$
$6x+3y=51$

5) $4x-6y=-2{\cdot}5$
$7x-5y=-0{\cdot}25$

6) $\frac{x}{2}+\frac{y}{3}=\frac{13}{6}$
$\frac{2x}{7}-\frac{y}{4}=\frac{5}{14}$

7) $\frac{x}{8}-y=-\frac{5}{2}$
$3x+\frac{y}{3}=13$

8) $\frac{x-2}{3}+\frac{y-1}{4}=\frac{13}{12}$
$\frac{2-x}{2}+\frac{3+y}{3}=\frac{11}{6}$

9) $\frac{x}{3}-\frac{y}{2}+1=0$
$6x+y+8=0$

10) $3x-4y=5$
$2x-5y=8$

11) $x-y=3$
$\frac{x}{5}-\frac{y}{7}=\frac{27}{35}$

12) $3x+4y=0$
$2x-2y=7$

PROBLEMS INVOLVING SIMULTANEOUS EQUATIONS

In problems which involve two unknowns it is first necessary to form two separate equations from the given data. The equations may then be solved as shown previously.

Examples

1) A bill for £74 was paid with £5 and £1 notes, a total of 50 notes being used. Find how many £5 notes were used.
Let x be the number of £5 notes and y be the number of £1 notes. Then

$x+y=50 \qquad \ldots(1)$

The total value of x £5 notes is £$5x$ and the value of y £1 notes is £y. Hence the total value of x £5 notes and y £1 notes is £$(5x+y)$ and this must equal £74. Hence,

$5x+y=74 \qquad \ldots(2)$

Subtracting equation (1) from equation (5),

$4x=24$
$x=6$

Therefore six £5 notes were used.

2) Find two numbers such that their sum is 108 and their difference is 54.
Let x and y be the two numbers.
Then their sum is $x+y$ and their difference is $x-y$. Hence,

$x+y=108 \qquad \ldots(1)$
$x-y=54 \qquad \ldots(2)$

adding equations (1) and (2),

$2x=162$
$x=81$

Substituting for x in (1),

$81+y=108$
$y=108-81$
$y=27$

3) A foreman and 7 men together earn £130 per week whilst two foremen and 17 men together earn £305 per week. Find the weekly wages of a foreman.
Let a foreman earn £x per week and a man earn £y per week.

$\therefore x+7y=130 \qquad \ldots(1)$
$2x+17y=305 \qquad \ldots(2)$

Multiplying equation (1) by 2,

$2x+14y=260 \qquad \ldots(3)$

Subtracting equation (3) from equation (2)

$3y=45$
$y=15$

Substituting for $y=15$ in equation (1),

$x+7\times 15=130$
$x+105=130$
$x=25$

Hence a foreman earns £25 per week.

Exercise 39

1) Find two numbers such that their sum is 27 and their difference is 3.

2) A bill for £123 was paid with £5 and £1 notes a total of 59 notes being used. Find how many £5 notes were used.

3) £x is invested at 6% and £y is invested at 8%. The annual income from these investments is £23·20. If £x had been invested at 8% and £y at 6% the annual income would have been £21·60. Find x and y.

4) An alloy containing 8 cm^3 of copper and 7 cm^3 of tin has a mass of 121 g. A second alloy contains 9 cm^3 of copper and 11 cm^3 of tin has a mass of 158 g. Find the densities of copper and tin in g/cm^3.

5) A motorist travels x km at 40 km/h and y km at 50 km/h. The total time taken is $2\frac{1}{2}$ hours. If the time taken to travel $6x$ km at 30 km/h and $4y$ km at 50 km/h is 14 hours find x and y.

6) 500 tickets were sold for a concert, some at 20p each and the remainder at $12\frac{1}{2}$ p each. The money received for the dearer tickets was £35 more than for the cheaper tickets. Find the number of dearer tickets which were sold.

7) The ages of A and B are in the ratio 4 : 3. In eight years time the ratio of their ages will be 9 : 7. Find their present ages. If n years ago, A was three times as old as B, find the value of n.

8) The organisers of a charity concert sold tickets at two different prices. If they had sold 112 of the dearer tickets and 60 of the cheaper ones they would have received £45·60 but if they had sold 96 of the dearer tickets and 120 of the cheaper ones they would have received £52·80. Find the price of the dearer tickets.

9) Two numbers are in the ratio 5 : 7. When 15 is added to each the ratio changes to 5 : 6. Calculate the two numbers.

10) A man bought a number of 3p stamps and also sufficient 4p stamps to make his total expenditure 120p. If, instead of the 3p stamps, he had bought three times as many 2p stamps he would have needed 9 fewer 4p stamps than before for his expenditure to be 120p. Find how many 3p stamps he bought.

SELF TEST 13

In the following questions state the letter (or letters) which correspond to the correct answer (or answers).

1) In the simultaneous equations:

$$2x+3y=17$$
$$3x+4y=24$$

x is equal to 4. Hence the value of y is

a $\frac{25}{3}$ b 75   c 3   d 4

2) In the simultaneous equations:

$$2x-3y=-16$$
$$5y-3x=25$$

x is equal to -5. Hence the value of y is

a 2   b -2 c $\frac{26}{3}$ d 0

3) By eliminating x from the simultaneous equations:

$$2x-5y=8$$
$$2x-3y=-7$$

the equation below is obtained

a $-8y=1$ b $-2y=15$

c $-8y=15$ c $-2y=1$

4) By eliminating y from the simultaneous equations:

$$3x-4y=-10$$
$$x+4y=8$$

the equation below is obtained

a $2x=-18$ b $4x=-18$

c $2x=-2$ d $4x=-2$

5) By eliminating x from the simultaneous equations:

$$3x+5y=2$$
$$x+3y=7$$

the equation below is obtained

a $4y=-19$ b $8y=9$

c $4y=19$ d $4y=5$

6) By eliminating y from the simultaneous equations:

$$2x-4y=3$$
$$3x+8y=7$$

the equation below is obtained:

a $7x=13$ b $x=-1$

c $x=1$ d $7x=10$

7) The solutions to the simultaneous equations

$$2x-5y=3$$
$$x-3y=1$$ are

a $x=4, y=1$ b $y=4, x=1$

c $y=4, x=13$ d $x=4, y=3$

8) The solutions to the simultaneous equations

$$3x-2y=5$$
$$4x-y=10$$

a $x=-3, y=22$ b $x=-3, y=-22$
c $x=3, y=2$ d $x=3, y=-2$

9) Two numbers, x and y, are such that their sum is 18 and their difference is 12. The equations below will allow x and y to be found:

a $x+y=18$, $y-x=12$ b $x+y=18$, $x-y=12$

c $x-y=18$, $x+y=12$ d $y-x=18$, $x+y=12$

10) A bill for 40 pence is paid by means of 5 pence and 10 pence pieces. Seven coins were used in all. If x is the number of 5 pence pieces used and y is the number of 10 pence pieces used then

a $x+y=7$, $x+2y=40$ b $x+y=7$, $5x+10y=40$

c $x-y=7$, $x+y=40$ d $x-y=7$, $5x+10y=40$

11) A motorist travels x km at 50 km/h and y km at 60 km/h. The total time taken is 5 hours. If his average speed is 56 km/h then

a $50x+60y=5$, $x+y=280$ b $6x+5y=1500$, $x+y=280$

c $\frac{x}{50}+\frac{y}{60}=5$, $\frac{x+y}{5}=56$ d $50x+60y=5$, $x+y=280$

12) 300 tickets were sold for a concert some at 20p each and the remainder at 30p each. The cash received for the cheaper tickets was £10 more than that received for the dearer tickets. Therefore,

a $x+y=300$, $20x-30y=10$ b $x+y=300$, $20x-30y=1000$

c $x+y=300$, $2x-3y=100$ d $x+y=300$, $2x+3y=1$

Miscellaneous Exercise

Exercise 40

These questions are of the type found in O Level papers.

1) If $p=2$, $q=0$ and $r=-3$, find the value of

$$\frac{pq-r^2}{p^2-qr}$$

2) If $v=u-at$ and $v^2=u^2+2as$ express s in terms of u, a and t.

3) Factorise

a) $9z^2-25$, b) $3a^2+2ab-12ac-8bc$

4) Simplify

$$\frac{3}{x+1}+\frac{2x-1}{(x+1)(x+2)}-\frac{2}{x+2}$$

5) Make f the subject of the formula

$$s=u+\frac{t}{2}\sqrt{f}$$

6) Simplify $\dfrac{1}{x+1}+\dfrac{x}{x-1}-1$

7) Add $5x-3y+2z$ and $3x-y-3z$ and subtract the total from $7x-2y+5z$

8) Factorise $7-63a^2$

9) Factorise $7t^2-14t$

10) If $a=b-\dfrac{c}{d}$ find d in terms of a, b and c.

11) Factorise $6x^2-x-70$

12) Simplify $(3x+y)^2-(9x-y)(x+y)$

13) Make x the subject of the formula

$$y=\frac{3x-1}{4x-5}$$

14) Simplify $\dfrac{x-3}{3}-\dfrac{x-7}{6}$

15) Copy and complete the following

$$25x^2-70x+\qquad=(5x\qquad)^2$$

16) If $y=2(x-1)(x-3)$, find the value of y when

i) $x=5$, ii) $x=1$, iii) $x=0$

17) Factorise a^2+4a+3

18) Factorise

a) $(a+b)^2-4c^2$

b) $2(x+2y)+ax+2ay$

19) Express $\dfrac{a+1}{3a}+\dfrac{a+3}{4a}+\dfrac{a-13}{12a}$

as a single fraction in its lowest terms.

20) If $x=\dfrac{2n-4}{n}$ express n in terms of x.

21) Simplify $\dfrac{1}{x^2-2x}+\dfrac{1}{2x}$

22) Factorise
i) $4x^2-y^2$, ii) $a^2+2ab+b^2-x^2$

23) Factorise i) $3x^2+5x-12$,
ii) $xy+y^2-2x-2y$, iii) $3x^2-12y^2$

24) Factorise completely a^4-b^4

25) Find the value of k (other than 0) for which $(3p-q)^2+kpq$ is a perfect square.

26) Given $x=\dfrac{2t}{1+t^2}$ and $y=\dfrac{1-t^2}{1+t^2}$ find

i) the values of x and y when $t=2$,
ii) the values of t and x when $y=0{\cdot}6$.

27) If $V=\pi h^2(R-\frac{1}{3}h)$, express R in terms of V, h (and π).

28) Factorise

i) $25x^2-81$
ii) $(x+2)^2-(y+1)^2$, iii) $5x^2+11x+2$,
iv) $(x+a)^2-3(x+a)+2$.

29) Find the value of $p^2+q^2+r^2-2qr$ when $p=2$, $q=3$ and $r=-4$.

30) Factorise completely $27x^2-48y^2$

31) From the formula $F=\dfrac{W(V-U)}{gt}$ calculate V when $F=8$, $g=32$, $t=3$, $W=24$ and $U=5$.

32) Factorise $2x^3-2x^2-4x$

33) Simplify $\dfrac{2}{x^2+4x+3}-\dfrac{1}{x^2+3x+2}$

34) Simplify

$$\frac{7}{x^2+3x-10}-\frac{2}{x^2+5x}-\frac{2}{x^2-2x}$$

35) Find the value of $(2p-3q)^2+12pq$ given that $p=4$ and $q=-1$.

36) Rearrange the formula $y=(x+a)(p+q)$ to express x in terms of a, p, q and y.

37) Factorise $ac-2ab-3bc+6b^2$

38) Factorise completely $24p^2-6p-9$

39) Factorise

i) $12x^2-3y^2$,
ii) $2x+xy+y^2+2y$,
iii) $3(x+2)^2+2(x+2)-8$.

40) Find the value of $\dfrac{a}{b^2}+\dfrac{b}{c^2}-\dfrac{c}{a^2}$ when $a=\frac{1}{2}$, $b=-3$ and $c=3$.

41) If $a=x+3y$ and $b=3x-y$ find the values in terms of x and y of

i) $3a-2b$, ii) a^2+b^2

42) Express

$$\frac{2-p}{2p}-\frac{3-2p}{3p}-\frac{p+2}{6p}$$

as a single fraction in its lowest terms.

43) If $a=-2$, $b=3$, $c=7$ and $d=0$ find the value of

i) $(b+2a)^3$, ii) $bc+ad$, iii) $a(b-c)$

44) Factorise

i) $2p^2-5p+2$ ii) $b(a-2c)+ad-2cd$

45) Express

$$\frac{2x-5}{5x}-\frac{3x+2}{4x}+\frac{7x+15}{10x}$$

as a single fraction in its lowest terms.

46) Simplify

$$\frac{8}{x^2-x-12}+\frac{x-1}{x(x+3)}-\frac{1}{x-4}$$

47) Express

$$\frac{2}{2x+1}+\frac{7}{(2x+1)(x-3)}$$

as a single fraction in its simplest form.

48) Use the formula $a=b-\dfrac{2}{c}$ to express c in terms of a and b.

49) Find the value of $ab(b-c)-2abc$ when $a=2$, $b=-5$ and $c=1$.

50) Factorise

i) $6p^2+5p-6$, ii) $(5-a)^2-49b^2$

51) Solve the equation

$$3(x+2)-\frac{4x-5}{4}=3\tfrac{1}{4}$$

52) Solve the equations

$3x-5y=44$
$5x+7y=12$

53) If $\dfrac{3a+b}{3b-2a}=4$, calculate the value of $\dfrac{a}{b}$

54) If $V^2=U^2+2as$ find

a) s in terms of a, U and V

b) the value of U when $a=4$, $s=45$ and $V=23$

55) If apples cost x pence per kilogramme find

a) the weight of apples that can be purchased for £y

b) the extra weight which can be purchased for £y when the cost per kilogramme decreases by 2 pence per kilogramme. Express your answer as a single fraction in its simplest form.

56) Simplify

$$\frac{2}{x^2+x-6}\times\frac{3}{2x^2+3x-2}\div\left(1+\frac{3-5x}{2x^2+5x-3}\right)$$

57) Potatoes costing £c per 100 kg are sold, at a profit, for f pence per kg. Write down the profit in pence made on each kilogramme sold.

58) The cost of travelling by taxi is worked out as follows: There is a hiring charge of 7 pence and in addition a charge of 8 pence per kilometre. Find the cost of travelling 5 km by taxi and also the length of a journey which costs £1·03. Write down the total cost, in pence, for a journey of n kilometres. Find n when the average cost per kilometer is 9 pence.

59) Solve the simultaneous equations

$$4x-5y=13$$
$$x+2y=0$$

60) Solve the equation

$$\frac{x-1}{2}-\frac{2(3-x)}{3}=1$$

61) Solve the equation

$$\frac{5(x-2)}{6}=x-4$$

62) If $\frac{1}{x}=\frac{1}{y}+\frac{2}{z}$, find the value of x when $y=\frac{1}{3}$ and $z=4$.

63) Solve the simultaneous equations

$$3p+2q=8$$
$$p-q=6$$

64) If $(x+11)$ is a factor of $2x^2+ax-187$ find the other factor and the value of a.

65) Solve the equation

$$\frac{x}{3}+\frac{3x-1}{4}=\frac{3x+7}{12}$$

66) If $\frac{a+3b}{4a+3b}=4$, find a in terms of b.

67) Evaluate $3x^2-7xy-5y^2$ when $x=2$ and $y=-3$.

68) Given that $y=\frac{1-2t}{1+3t}$ find t in terms of y.

69) Factorise $12a^2+7a-49$.

70) Find the value of

$$\frac{a}{b^2}+\frac{b}{c^2}-\frac{c}{a^2}$$

when $a=\frac{1}{4}$, $b=-3$ and $c=3$.

71) If $c=\sqrt{\frac{4b(a-b)}{5}}$ express a in terms of b and c.

72) Solve the equation

$$\frac{2x}{x+2}=\frac{3x}{x+5}-1$$

73) Solve the simultaneous equations

$$\frac{6}{x}-\frac{2}{y}=\frac{1}{2}$$
$$\frac{4}{x}-\frac{3}{y}=0$$

(Hint: let $X=\frac{1}{x}$ and $Y=\frac{1}{y}$)

74) If $A(x-1)+B(x+1)=3x+5$ for all values of x, find the values of A and B.

75) A shopkeeper purchased a number of articles at x pence each and added 4 pence to the cost price of each article to form his selling price. He sold some of these articles

for a total of $4a$ pence. Write down an expression for the number of articles sold.

76) Solve the equation

$$\frac{3+2x}{3-2x}=-5$$

77) A clock loses x minutes in y hours. How many minutes will it lose in z minutes?

78) Solve the simultaneous equations

$$3x-4y=8$$
$$5x+6y=-7$$

Chapter 14 Squares, Square Roots and Reciprocals

SQUARE ROOTS

The square root of a number is the number whose square equals the given number.Thus since

$$5^2=25, \quad \sqrt{25}=5$$

and

$$9^2=81, \quad \sqrt{81}=9$$

The square root of any number can usually be found to sufficient accuracy by using the printed tables of square roots. There are two of these tables, one giving square roots of numbers from 1 to 10 and the other square roots of numbers from 10 to 100. The reason for having two sets of tables is as follows:

$$\sqrt{8{\cdot}1}=2{\cdot}846 \quad \text{but} \quad \sqrt{81}=9$$

Thus there are two quite different square roots for the same figures and the square root of a number depends upon the position of the decimal point.

USE OF SQUARE ROOT TABLES

1) *The square root of a number having two significant figures.* To find $\sqrt{1{\cdot}8}$ find 1·8 in the first column of the table. Move along the row to the number under the column headed 0. We find the number to be 1342. Hence

$$\sqrt{1{\cdot}8}=1{\cdot}342$$

Similarly $\sqrt{2{\cdot}7}=1{\cdot}643$

2) *The square root of a number having three significant figures.* To find $\sqrt{1{\cdot}87}$ find 1·8 in the first column of the table. Move along the row to the number under the column headed 7. We find the number to be 1·367. Hence

$$\sqrt{1{\cdot}87}=1{\cdot}367$$

Similarly $\sqrt{2{\cdot}74}=1{\cdot}655$

3) *The square root of a number having four significant figures.* To find $\sqrt{1{\cdot}873}$ find 1·8 in the first column. Move along the row to the number under the column headed 7 to find the number 1·367. Now move along the same row to the number under the column headed 3 of the proportional parts and read 1. Add this 1 (that is 0·001) to 1·367 to give 1·368. Thus,

$$\sqrt{1{\cdot}873}=1{\cdot}368$$

Similarly $\sqrt{2{\cdot}748}=1{\cdot}657$

Square roots of numbers between 10 and 100 may be found in a similar way by using the square root tables for numbers between 10 and 100. Thus

$$\sqrt{92{\cdot}65}=9{\cdot}626$$

$$\sqrt{33{\cdot}28}=5{\cdot}769$$

SQUARE ROOTS OF NUMBERS OUTSIDE THE RANGE OF THE SQUARE ROOT TABLES

Although the two sets of square root tables only give the square root of numbers between 1 and 100 we can use the tables to find the square roots of numbers outside this range. We must be sure that we use the correct tables and that the decimal point in the answer is positioned correctly. The methods shown in the following examples are recommended.

Examples

1) To find $\sqrt{836 \cdot 3}$

Mark off the figures in pairs to the *left* of the decimal point. Each pair of figures is called a period. Thus 836·3 becomes 8′ 36·3. The first period is 8 so we use the table of numbers from 1 to 10 and look up $\sqrt{8 \cdot 363} = 2 \cdot 892$. To position the decimal point in the final answer, remember that for *each period to the left* of the decimal point there will be *one figure to the left* of the decimal point in the answer. Thus,

```
2  8 · 92
┌──┐
8′ 36 · 3
```

$\therefore \quad \sqrt{836 \cdot 3} = 28 \cdot 92$

2) To find $\sqrt{173\,900}$

Marking off in periods 173 900 becomes 17′ 39′ 00. The first period is 17 so we use the table of numbers from 10 to 100 and look up

$\sqrt{17 \cdot 3900} = 4 \cdot 170$

```
 4  1  7 · 0
┌──────┐
17′ 39′ 00
```

$\therefore \quad \sqrt{173\,900} = 417 \cdot 0$

3) To find $\sqrt{0 \cdot 000\,094\,31}$

In the case of numbers less than 1, mark off the periods to the *right* of the decimal point. 0·000 094 31 becomes 00′ 00′ 94′ 31. Apart from the zero pairs the first period is 94 so we use the table of numbers from 10 to 100 and look up

$\sqrt{94 \cdot 31} = 9 \cdot 712$

For *each zero pair* in the original number there is *one zero* following the decimal point in the answer.

```
0 ·  0   0   9   7 1 2
  ┌──────────────┐
0 · 00′ 00′ 94′ 31
```

$\therefore \quad \sqrt{0 \cdot 000\,094\,31} = 0 \cdot 009\,712$

4) To find $\sqrt{0 \cdot 073\,65}$

Marking off in periods to the *right* of the decimal point 0·073 65 becomes 07′ 36′ 50. Since the first period is 07 we use the table of numbers from 1 to 10 and look up

$\sqrt{7 \cdot 365} = 2 \cdot 714$

```
0 ·  2   7   1 4
  ┌─────────┐
0 · 07′ 36′ 50
```

$\therefore \quad \sqrt{0 \cdot 073\,65} = 0 \cdot 271\,4$

Exercise 41

Find the square roots of the following numbers:

1) 3·4

2) 8·19

3) 5·264

4) 9·239

5) 7·015

6) 3·009

7) 35

8) 89·2

9) 53·17

10) 82·99

11) 79·23

12) 50·01

13) 900

14) 725·3

15) 7142

16) 89 000

17) 3945

18) 893 400 000

19) 0·1537

20) 0·001 698

21) 0·000 007 1

22) 0·039 47

23) 0·000 783 1

24) 0·001 978

SQUARES OF NUMBERS

When a number is multiplied by itself the result is called the square of the number. Thus,

the square of $15 = 15 \times 15 = 225$

Instead of writing 15×15 we often write 15^2 which is read as the square of 15. Thus,

$12^2 = 12 \times 12 = 144$

and

$2 \cdot 27^2 = 2 \cdot 27 \times 2 \cdot 27 = 5 \cdot 1529$

USING TABLES TO FIND THE SQUARE OF A NUMBER

If available, a table of squares of numbers may be used. However many books of tables do not contain a table of squares of numbers in which case the square of a number may be found by using the square root tables.

1) To find $(1{\cdot}806)^2$. In the body of the square root tables we search until we find the number 1·806. This is in the row 3·2 and the column headed 6. Thus $(1{\cdot}806)^2 = 3{\cdot}26$.

2) To find $(6{\cdot}682)^2$ search in the body of the square root tables until a number as close as possible to 6·682 is found. This is 6·678 which is in the row 44 and the column 6. Thus, $(6{\cdot}678)^2 = 44{\cdot}6$. The difference between 6·682 and 6·678 is 0·004 (or 4 in the table of proportional parts) which occurs in the column headed 5. Hence,

$$(6{\cdot}682)^2 = 44{\cdot}65.$$

3) To find $(468{\cdot}8)^2$.

$$(468{\cdot}8)^2 = (4{\cdot}688 \times 100) \times (4{\cdot}688 \times 100)$$
$$= 4{\cdot}688^2 \times 100^2$$

From the square root tables $4{\cdot}688^2 = 21{\cdot}97$
Hence

$$(468{\cdot}8)^2 = 21{\cdot}97 \times 100^2 = 219\,700$$

4) To find $(0{\cdot}2388)^2$

$$(0{\cdot}2388)^2 = \frac{2{\cdot}388}{10} \times \frac{2{\cdot}388}{10} = \frac{2{\cdot}388^2}{100}$$

From the square root tables $2{\cdot}388^2 = 5{\cdot}703$
Hence

$$(0{\cdot}2388)^2 = \frac{5{\cdot}703}{100} = 0{\cdot}05703$$

Exercise 42

Find the square of the following numbers:

1) 1·5 2) 2·1
3) 8·6 4) 3·15
5) 7·68 6) 5·23
7) 4·263 8) 7·916
9) 8·017 10) 8·704
11) 23 12) 40·6
13) 3093 14) 112·3
15) 98·12 16) 0·019
17) 0·7292 18) 0·004219
19) 0·2834 20) 0·000 578 4

Find the values of the following:

21) $(2{\cdot}58)^2 - (0{\cdot}89)^2$

22) $(17{\cdot}36)^2 + (14{\cdot}18)^2$

23) $(11{\cdot}412 - 8{\cdot}097)^2$

24) $(125{\cdot}3 + 17{\cdot}81)^2$

25) $81^2 - (17{\cdot}53 - 11{\cdot}60)^2$

26) $(1{\cdot}531 - 0{\cdot}968)^2 + (5{\cdot}261)^2$

27) $(3{\cdot}218)^2 + (7{\cdot}361)^2 - (6{\cdot}153)^2$

28) $(3{\cdot}1 \times 2{\cdot}2)^2 - (5{\cdot}71)^2$

RECIPROCALS OF NUMBERS

The reciprocal of a number is $\dfrac{1}{\text{number}}$.

Thus the reciprocal of 5 is $\dfrac{1}{5}$ and the reciprocal of 21·34 is $\dfrac{1}{21{\cdot}34}$.

The tables of reciprocals are used in much the same way as the tables of squares but the proportional parts are subtracted *not* added. The tables give the reciprocals of numbers from 1 to 10 in decimal form. From the tables:

the reciprocal of 6 = 0·1667
the reciprocal of 3·157 = 0·3168

The method of finding the reciprocals of numbers less than 1 or greater than 10 is shown in the following examples.

Examples

1) Find the value of $\frac{1}{639\cdot2}$

$$\frac{1}{639\cdot2} = \frac{1}{6\cdot392} \times \frac{1}{100}$$

$$= \frac{0\cdot1565}{100} = 0\cdot001\,565$$

2) Find the value of $\frac{1}{0.03982}$

$$\frac{1}{0\cdot03\,982} = \frac{1}{3\cdot982} \times \frac{100}{1}$$

$$= 0\cdot2512 \times 100 = 25\cdot12$$

Exercise 43

Find the reciprocals of the following:

1) 3·4 2) 8·19

3) 5·264 4) 9·239

5) 7·015 6) 35

7) 89·2 8) 53·17

9) 900 10) 7142

11) 0·1537 12) 0·001 698

13) 0·039 47 14) 0·000 783 1

15) 0·001 978

Find the values of the following:

16) $\frac{1}{(15\cdot28)^2}$ 17) $\frac{1}{(0\cdot1372)^2}$

18) $\frac{1}{(250)^2}$ 19) $\frac{1}{\sqrt{8\cdot406}}$

20) $\frac{1}{\sqrt{18\cdot73}}$ 21) $\frac{1}{\sqrt{0\cdot01798}}$

22) $\frac{1}{\sqrt{(30\cdot15)^2+(8\cdot29)^2}}$

23) $\frac{1}{\sqrt{(11\cdot26)^2-(8\cdot18)^2}}$

24) $\frac{1}{8\cdot2} + \frac{1}{9\cdot9}$

25) $\frac{1}{0\cdot7325} + \frac{1}{0\cdot9817}$

26) $\frac{1}{18\cdot27} - \frac{1}{97\cdot58}$

27) $\frac{1}{\sqrt{7\cdot517}} + \frac{1}{(0\cdot829)^2} + \frac{1}{0\cdot0749}$

28) $\frac{1}{71\cdot36} + \frac{1}{\sqrt{863\cdot5}} + \frac{1}{(7\cdot589)^2}$

SELF TEST 14

In the following questions state the letter (or letters) corresponding to the correct answer (or answers).

1) 80^2 is equal to

a 64 b 640 c 6400

2) 700^2 is equal to

a 490 000 b 49 000 c 4900

3) $0\cdot8^2$ is equal to

a 6·4 b 0·64 c 0·064

4) $0\cdot09^2$ is equal to

a 0·081 b 0·0081 c 0·81

5) $\sqrt{0\cdot25}$ is equal to

a 0·5 b 0·05 c 0·158

6) $\sqrt{0\cdot036}$ is equal to

a 0·6 b 0·06 c 0·1897

7) $\sqrt{0\cdot0049}$ is equal to

a 0·7 b 0·07 c 0·2214

8) $\sqrt{1690}$ is equal to

a 130 b 13 c 41·11

9) $\sqrt{810}$ is equal to

a 28·46 b 90 c 9

10) $\sqrt{12\,100}$ is equal to

a 1100 b 110 c 347·9

11) $\frac{1}{12 \cdot 5}$ is equal to

a 8 b 0·08 c 0·8

12) $\frac{1}{0 \cdot 25}$ is equal to

a 40 b 4 c 0·4

13) $\frac{1}{0 \cdot 020}$ is equal to

a 5 b 50 c 500

14) $\frac{1}{250}$ is equal to

a 0·4 b 0·04 c 0·004

15) $18 \cdot 3^2 - 8 \cdot 3^2$ is equal to

a 100 b 266 c 26·6

16) If $h = \frac{1}{u} + \frac{1}{v}$ then when $u = 37 \cdot 17$ and $v = 1 \cdot 477$ the value of h is

a 0·7039 b 0·9457 c 0·09 457

17) If $p = b\sqrt{\frac{c}{q}}$ then when $b = 12$, $c = 4$ and $q = 16$ then p is equal to

a 7·59 b 75·9 c 6

18) An approximate value of $\sqrt{2562 \cdot 8}$ is

a 16·2 b 50·6 d 520

19) $\frac{1}{0 \cdot 3128}$ is equal to

a 0·3197 b 3·197 c 31·97

20) $(30 \cdot 16)^2 + \frac{1}{0 \cdot 0478}$ is equal to

a 93·06 b 111·89 c 930·5

Chapter 15 Indices and Logarithms

LAWS OF INDICES

Multiplication

$$a^4 \times a^3 = (a \times a \times a \times a) \times (a \times a \times a)$$
$$= a \times a \times a \times a \times a \times a \times a = a^7$$

It will be noticed that $a^4 \times a^3 = a^{4+3} = a^7$; that is we have **added** the two indices together in order to obtain the power of a in the result. In general,

$$a^m \times a^n = a^{m+n}$$

and the law is:

When multiplying powers of the same number together add the indices.

Examples

1) $5^2 \times 5^3 = 5^{2+3} = 5^5$

2) $x^6 \times x^7 = x^{6+7} = x^{13}$

3) $y^2 \times y^3 \times y^4 \times y^5 = y^{2+3+4+5} = y^{14}$

Division

$$\frac{a^5}{a^2} = \frac{a \times a \times a \times a \times a}{a \times a} = a^3$$

It will be noticed that $\frac{a^5}{a^2} = a^{5-2} = a^3$; that is we have **subtracted** the index of the denominator from the index of the numerator in order to obtain the power of a in the result. In general,

$$\frac{a^m}{a^n} = a^{m-n}$$

and the rule is:

When dividing powers of the same number subtract the index of the denominator from the index of the numerators.

Examples

1) $\frac{x^5}{x^2} = x^{5-2} = x^3$

2) $\frac{5^7}{5^3} = 5^{7-3} = 5^4$

3) $\frac{a^3 \times a^4 \times a^8}{a^5 \times a^7} = \frac{a^{3+4+8}}{a^{5+7}} = \frac{a^{15}}{a^{12}} = a^{15-12} = a^3$

4) $\frac{3y^2 \times 2y^5 \times 5y^4}{6y^3 \times 4y^4} = \frac{30y^{2+5+4}}{24y^{3+4}} = \frac{30y^{11}}{24y^7}$

$$= \frac{5y^{11-7}}{4} = \frac{5y^4}{4}$$

Powers

$$(a^3)^2 = a^3 \times a^3 = a^{3+3} = a^6$$

It will be noticed that $(a^3)^2 = a^{3 \times 2} = a^6$; that is, we have **multiplied** the indices together. In general,

$$(a^m)^n = a^{mn}$$

and the rule is:

When raising the power of a number to a power, multiply the indices together.

Examples

1) $(5^4)^3 = 5^{4 \times 3} = 5^{12}$

2) $(3x)^3 = 3^{1 \times 3} \times x^{1 \times 3} = 3^3x^3 = 27x^3$

3) $(a^2b^3c^4)^2 = a^{2 \times 2}b^{3 \times 2}c^{4 \times 2} = a^4b^6c^8$

4) $\left(\frac{3m^3}{5n^2}\right)^2 = \frac{3^2m^{3 \times 2}}{5^2n^{2 \times 2}} = \frac{9m^6}{25n^4}$

Negative indices

$$\frac{a^3}{a^5} = \frac{\not{a} \times \not{a} \times \not{a}}{\not{a} \times \not{a} \times \not{a} \times a \times a} = \frac{1}{a \times a} = \frac{1}{a^2}$$

Now by the law for division,

$$\frac{a^3}{a^5}=a^{3-5}=a^{-2}$$

Thus, $a^{-2}=\frac{1}{a^2}$

A negative index therefore indicates the reciprocal of the quantity.

Generally speaking,

$$a^{-m}=\frac{1}{a^m}$$

Examples.

1) $a^{-1}=\frac{1}{a}$

2) $5x^{-3}=\frac{5}{x^3}$

3) $(8x^2)^{-3}=\frac{1}{(8x^2)^3}=\frac{1}{8^3x^6}=\frac{1}{512x^6}$

4) $a^2b^{-2}c^{-3}=\frac{a^2}{b^2c^3}$

5) $\frac{1}{a^{-3}}=(a^{-3})^{-1}=a^3$

6) $\frac{(27a^2)^{-2}}{(81a^3)^{-3}}=\frac{(81a^3)^3}{(27a^2)^2}=\frac{(3^4a^3)^3}{(3^3a^2)^2}$

$$=\frac{3^{12}a^9}{3^6a^4}=3^6a^5=729a^5$$

Fractional Indices

To find a meaning for $a^{1/3}$

$$\sqrt[3]{a}\times\sqrt[3]{a}\times\sqrt[3]{a}=a$$

also,

$$a^{1/3}\times a^{1/3}\times a^{1/3}=a$$

$$\therefore\ a^{1/3}=\sqrt[3]{a}$$

To find a meaning for $a^{2/3}$

$$\sqrt[3]{a^2}\times\sqrt[3]{a^2}\times\sqrt[3]{a^2}=a^2$$

$$a^{2/3}\times a^{2/3}\times a^{2/3}=a^2$$

$$\therefore\ a^{2/3}=\sqrt[3]{a^2}$$

Generally speaking,

$$a^{n/m}=\sqrt[m]{a^n}$$

The numerator of the index indicates the power to which the number must be raised; the denominator denotes the root which is to be taken.

Examples

1) $3^{5/4}=\sqrt[4]{3^5}$

2) $x^{2/3}=\sqrt[3]{x^2}$

3) $ab^{3/4}=a\times\sqrt[4]{b^3}$

4) $\sqrt{a}=a^{1/2}$

(Note that for square roots the number indicating the root which is to be taken is usually omitted).

5) $\sqrt{64a^6}=(64a^6)^{1/2}=(8^2a^6)^{1/2}$
$=8^{2\times 1/2}a^{6\times 1/2}=8a^3$

6) $(32)^{2/5}=(2^5)^{2/5}=2^{5\times 2/5}=2^2$

Zero Index

$$\frac{a^n}{a^n}=a^{n-n}=a^0$$

also, if a is not zero,

$$\frac{a^n}{a^n}=1$$

$$\therefore\ a^0=1$$

It should be noted that a represents any non-zero number and hence *any non-zero number raised to the power* 0 *is equal to* 1.

Examples

1) $(20)^0=1$

2) $(0.56)^0=1$

3) $\left(\frac{1}{4}\right)^0=1$

MISCELLANEOUS EXAMPLES ON INDICES

1) $\left(\frac{1}{2}\right)^{-3}=\frac{1^{-3}}{2^{-3}}=\frac{2^3}{1^3}=8$

2) $4^{3/2}=(2^2)^{3/2}=2^{2\times 3/2}=2^3=8$

3) $\sqrt{25x^{16}}=(5^2x^{16})^{1/2}=5^{2\times 1/2}x^{16\times 1/2}=5x^8$

4) $9{\cdot}5\times 10^2\times 1{\cdot}2\times 10^{-5}$

$=(9{\cdot}5\times 1{\cdot}2)\times 10^2\times 10^{-5}$

$=11{\cdot}40\times 10^{-3}$

$=\frac{11{\cdot}40}{10^3}=\frac{11{\cdot}40}{1000}$

$=0{\cdot}0114$

5) $16^{3/4}\times 125^{1/3}\times 27^{-1/3}$

$=(2^4)^{3/4}\times(5^3)^{1/3}\times(3^3)^{-1/3}$

$=2^3\times 5\times 3^{-1}$

$=\frac{8\times 5}{3}=\frac{40}{3}=13\frac{1}{3}$

Exercise 44

Simplify the following:

1) $a^5\times a^6$

2) $z^4\times z^7$

3) $y^3\times y^4\times y^5$

4) $2^3\times 2^5$

5) $3\times 3^2\times 3^5$

6) $\frac{1}{2}a\times\frac{1}{4}a^2\times\frac{3}{4}a^3$

7) $a^5\div a^2$

8) $m^{12}\div m^5$

9) $2^8\div 2^4$

10) $x^{20}\div x^5$

11) $a^5\times a^3\div a^4$

12) $q^7\times q^6\div q^5$

13) $\frac{m^5}{m^3}\times\frac{m}{m^2}$

14) $\frac{l^5\times l^6}{l^2\times l^7}$

15) $\frac{aL^4}{aL^2}$

16) $(x^3)^4$

17) $(a^5)^3$

18) $(3x^4)^2$

19) $(2^3)^2$

20) $(10^3)^2$

21) $(ab^2)^3$

22) $(ab^2c^3)^4$

23) $(2x^2y^3z)^5$

24) $\left(\frac{3m^2}{4n^3}\right)^5$

25) Find the values of 10^{-1}, 2^{-2}, 3^{-4}, 5^{-2}.

26) Find the values of $2^4\times 2$, $5^2\times 5^{1/2}\times 5^{3/2}$, $8^{1/3}$, $27^{1/3}$.

27) Express as powers of 3: 9^3, 27^5, 81^3, $9^4\times 27^3$

28) Express as powers of a: $\sqrt[5]{a}$, $\sqrt[3]{a^2}$, $\sqrt[7]{a^4}$, $\sqrt{a^6}$

29) Find the values of: $2^4\times 2^2$, $(10^2)^3$, $(2^5)^{2/5}$, $64^{1/2}$

30) Evaluate: $32^{2/5}$, $16^{3/4}$, $(81)^{-1/4}$, $9^{-1/2}$

31) Find the value of $32^{3/5}\times 25^{1/2}\times 64^{-1/3}$

32) Find the value of $4^{-1/2}+\left(\frac{1}{27}\right)^{1/3}$

33) Find the values of $\left(\frac{1}{5}\right)^0$, $(125)^{-1/3}$, $(1\,000\,000)^{5/6}$

34) Simplify: $(9x^4)^{-1/2}$, $(27x^6)^{-1/3}$, $(25a^8)^{-1/2}$

35) Simplify: $(64a^6)^{1/2}\div(64a^6)^{-1/3}$

36) Simplify: $\frac{(x^2y^3)^2+(x^3y^2\times x^2y^3)}{xy+y^2}$

37) Evaluate: $\sqrt{\frac{p}{q}}$ when $p=64^{2/3}$ and $q=3^{-2}$

NUMBERS IN STANDARD FORM

Any number can be expressed as a value between 1 and 10 multiplied by a power of 10. A number expressed in this way is said to be in standard form. The repeating of zeros in large and small numbers often leads to errors. Stating the numbers in standard form often avoids these errors.

Examples

1) $49{\cdot}4=4{\cdot}94\times 10$

2) $385 \cdot 3 = 3 \cdot 853 \times 100 = 3 \cdot 853 \times 10^2$

3) $20\,000\,000 = 2 \times 10\,000\,000 = 2 \times 10^7$

4) $0 \cdot 596 = \dfrac{5 \cdot 96}{10} = 5 \cdot 96 \times 10^{-1}$

5) $0 \cdot 000\,478 = \dfrac{4 \cdot 78}{10\,000} = \dfrac{4 \cdot 78}{10^4} = 4 \cdot 78 \times 10^{-4}$

The convenience of using numbers in standard form is shown in the following examples.

6) Evaluate $\dfrac{27\,000\,000 \times 0 \cdot 000\,3}{18\,000}$

Putting each number in standard form:

$27\,000\,000 = 2 \cdot 7 \times 10^7$

$0 \cdot 000\,3 = 3 \times 10^{-4}$

$18\,000 = 1 \cdot 8 \times 10^4$

$$\frac{27\,000\,000 \times 0 \cdot 000\,3}{18000} = \frac{2 \cdot 7 \times 10^7 \times 3 \times 10^{-4}}{1 \cdot 8 \times 10^4}$$

$$= \frac{8 \cdot 1 \times 10^3}{1 \cdot 8 \times 10^4}$$

$$= \frac{8 \cdot 1}{1 \cdot 8} \times 10^{-1}$$

$$= 4 \cdot 5 \times 10^{-1}$$

7) Evaluate $\dfrac{\sqrt{P}}{Q}$ when $P = 1 \cdot 44 \times 10^{-6}$ and $Q = 4 \cdot 8 \times 10^{-6}$ putting the answer in standard form.
Substituting the given values,

$$\frac{\sqrt{P}}{Q} = \frac{\sqrt{1 \cdot 44 \times 10^{-6}}}{4 \cdot 8 \times 10^{-6}} = \frac{(1 \cdot 44 \times 10^{-6})^{1/2}}{4 \cdot 8 \times 10^{-6}}$$

$$= \frac{1 \cdot 2 \times 10^{-3}}{4 \cdot 8 \times 10^{-6}} = \frac{1 \cdot 2}{4 \cdot 8} \times 10^{-3} \times 10^6$$

$$= 0 \cdot 25 \times 10^3 = 2 \cdot 5 \times 10^2$$

Exercise 45

Express the following numbers in standard form:

1) 19·6 2) 385

3) 56 4) 59 876

5) 189·7 6) 15 000 000

7) 0·013 8) 0·000 698

9) 0·003 85 10) 0·697

Find the values of the following expressing the answer in standard form to the number of significant figures stated.

11) $\dfrac{800 \times 60 \times 12}{4000}$ to 3 sig. figures

12) $\dfrac{704 \times 80 \times 1728}{500 \times 6}$ to 2 sig. figures

13) $0 \cdot 003\,6 \times 0 \cdot 143 \times 51$ to 3 sig. figures

14) $\dfrac{18 \cdot 15}{0 \cdot 000\,73 \times 30 \times 0 \cdot 17}$ to 2 sig. figures

15) $\dfrac{0 \cdot 001\,8 \times 86\,000 \times 0 \cdot 025}{19 \cdot 0 \times 0 \cdot 000\,78}$ to 2 sig. figures

16) $(0 \cdot 3)^2 \times (0 \cdot 2)^3$

17) $\dfrac{0 \cdot 027 \times 130}{0 \cdot 17}$ to the nearest whole number

18) In the formula $y = x(1 + pt)$ the following values are given: $y = 20 \cdot 06$, $x = 20$ and $t = 200$. Calculate the exact value of p giving the answer in standard form.

19) In the formula $q = \dfrac{t}{\sqrt{c}}$ calculate the value of q when $t = 15 \times 10^{-5}$ and $c = 125 \times 10^{-8}$ expressing the answer in standard form.

20) In the formula $d = \sqrt{\dfrac{D^2 S}{V}}$ find d when $D = 4$, $S = 300$ and $V = 20\,000$ expressing the answer in standard form.

LOGARITHMS

Any positive number can be expressed as a power of 10. For example $100 = 10^2$ and

$86 = 10^{1 \cdot 9345}$. These powers of 10 are called **logarithms to the base 10**. That is,

$$\text{number} = 10^{\text{power}} = 10^{\text{logarithm}}$$

We have seen above that $86 = 10^{1 \cdot 9345}$ and we write

$$\log_{10} 86 = 1{\cdot}9345$$

The base 10 is indicated as shown but it is frequently omitted and we write

$$\log\ 86 = 1{\cdot}934\,5.$$

The log tables give the logarithms of numbers between 1 and 10. Thus,

$$\log 5 = 0{\cdot}6990 \quad \text{and} \quad \log 8{\cdot}293 = 0{\cdot}9188$$

To find the logarithms of numbers outside this range we make use of numbers in standard form. Then by using the multiplication law of indices and the log tables we find the complete logarithm. For example:

To find log 249·3

$$249{\cdot}3 = 2{\cdot}493 \times 100 = 2{\cdot}493 \times 10^2$$

From the log. tables

$$\log 2{\cdot}493 = 0{\cdot}396\,7$$
$$249{\cdot}3 = 10^{0 \cdot 3967} \times 10^2 = 10^{2 \cdot 3967}$$
$$\log 249{\cdot}3 = 2{\cdot}396\,7$$

A logarithm therefore consists of two parts:
i) a whole number part called the **characteristic**,
ii) a decimal part called the **mantissa**.
As can be seen from the above example the **characteristic** depends on the size of the number. It is found by **subtracting** 1 from the number of figures which occur to the left of the decimal point in the given number. The mantissa is found directly from the log. tables.

Examples

1) In log 8293 the characteristic is 3

$\therefore \log 8293 = 3{\cdot}918\,8$

2) In log 829·3 the characteristic is 2

$\log 829{\cdot}3 = 2{\cdot}918\,8$

3) In log 82·93 the characteristic is 1

$\log 82{\cdot}93 = 1{\cdot}918\,8$

4) In log 8·293 the characteristic is 0

$\log 8{\cdot}293 = 0{\cdot}918\,8$

Numbers which have the same set of significant figures have the same mantissa in their logarithms.

ANTI-LOGARITHMS

The table of anti-logarithms contains the *numbers which correspond to the given logarithms*. The tables are used in a similar way to the log. tables but it must be remembered that:

i) the mantissa (or decimal part) of the logarithm only is used in the table,
ii) the number of figures to the left of the decimal point is found by **adding 1** to the characteristic of the logarithm.

Example
To find the number whose log is 2·182 5. Using the mantissa ·182 5 we find from the anti-log tables that the number corresponding is 1 523. Since the characteristic is 2, the number must have three figures to the left of the decimal point. The number is therefore 152·3. (note that log 152·3 = 2·182 5).

RULES FOR THE USE OF LOGARITHMS

It has been shown that logarithms are indices. When using logarithms, the laws of indices must be observed.

Multiplication

To find $39{\cdot}27 \times 6{\cdot}127$

$\log 39{\cdot}27 = 1{\cdot}5921$; $\log 6{\cdot}127 = 0{\cdot}7873$

$$\therefore\ 39{\cdot}27 \times 6{\cdot}127 = 10^{1\cdot5921} \times 10^{0\cdot7873}$$
$$= 10^{1\cdot5921+0\cdot7873}$$
$$= 10^{2\cdot3814}$$

By finding the anti-log of 2·3814,

$39{\cdot}27 \times 6{\cdot}127 = 240{\cdot}6$

From this example we see that the rule for multiplication is:

Find the logarithms of the numbers to be multiplied and **add** *them together. The required answer is found by taking the anti-log of the sum.*

The method shown above is not very convenient and a better way is shown below.

	number	log
	39·27	1·5941
	6·127	0·7873
Answer =	240·6	2·3814

Division

To find $\dfrac{293{\cdot}6}{18{\cdot}78}$

$\log 293{\cdot}6 = 2{\cdot}4678$; $\log 18{\cdot}78 = 1{\cdot}2737$

$$\therefore\ \frac{293{\cdot}6}{18{\cdot}78} = \frac{10^{2\cdot4678}}{10^{1\cdot2737}}$$

$$= 10^{2\cdot4678-1\cdot2737} = 10^{1\cdot1941}$$

By finding the anti-log of 1·1941,

$$\frac{293{\cdot}6}{18{\cdot}78} = 15{\cdot}63$$

Therefore the rule for division is:

Find the logarithm of each number. **Subtract** *the logarithm of the denominator from the logarithm of the numerator. The answer is found by taking the anti-log of the difference.*

A better way of performing the process is shown below.

	number	log
	293·6	2·4678
	18·78	1·2737
Answer =	15·63	1·1941

Example

Find the value of $\dfrac{783{\cdot}9 \times 2{\cdot}023}{2{\cdot}168 \times 39{\cdot}47}$

	number	log	number	log
	783·9	2·8943	2·168	0·3361
	2·023	0·3060	39·47	1·5963
	numerator	3·2003	denominator	1·9324
	denominator	1·9324		
Ans. =	18·53	1·2679		

Powers

To find $(3{\cdot}968)^3$

$\log 3{\cdot}968 = 0{\cdot}5986 \quad \therefore\ 3{\cdot}968 = 10^{0\cdot5986}$

$(3{\cdot}968)^3 = (10^{0\cdot5986})^3 = 10^{0\cdot5986\times3} = 10^{1\cdot7958}$

By finding the anti-log of 1·7958,

$(3{\cdot}968)^3 = 62{\cdot}48$

The rule for finding the powers of numbers is therefore:

Find the logarithm of the number and **multiply** *it by the index denoting the power. The value of the number raised to the given power is found by taking the anti-log. of the product.*

Example
Find the value of $(11{\cdot}63 \times 2{\cdot}87)^4$

number	log
11·63	1.0656
2·87	0·4579
11·63 × 2·87	1·5235
	×4
Answer = 1 242 000	6·0940

Roots

To find $\sqrt[4]{70{\cdot}35}$

$\log 70{\cdot}35 = 1{\cdot}8473 \quad \therefore \; 70{\cdot}35 = 10^{1{\cdot}8473}$

$$\sqrt[4]{70{\cdot}35} = (70{\cdot}35)^{1/4}$$
$$= (10^{1{\cdot}8473})^{1/4}$$
$$= 10^{1{\cdot}8473/4} = 10^{0{\cdot}4618}$$

By finding the anti-log of 0·4618,

$$\sqrt[4]{70{\cdot}35} = 2{\cdot}896$$

The rule for finding the root of a number is:

Find the logarithm of the number and **divide** *it by the number denoting the root. The result obtained by this division is the log of the required root and the anti-log is the required root.*

Example
Find the value of $\sqrt[3]{(1{\cdot}832)^2 \times 6{\cdot}327}$

number	log
1·832	0·2630
	×2
$(1{\cdot}832)^2$	0·5260
6·327	0·8012
$(1{\cdot}832)^2 \times 6{\cdot}327$	1·3272
	÷3
Answer = 2·770	0·4424

Exercise 46

1) Write down the characteristics of the following:

23, 2300, 2·3, 17 970, 983, 950 000, 55·27, 1·794, 333·4, 2893, 1988, 17·25, 390·1, 78·16

2) Find the logarithms of the following:
a) 7, 70, 700, 7000, 70 000
b) 3·1, 31, 310, 3100, 3 100 000
c) 48·3, 483 000, 4·83, 483
d) 7895, 7·895, 78·95, 78 950
e) 1·003, 10·03, 1003, 100·3

3) Write down the anti-log of the following:
a) 0·32, 2·32, 4·32, 1·32
b) 3·275, 0·275, 4·275, 6·275
c) 0·5987, 1·5987, 4·5987, 2·5987
d) 7·8949, 0·8949, 2·8949, 4·8949

Use logs to find the values of the following:

4) $17{\cdot}63 \times 20{\cdot}54$

5) $328{\cdot}4 \times 54{\cdot}7$

6) $6819 \times 1{\cdot}285 \times 17$

7) $305{\cdot}2 \times 1{\cdot}003 \times 12{\cdot}36$

8) $25{\cdot}14 \div 12{\cdot}95$

9) $8{\cdot}165 \div 3{\cdot}142$

10) $128{\cdot}3 \div 12{\cdot}95$

11) $1{\cdot}975 \div 1{\cdot}261$

12) $\dfrac{95{\cdot}83 \times 6{\cdot}138}{8{\cdot}179}$

13) $\dfrac{9{\cdot}125 \times 123}{120{\cdot}2}$

14) $\dfrac{7658}{26{\cdot}28 \times 43{\cdot}93}$

15) $\dfrac{42{\cdot}7 \times 16{\cdot}15 \times 3{\cdot}298}{11{\cdot}69 \times 7{\cdot}58}$

16) $(7{\cdot}326)^3$

17) $(29{\cdot}38)^2$

18) $(1{\cdot}098)^5$

19) $(2{\cdot}998)^2 \times 11{\cdot}35$

20) $(16{\cdot}29)^3 \div 86{\cdot}76$

21) $73{\cdot}25 \div (3{\cdot}924)^3$

22) $\dfrac{(7{\cdot}36)^2 \times (1{\cdot}088)^3}{42{\cdot}35}$

23) $\dfrac{45\,827}{(56{\cdot}3)^2 \times (1{\cdot}82)^3}$

24) $\sqrt[3]{15{\cdot}38}$

25) $\sqrt[4]{1{\cdot}295}$

26) $\sqrt[3]{(2{\cdot}593)^2}$

27) $\sqrt[5]{1{\cdot}637 \times 11{\cdot}87}$

28) $\sqrt{61{\cdot}5 \times (19{\cdot}27)^3}$

LOGARITHMS OF NUMBERS BETWEEN 0 AND 1

$$0{\cdot}1 = \frac{1}{10} = 10^{-1}$$

$$\therefore\ \log 0{\cdot}1 = -1$$

$$0{\cdot}01 = \frac{1}{100} = \frac{1}{10^2} = 10^{-2}$$

$$\log 0{\cdot}01 = -2$$

From these results we may deduce that:

The logarithms of numbers between 0 *and* 1 *are negative.*

To find the logarithm of any number between 0 and 1, we first express the number in standard form. Then by using the multiplication law of indices the logarithm may be found.

Example

To find the logarithm of 0·378 3.

$$0{\cdot}378\,3 = \frac{3{\cdot}783}{10} = 3{\cdot}783 \times 10^{-1}$$

$$= 10^{0{\cdot}5778} \times 10^{-1}$$

$$= 10^{-1+0{\cdot}5778}$$

$$\therefore\ \log 0{\cdot}378\,3 = -1 + 0{\cdot}577\,8$$

The characteristic is therefore -1 and the mantissa is 0·577 8. In the case of numbers greater than 1, the mantissa remains the same when the numbers are multiplied or divided by powers of 10. That is, with the same set of significant figures we have the same mantissa. It would be advantageous if we could do the same thing for the logarithms of numbers less than 1. This can be done if we retain the negative characteristic as shown above, but to write log 0·378 3 as $-1+0{\cdot}577\,8$ would be awkward so we adopt the notation $\bar{1}{\cdot}5778$. The minus sign is written above the characteristic. It must be clearly understood that:

$$\bar{1}{\cdot}577\,8 = -1 + 0{\cdot}577\,8$$
$$\bar{2}{\cdot}609\,3 = -2 + 0{\cdot}609\,3 \text{ and so on.}$$

A logarithms written in this way has a *negative characteristic and a positive mantissa.* Using this notation:

$$\log 0{\cdot}426\,3 = \bar{1}{\cdot}629\,7$$
$$\log 0{\cdot}042\,63 = \bar{2}{\cdot}629\,7$$
$$\log 0{\cdot}004\,263 = \bar{3}{\cdot}629\,7$$

The negative characteristic is numerically one more than the number of zeros which follow the decimal point in the given number.

Examples

1) Find the value of $0{\cdot}01 \times 0{\cdot}0001$

	number	log
	0·01	$\bar{2}{\cdot}0000$
	0·0001	$\bar{4}{\cdot}0000$
Answer =	0·000 001	$\bar{6}{\cdot}0000$

Note:- $(-2)+(-4) = -6$ and $\bar{2}+\bar{4}=\bar{6}$

2) Add together the following logarithms:

$\bar{1}{\cdot}7318$
$\bar{1}{\cdot}8042$
$\bar{2}{\cdot}7658$
$0{\cdot}5823$

Answer = $\bar{2}{\cdot}8841$

The addition of the decimal parts gives 2·8841. Thus the addition of the characteristic becomes:

$$(-1)+(-1)+(-2)+0+2=-2$$

3) Substract the following logarithms:

$$\begin{array}{r} \bar{3}\cdot5903 \\ \bar{2}\cdot4061 \\ \hline \end{array}$$

Answer $=\bar{1}\cdot1842$

The characteristic of the answer becomes:

$$-3-(-2)=-3+2=-1$$

4) Subtract the following logarithms

$$\begin{array}{r} \bar{3}\cdot2584 \\ 1\cdot5789 \\ \hline \end{array}$$

Answer $=\bar{5}\cdot6795$

We cannot take 0·5789 from 0·2584 so we borrow 1 from -3 making it -4. We now take 0·5789 from 1·2584. The characteristic becomes $(-4)-1=-5$

5) Subtract the following logarithms:

$$\begin{array}{r} \bar{4}\cdot1179 \\ \bar{6}\cdot6218 \\ \hline \end{array}$$

Answer $=1\cdot4961$

We cannot take 0·6218 from 0·1179 so we borrow 1 from -4 thereby making it -5. We now take 0·6218 from 1·1179. The characteristic becomes $-5-(-6)=-5+6=1$

6) Multiply $\bar{2}\cdot6192$ by 4.

$$\begin{array}{r} \bar{2}\cdot6192 \\ \times 4 \\ \hline \bar{6}\cdot4768 \end{array}$$

$0\cdot6192\times4=2\cdot4768$. The characteristic then becomes $(-2)\times4+2=-8+2=-6$

7) Divide $\bar{5}\cdot8293$ by 3
We must make the negative characteristic exactly divisible by 3, so we write:

$$\begin{aligned} \bar{5}\cdot8293\div3 &= (\bar{6}+1\cdot8293)\div3 \\ &= \bar{2}+0\cdot6098=\bar{2}\cdot6098 \end{aligned}$$

The work is best set out as follows:

$$\begin{array}{l} 3\overline{)\bar{6}+1\cdot8293} \\ \quad\bar{2}+0\cdot6098=\bar{2}\cdot6098 \end{array}$$

8) Find the value of $\sqrt[5]{0\cdot0139}$
It will be remembered that to find the root of a number we find the log. of the number and divide it by the number denoting the root.

$$\log 0\cdot0139=\bar{2}\cdot1430$$

$$\begin{array}{l} 5\overline{)\bar{5}+3\cdot1430} \\ \quad\bar{1}+0\cdot6286=\bar{1}\cdot6286 \end{array}$$

By finding the anti-log of $\bar{1}\cdot6286$,

$$\sqrt[5]{0\cdot0139}=0\cdot4252$$

Exercise 47

1) Write down the logs of the following numbers:
   a) 2.817, 0·2817, 0·02817, 0·002 817
   b) 4·597, 0·4597, 0·004 597, 0·000 045 97
   c) 0·09768, 0·000 976 8, 0·9768
   d) 0·000 058 75, 0·05875, 0·000 587 5

2) Find the numbers whose logs are:
   a) $\bar{1}\cdot4337$
   b) $\bar{3}\cdot8199$
   c) $\bar{4}\cdot5486$
   d) $\bar{2}\cdot4871$
   e) $\bar{8}\cdot5319$
   f) $\bar{1}\cdot0218$

3) Add the following:
   $1+\bar{1}$, $3+\bar{2}$, $\bar{1}+\bar{3}$, $\bar{2}+\bar{2}$, $\bar{3}+2$, $0+\bar{2}$, $\bar{3}+0$, $\bar{5}+\bar{4}$, $\bar{6}+3$, $2+\bar{4}$

4) Add the following:
   $\bar{2}\cdot7+1\cdot4$, $\bar{1}\cdot2+3\cdot1$, $0\cdot6+\bar{2}\cdot3$, $2\cdot7+\bar{3}\cdot4$, $2\cdot1+\bar{1}\cdot0$, $\bar{1}\cdot3+\bar{1}\cdot4$, $\bar{2}\cdot0+\bar{2}\cdot1$, $\bar{2}\cdot4+\bar{1}\cdot6$, $1\cdot2+\bar{1}\cdot9$, $\bar{3}\cdot7+1\cdot5$, $\bar{2}\cdot8+\bar{3}\cdot7$, $\bar{1}\cdot9+4\cdot5$

5) Add the following:
   $1\cdot5176+1\cdot8973+\bar{5}\cdot4398+0\cdot0625$
   $\bar{3}\cdot3785+2\cdot7827+1\cdot6879+\bar{2}\cdot8898$
   $3\cdot1189+\bar{2}\cdot7615+\bar{5}\cdot2319+\bar{6}\cdot0527$

6) Subtract the following:
   $2-3$, $2-5$, $0-3$, $\bar{2}-1$, $1-\bar{2}$, $\bar{3}-2$, $\bar{2}-\bar{2}$, $\bar{1}-\bar{4}$, $\bar{4}-\bar{1}$, $0-\bar{3}$

7) Subtract the following:
$3{\cdot}8-\bar{2}{\cdot}7$, $\bar{2}{\cdot}6-1{\cdot}4$, $\bar{1}{\cdot}7-\bar{1}{\cdot}3$, $\bar{1}{\cdot}8-\bar{3}{\cdot}5$,
$1{\cdot}7-3{\cdot}2$, $2{\cdot}8-\bar{2}{\cdot}6$, $\bar{3}{\cdot}5-\bar{1}{\cdot}4$, $2{\cdot}5-3{\cdot}6$,
$1{\cdot}3-1{\cdot}8$, $\bar{2}{\cdot}3-1{\cdot}8$, $\bar{1}{\cdot}5-\bar{1}{\cdot}7$, $\bar{1}{\cdot}3-\bar{3}{\cdot}5$

8) Subtract the following:
$3{\cdot}2973-\bar{4}{\cdot}3879$, $0{\cdot}4973-0{\cdot}8769$,
$\bar{2}{\cdot}5321-1{\cdot}9897$, $\bar{3}{\cdot}0036-\bar{6}{\cdot}8798$

9) Simplify:
a) $\bar{1}{\cdot}4\times 2$
b) $\bar{3}{\cdot}1\times 3$
c) $\bar{1}{\cdot}7\times 2$
d) $\bar{2}{\cdot}8\times 3$
e) $\bar{1}{\cdot}8\times 5$

10) Simplify:
a) $\bar{2}{\cdot}6\div 2$
b) $\bar{3}{\cdot}9\div 3$
c) $\bar{1}{\cdot}2\div 2$
d) $\bar{3}{\cdot}5\div 5$
e) $\bar{4}{\cdot}1\div 3$

Use log tables to evaluate the following putting the answers in standard form:

11) $\dfrac{0{\cdot}3786\times 0{\cdot}03972}{31{\cdot}67}$

12) $\dfrac{97{\cdot}61\times 0{\cdot}00046}{0{\cdot}09174}$

13) $\dfrac{0{\cdot}86\times 298{\cdot}7\times 0{\cdot}683\times 15}{39{\cdot}4}$

14) $\dfrac{0{\cdot}0146\times 0{\cdot}798\times 643}{33\,000\times 11{\cdot}8}$

15) $\dfrac{9{\cdot}87\times 30\times 10^6\times 0{\cdot}048}{70^2}$

16) $(0{\cdot}03614)^2$

17) $(0{\cdot}785)^3$

18) $(0{\cdot}00153)^4$

19) $\sqrt{0{\cdot}2569}$

20) $\sqrt[3]{0{\cdot}06987}$

21) $\sqrt[3]{0{\cdot}000\,781\,6}$

22) $\sqrt[5]{0{\cdot}6978}$

USE OF LOGARITHMS IN EVALUATING FORMULAE

Examples

1) A positive number is given by the formula

$$y=\frac{3t}{\sqrt{c}}$$

i) Use tables to calculate y when $t=7{\cdot}32$ and $c=205$.

ii) If t may take any value from 5 to 8 (inclusive) and c may take any value from 100 to 225 (inclusive), calculate the greatest possible value of y.

i) Substituting the given values $t=7{\cdot}32$ and $c=205$,

$$y=\frac{3\times 7{\cdot}32}{\sqrt{205}}=\frac{21{\cdot}96}{14{\cdot}32}$$

by using the square root tables.
Using the log. tables

	number	log
	21·96	1·3416
	14·32	1·1559
Answer =	1·533	0·1857

Hence $y=1{\cdot}5333$.

ii) The greatest possible value of y will occur when t has its greatest possible value and c has its least possible value. Hence, when $t=8$ and $c=100$,

$$y=\frac{3\times 8}{\sqrt{100}}=\frac{24}{10}=2{\cdot}4$$

2) Calculate the value of y from the formula

$$y^3=\frac{ab}{a-b}$$

when $a=0{\cdot}649$ and $b=0{\cdot}022$.

Since $y^3=\dfrac{ab}{a-b}$

$$y=\sqrt[3]{\frac{ab}{a-b}}$$

Substituting the given values $a=0{\cdot}649$ and $b=0{\cdot}022$,

$$y=\sqrt[3]{\frac{0{\cdot}649\times 0{\cdot}022}{0{\cdot}649-0{\cdot}022}}=\sqrt[3]{\frac{0{\cdot}649\times 0{\cdot}022}{0{\cdot}627}}$$

Using the log. tables:

number	log
0·649	$\bar{1}$·8122
0·022	$\bar{2}$·3424
0·649 × 0·022	$\bar{2}$·1546
0·627	$\bar{1}$·7973
$\frac{0·649 \times 0·022}{0·627}$	$\bar{2}$·3573 ÷ 3
Answer = 0·2834	$\bar{1}$·4524

Hence $y = 0·2834$.
(Note that we can only use logs. when numbers are to be multiplied or divided. They must never be used when numbers are to be added or subtracted)

Exercise 48

1) Evaluate the formula $1·73 \times \sqrt[3]{d^2}$ when $d = 2·8$.

2) Find D from the formula $D = 1·2 \times \sqrt{dL}$ when $d = 12$ and $L = 0·756$.

3) Find the value of the expression $0·25 \times (d - 0·5)^2 \times \sqrt{S}$ when $d = 4·33$ and $S = 5·12$.

4) Find the value of the expression

$$\frac{p_1 v_1 - p_2 v_2}{c - 1}$$

when $v_1 = 28·6$, $v_2 = 32·2$, $p_1 = 18·5$, $p_2 = 13·5$ and $c = 1·42$.

5) Find P from the formula

$$P = \sqrt{\frac{x^2 - y^2}{2xy}}$$

when $x = 5·531$ and $y = 3·469$.

6) If $A = PV^n$ find A when $P = 0·9314$, $V = 0·6815$ and $n = \frac{1}{2}$.

7) Calculate M from the formula

$$M^3 = \frac{3x^2 w}{p}$$

when $x = 0·3512$, $w = 1·664$ and $p = 2·308$.

8) Given that $x = \frac{dh}{D - d}$ find x when $d = 0·638$, $h = 0·516$ and $D = 0·721$.

9) Find the value of R from the equation

$$R = k\sqrt{\frac{P}{H^2}}$$

when $k = 65·2$, $P = 81·3$ and $H = 22·7$.

10) Transpose the formula

$$y = \frac{Wl^3}{48EI}$$

to make I the subject and find its value when $y = 0·346$, $W = 10\,300$, $l = 122$ and $E = 30 \times 10^6$.

SELF TEST 15

1) If $2^x = 32$ then x is equal to

a 5 b $\sqrt{32}$ c 16 d $\frac{1}{5}$

2) The product of $2·5 \times 10^4$ and 8×10^{-5} is

a 1·8 b 2·0 c 5·5 d 105

3) $(x^{1/2})^3 \times \sqrt{x^9}$ is equal to

a $x^{9/2}$ b x^5 c $x^{11/2}$ d x^9

4) The cube root of 0·036 03 is

a 0·3303 b 0·03303 c 0·6002 d 6·002

5) $(8x^3)^{-1/3}$ is equal to

a $8x$ b $\frac{8}{x}$ c $2x$ d $\frac{1}{2x}$

6) $(3x^3)^2$ is equal to

a $3x^5$ b $9x^5$ c $3x^6$ d $9x^6$

7) $\sqrt{9p^4} \div \frac{1}{2}p^2$ is equal to

a 6 b 3 c $6p^4$ d $3p^4$

8) Write down 0·09763 in standard form.

a $97{\cdot}63 \times 10^3$ b $9{\cdot}763 \times 10^2$
c $9{\cdot}763 \times 10^{-2}$ d 97.63×10^{-3}

9) If $a = 1{\cdot}2 \times 10^7$ and $b = 3{\cdot}2 \times 10^6$ then $\sqrt{a^2 + b^2}$ is equal to

a $1{\cdot}242 \times 10^6$ b $1{\cdot}242 \times 10^7$
c $2{\cdot}098 \times 10^7$ d $2{\cdot}098 \times 10^6$

10) If $\sqrt{3} = 1{\cdot}732$ correct to three decimal places then $\sqrt{27}$, correct to two decimal places, is equal to

a 2·99 b 5·19 c 5·20 d 8·98

11) The cube root of 27×10^{-6} is

a 3×10^{-3} b $5{\cdot}20 \times 10^{-3}$
c 3×10^{-2} d $5{\cdot}20 \times 10^{-2}$

12) $(5 \times 7^3) \times (3 \times 7^5)$ is equal to

a 8×7^8 b 15×7^8 c 8×7^{15}
d 15×7^{15}

13) $\sqrt{16a^4b^{16}}$ is equal to

a $8a^2b^4$ b $4a^2b^4$ c $4a^2b^8$
d $8a^2b^8$

14) If $\log y = -2$ then y is equal to

a -100 b $-0{\cdot}01$ c 0·1
d 0·01

15) The largest of the numbers $\frac{1}{7}$, $1{\cdot}3 \times 10^{-1}$, 0·12 and $1{\cdot}4 \times 10^{-2}$ is

a $\frac{1}{7}$ b $1{\cdot}3 \times 10^{-1}$ c 0·12
d $1{\cdot}4 \times 10^{-2}$

16) The value of $32^{-3/5}$ is

a 19·2 b $\frac{1}{8}$ c 8 d $-19{\cdot}2$

17) $\dfrac{5 \times 10^{-6}}{0{\cdot}001}$ is equal to

a 5×10^{-9} b 5×10^{-4} c 5×10^{-3}
d 5×10^{-2}

18) The value of $\left(\frac{1}{9}\right)^{-1/2}$ is

a -9 b $-\frac{1}{18}$ c $-\frac{1}{3}$ d 3

19) $8 \times 10^{-3} \times 1{\cdot}25 \times 10^4$ is equal to

a 1 b 10 c 100 d 2

20) The value of $10^{1{\cdot}3243}$ is

a 13·243 b 0·13243 c 2·110
d 21·10

21) The value of $10^{0{\cdot}3010} \times 10^{0{\cdot}4771}$ is

a 6 b 60 c 0·7781 d 7·781

22) The value of $(10^{1{\cdot}4314})^{1/3}$ is

a 4·771 b 0·4771 c 3 d 30

23) The value of $10^{-1/7709}$ is

a $-0{\cdot}17709$ b $-17{\cdot}709$
c 0·1694 d 0·01694

24) $\bar{2}{\cdot}8 \div 6$ is equal to

a $\bar{2}{\cdot}6$ b $-2{\cdot}6$ c $\bar{1}{\cdot}8$ d $-1{\cdot}8$

25) $\bar{8}{\cdot}2 \div 3$ is equal to

a $\bar{2}{\cdot}4$ b $\bar{3}{\cdot}4$ c $-3{\cdot}4$ d $-2{\cdot}4$

Chapter 16 Quadratic Equations

An equation of the type $ax^2+bx+c=0$ is called a *quadratic equation*. The constants a, b and c can have any numerical values. Thus,

$x^2-36=0$
in which $a=1$, $b=0$ and $c=-36$
$5x^2+7x+8=0$
in which $a=5$, $b=7$ and $c=8$
$2{\cdot}5x^2-3{\cdot}1x-2=0$
in which $a=2{\cdot}5$, $b=-3{\cdot}1$ and $c=-2$

are all examples of quadratic equations. A quadratic equation may contain only the square of the unknown quantity, as in the first of the above equations, or it may contain both the square and the first power as in the remaining two equations.

EQUATIONS OF THE TYPE $ax^2 + c = 0$

When $b=0$, the standard quadratic equation $ax^2+bx+c=0$ becomes $ax^2+c=0$. The methods of solving such equations are shown in the examples which follow.

Examples

1) Solve the equation $x^2-16=0$

Since $x^2-16=0$

$x^2=16$

Taking the square root of both sides

$x=\pm\sqrt{16}$

and $x=\pm 4$

It is necessary to insert the double sign before the value obtained for x since $+4$ and -4 when squared both give 16. This means that there are two solutions which will satisfy the given equation. The solution $x=\pm 4$ means that either $x=+4$ or $x=-4$.

The solution to a quadratic equation always consists of a pair of numbers (i.e. there are always two solutions although it is possible for both solutions to have the same numerical value).

2) Solve the equation $2x^2-12=0$

Since $2x^2-12=0$

$2x^2=12$

$x^2=6$

$x=\pm\sqrt{6}$

and $x=\pm 2{\cdot}45$

(by using the square root tables to find $\sqrt{6}$).

3) Solve the equation $2x^2+18=0$

Since $2x^2+18=0$

$2x^2=-18$

$x^2=-9$

and $x=\pm\sqrt{-9}$

The square root of a negative quantity has no arithmetic meaning and it is called an imaginary number. The reason is as follows:

$(-3)^2=9 \qquad (+3)^2=9$

$\therefore\ \sqrt{9}=\pm 3$

Hence it is not possible to give a meaning to $\sqrt{-9}$. The equation $2x^2+18=0$ is said, therefore, to have imaginary roots.

SOLUTION BY FACTORS

If the product of two factors is zero, then one factor or the other factor must be zero or they may both be zero. Thus if $ab=0$, then either $a=0$ or $b=0$ or both $a=0$ and $b=0$.

We make use of this fact in solving quadratic equations.

Examples

1) Solve the equation $(2x+3)(x-5)=0$
Since the product of the two factors $(2x+3)$ and $(x-5)$ is zero then,

either $2x+3=0$ giving $x=-\frac{3}{2}$

or $x-5=0$ giving $x=5$

The solutions are $x=-\frac{3}{2}$ or $x=5$

2) Solve the equation $x^2-5x+6=0$

Factorising, $(x-3)(x-2)=0$

either $x-3=0$ giving $x=3$
or $x-2=0$ giving $x=2$

The solutions are $x=3$ or $x=2$

3) Solve the equation $6x^2+x-15=0$

Factorising $(2x-3)(3x+5)=0$

either $2x-3=0$ giving $x=\frac{3}{2}$

or $3x+5=0$ giving $x=-\frac{5}{3}$

The solutions are $x=\frac{3}{2}$ or $x=-\frac{5}{3}$

4) Solve the equation $x^2-5x=0$
Factorising $x(x-5)=0$

either $x=0$

or $x-5=0$ giving $x=5$

The solutions are $x=0$ and $x=5$ (note that it is incorrect to say that the solution is $x=5$. The solution $x=0$ must be stated also).

Exercise 49

Solve the following quadratic equations:

1) $x^2-25=0$
2) $x^2-8=0$
3) $x^2-16=0$
4) $3x^2-48=0$
5) $5x^2-80=0$
6) $7x^2-21=0$
7) $(x-5)(x-2)=0$
8) $(3x-4)(x+3)=0$
9) $x(x+7)=0$
10) $3x(2x-5)=0$
11) $m^2+4m-32=0$
12) $x^2+9x+20=0$
13) $m^2=6m-9$
14) $x^2+x-72=0$
15) $3x^2-7x+2=0$
16) $14q^2=29q-12$
17) $9x+28=9x^2$
18) $x^2-3x=0$
19) $y^2+8y=0$
20) $4a^2-4a-3=0$

SOLUTION BY FORMULA

The standard form of the quadratic equation is

$$ax^2+bx+c=0$$

It can be shown that the solution of this equation is

$$x=\frac{-b\pm\sqrt{b^2-4ac}}{2a}$$

Note that the whole of the numerator including $-b$ is divided by $2a$. The formula is used when factorisation is not possible.

Examples

1) Solve the equation $3x^2-8x+2=0$.
Comparing with $ax^2+bx+c=0$ we have $a=3$, $b=-8$ and $c=2$. Substituting these values in the formula,

$$x=\frac{-(-8)\pm\sqrt{(-8)^2-4\times3\times2}}{2\times3}$$

$$=\frac{8\pm\sqrt{64-24}}{6}$$

$$=\frac{8\pm\sqrt{40}}{6}$$

$$=\frac{8\pm6{\cdot}325}{6}$$

either $x=\dfrac{8+6\cdot325}{6}$ or $\dfrac{8-6\cdot325}{6}$

$=\dfrac{14\cdot325}{6}$ or $\dfrac{1\cdot675}{6}$

$=2\cdot39$ or $0\cdot28$

2) Solve the equation $-2x^2+3x+7=0$. Where the coefficient of x^2 is negative it is best to make it positive by multiplying both sides of the equation by (-1). This is equivalent to changing the sign of each of the terms. Thus,

$$2x^2-3x-7=0$$

This gives $a=2$, $b=-3$ and $c=-7$

$$x=\frac{-(-3)\pm\sqrt{(-3)^2-4\times2\times(-7)}}{2\times2}$$

$$=\frac{3\pm\sqrt{9+56}}{4}$$

$$=\frac{3\pm\sqrt{65}}{4}$$

$$=\frac{3\pm8\cdot063}{4}$$

either $x=\dfrac{11\cdot063}{4}$ or $x=\dfrac{-5\cdot063}{4}$

$=2\cdot766$ or $-1\cdot266$

3) Solve the equation $\dfrac{3}{2x-3}-\dfrac{2}{x+1}=5$.

The L.C.M. of the denominators is $(2x-3)(x+1)$. Multiplying each side of the equation by this gives,

$$3(x+1)-2(2x-3)=5(2x-3)(x+1)$$
$$3x+3-4x+6=5(2x^2-x-3)$$
$$-x+9=10x^2-5x-15$$
$$10x^2-4x-24=0$$

Here $a=10$, $b=-4$ and $c=-24$

$$x=\frac{-(-4)\pm\sqrt{(-4)^2-4\times10\times(-24)}}{2\times10}$$

$$=\frac{4\pm\sqrt{16+960}}{20}$$

$$=\frac{4\pm\sqrt{976}}{20}$$

$$=\frac{4\pm31\cdot24}{20}$$

either $x=\dfrac{4+31\cdot24}{20}$ or $\dfrac{4-31\cdot24}{20}$

$=\dfrac{35\cdot24}{20}$ or $\dfrac{-27\cdot24}{20}$

$=1\cdot762$ or $-1\cdot362$

Exercise 50

Solve the following equations:

1) $4x^2-3x-2=0$

2) $x^2-x-1=0$

3) $3x^2+7x-5=0$

4) $7x^2+8x-2=0$

5) $5x^2-4x-1=0$

6) $2x^2-7x=3$

7) $x(x+4)+2x(x+3)=5$

8) $5x(x+1)-2x(2x-1)=20$

9) $\dfrac{2}{x+2}+\dfrac{3}{x+1}=5$

10) $\dfrac{x+2}{3}-\dfrac{5}{x+2}=4$

11) $\dfrac{3x-5}{4}=\dfrac{x^2-2}{x}$

12) $x(x+5)=66$

13) $(2x-3)^2=13$

14) $\dfrac{12}{x+2}-\dfrac{1}{x}=2$

EQUATIONS GIVING RISE TO QUADRATIC EQUATIONS

Examples

1) Solve the simultaneous equations:

$3x-y=4$ ...(1)

$x^2-3xy+8=0$ ...(2)

From equation (1),

$y=3x-4$

Substituting for y in equation (2),

$$x^2-3x(3x-4)+8=0$$
$$x^2-9x^2+12x+8=0$$
$$-8x^2+12x+8=0$$
$$8x^2-12x-8=0$$

Dividing throughout by 4,

$$2x^2-3x-2=0$$
$$(2x+1)(x-2)=0$$

either $2x+1=0$ giving $x=-\frac{1}{2}$

or $x-2=0$ giving $x=2$

when $x=-\frac{1}{2}$, $y=3\times\left(-\frac{1}{2}\right)-4=-5\frac{1}{2}$

when $x=2$, $y=3\times 2-4=2$

Thus the solutions are,

$$x=-\frac{1}{2},\ y=-5\tfrac{1}{2} \quad \text{or} \quad x=2,\ y=2$$

2) Solve the simultaneous equations

$x-6y-5=0$ ...(1)

$xy-6=0$ ...(2)

From equation (1),

$x=6y+5$

Substituting for x in equation (2),

$$(6y+5)y-6=0$$
$$6y^2+5y-6=0$$
$$(3y-2)(2y+3)=0$$

either $3y-2=0$ giving $y=\frac{2}{3}$

or $2y+3=0$ giving $y=-\frac{3}{2}$

when $y=\frac{2}{3}$, $x=6\times\frac{2}{3}+5=9$

when $y=-\frac{3}{2}$, $x=6\times-\frac{3}{2}+5=-4$

Thus the solutions are

$$x=9,\ y=\frac{2}{3} \quad \text{or} \quad x=-4,\ y=-\frac{3}{2}$$

Exercise 51

Solve the following simultaneous equations:

1) $x+y=3$
$xy=2$

2) $x-y=3$
$xy+10x+y=150$

3) $x^2+y^2-6x+5y=21$
$x+y=9$

4) $x+y=12$
$2x^2+3y^2=7xy$

5) $2x^2-3y^2=20$
$2x+y=6$

6) $x^2+y^2=34$
$x+2y=13$

7) $3x+2y=13$
$xy=2$

8) $-3x+y+15=0$
$2x^2+4x+y=0$

PROBLEMS INVOLVING QUADRATIC EQUATIONS

1) The area of a rectangle is 6 square metres. If the length is 1 metre longer than the width find the dimensions of the rectangle.
Let x metres be the width of the rectangle.

Then the length of the rectangle is $(x+1)$ metres since the area is length × breadth, then

$$x(x+1)=6$$
$$x^2+x-6=0$$
$$(x+3)(x-2)=0$$

either $x+3=0$ giving $x=-3$

or $x-2=0$ giving $x=2$

The solution cannot be negative and hence $x=2$. Hence the width is 2 metres and the length is $(2+1)=3$ metres.

2) Two square rooms have a total floor area of 208 square metres. One room is 4 metres longer each way than the other. Find the floor dimensions of each room.

Let the smaller room have sides of x metres. The area of this room is then x^2 square metres.
The larger room will then have sides of $(x+4)$ metres and its floor area is $(x+4)(x+4)=(x+4)^2$ square metres.

Hence $$x^2+(x+4)^2=208$$
$$x^2+x^2+8x+16=208$$
$$2x^2+8x-192=0$$

Dividing through by 2,

$$x^2+4x-96=0$$
$$(x+12)(x-8)=0$$

either $x+12=0$ giving $x=-12$

or $x-8=0$ giving $x=8$

The negative value of x is not possible hence $x=8$.
The floor dimensions of the two rooms are 8 metres by 8 metres and 12 metres by 12 metres.

3) A rectangular room is 4 metres wider than it is high and it is 8 metres longer than it is wide. The total area of the walls is 512 square metres. Find the width of the room.
Let the height of the room be x metres.
Then the width of the room is $(x+4)$ metres and the length of the room is $x+4+8=(x+12)$ metres.

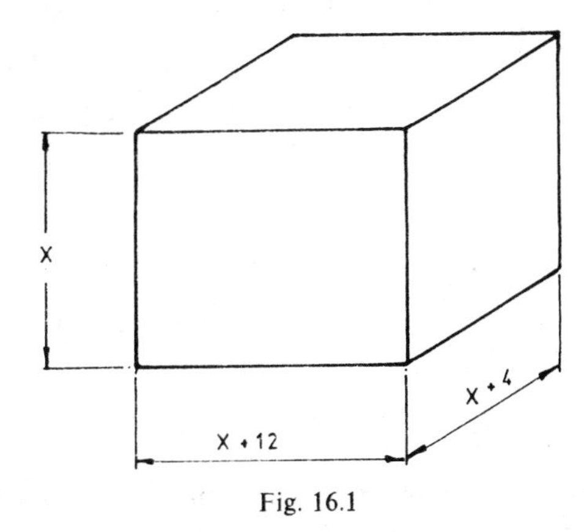

Fig. 16.1

These dimensions are shown in Fig. 16.1.
The total wall area is $2x(x+12)+2x(x+4)$.

Hence $2x(x+12)+2x(x+4)=512$

Dividing both sides of the equation by 2,

$$x(x+12)+x(x+4)=256$$
$$x^2+12x+x^2+4x=256$$
$$2x^2+16x=256$$

Dividing both sides of the equation by 2 again,

$$x^2+8x=128$$
$$x^2+8x-128=0$$
$$(x+16)(x-8)=0$$

either $x+16=0$ giving $x=-16$

or $x-8=0$ giving $x=8$

Thus the height of the room is 8 metres. Its width is $(x+4)=8+4=12$ metres.

4) The smallest of three consecutive positive numbers is m. Three times the square of the largest is greater than the sum of the squares of the other two numbers by 67. Find m.
The three numbers are m, $m+1$ and $m+2$.
Three times the square of the larger number is $3(m+2)^2$.
The sum of the squares of the other two numbers is $m^2+(m+1)^2$

$$\therefore \quad 3(m+2)^2-[m^2+(m+1)^2]=67$$
$$3(m^2+4m+4)-[m^2+m^2+2m+1]=67$$
$$3m^2+12m+12-[2m^2+2m+1]=67$$
$$3m^2+12m+12-2m^2-2m-1=67$$

$$m^2+10m+11=67$$
$$m^2+10m-56=0$$
$$(m-4)(m+14)=0$$

either $m-4=0$ giving $m=4$

or $m+14=0$ giving $m=-14$

Since m must be positive its value is 4.

Exercise 52

1) Find the number which when added to its square gives a total of 42.

2) A rectangle is 72 square metres in area and its perimeter is 34 metres. Find its length and breadth.

3) Two squares have a total area of 274 square centimetres and the sum of their sides is 88 centimetres. Find the side of the larger square.

4) The area of a rectangle is 4 square metres and its length is 3 metres longer than its width. Find the dimensions of the rectangle.

5) Part of a garden consists of a square lawn with a path 1·5 metres wide around its perimeter. If the lawn area is two-thirds of the total area find the length of a side of the lawn.

6) The largest of three consecutive positive numbers is n. The square of this number exceeds the sum of the other two numbers by 38. Find the three numbers.

7) The length of a rectangle exceeds its breadth by 4 centimetres. If the length were halved and the breadth increased by 5 cm the area would be decreased by 35 square centimetres. Find the length of the rectangle.

8) In a certain fraction the denominator is greater than the numerator by 3. If 2 is added to both the numerator and denominator, the fraction is increased by $\frac{6}{35}$. Find the fraction.

9) A piece of wire which is 18 metres long is cut into two parts. The first part is bent to form the four sides of a square. The second part is bent to form the four sides of a rectangle. The breadth of the rectangle is 1 metre and its length is x metres. If the sum of the areas of the square and rectangle is A square metres show that:

$$A=16-3x+\frac{x^2}{4}$$

If $A=9$, calculate the value of x.

10) One side of a rectangle is d cm long. The other side is 2 cm shorter. The side of a square is 2 cm shorter still. The sum of the areas of the square and rectangle is 148 square centimetres. Find an equation for d and solve it.

11) The exchange rate in 1964 was x dollars to £1. An American tourist remembered that in 1952 he needed $1\frac{1}{2}$ dollars more for each £1 he received in exchange. Using these facts only write down expressions for the number of pounds he received for a 100 dollar note a) in 1964 b) in 1952.
In 1952 he received £12 less for his 100 dollars than in 1964. Form an equation in x and show that it may be reduced to the form $2x^2+3x-25=0$. Solve this equation for x.

12) A garage owner bought a certain number of litres of petrol for £156. If petrol costs x pence per litre write down an expression for the number of litres of petrol he received.

When the price per litre was increased by a penny he found that he received 100 litres fewer for the same sum of money. Form an equation for x and show it reduces to $x^2+x-156=0$. Calculate the original price of the petrol per litre.

13) Two rectangular rooms each have an area of 240 square metres. If the length of one of the rooms is x metres and the other room is 4 metres longer, write down the width of each room in terms of x. If the widths of the rooms differ by 3 metres form an equation in x and show that this reduces to $x^2+4x-320=0$. Solve this equation and hence find the difference between the perimeters of the rooms.

SELF TEST 16

In the following questions state the letter (or letters) corresponding to the correct answer (or answers).

1) If $(2x-3)(3x+4)=0$ then x is equal to

a $-\frac{3}{2}$ or $\frac{4}{3}$ b $-\frac{2}{3}$ or $\frac{3}{4}$

c $\frac{2}{3}$ or $-\frac{3}{4}$ d $\frac{3}{2}$ or $-\frac{4}{3}$

2) If $(5x+2)(3x-2)=0$ then x is equal to

a $-\frac{2}{5}$ or $\frac{2}{3}$ b $-\frac{2}{5}$ or $-\frac{2}{3}$

c $\frac{5}{2}$ or $-\frac{3}{2}$ d $\frac{2}{5}$ or $-\frac{2}{3}$

3) If $(2x+5)(4x-7)=0$ then x is equal to

a $-\frac{2}{5}$ or $\frac{4}{7}$ b $-\frac{5}{2}$ or $\frac{7}{4}$

c $-\frac{5}{2}$ or $-\frac{7}{4}$ d $\frac{2}{5}$ or $-\frac{4}{7}$

4) If $x(2x-5)=0$ then x is equal to

a $\frac{5}{2}$ b $-\frac{5}{2}$ c 0 or $\frac{5}{2}$

d 0 or $-\frac{5}{2}$

5) If $x^2-25=0$ then x is equal to

a 0 b 5 c ± 5

6) If $3x^2-27=0$ then x is equal to

a 9 b 3 c ± 3

7) If $x^2-5x-2=0$ then x is equal to

a $\frac{-5\pm\sqrt{33}}{2}$ b $\frac{5\pm\sqrt{33}}{2}$

c $\frac{-5\pm\sqrt{17}}{2}$ d $\frac{5\pm\sqrt{17}}{2}$

8) If $3x^2+2x-3=0$ then x is equal to

a $\frac{2\pm\sqrt{40}}{6}$ b $\frac{-2\pm\sqrt{40}}{6}$

c $\frac{2\pm\sqrt{32}}{6}$ d $\frac{-2\pm\sqrt{36}}{6}$

9) If $x^2+9x+7=0$ then x is equal to

a $\frac{9\pm\sqrt{109}}{2}$ b $\frac{9\pm\sqrt{53}}{2}$

c $\frac{-9\pm\sqrt{109}}{2}$ d $\frac{-9\pm\sqrt{53}}{2}$

10) If $x^2-7x+3=0$ then x is equal to

a $\frac{7\pm\sqrt{37}}{2}$ b $\frac{7\pm\sqrt{61}}{2}$

c $\frac{-7\pm\sqrt{37}}{2}$ d $\frac{-7\pm\sqrt{61}}{2}$

11) If $\frac{3}{x-3}-\frac{2}{x-1}=5$ then

a $5x^2-21x+24=0$

b $5x^2-21x+12=0$

c $x-14=0$ d $x-2=0$

12) If $\frac{2}{x+1}-3=\frac{1}{x-2}$ then

a $3x^2+2x+11=0$

b $-3x^2-2x-11=0$

c $3x^2-4x-1=0$

d $-3x^2+4x+1=0$

Miscellaneous Exercise

Exercise 53

These questions are of the type found in O Level papers

1) Solve the equation $x-10+\frac{9}{x}=0$

2) Solve the equation $5x(x+1)=3$ giving your answers correct to two places of decimals.

3) If $\frac{6x^2-5y^2}{x^2+2y^2}=2$ find the two possible values of $\frac{x}{y}$.

4) Without using tables find the value of $(2)^{-1}\times 8^{2/3}$.

5) Use logarithms to find the fifth root of 0·003148.

6) Find, correct to two decimal places, the roots of the quadratic equation

$$3x^2-6x+2=0$$

7) If $\frac{(a^{1/2}b^{-3/4})^2\times(a^2b^2)^{-1/2}}{a^{1/3}b^{-1/2}}=a^mb^n$ find the values of m and n.

8) If $T=2\pi\sqrt{\frac{k^2+h^2}{gh}}$ find the value of T correct to three significant figures when $k=50{\cdot}17$, $h=42{\cdot}76$ and $g=981{\cdot}1$. (Take $\pi=3{\cdot}142$)

9) Solve the equation $y-9y^2=0$

10) Solve the equation $3x^2-4x=5$, giving the roots correct to two places of decimals.

11) Use logarithms to evaluate

a) $\frac{0{\cdot}0759\times 41{\cdot}19}{0{\cdot}00593}$ b) $\sqrt[3]{0{\cdot}08135}$

12) Express in its simplest form with positive indices

$$\sqrt{\frac{a^{3/2}b^{1/4}}{a^{-3/2}b^{-5/2}}}$$

13) Use logarithms to evaluate $\sqrt[3]{0{\cdot}03791}$

14) Solve the equation $x(x+5)=66$

15) A shopkeeper purchased a number of articles at x pence each and added 4 pence to the cost price of each article to form his selling price. He sold some of the articles for a total of $4a$ pence. Write down an expression for the number of articles sold. In order to dispose of the remainder he had to reduce his selling price per article to one penny below his original cost price of x pence. He then managed to sell all the remaining articles for a total of a pence. Prove that the number of articles bought originally was

$$\frac{5ax}{(x+4)(x-1)}$$

If $a=360$ and the number of articles originally bought was 96, calculate the value of x.

16) Find x given that $\log x^3=\bar{2}{\cdot}7256$

17) If $4x^2-12xy+5y^2=0$ find the possible values of $\frac{x}{y}$.

18) Evaluate $4^{-1/2}+\left(\frac{1}{27}\right)^{1/3}$

19) From the formula $v^2=2k\left(\frac{1}{x}-\frac{1}{a}\right)$ find the value of v when $k=1{\cdot}73$, $x=0{\cdot}275$ and $a=1{\cdot}43$.

20) Solve the equation $\frac{12}{x+2}-\frac{1}{x}=2$ giving each answer correct to two decimal places.

21) The area of metal used to make a fruit can is $\frac{44}{7}\cdot R(R+H)$ where R is the radius and H the height. If the height is 8 cm and the area of metal 660 square centimetres, find the radius.

22) Write down using tables the values of

a) $\frac{1}{0\cdot 3057}$ b) $(237\cdot 5)^2$

23) Find the value of $100^{4/5}$

24) If $A=5\cdot 7\times 10^5$ and $B=3\cdot 8\times 10^4$ find, without using tables, the values of $\frac{A}{B}$ and $\frac{A-B}{1000}$.

25) Find, correct to two decimal places, the values of x which satisfy $(7x-1)(x+3)=12$

26) Use tables to find the values of

a) $\sqrt{\frac{18\cdot 25}{197}}$ b) $(2\cdot 85)^2+\frac{1}{1\cdot 85}$

27) a) Use logarithms to calculate the value of $\sqrt[3]{0\cdot 6025)^2+0\cdot 3}$

b) Find, without using tables, the value of $64^{2/3}$.

28) Verify that $\frac{2}{3}$ is a solution of the equation

$$\frac{x+2}{2x}-\frac{7x-2}{x-2}=4$$

Find the other solution.

29) Find correct to three significant figures the value of:

a) $\frac{18\cdot 57\times 3\cdot 981}{12\cdot 38}$ b) $\sqrt{(12\cdot 65)^2-(7\cdot 35)^2}$

c) $\left(\frac{1}{31\cdot 25}\right)^2$

30) Find the values of

a) $(6\cdot 79)^2$ b) $\frac{1}{0\cdot 128}$ c) $\sqrt{0.327}$

31) Simplify $\frac{(10^2)^3\times 10^5}{10^7}$

32) Using the formula $T=\sqrt{\frac{x^2+y^2}{kx}}$ find the value of k correct to two significant figures if $T=0\cdot 42$, $x=1\cdot 5$ and $y=2\cdot 5$. Show that when $T=0\cdot 5$, $k=32$ and $y=2$, $x^2-8x+4=0$. Solve this equation giving the answers correct to two significant figures.

33) Using the formula $R=W(5\cdot 6+0\cdot 009V^2)$ find the value of V when $R=2000$ and $W=100$.

34) Solve the equation $\frac{2}{x+1}-\frac{1}{2x-1}=\frac{1}{x}$ giving the answers correct to two decimal places.

35) Simplify

a) $(x^2)^3$ b) $y^2\times y^{-1/2}$ c) $\sqrt{z^{16}}$

36) Without using tables find the value of

a) $9^{3/2}$ b) $8^{-2/3}$

37) Using tables find correct to three significant figures the value of

a) $\frac{20\cdot 42\times 3\cdot 675}{54\cdot 36}$ b) $\sqrt{\frac{4\cdot 796}{24\cdot 86}}$

38) Given $x=\frac{2t}{1+t^2}$ and $y=\frac{1-t^2}{1+t^2}$ find

a) the values of x and y when $t=2$
b) the values of t and x when $y=0\cdot 6$

39) State the time taken in minutes for a journey of d kilometres at V kilometres per hour. A man cycles at a speed of 10 kilometres per hour faster than his walking speed. He completes a journey, during which he cycled 20 kilometres and walked 2 kilometres in 1 hour 35 minutes. Find the speed at which he walked.

40) Evaluate a) $\sqrt{547{\cdot}5}$ b) $35^{0{\cdot}6}$

41) a) Factorise $3x^2-4x-32$ and solve the equation $3x^2-4x-32=0$

b) Solve correct to 2 decimal places $5x^2+2x-2=0$

42) Using tables find, correct to three significant figures the value of

a) $\dfrac{45{\cdot}67\times0{\cdot}02981}{0{\cdot}7345}$

b) $\sqrt{(2{\cdot}37)^2+(0{\cdot}856)^2}$

c) $\dfrac{1}{3{\cdot}65}+\dfrac{1}{0{\cdot}904}$

43) Solve the equations:

a) $\dfrac{x+3}{3}-\dfrac{x-2}{4}=\dfrac{7}{4}$

b) $x^2-5x+3=0$

giving your answers correct to two significant figures.

44) a) Use logarithms to evaluate

i) $\sqrt[3]{0{\cdot}0822}$ ii) $\dfrac{47{\cdot}28\times(0{\cdot}8129)^2}{723{\cdot}9}$

b) Find the product of $4{\cdot}47\times10^7$ and $5{\cdot}27\times10^{-3}$. Give your answer in the form $A\times10^n$, where A is a number between 1 and 10 and n is a positive integer.

Chapter 17 Mensuration

SI UNITS

The Systeme International d'Unites (the international system of units) is essentially a metric system. It is based upon six fundamental units which are:

Length—the metre (abbreviation m)
Mass—the kilogram (kg)
Time—the second (s)
Electric current—the ampere (A)
Luminous Intensity—the candela (cd)
Temperature—the Kelvin (K)

For many applications some of the above units are too small or too large and hence multiples and sub-multiples are often needed. These multiples and sub-multiples are given special names which are as follows:

MULTIPLICATION FACTOR		PREFIX	SYMBOL
1 000 000 000 000	10^{12}	tera	T
1 000 000 000	10^{9}	giga	G
1 000 000	10^{6}	mega	M
1 000	10^{3}	kilo	k
100	10^{2}	hecto	h
10	10^{1}	deca	da
0·1	10^{-1}	deci	d
0·01	10^{-2}	centi	c
0·001	10^{-3}	milli	m
0·000 001	10^{-6}	micro	μ
0·000 000 001	10^{-9}	nano	n
0·000 000 000 001	10^{-12}	pico	p
0·000 000 000 000 001	10^{-15}	femto	f
0·000 000 000 000 000 001	10^{-18}	atto	a

Where possible, multiples and sub-multiples should be of the form 10^{3n} where n is an integer. Thus 5000 metres should be written as 5 kilometres and not as 50 hectometres. Double prefixes are not permitted in the SI system. For example 1000 km cannot be written as 1 kkm but only as 1 Mm. Again, 0·000006 km cannot be written as 6 μkm but only as 6 mm.

Examples

1) Express 203 560 kg as the highest multiple possible.

$$\begin{aligned} 203\,560 \text{ kg} &= 203\,560 \times 10^3 \text{ gramme} \\ &= 203{\cdot}560 \times 10^3 \times 10^3 \text{ gramme} \\ &= 203{\cdot}560 \times 10^6 \text{ gramme} \\ &= 203{\cdot}560 \text{ mega gramme} \\ &= 203{\cdot}560 \text{ Mg} \end{aligned}$$

(It is usually better to use 203·560 Mg rather than 0·203 560 Gg).

2) A measurement is taken as 0·000 000 082 m. Express this measurement as a standard sub-multiple of a metre.

$$0{\cdot}000\,000\,082 \text{ m} = \frac{82}{1\,000\,000\,000} \text{ m}$$

$$= \frac{82}{10^9} \text{ m} = 82 \times 10^{-9} \text{m}$$

$$= 82 \text{nm}$$

(It is better to use 82 nm rather than 0·082 pm).

Exercise 54

Express each of the following as a standard multiple or sub-multiple.

1) 8000 m
2) 15 000 kg
3) 3800 km
4) 1 891 000 kg
5) 0·007 m
6) 0·000 001 3 m
7) 0·028 kg
8) 0·000 36 km
9) 0·000 064 kg
10) 0·003 6 A

Express each of the following in the form $A \times 10^n$ where A is a number between 1 and 10 and n is an integer.

11) 53 km

12) 18 kg

13) 3·563 Mg

14) 18·76 Gg

15) 70 mm

16) 78 mg

17) 358 pm

18) 18·2 μm

19) 270·6 Tm

20) 253 μg

DERIVED UNITS

Derived units are obtained by using two or more of the six fundamental units. For instance speed is defined as $\frac{\text{change of distance}}{\text{change of time}}$. Since the fundamental unit of distance is the metre and the fundamental unit of time is the second, the unit of speed is the $\frac{\text{unit of distance}}{\text{unit of time}}$ which is $\frac{\text{metre}}{\text{second}}$. This is written metre/second and abbreviated to m/s.
Density is defined as mass per unit volume. The unit of mass is the kilogramme (kg) and since volume = length × length × length the unit of volume is metre × metre × metre or cubic metre which is abbreviated to m^3.

$$\text{Now density} = \frac{\text{mass}}{\text{volume}}$$

Unit of density

$$= \frac{\text{unit of mass}}{\text{unit of volume}} = \frac{\text{kg}}{\text{m}^3} \text{ or kg/m}^3$$

UNIT OF AREA

The area of a plane figure is measured by seeing how many square units it contains. A square metre is the area inside a square which has a side of 1 metre. Similarly a square millimetre is the area inside a square which has a side of 1 millimetre.
The standard abbreviations for units of area are

square metre	m^2
square centimetre	cm^2
square millimetre	mm^2

In the SI system a prefix, when present, is raised to the same power as the unit. For example, mm^3 is actually $(mm)^3$ and *not* $m(m)^3$.

Examples

1) How many square centimetres are there in 1 square metre?

$$1 \text{ m} = 10^2 \text{ cm}$$
$$1 \text{ m}^2 = (10^2 \text{ cm})^2 = 10^4 \text{ cm}^2 = 10\,000 \text{ cm}^2$$

2) An area is 300 000 mm^2. How many square metres is this?

$$1 \text{ mm} = 10^{-3} \text{ m}$$
$$1 \text{ mm}^2 = (10^{-3} \text{ m})^2 = 10^{-6} \text{ m}^2$$

$$\therefore\ 300\,000 \text{ mm}^2 = 300\,000 \times 10^{-6} \text{ m}^2$$
$$= 3 \times 10^5 \times 10^{-6} \text{ m}^2$$
$$= 3 \times 10^{-1} \text{ m}^2$$
$$= \frac{3}{10} = 0{\cdot}3 \text{ m}^2$$

Exercise 55

Convert the following areas into the units stated:

1) 5000 mm^2 into cm^2

2) 52 600 mm^2 into m^2

3) 82 000 cm^2 into m^2

4) 2·65 m^2 into cm^2

5) 12·38 m^2 into cm^2

6) 0·78 m^2 into mm^2

AREAS AND PERIMETERS

The following table gives the areas and perimetres of some simple geometrical shapes.

Figure | *Diagram* | *Formulae*

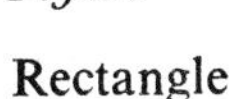

Rectangle

area $= l \times b$

Perimeter $= 2l + 2b$

Parallelogram

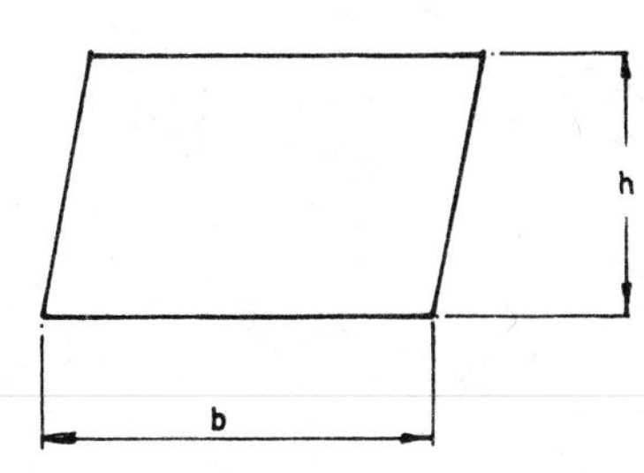

area $= b \times h$

Triangle

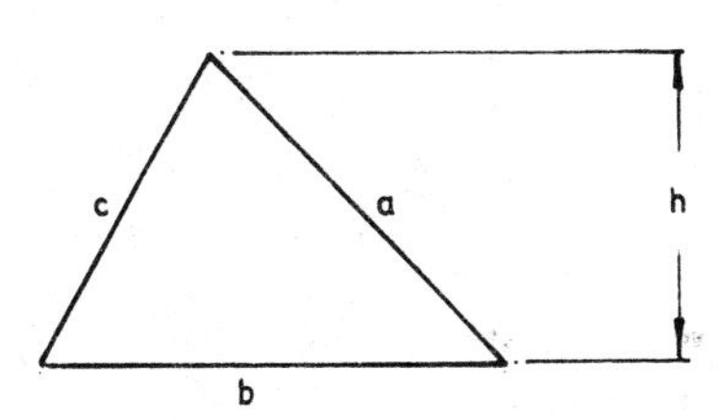

area $= \dfrac{1}{2} \times b \times h$

area $= \sqrt{s(s-a)(s-b)(s-c)}$

where $s = \dfrac{a+b+c}{2}$

Trapezium

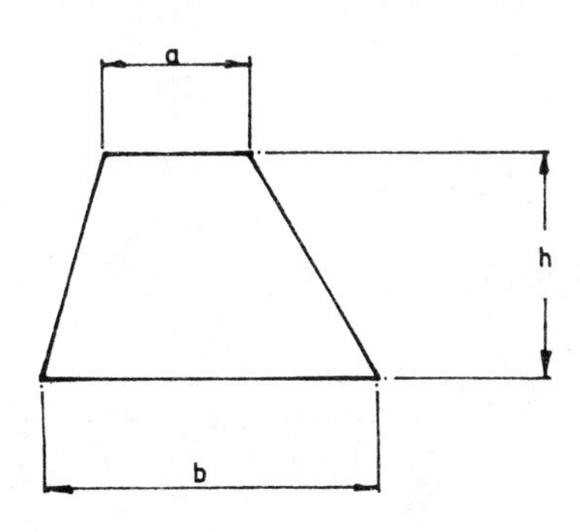

area $= \dfrac{1}{2} \times h \times (a+b)$

Circle

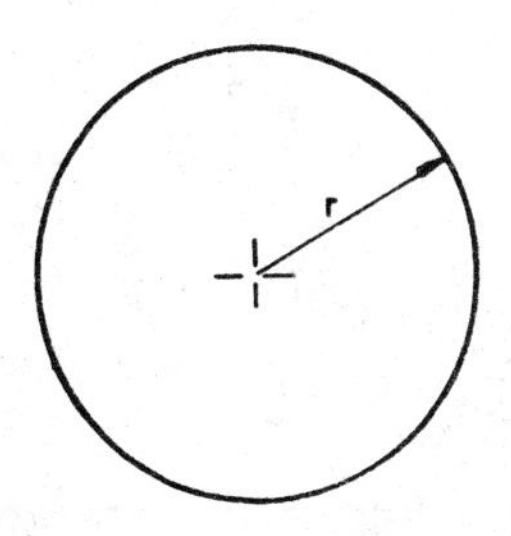

area $= \pi r^2$

circumference $= 2\pi r$

Figure	*Diagram*	*Formulae*
Sector of a circle	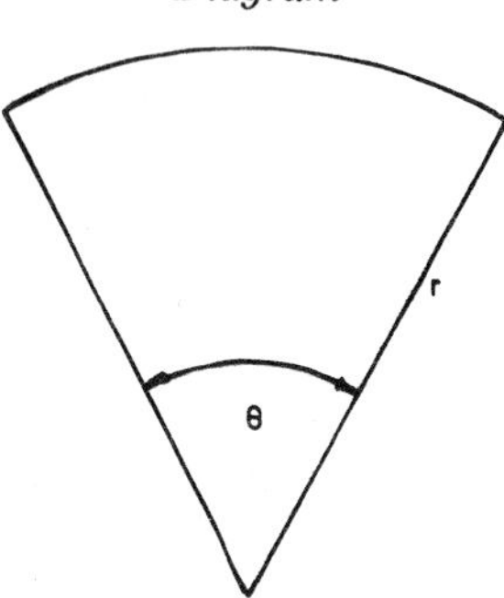	area $=\pi r^2 \times \dfrac{\theta}{360}$ Length of arc $=2\pi r \times \dfrac{\theta}{360}$

Examples

1) Figure 17.1 shows the cross-section of a girder. Find its area in square centimetres. The section may be split up into two rectangles and a parallelogram as shown.

area of rectangle $=6 \times 1{\cdot}3=7{\cdot}8$ cm^2
area of parallelogram $=1{\cdot}5 \times 5{\cdot}2=7{\cdot}8$ cm^2
area of section $=7{\cdot}8+7{\cdot}8+7{\cdot}8=23{\cdot}4$ cm^2

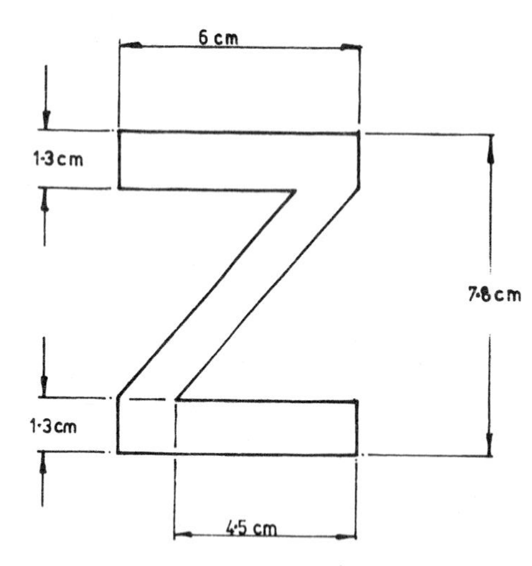

Fig. 17.1

2) A quadrilateral has the dimensions shown in Figure 17.2. Find its area.
The quadrilateral is made up to the triangles ABC and ACD.
To find the area of Δ ABC,

$$s=\frac{5+7+10}{2}=11$$

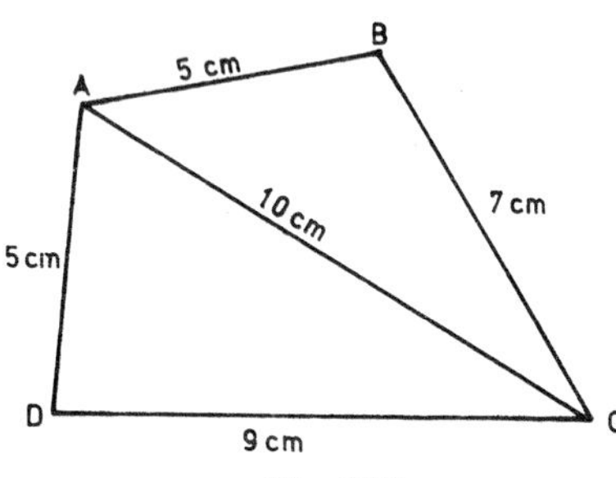

Fig. 17.2

area of Δ ABC

$$=\sqrt{s(s-a)(s-b)(s-c)}$$
$$=\sqrt{11 \times (11-5) \times (11-7) \times (11-10)}$$
$$=\sqrt{11 \times 6 \times 4 \times 1}$$
$$=\sqrt{264}=16{\cdot}25 \text{ cm}^2$$

To find the area of Δ ACD,

$$s=\frac{5+9+10}{2}=12$$

area of Δ ACD

$$=\sqrt{s(s-a)(s-b)(s-c)}$$
$$=\sqrt{12 \times (12-5) \times (12-9) \times (12-10)}$$
$$=\sqrt{12 \times 7 \times 3 \times 2}$$
$$=\sqrt{504}=22{\cdot}45 \text{ cm}^2$$

∴ area of quadrilateral

$$=\text{area of } \Delta \text{ ABC}+\text{area of } \Delta \text{ ACD}$$
$$=16{\cdot}25+22{\cdot}45=38{\cdot}70 \text{ cm}^2$$

3) The cross-section of a block of metal is shown in Fig. 17.3. Find its area.

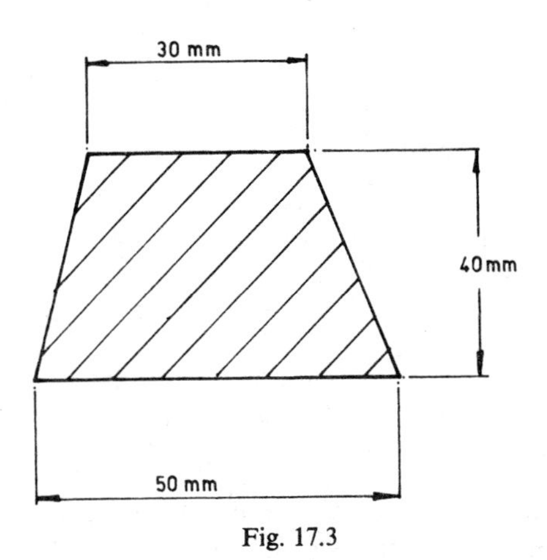

Fig. 17.3

$$\text{area of trapezium} = \frac{1}{2} \times 40 \times (30+50)$$

$$= \frac{1}{2} \times 40 \times 80$$

$$= 1\,600 \text{ mm}^2$$

4) A hollow shaft has an outside diameter of 3·25 cm and an inside diameter of 2·5 cm. Calculate the cross-sectional area of the shaft (Fig. 17.4).

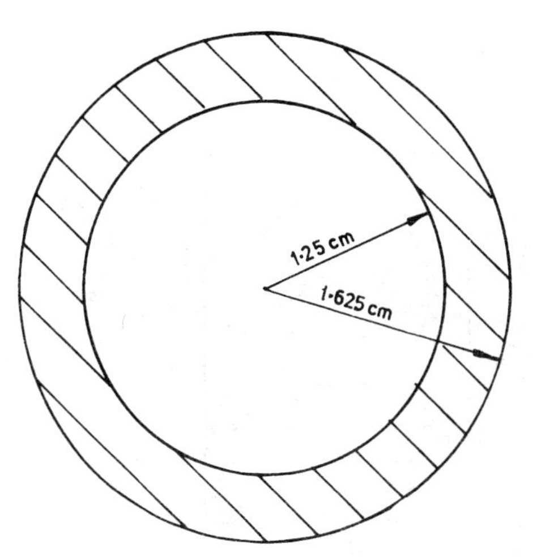

Fig. 17.4

area of cross-section

= area of outside circle − area of inside circle

$= \pi \times 1{\cdot}625^2 - \pi \times 1{\cdot}25^2$

$= \pi(1{\cdot}625^2 - 1{\cdot}25^2)$

$= 3{\cdot}142 \times (2{\cdot}640 - 1{\cdot}563)$

$= 3{\cdot}142 \times 1{\cdot}077$

$= 3{\cdot}388 \text{ cm}^2$

5) Calculate a) the length of arc of a circle whose radius is 8 m and which subtends an angle of 56° at the centre, and b) the area of the sector so formed.

$$\text{Length of arc} = 2\pi r \times \frac{\theta^\circ}{360}$$

$$= 2 \times \pi \times 8 \times \frac{56}{360} = 7{\cdot}82 \text{ m}$$

$$\text{area of sector} = \pi r^2 \times \frac{\theta^\circ}{360}$$

$$= \pi \times 8^2 \times \frac{50}{360} = 31{\cdot}28 \text{ m}^2$$

Exercise 56

1) The area of a rectangle is 220 mm². If its width is 25 mm find its length.

2) A sheet metal plate has a length of 147·5 mm and a width of 86·5 mm. Find its area in m².

3) Find the areas of the sections shown in Fig. 17.5.

4) Find the area of a triangle whose base is 7·5 cm and whose altitude is 5·9 cm.

5) A triangle has sides 4 cm, 7 cm and 9 cm long. What is its area?

6) A triangle has sides 37 mm, 52 mm and 63 mm long. What is its area in cm²?

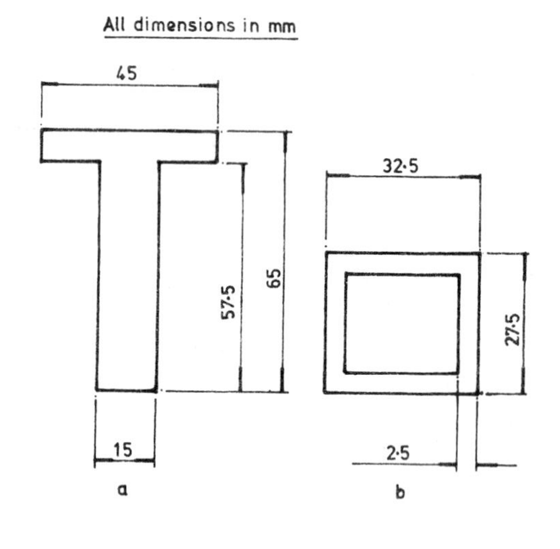

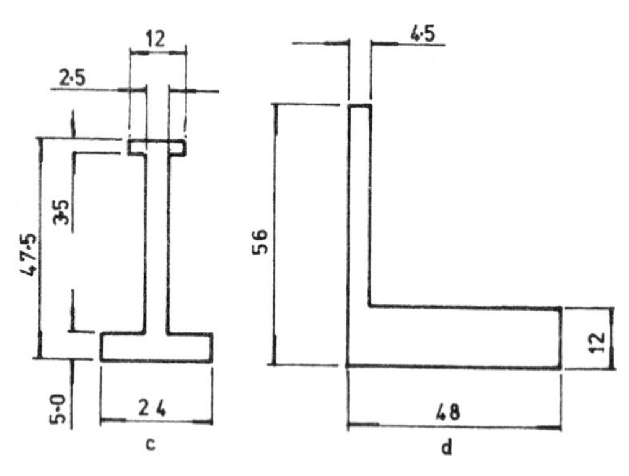

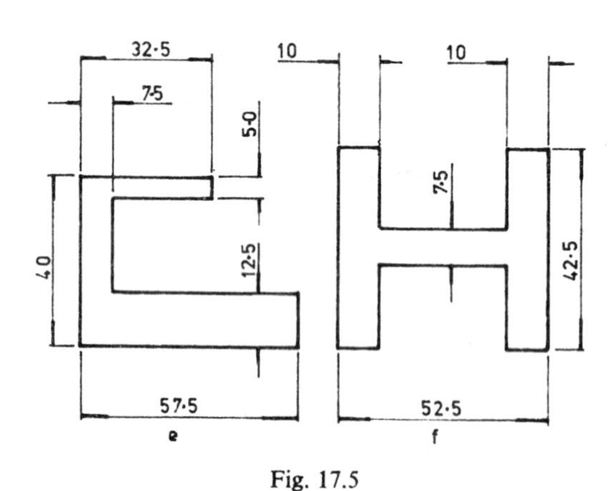

Fig. 17.5

7) Find the area of the shape shown in Fig. 17.6.

8) Find the areas of the quadrilateral shown in Fig. 17.7.

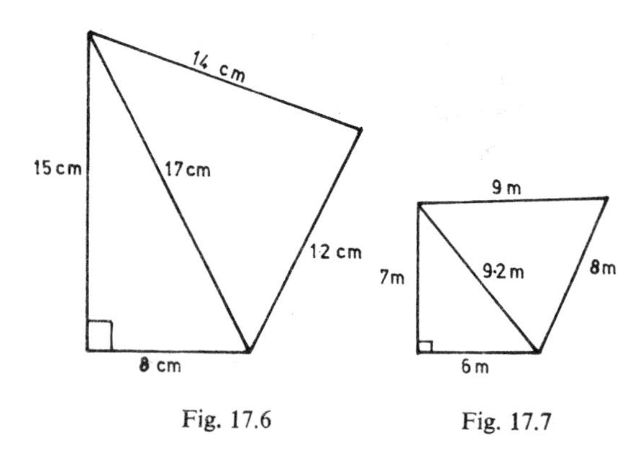

Fig. 17.6 Fig. 17.7

9) What is the area of a parallelogram whose base is 7 cm long and whose vertical height is 4 cm?

10) Determine the length of the side of a square whose area is equal to that of a parallelogram with a base of 3 m and a vertical height of 1·5 m.

11) Find the area of a trapezium whose parallel sides are 75 mm and 82 mm long respectively and whose vertical height is 39 mm.

12) The parallel sides of a trapezium are 12 cm and 16 cm long. If its area is 220 cm^2, what is its altitude?

13) Find the areas of the shaded portions in each of the diagrams of Fig. 17.8.

14) Find the circumference of circles whose radii are:

a) 3·5 mm b) 13·8 m c) 4·2 cm

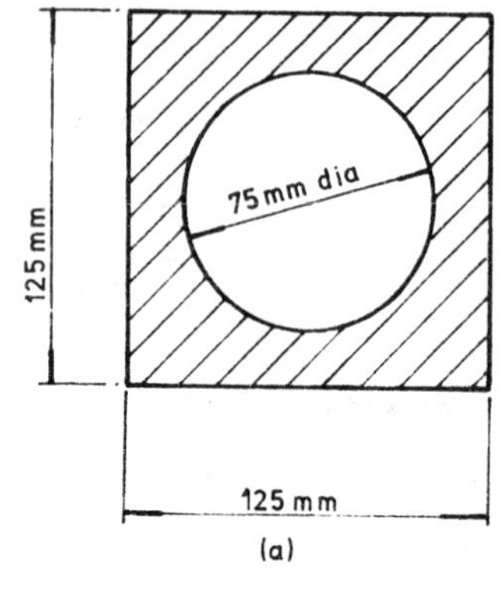

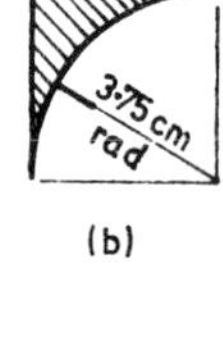

Fig. 17.8

15) Find the diameter of circles whose circumferences are:

a) 34·4 mm b) 18·54 cm c) 195·2 m

16) A ring has an outside diameter of 3·85 cm and an inside diameter of 2·63 cm. Calculate its area.

17) A hollow shaft has a cross-sectional area of 8·68 cm^2. If its inside diameter is 0·75 cm, calculate its outside diameter.

18) Find the area of the plate shown in Fig. 17.9.

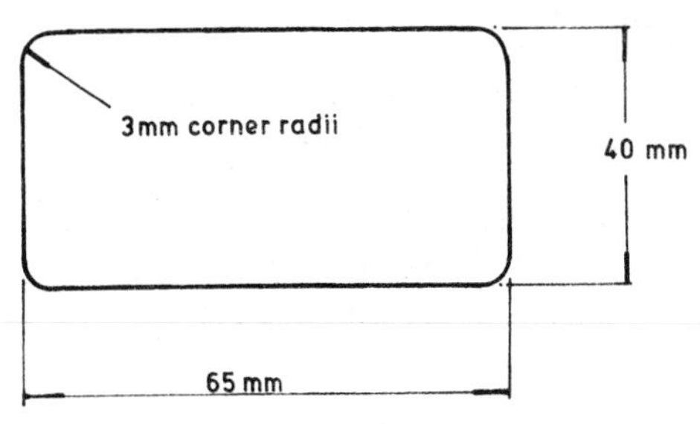

Fig. 17.9

19) How many revolutions will a wheel make in travelling 2 km if its diameter is 700 mm?

20) If r is the radius and θ is the angle subtended at the centre by an arc find the length of arc when

a) $r=2$ cm, $\theta=30°$
b) $r=3{\cdot}4$ cm, $\theta=38°40'$.

21) If l is the length of an arc, r is the radius and θ is the angle subtended by the arc, find θ when

a) $l=9{\cdot}4$ cm, $r=4{\cdot}5$ cm
b) $l=14$ mm, $r=79$ mm.

22) If an arc 7 cm long subtends an angle of 45° at the centre what is the radius of the circle?

23) Find the areas of the following sectors of circles:

a) radius 3 m, angle of sector 60°.
b) radius 2·7 cm, angle of sector 79°45'.
c) radius 7·8 cm, angle of sector 143°42',

24) Calculate the area of the cross-section shown in Fig. 17.10.

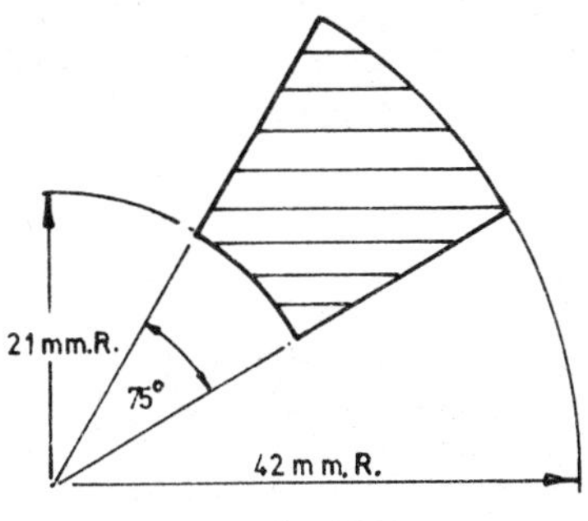

Fig. 17.10

UNIT OF VOLUME

The volume of a solid figure is measured by seeing how many cubic units it contains. A cubic metre is the volume inside a cube which has a side of 1 metre. Similarly a cubic centimetre is the volume inside a cube which has a side of 1 centimetre. The standard abbreviations for units of volume are:

cubic metre	m^3
cubic centimetre	cm^3
cubic millimetre	mm^3

Examples

1) How many cubic centimetres are contained in 1 cubic metre?

$$1 \text{ m} = 10^2 \text{ cm}$$
$$1 \text{ m}^3 = (10^2 \text{ cm})^3 = 10^6 \text{ cm}^3 = 1\,000\,000 \text{ cm}^3$$

2) A tank contains 84 000 000 cubic millimetres of liquid. How many cubic metres does it contain?

$$1 \text{ mm} = 10^{-3} \text{ m}$$
$$1 \text{ mm}^3 = (10^{-3} \text{ m})^3 = 10^{-9} \text{ m}^3$$
$$84\,000\,000 \text{ mm}^3 = 84\,000\,000 \times 10^{-9} \text{ m}^3$$
$$= 8{\cdot}4 \times 10^7 \times 10^{-9} \text{ m}^3$$
$$= 8{\cdot}4 \times 10^{-2} \text{ m}^3$$
$$= \frac{8{\cdot}4}{10^2} = 0{\cdot}084 \text{ m}^3$$

Sometimes the capacity of a container is measured in litres (abbreviation: l). Strictly speaking a litre is not used in precise measurement since 1 litre = 1000·028 cm^3. However, for most purposes it is near enough to take

1 litre = 1000 cm^3

Example

A tank contains 30 000 litres of liquid. How many cubic metres does it contain?

$$30\,000 \text{ litres} = 30\,000 \times 1\,000 \text{ cm}^3 = 3 \times 10^7 \text{ cm}^3$$

$$1 \text{ cm} = 10^{-2} \text{ m}$$

$$1 \text{ cm}^3 = (10^{-2} \text{ m})^3 = 10^{-6} \text{ m}^3$$

$$\therefore\ 3 \times 10^7 \text{ cm}^3 = 3 \times 10^7 \times 10^{-6} \text{ m}^3 = 3 \times 10 = 30 \text{ m}^3$$

Exercise 57

Convert the following volumes into the units stated:

1) 5 m^3 into cm^3

2) 0·08 m^3 into mm^3

3) 18 m^3 into mm^3

4) 830 000 cm^3 into m^3

5) 850 000 mm^3 into m^3

6) 78 500 cm^3 into m^3

7) A tank contains 5000 litres of petrol. How many cubic metres of petrol does it contain?

8) A small vessel contains 2500 mm^3 of oil. How many litres does it contain?

9) A tank holds, when full, 827 m^3 of water. How many litres does it hold?

10) A container holds 8275 cm^3 when full. How many litres does it hold?

VOLUMES AND SURFACE AREAS

Figure	*Volume*	*Surface area*
Any solid having a uniform cross-section.	Cross-sectional area × length of solid	Curved surface + ends. *i.e.* (perimeter of cross-sections × length of solid) + (total area of ends).
Cylinder 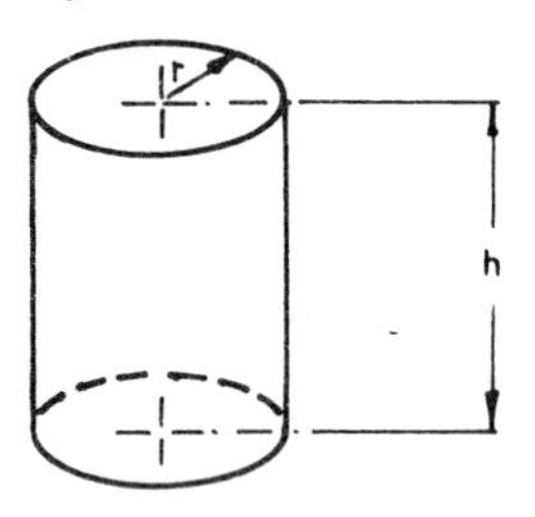	$\pi r^2 h$	$2\pi rh + 2\pi r^2 = 2\pi r(h+r)$
Cone 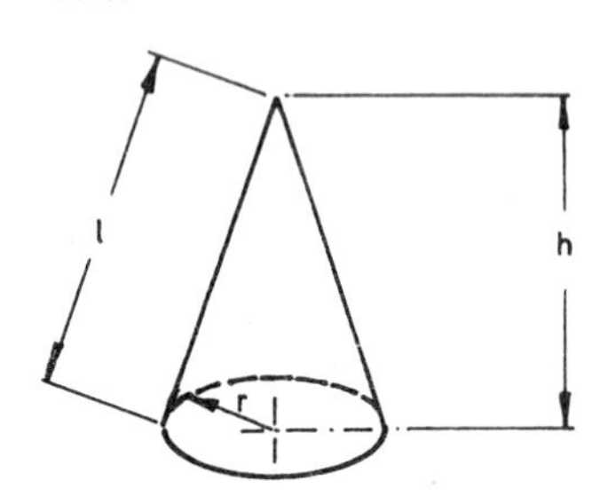	$\frac{1}{3}\pi r^2 h$ (*h* is the vertical height)	$\pi rl + \pi r^2$ (*l* is the slant height)

Figure	*Volume*	*Surface area*
Sphere 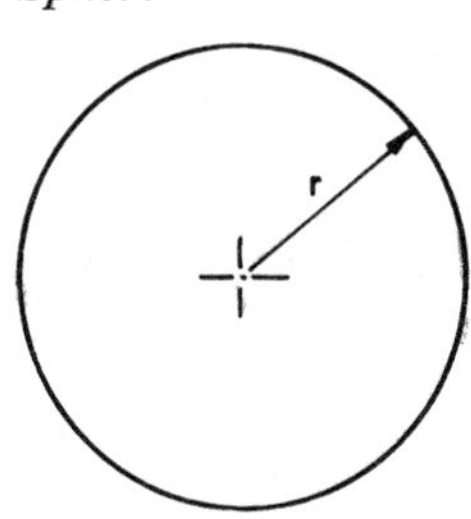	$\frac{4}{3}\pi r^3$	$4\pi r^2$
Pyramid 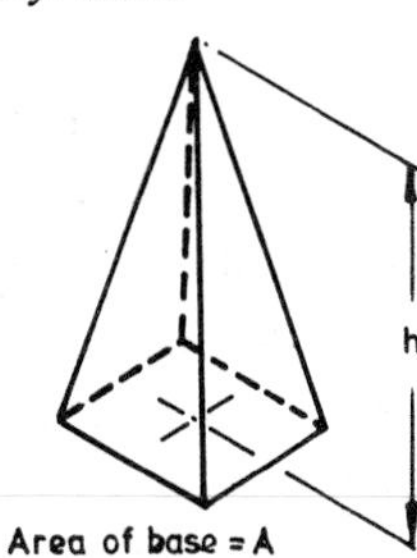	$\frac{1}{3}Ah$	Sum of the areas of the triangles plus the area of the base.
Prism (any solid with two faces parallel and a constant cross-section. The end faces must be triangles, quadrilaterals or polygons).	Cross-sectional area $\times$ length of prism.	

Examples

1) A steel section has the cross-section shown in Fig. 17.11. If it is 9 m long claculate its volume and total surface area.

The section is made up of a rectangle whose length is 100 mm and whose breadth is 150 mm, and a semi-circle whose radius is 75 mm.

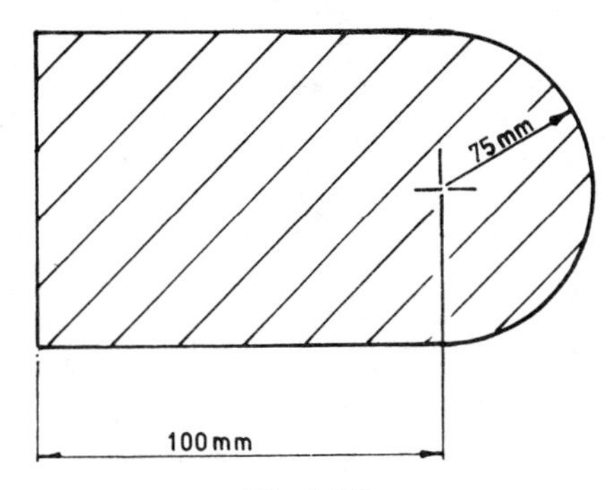

Fig. 17.11

To find the volume:

area of cross-section

$$= 100 \times 150 + \frac{1}{2} \times \pi \times 75^2$$

$$= 23\,836 \text{ mm}^2$$

$$= \frac{23\,836}{(1000)^2} = 0{\cdot}023\,836 \text{ m}^2$$

$$\text{Volume of solid} = 0{\cdot}023\,836 \times 9$$

$$= 0{\cdot}214\,5 \text{ m}^3$$

To find the surface area:

Perimeter of cross-section

$$= \pi \times 75 + 2 \times 100 + 150$$

$$= 585{\cdot}5 \text{ mm}$$

$$= \frac{585{\cdot}5}{1000} = 0{\cdot}585\,5 \text{ m}$$

Lateral surface area

$=9\times 0{\cdot}585\,5=5{\cdot}270\ \text{m}^2$

Surface area of ends

$=2\times 0{\cdot}024=0{\cdot}048\ \text{m}^2$

Total surface area

$=5{\cdot}270+0{\cdot}048=5{\cdot}318\ \text{m}^2$

2) A metal bar of length 200 mm and diameter 75 mm is melted down and cast into washers 2·5 mm thick with an internal diameter of 12·5 mm and an external diameter of 25 mm. Calculate the number of washers obtained assuming no loss of metal.

Volume of original bar of metal

$=\pi\times 37{\cdot}5^2\times 200$
$=883\,500\ \text{mm}^3$

Volume of one washer

$=\pi\times(12{\cdot}5^2-6{\cdot}25^2)\times 2{\cdot}5$
$=\pi\times 117{\cdot}2\times 2{\cdot}5$
$=920{\cdot}4\ \text{mm}^3$

Number of washers obtained

$$=\frac{883\,500}{920{\cdot}4}=960$$

Exercise 58

1) A steel ingot whose volume is 2 m^3 is rolled into a plate 15 mm thick and 1·75 m wide. Calculate the length of the plate in m.

2) A block of lead 1·5 m × 1 m × 0·75 m is hammered out to make a square sheet 10 mm thick. What are the dimensions of the square?

3) Calculate the volume of a metal tube whose bore is 50 mm and whose thickness is 8 mm if it is 6 m long.

4) The volume of a small cylinder is 180 cm^3. If the radius of the cross-section is 25 mm find its height.

5) A steel ingot is in the shape of a cylinder 1·5 m diameter and 3·5 m long. How many metres of square bar of 50 mm side can be rolled from it.

6) A cone has a diameter of 70 mm and a height of 100 mm. What is its volume?

7) Calculate the diameter of a cylinder whose height is the same as its diameter and whose volume is 220 cm^3.

8) An ingot whose volume is 2 m^3 is to be made into ball bearings whose diameters are 12 mm. Assuming 20% of the metal in the ingot is wasted, how many ball bearings will be produced from the ingot?

9) The washer shown in Fig. 17.12 has a square of side l cut out of it. If its thickness is t find an expression for the volume, V, of the washer. Hence find the volume of a washer when $D=6$ cm, $t=0{\cdot}2$ cm and $l=4$ cm.

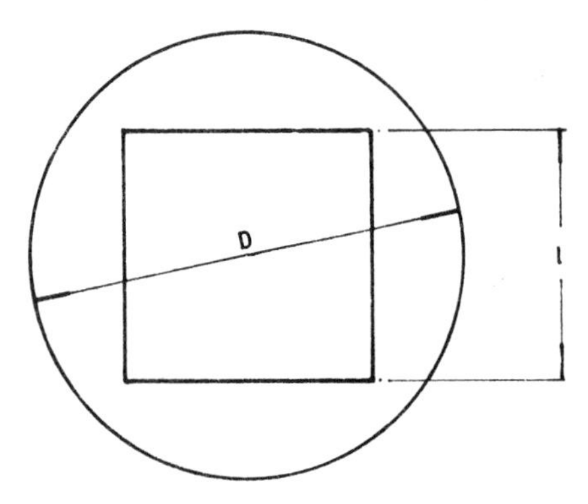

Fig. 17.12

10) A water tank with vertical sides has a horizontal base in the shape of a rectangle with semi-circular ends as illustrated in Fig. 17.13. The total inside length of the tank is 7 m, its width 4 m and its height 2 m.

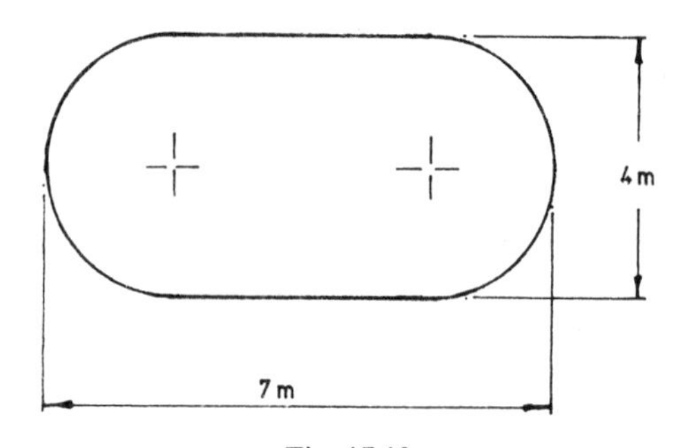

Fig. 17.13

Calculate

a) the surface area of the vertical walls of the tank in m^2,

b) the area of the base in m^2,

c) the number of litres of water in the tank when the depth of water is 1·56 m.

11) A tank 1 m long and 60 cm wide, internally, contains water to a certain depth. An empty tank 40 cm long, 30 cm wide and 25 cm deep, internally, is filled with water from the first tank. If the depth of water in the first tank is now 35 cm, what was the depth at first?

12) Figure 17.14 represents a bird cage in the form of a cylinder surmounted by cone. The diameter of the cylinder is 35 cm and its height is 25 cm. The total volume of the bird cage is 31 178 cm^3. Calculate

a) the total height of the bird cage,

b) the surface area of the cover for the cage (except the base).

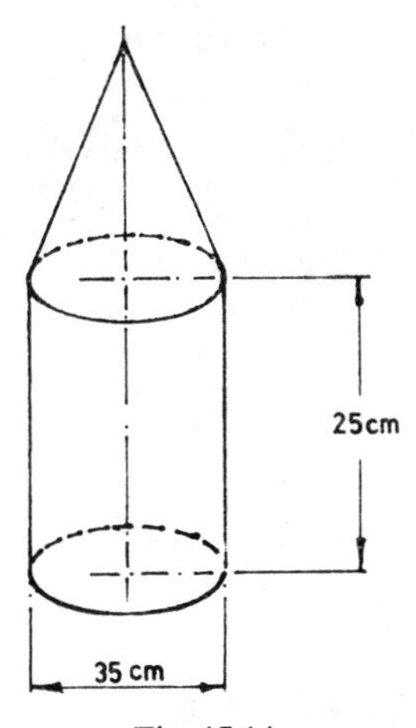

Fig. 17.14

13) A solid iron cone is 12 cm in height and the radius of the circular base is 4 cm. It is placed on its base in a cylindrical vessel of internal radius 5 cm. Water is poured into the cylinder until the depth of water is 16 cm. The cone is then removed. Find by how much the water level falls.

14) A rolling mill produces steel sheet 1·2 m wide and 3 mm thick. If the length of sheeting produced per hour is 2 km and the steel has a density of 7·75 grams per cubic centimetre, find in kilograms the mass of steel produced per hour.

15) A measuring jar is in the form of a vertical cylinder which is graduated so that the volume of liquid in the jar can be read directly in cubic centimetres. The internal radius of the jar is 2·4 cm. Find to the nearest millimetre the distance between the two marks labelled 200 cm^3 and 300 cm^3. When the jar is partly full of water, a steel sphere of radius 1·8 cm is lowered into the jar and completely immersed in the water without causing the water to overflow. Find in millimeters the distance the water-level rises in the jar.

16) A cylindrical tank, open at the top, is made of metal 75 mm thick. The internal radius of the tank is 1·2 m and the internal depth of the tank is 1·9 m. The tank stands with its plane base horizontal. Calculate

a) the number of litres of liquid in the tank when it is $\frac{4}{5}$ full,

b) the area of the external curved surface of the tank,

c) the area of the plane surface of metal at the base of the tank.

17) A cylindrical can whose height is equal to its diameter has a capacity of 9 litres. Find the height of the can. If the diameter is halved and the height altered so that the can still has a capacity of 9 litres, find the ratio of the original curved surface area to the final curved surface area.

18) A cheese is made in the form of a cylinder of radius 21 cm and height 45 cm. The slice shown in Fig. 17.15 (where AB is 13·5 cm and lies along the axis of the cylinder and where $\angle XAY = 30°$) has a mass of 1·3 kg. Calculate the mass of the whole cheese in kg.

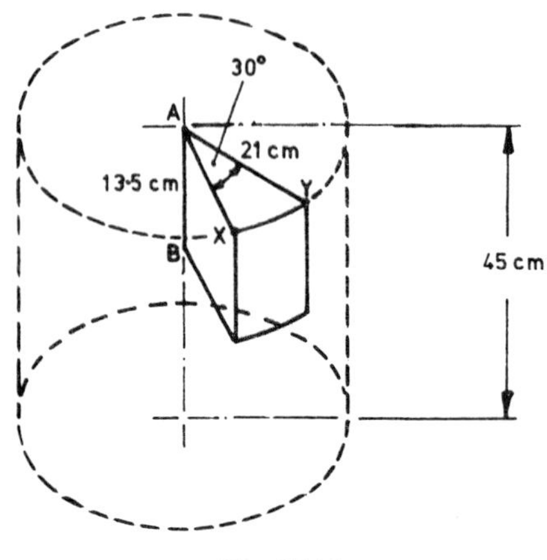

Fig. 17.15

19) The diagram (Fig. 17.16) shows a section through a chemical flask consisting of a spherical body of internal radius r cm with a cylindrical neck of internal radius $\frac{1}{6}r$ cm and a length r cm. Show that when filled to the brim the flask will hold approximately $\frac{77r^3}{18}$ cm^3. (The volumes enclosed by the sphere and cylinder overlap—between the dotted lines—but this fact may be ignored).

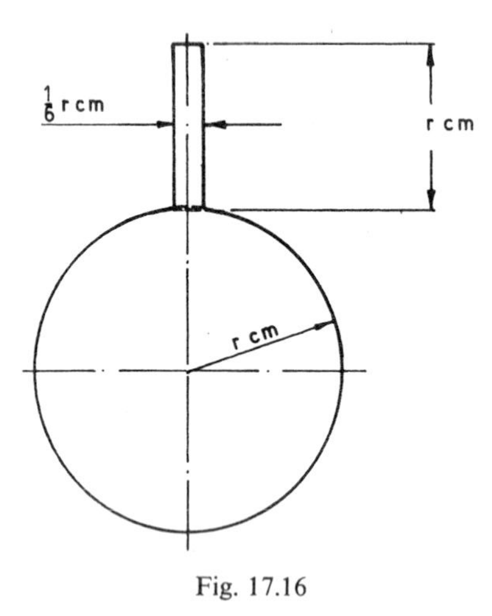

Fig. 17.16

20) A cylindrical can is filled with water. It has a capacity of 300 cm^3 and is 6·5 cm high. Calculate its radius. The water is now poured into another container which has a square horizontal base of side 8 cm and vertical sides. Calculate the depth of water in this container. Calculate also the area of this container which is in contact with the water.

DENSITY AND SPECIFIC GRAVITY

The density of a substance is its mass per unit volume. Thus

$$\text{density} = \frac{\text{mass}}{\text{volume}}$$

or $\quad \text{mass} = \text{volume} \times \text{density}$

Examples

1) A block of lead has a volume of 800 cm^3. What is its mass if the density of lead is 11·4 g/cm^3.

$$\text{mass} = \text{volume} \times \text{density}$$
$$= 800 \times 11{\cdot}4 = 9120 \text{ g} = 9{\cdot}12 \text{ kg}$$

2) A steel pipe 3 m long has an internal radius of 15 mm and an external radius of 18 mm. If steel has a density of 7·7 g/cm^3 calculate (to the nearest gram) the mass of the pipe.

Converting all the dimensions to centimetres, internal radius = 1·5 cm, external radius = 1·8 cm and the length = 3 000 cm.

Volume of pipe
$$= \text{cross-sectional area} \times \text{length}$$
$$= \pi(1{\cdot}8^2 - 1{\cdot}5^2) \times 3000$$
$$= \pi \times 0{\cdot}99 \times 3000$$
$$= 9330 \text{ cm}^3$$

$$\text{mass} = \text{volume} \times \text{density}$$
$$= 9330 \times 7{\cdot}7$$
$$= 71\,841 \text{ g}$$
$$= 71{\cdot}841 \text{ kg}$$

The specific gravity is the number of times a substance is heavier than water, volume for volume.

$$\text{Specific gravity} = \frac{\text{density of the substance}}{\text{density of water}}$$

Since the density of water is 1 g/cm^3 the density of a substance in g/cm^3 is numerically equal to the specific gravity.

Examples

1) The flask shown in Fig. 17.17 is completely filled with oil whose specific gravity is 0·8. Find the mass of oil in the flask

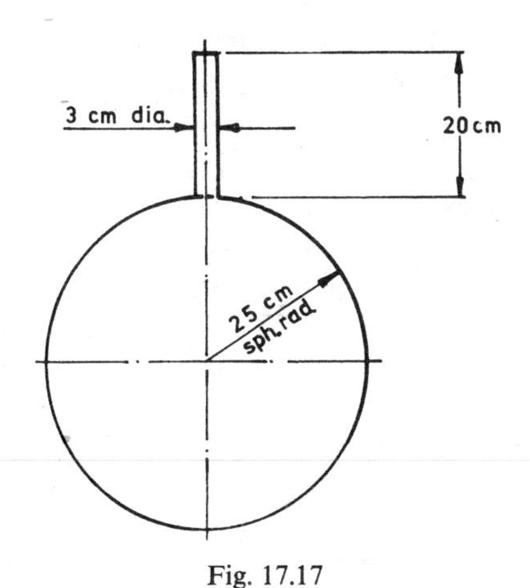

Fig. 17.17

Volume of the sphere

$$= \frac{4}{3}\pi \times 25^3 = 65\,450 \text{ cm}^3$$

Volume of the cylinder

$$= \pi \times 1{\cdot}5^2 \times 20 = 141 \text{ cm}^3$$

Volume of the flask

$$= 65\,450 + 141 = 65\,591 \text{ cm}^3$$

Since the specific gravity is 0·8, the density of the oil is 0·8 g/cm^3. Hence

$$\begin{aligned} \text{mass of the oil} &= \text{volume} \times \text{density} \\ &= 65\,591 \times 0{\cdot}8 \\ &= 52\,473 \text{ g} \\ &= 52{\cdot}473 \text{ kg} \end{aligned}$$

Exercise 59

1) The density of aluminium is 2590 kg/m^3. Find the mass of a piece of aluminium which has a volume of 2 m^3.

2) A lead sheet is 1·5 m × 0·75 m × 3 mm thick. If lead has a density of 11·4 g/cm^3 find the mass of the sheet.

3) A cast iron pipe has an external diameter of 75 mm and an internal diameter of 63 mm. If it is 5 m long and the density of cast iron is 7·5 g/cm^3, what is the mass of the pipe?

4) The diagram (Fig. 17.18) represents the vertical cross-section of a horizontal feeding trough 2 m long and closed at both ends. The trough is made from sheet metal. Calculate in square metres the total area of sheet metal required . If the metal is 4 mm thick and it has a density of 7·7 g/cm^3 find the mass of the sheet metal.

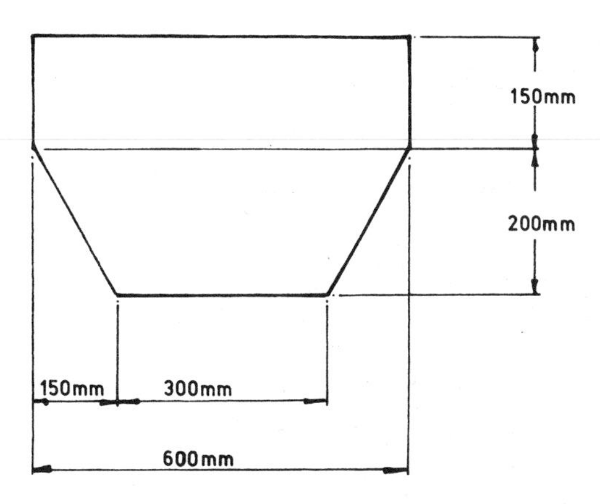

Fig. 17.18

5) The diagram (Fig. 17.19) shows the end view of a metal block of uniform cross-section. The block is 90 mm long and has a mass of 2·898 kg. Calculate the volume of the block and hence find the density in g/cm^3 of the metal from which it is made.

6) A porcelain crucible is in the form of a thick hemi-spherical shell. The radius of the internal hemi-sphere is 5 cm and the thickness of the shell is 1 cm. If the crucible weighs 720 g calculate the density of porcelain.

7) Figure 17.20 shows the cross-section of a steel girder 2 m long. If steel has a density of 7·7 g/cm^3 calculate the mass of the girder in kg.

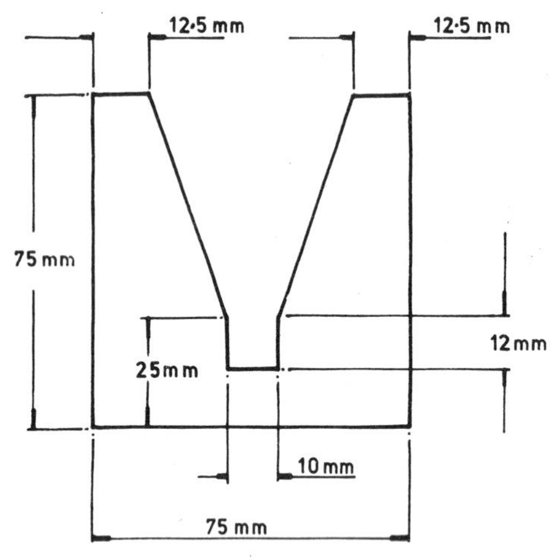

Fig. 17.19

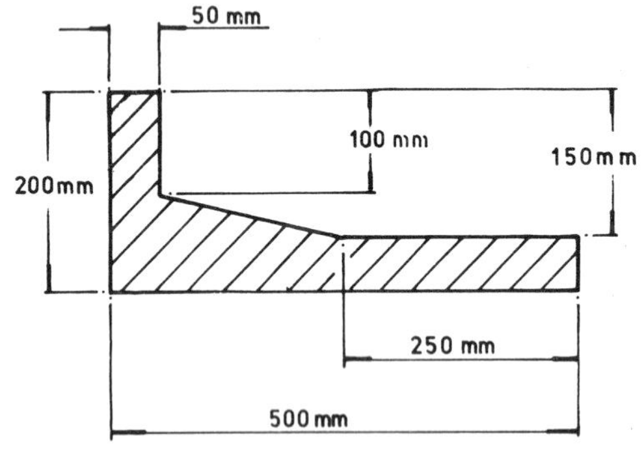

Fig. 17.20

8) A test tube consists of a cylindrical part of internal diameter 2·4 cm and a hemispherical base of internal diameter 2·4 cm. It is placed so that the axis of the cylinder is vertical and liquid is poured into the test tube so that the greatest depth of liquid is 9·5 cm. If the specific gravity of the liquid is 0·7 find the mass of liquid in the test tube.

9) An iron weight is found to be 15 g too light and to correct this a cylindrical hole of diameter 1·60 cm is made in the base and the iron removed is replaced by lead. Given that 1 cm^3 of iron has a mass of 7·14 g and 1 cm^3 of lead has a mass of 11·34 g, calculate the depth of the hole, giving your answer correct to the nearest mm.

10) A bowl is made by cutting in half a hollow sphere of external diameter 50 cm and made of metal 2·5 cm thick. If the bowl is filled with liquid whose specific gravity is 0·95 calculate the total mass of liquid in the bowl. The bowl when empty has a mass of 97 kg. Find the specific gravity of the metal from which the bowl is made.

THE FLOW OF WATER

Suppose that water is flowing through a pipe whose cross-sectional area is A square meters at a speed of v metres per second. If the pipe is running full, the discharge from the pipe per second is the volume of a cylinder of cross-section A and length v. That is, the discharge per second is Av cubic metres per second.

Examples

1) Water is flowing through a pipe whose bore is 75 mm at a speed of 2 metres per second. Calculate the discharge from the pipe a) in cubic metres per second b) in litres per minute.

a) Bore of pipe = 75 mm = 0·075 m
Area of pipe $= \pi \times (0{\cdot}037\,5)^2 = 0{\cdot}004\,4$ m^3
Discharge from pipe
= speed of flow × area of pipe
$= 2 \times 0{\cdot}004\,4 = 0{\cdot}008\,8$ m^3/s

b) Since 1m = 100 cm
Discharge from pipe $= 0{\cdot}008\,8 \times 100^3$
$= 8800$ cm^3/s
Since 1 l = 1000 cm^3
Discharge from pipe

$$= \frac{8800}{1000} = 8{\cdot}8 \text{ l/s}$$

$$= 8{\cdot}8 \times 60 = 528 \text{ l/min}$$

2) Water is being pumped through a pipe of 10 cm diameter so that it discharges 1250 litres per minute. Calculate the speed of flow of the water in metres per second. The pipe is used to empty a swimming bath containing 800 cubic metres of water. How long does it take, in hours, to empty the bath?

Area of the pipe $= \pi \times 5^2 = 25\pi$ cm^2
Volume of water discharged per second
$= 1250$ l $= 1\,250\,000$ cm^3

$$\text{Speed of flow} = \frac{\text{discharge}}{\text{area}} = \frac{1\,250\,000}{25\pi}$$

$$= 15\,920 \text{ cm/min}$$

$$= 159{\cdot}2 \text{ m/min}$$

$$= \frac{159{\cdot}2}{60} = 2{\cdot}65 \text{ m/s}$$

$$\text{Time taken to empty bath} = \frac{\text{Volume}}{\text{Discharge}}$$

Volume $= 800$ m^3
Discharge $= 1\,250\,000$ cm^3/min
$= 1{\cdot}25$ m^3/min
Time taken to empty bath

$$= \frac{800}{1{\cdot}25} = 640 \text{ min}$$

$$= \frac{640}{60} = 10{\cdot}67 \text{ hours}$$

Exercise 60

1) Water is flowing through a pipe whose bore is 15 cm at a speed of 3 metres per second. Calculate the discharge from the pipe in cubic metres per second.

2) The discharge of water from a pipe is 500 cubic centimetres per second. The bore of the pipe is 10 cm. Calculate the speed of flow of the water.

3) Water is pumped through a pipe so that it discharges 10 litres per minute. It is used to empty a tank containing 3 m^3 of water. How long does it take to empty the tank?

4) A rectangular swimming bath with vertical sides is 25 m long and 10 m wide and its rectangular base slopes uniformly from a depth of 1 m at the shallow end to 3 m at the deep end. If the bath contains 400 cubic metres of water find the distance of the water level below the top of the bath.
The water is emptied from the bath through a pipe at a rate of 1·5 cubic metres per second. Find the time taken to empty the bath.

5) Water is poured into a cylindrical reservoir 10 m in diameter at the rate of 3000 litres per minute. Find at what rate the level of the water in the reservoir rises.

6) A large reservoir was replenished with 28 000 m^3 of water flowing through three inlet pipes whose diameters were 0·8 m, 1 m and 1·3 m, and the speeds at which water flowed through them were 2 m/s, 1·5 m/s and 1 m/s. Calculate the time taken to replenish the reservoir.

7) A rectangular block of lead is 40 cm long, 35 cm wide and 25 cm high. The metal is to be used in the manufacture of lead pipe of internal diameter 5 cm and external diameter 10 cm. Calculate in metres, the length of pipe manufactured. If water is to be pumped through the pipe at a rate of 15 000 litres per minute calculate, in m/s, the speed of water flowing through the pipe.

8) A swimming bath is 20 m long and 7 m wide. The depth of water increases uniformly from 1·2 m at the shallow end to 2·2 m at the deep end. Calculate the number of cubic metres of water in the bath. This water is pumped into the cleaning plant through a cylindrical pipe whose internal diameter is 20 cm at 4000 litres per minute. Calculate the speed in metres per second at which the water is moving through the pipe.

SIMILAR SOLIDS

Two solids are similar if the ratios of their corresponding linear dimensions are equal. The two cones shown in Fig. 17.21 are similar if

$$\frac{h_1}{h_2} = \frac{r_1}{r_2}$$

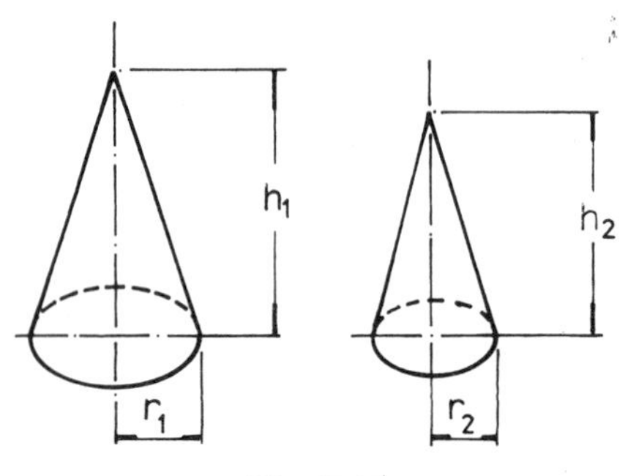

Fig. 17.21

The two cylinders shown in Fig. 17·22 are similar since

$$\frac{150}{75}=\frac{100}{50}$$

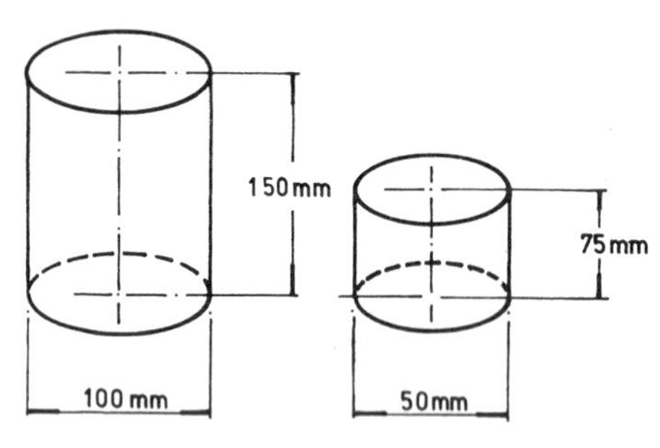

Fig. 17.22

The surface areas of similar solids are proportional to the squares of their linear dimensions.

i) If two spheres having radii r_1 and r_2 have surface areas A_1 and A_2 respectively, then

$$\frac{A_1}{A_2}=\left(\frac{r_1}{r_2}\right)^2$$

ii) The ratio of the surface areas of the two cones shown in Fig. 17.21 is:

$$\frac{A_1}{A_2}=\frac{{r_1}^2}{{r_2}^2}=\frac{{h_1}^2}{{h_2}^2}$$

iii) The ratio of the surface areas of the two cylinders shown in Fig. 17.22 is:

$$\frac{A_1}{A_2}=\frac{100^2}{50^2}=\frac{150^2}{75^2}=\frac{4}{1}$$

Example

Find the surface area of a sphere 120 mm radii. What is the surface area of a sphere 60 mm radii?

Surface area of 60 mm radii sphere
$=4\pi r^2=4\pi\times 60^2=45\,239\text{ mm}^2$

If A_1 = surface area of 120 mm sphere and A_2 = surface area of 60 mm sphere

$$\frac{A_2}{A_1}=\frac{60^2}{120^2}=\frac{1}{4}$$

$$A_2=\frac{1}{4}\times A_1=\frac{45\,239}{4}=11\,310\text{ mm}^2$$

The volumes of similar solids are proportional to the cubes of their corresponding linear dimensions.

i) If two spheres of radii r_1 and r_2 have volumes V_1 and V_2 respectively, then

$$\frac{V_1}{V_2}=\frac{{r_1}^3}{{r_2}^3}$$

ii) The ratio of the volumes of the two cones shown in Fig. 17.21 is

$$\frac{V_1}{V_2}=\frac{{r_1}^3}{{r_2}^3}=\frac{{h_1}^3}{{h_2}^3}$$

iii) The ratio of the volumes of the two cylinders shown in Fig. 17.22 is

$$\frac{V_1}{V_2}=\frac{100^3}{50^3}=\frac{150^3}{75^3}=8$$

Since mass is proportional to volume

$$\frac{M_1}{M_2}=\frac{V_1}{V_2}$$

Where M_1 and M_2 are the masses of two similar solids made of material of the same density and whose volumes are V_1 and V_2 respectively.

Examples

1) The volume of a cone of height 135 mm is 1090 mm^3. Find the volume of a cone whose height is 72 mm.

Let V_1 and V_2 be the volumes of the two cones. Then

$$\frac{V_2}{V_1} = \frac{h_2^3}{h_1^3}$$

$$\frac{V_2}{1090} = \frac{72^3}{135^3}$$

$$V_2 = 1090 \times \frac{72^3}{135^3} = 165{\cdot}4\ \text{mm}^3$$

2) A cone 90 mm high has a mass of 8 kg. A frustum of this cone is formed by cutting off the top 20 mm of the cone as shown in Fig. 17.23. Find the mass of the frustum.

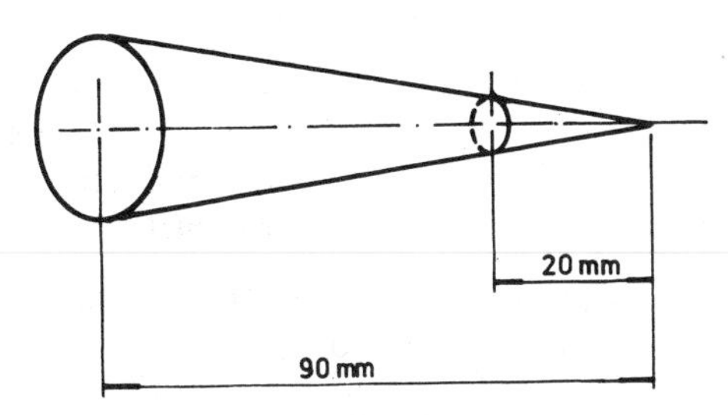

Fig. 17.23

To solve this problem we note that:

mass of frustum = mass of original cone − mass of piece cut off

Let M_1 be mass of original cone and M_2 be mass of piece cut off.

$$\frac{M_2}{M_1} = \frac{20^3}{90^3}$$

$$M_2 = \frac{20^3}{90^3} \times M_1 = \frac{20^3}{90^3} \times 8 = 0{\cdot}088\ \text{kg}$$

mass of frustum $= 8 - 0{\cdot}088 = 7{\cdot}912$

Exercise 61

1) Two spheres have radii 3 cm and 5 cm respectively. Find their volumes.

2) A spherical cap has a height of 2 cm and a volume of 8 cm^3. A similar cap has a height of 3 cm. What is its volume?

3) The volume of a cone of height 14·2 cm is 210 cm^3. Find the height of a similar cone whose volume is 60 cm^3.

4) Find the surface area of a metal sphere whose radius is 73 mm. What is the area of a sphere whose radius is 29 mm?

5) The curved surface of a cone has an area of 20·5 cm^2. What is the curved surface area of a similar cone whose height is 1·5 times as great as the first cone?

6) Find the mass of a hemi-spherical bowl of copper whose external and internal diameters are 24 cm and 16 cm respectively. The density of copper is 8·9 g/cm^3. What is the mass of a similar bowl whose external diameter is 20 cm?

7) A metal pyramid has a square base of side 3 cm and it has a mass of 70 kg. What is the mass of a similar pyramid whose base is a square of side 6 cm. If the height of the first pyramid is 8 cm what is the height of the second pyramid?

8) A solid cylinder has a radius of r cm. The length of the cylinder is $3r$ cm and the **total** surface area is 308 cm^2. Taking π to be $\frac{22}{7}$ calculate the value of r. A second cylinder has a radius of $2x$ cm and a height of $6x$ cm. If its surface area is 254 cm^2 what is the value of x?

9) A cone is 100 mm high and has a radius of 20 mm. Find its volume. A frustum of this cone is formed by cutting off the top 20 mm. Find the volume of the frustum.

10) In a scale model of a school the area of the assembly hall is $\frac{1}{100}$ of the actual area. Calculate the ratio of the volume of the model hall to the volume of the actual hall.

SELF TEST 17

1) An area is 30 000 square centimeters. Hence it is

a 3000 m^2 b 300 m^2 c 3 m^2
d 30 m^2

2) An area is 5 m^2. Hence it is
a 50 cm^2 b 500 cm^2 c 5000 cm^2
d 50 000 cm^2

3) An area is 2000 mm^2. Hence it is
a 2 cm^2 b 20 cm^2 c 0·2 cm^2
d 200 cm^2

4) An area is 600 000 mm^2. Hence it is
a 6000 m^2 b 600 m^2 c 6 m^2
d 0·6 m^2

5) An area is 0·3 m^2. Hence it is
a 30 mm^2 b 300 mm^2
c 30 000 mm^2 d 300 000 mm^2

6) An area is 20 km^2. It is therefore
a 2000 m^2 b 20 000 m^2
c 20 000 000 m^2 d 200 000 m^2

7) A rectangular plot of ground is 4 km long and 8 km wide. Its area is therefore
a 32 Mm^2 b 32 km^2
c 32 000 m^2 d 0·32 Mm^2

8) A triangle has an altitude of 100 mm and a base of 50 mm. Its area is
a 2500 mm^2 b 5000 mm^2
c 25 cm^2 d 50 cm^2

9) A parallelogram has a base 10 cm long and a vertical height of 5 cm. Its area is
a 25 cm^2 b 50 cm^2 c 2500 mm^2
d 5000 mm^2

10) A trapezium has parallel sides whose lengths are 18 cm and 22 cm. The distance between the parallel sides is 10 cm. Hence the area of the trapezium is
a 400 cm^2 b 200 cm^2
c 3960 cm^2 d 495 cm^2

11) The area of a circle is given by the formula
a $2\pi r^2$ b $2\pi r$ c πr^2 d πr

12) The circumference of a circle is given by the formula
a πr^2 b $2\pi r$ c πr d πd

13) A ring has an outside diameter of 8 cm and an inside diameter of 4 cm. Its area is therefore
a $\pi(8^2-4^2)$ b $8\pi-4\pi$
c $\pi(8+4)(8-4)$ d $\pi(4^2-2^2)$

14) A wheel has a diameter of 70 cm. The number of revolutions it will make in travelling 55 km is
a 25 000 b 50 000 c 5000
d 2500

15) An arc of a circle is 22 cm and the radius of the circle is 140 cm. The angle subtended by the arc is
a 90° b 9° c 180° d 18°

16) A sector of a circle subtends an angle of 120°. If the radius of the circle is 42 cm then the area of the sector is
a 88 cm^2 b 1848 cm^2
c 3696 cm^2 d 176 cm^2

17) A tank has a volume of 8 m^3. Hence the Volume of the tank is also
a 800 cm^3 b 8000 cm^3
c 80 000 cm^3 d 8 000 000 cm^3

18) A solid has a volume of 200 000 mm^3. Hence the volume of the solid is also
a 2000 cm^3 b 200 cm^3 c 20 cm^3
d 20 000 cm^3

19) The capacity of a container is 50 litres. Hence its capacity is also
a 50 000 cm^3 b 5000 cm^3
c 0·5 m^3 d 0·05 m^3

20) The area of the curved surface of a cylinder of radius r and height h is
a $2\pi rh$ b $2\pi r^2h$ c πrh
d πr^2h

21) The volume of a cylinder of radius r and height h is
a $2\pi rh$ b $2\pi r^2h$ c πrh
d πr^2h

22) The total surface area of a closed cylinder whose radius is r and whose height is h is
a $\pi rh+2\pi r^2$ b $\pi r(h+2r)$
c $2\pi rh+2\pi r^2$ d $2\pi r(h+r)$

23) A small cylindrical container has a diameter of 280 mm and a height of 50 mm. It will hold
a 3·08 l b 30·8 l c 6·16 l
d 61·6 l

24) A cone has height of 90 mm and a diameter of 140 mm. Hence, the volume of the cone is

a 462 cm^3 b 19 800 mm^3

c 462 000 mm^3 d 19·8 cm^3

25) A test tube whose overall length is h and whose radius is r has a hemi-spherical end. A formula for its volume is

a $\pi r^2\left(\frac{2}{3}r+h\right)$ b $\pi r^2\left(h-\frac{1}{3}r\right)$

c $2\pi r(r+h)$ d $\pi r^2(2+h-r)$

26) The mass of an object is

a $\frac{\text{density}}{\text{volume}}$ b $\frac{\text{volume}}{\text{density}}$

c volume × density

27) The density of a material is

a $\frac{\text{volume}}{\text{mass}}$ b $\frac{\text{mass}}{\text{volume}}$

c mass × volume

28) A block of lead has a volume of 880 cm^3. If its mass is 80 g, the density of lead is

a 0·09 g/cm^3 b 11 g/cm^3

c 90 kg/m^3 d 11 000 kg/m^3

29) The specific gravity of a substance is 0·8. Hence its density is

a 0·8 g/cm^3 b 8 g/mm^3

c 800 kg/m^3 d 80 kg/m^3

30) A flask contains 500 litres of oil whose specific gravity is 0·7. The mass of oil in the flask is

a 350 kg b 35 kg c 3500 kg

d 3·5 kg

31) Water is flowing through a pipe at a speed of 5 m/s. If the bore of the pipe has an area of 2000 cm^2, the discharge from the pipe per second is

a 10 000 cm^3 b 1 000 000 cm^3

c 0·010 m^3 d 1000 l

32) A tank contains 2000 litres of water. It is emptied by means of a pipe through which the water discharges at 4 m^3/min. The time taken to empty the tank is

a 30 seconds b 500 min c 5 min

d 8 min

33) Two cylinders are similar. Cylinder A has a radius of 8 cm and cylinder B has a radius of 4 cm. Therefore

a $\frac{\text{Surface area of A}}{\text{Surface area of B}}=\frac{4}{1}$

b $\frac{\text{Surface area of A}}{\text{Surface area of B}}=\frac{2}{1}$

c $\frac{\text{Volume of A}}{\text{Volume of B}}=\frac{4}{1}$

d $\frac{\text{Volume of A}}{\text{Volume of B}}=\frac{8}{1}$

34) Figure 17.24 shows two cylinders. Therefore

a $\frac{\text{Volume of A}}{\text{Volume of B}}=\frac{8}{1}$

b $\frac{\text{Volume of A}}{\text{Volume of B}}=\frac{12}{1}$

c $\frac{\text{Volume of A}}{\text{Volume of B}}=\frac{27}{1}$

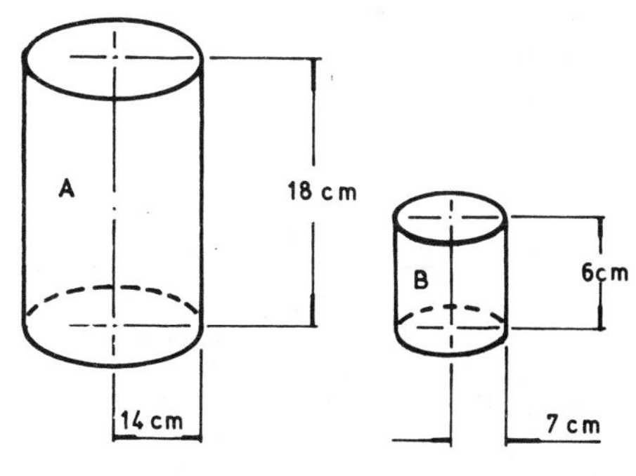

Fig. 17.24

35) A metal hemisphere has a diameter of 50 cm. A second hemisphere made of the same metal has a diameter of 25 cm. If the second hemisphere has a mass of 600 kg, the mass of the first hemisphere is

a 1200 kg b 2400 kg c 4800 kg

Chapter 18 **Graphs**

In newspapers, business reports and government publications use is made of pictorial illustrations to present and compare quantities of the same kind. These diagrams help the reader to understand what deductions can be drawn from the quantities represented in the diagrams. The most common form of diagram is the graph.

Axes of Reference

To plot a graph we take two lines at right angles to each other (Fig. 18.1). These lines are called the axes of reference. Their intersection, the point O, is called the origin.

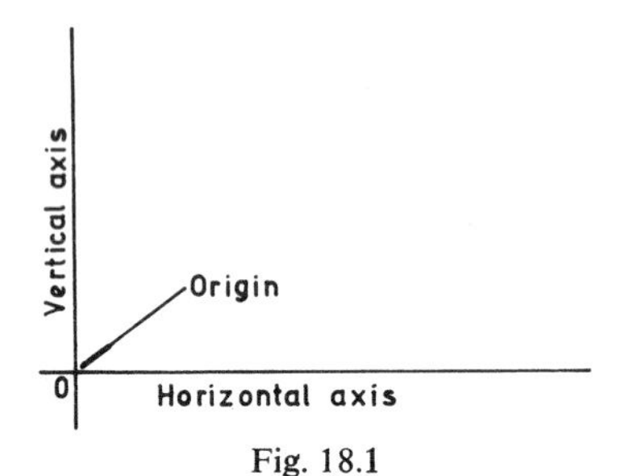

Fig. 18.1

Scales

The number of units represented by a unit length along an axis is called the scale. For instance 1 cm could represent 2 units. The scales need not be the same on both axes.

Co-ordinates

Co-ordinates are used to mark the points of a graph. In Fig. 18.2 values of x are to be plotted against values of y. The point P has been plotted so that $x=8$ and $y=10$. The values of 8 and 10 are said to be the rectangular co-ordinates of the point P. We then say that P is the point (8, 10).

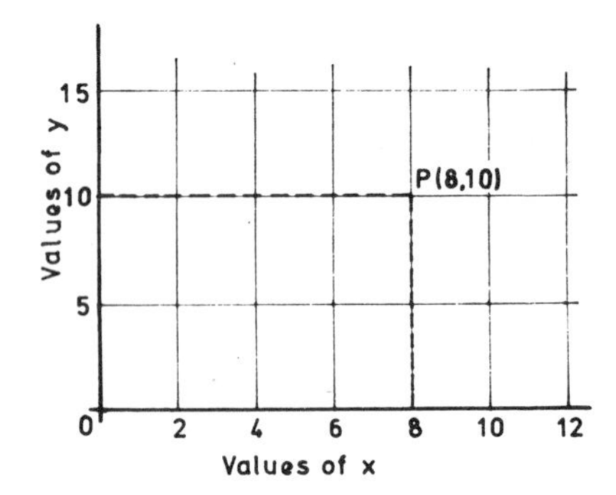

Fig. 18.2

DRAWING A GRAPH

Every graph shows a relation between two sets of numbers. The table below gives corresponding values of x and y.

x	0	2	4	6	8
y	0	4	16	36	64

To plot the graph we first draw the two axes of reference. Values of x are always plotted along the horizontal axis and values of y along the vertical axis. We next choose suitable scales. In Fig. 18.3 we have chosen 1 cm = 2 units along the horizontal axis and 1 cm = 10 units along the vertical axis. On plotting the graph we see that it is a smooth curve which passes through all the plotted points.

When a graph is either a straight line or a smooth curve we can use the graph to deduce corresponding values of x and y between those given in the table.

To find the value of y corresponding to $x=3$, find 3 on the horizontal axis and draw a vertical line to meet the graph at point P

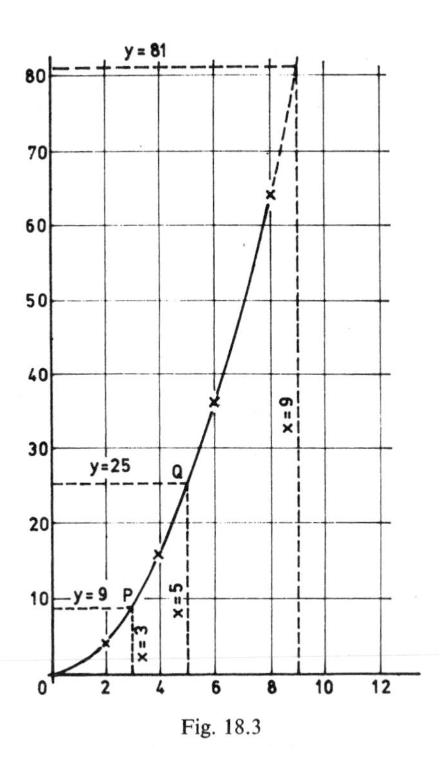

Fig. 18.3

(Fig. 18.3). From P draw a horizontal line to meet the vertical axis and read off the value which is 9. Thus when $x=3$, $y=9$. To find the value of x corresponding to $y=25$ find 25 on the vertical axis and draw a horizontal line to meet the graph at point Q. From Q draw a vertical line to meet the horizontal axis and read off this value which is 5. Thus when $y=25$, $x=5$.

Using a curve in this way to find values which are not given in the table is called *interpolation.* If we extend the curve so that it follows the general trend we can estimate values of x and y which lie *just beyond* the range of the given values. Thus in Fig. 18.3 by extending the curve we can find the probable value of y when $x=9$. This is found to be 81. Finding a probable value in this way is called *extrapolation.* An extrapolated value can usually be relied upon, but in certain cases it may contain a substantial amount of error. Extrapolated values must therefore be used with care. It must be clearly understood that interpolation can only be used if the graph is a smooth curve or a straight line. It is no use applying interpolation in the graph of the next example.

Example

The table below gives the temperature at 12.00 noon on seven successive days. Plot a graph to illustrate this information.

Day	June	1	2	3	4	5
Temp. °C		16	20	16	18	22

Day	June	6	7
Temp. °C		15	16·5

As before we draw two axes at right-angles to each other, indicating the day on the horizontal axis. Since the temperatures range from 15°C to 22°C we can make 14°C (say) our starting point on the vertical axis. This will allow us to use a larger scale on that axis which makes for greater accuracy in plotting the graph.

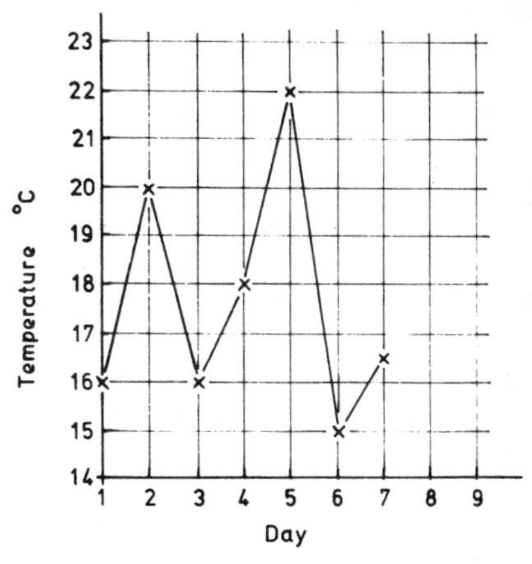

Fig. 18.4

On plotting the points (Fig. 18.4) we see that it is impossible to join the points by means of a smooth curve. The best we can do is to join the points by means of a series of straight lines. The rise and fall of temperatures do not follow any mathematical law and the graph shows this by means of the erratic line obtained. However the graph does present in pictorial form the variations in temperature and at a glance we can see that the 1st, 3rd and 6th June were cool days whilst the 2nd and 5th were warm days.

Exercise 62

1) The table below gives particulars of the amount of steel delivered to a factory during successive weeks. Plot a graph to show this with the week number on the horizontal axis.

Week number	1	2	3
Amount delivered (kg)	25 000	65 000	80 000

Week number	4	5
Amount delivered (kg)	30 000	50 000

2) The table below gives corresponding values of x and y. Plot a graph and from it estimate the value of y when $x=1{\cdot}5$ and the value of x when $y=30$.

x	0	1	2	3	4	5
y	3	5	11	21	35	63

3) The areas of circles for various diameters is shown in the table below. Plot a graph with diameter on the horizontal axis and from it estimate the area of a circle whose diameter is 18 cm.

Diameter (cm)	5	10	15	20	25
Area (cm^2)	19·6	78·5	176·6	314·2	492·2

4) The values in the table below are corresponding values of two quantities i and v.

v	15	25	35	50	70
i	1·1	2·0	2·5	3·2	3·9

Plot a graph with i horizontal and find v when $i=3{\cdot}0$.

5) An electric train starts from A and travelled to its next stop 6 km from A. The following readings were taken of the time since leaving A (in minutes) and the distance from A (in km).

Time	$\frac{1}{2}$	1	$1\frac{1}{2}$	2	$2\frac{1}{2}$	3
Distance	0·10	0·34	0·8	1·46	2·46	3·50

Time	$3\frac{1}{2}$	4	$4\frac{1}{2}$	5	$5\frac{1}{2}$	6
Distance	4·34	5·0	5·44	5·74	5·92	6

Draw a graph of these values taking time horizontally. From the graph estimate the time taken to travel 2 km from A.

GRAPHS OF SIMPLE EQUATIONS

Consider the equation:

$$y=2x+5$$

We can give x any value we please and so calculate a corresponding value for y. Thus,

when $x=0$ $\quad y=2\times0+5=5$
when $x=1$ $\quad y=2\times1+5=7$
when $x=2$ $\quad y=2\times2+5=9$ and so on.

The value of y therefore depends on the value allocated to x. We therefore call y the

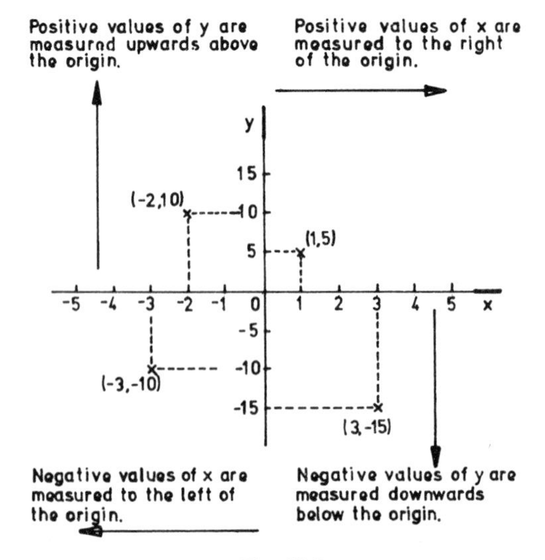

Fig. 18.5

dependent variable. Since we can give x any value we please, we call x the *independent variable*. It is usual to mark the values of the independent variable along the horizontal axis and this axis is frequently called the x-axis. The values of the dependent variable are then marked off along the vertical axis which is often called the y-axis.
In plotting graphs representing equations we may have to include co-ordinates which are positive and negative. To represent these on a graph we make use of the number scales used in directed numbers (Fig. 18.5).

Examples

1) Draw the graph of $y=2x-5$ for values of x between -3 and 4.
Having decided on some values for x we calculate the corresponding values for y by substituting in the given equation. Thus,

when $x=-3$,

$$y=2\times(-3)-5=-6-5=-11$$

For convenience the calculations are tabulated as shown below.

x	-3	-2	-1	0
$2x$	-6	-4	-2	0
-5	-5	-5	-5	-5
$y=2x-5$	-11	-9	-7	-5

x	1	2	3	4
$2x$	2	4	6	8
-5	-5	-5	-5	-5
$y=2x-5$	-3	-1	1	3

A graph may now be plotted using these values of x and y (Fig. 18.6). The graph is a straight line. Equations of the type $y=2x-5$, where the highest powers of the variables, x and y, is the first are called equations of the *first degree*. All equations of this type give graphs which are straight lines and hence they are often called *linear equations*. In order to draw graphs of linear equations we need only take two points. It is safer, however, to take three points, the third point acting as a check on the other two.

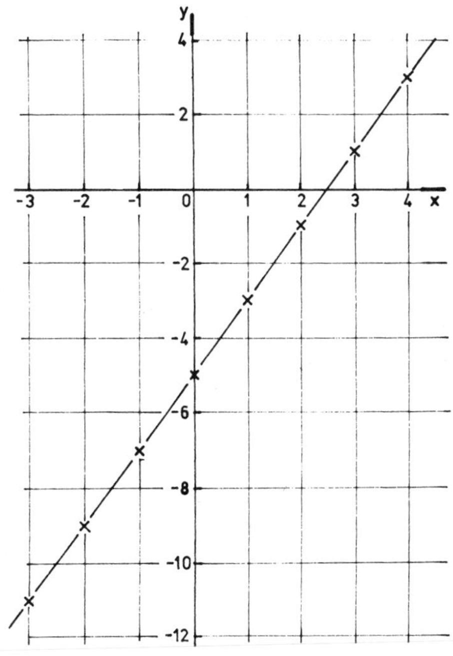

Fig. 18.6

2) By means of a graph show the relationship between x and y in the equation $y=5x+3$. Plot the graph between $x=-3$ and $x=3$.
Since this is a linear equation we need only take three points.

x	-3	0	$+3$
$y=5x+3$	-12	3	$+18$

The graph is shown in Fig. 18.7.

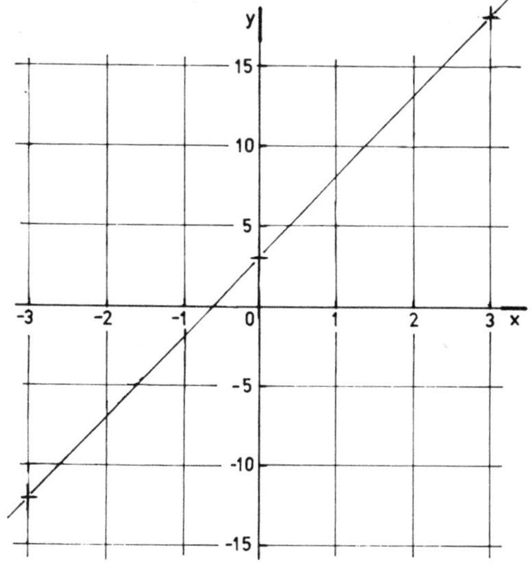

Fig. 18.7

THE EQUATION OF A STRAIGHT LINE

Every linear equation may be written in the standard form:

$$y=mx+c$$

Hence $y=2x-5$ is in the standard form with $m=2$ and $c=-5$.
The equation $y=4-3x$ is in standard form if we rearrange it to give $y=-3x+4$ so that we see $m=-3$ and $c=4$.

THE MEANING OF m AND c IN THE EQUATION OF A STRAIGHT LINE

The point B is any point on the straight line shown in Fig. 18.8 and it has the co-ordinates x and y. Point A is where the line cuts the y-axis and it has co-ordinates $x=0$ and $y=c$.

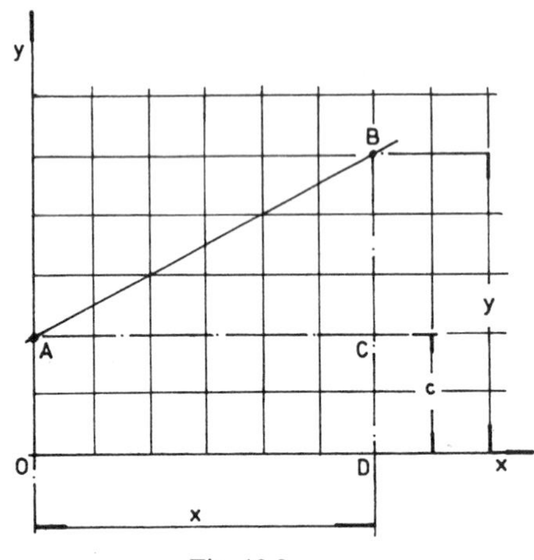

Fig. 18.8

$\frac{BC}{AC}$ is called the gradient of the line

now

$$BC=\frac{BC}{AC}\times AC=AC\times \text{gradient of the line}$$

$$\begin{aligned}y&=BC+CD=BC+AO\\&=AC\times \text{gradient of the line}+AO\\&=x\times \text{gradient of the line}+c\end{aligned}$$

But $y=mx+c$
Hence it can be seen that:

$m=$ gradient of the line
$c=$ intercept on the y-axis.

Figure 18.9 shows the difference between positive and negative gradients.

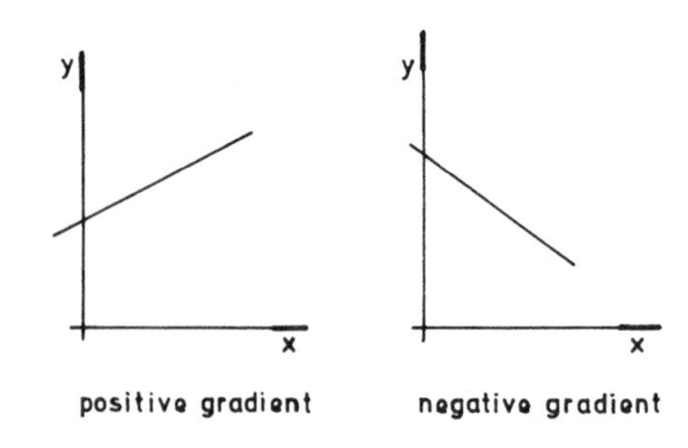

Fig. 18.9

Examples
1) Find the law of the straight line shown in Fig. 18.10.

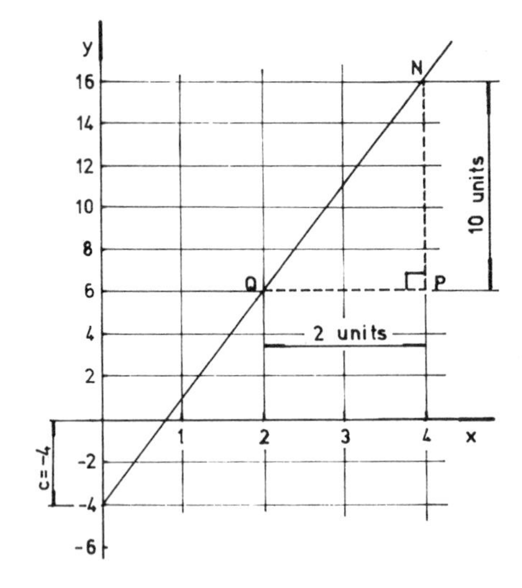

Fig. 18.10

Since the origin is at the intersection of the axes, c is the intercept on the y axis. From Fig. 18.10 it will be seen that $c=-4$. We now have to find m. Since this is the gradient of the line we draw Δ QNP making the sides reasonably long since a small triangle will give very inaccurate results. Using the scales of x and y we see that $QP=2$ units and $PN=10$ units.

$$\therefore \; m=\frac{NP}{QP}=\frac{10}{2}=5$$

$\therefore$ The standard equation of a straight line $y=mx+c$ becomes $y=5x-4$

2) Find the values of m and c if the straight line $y=mx+c$ passes through the point $(-1, 3)$ and has a gradient of 6.
Since the gradient is 6 we have $m=6$

$$\therefore \; y=6x+c$$

Since the line passes through the point $(-1, 3)$ we have $y=3$ when $x=-1$. By substitution,

$$3=6\times(-1)+c$$
$$3=-6+c$$
$$\therefore \; c=9$$

Hence $y=6x+9$

3) Find the law of the straight line shown in Fig. 18.11.

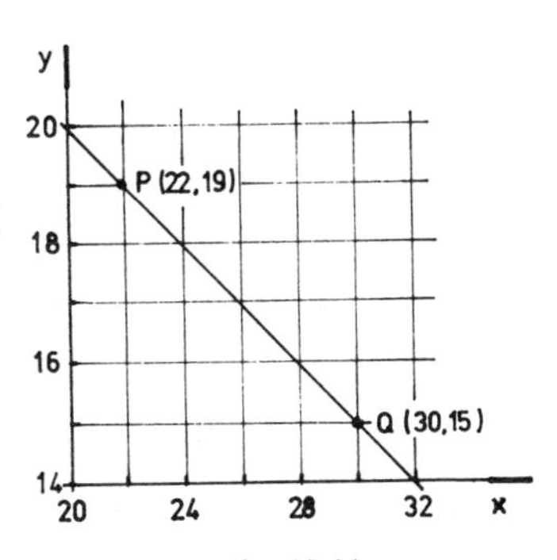

Fig. 18.11

It will seem from Fig. 18.11 that the *origin is not at the intersection of the axes.* In order to determine the law of the straight line we use two simultaneous equations as follows: Choose two convenient points P and Q and find their co-ordinates (these two points should be as far apart as possible to get maximum accuracy). If a point lies on a line then the x and y values of that point must satisfy the equation:

$$y=mx+c$$

at point P, $x=22$ and $y=19$

$$\therefore \; 19=22m+c \qquad ...(1)$$

at point Q, $x=30$ and $y=15$

$$15=30m+c \qquad ...(2)$$

Subtracting equation (2) from equation (1),

$$4=-8m$$

$$\therefore \; m=\frac{-8}{4}$$

$$m=-0{\cdot}5$$

Substituting $m=-0{\cdot}5$ in equation (1),

$$19=22\times(-0{\cdot}5)+c$$
$$19=-11+c$$
$$c=30$$

Thus the equation of the line shown in Fig. 18.11 is

$$y=-0{\cdot}5x+30$$

4) Find the values of m and c if the straight line $y=mx+c$ passes through the points $(3, 4)$ and $(7, 10)$.

$$y=mx+c$$

The first point has co-ordinates $x=3$, $y=4$. Hence

$$4=3m+c \qquad ...(1)$$

The second point has co-ordinates $x=7$, $y=10$. Hence

$$10=7m+c \qquad ...(2)$$

Subtracting equation (1) from equation (2),

$$6=4m$$
$$\therefore \; m=1{\cdot}5$$

Substituting for $m=1{\cdot}5$ in equation (1),

$$4=4{\cdot}5+c$$
$$\therefore \; c=-0{\cdot}5$$

The equation of the straight line is

$$y=1{\cdot}5x-0{\cdot}5$$

EXPERIMENTAL DATA

One of the most important applications of the straight-line equation is the determination of an equation connecting two quantities when values have been obtained from an experiment.

Example

In an experiment carried out with a lifting machine the effort E and the load W were found to have the values given in the table below:

W (kg)	15	25	40	50	60
E (kg)	2·75	3·80	5·75	7·00	8·20

Plot these results and obtain the equation connecting E and W which is thought to be of the type $E=aW+b$.

If E and W are connected by an equation of the type $E=aW+b$ then the graph must be a straight line. Note that when plotting the graph, W is the independent variable and must be plotted on the horizontal axis. E is the dependent variable and must be plotted on the vertical axis.

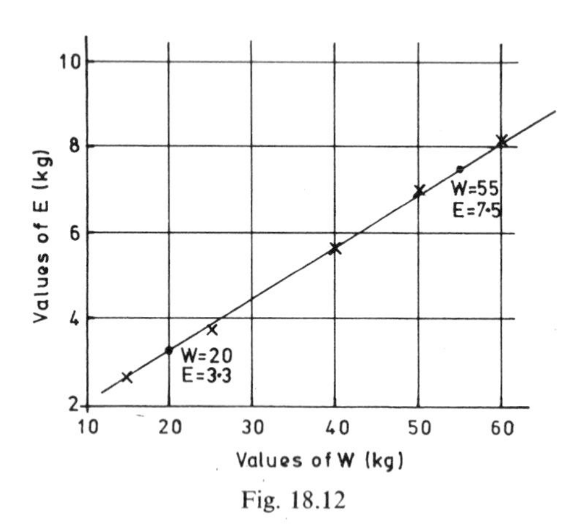

Fig. 18.12

On plotting the points (Fig. 18.12) it will be noticed that they deviate only slightly from a straight line. Since the data are experimental we must expect errors in measurement and observation and hence slight deviations from a straight line must be expected. Although the straight line will not pass through some of the points an attempt must be made to ensure an even spread of the points above and below the line.

To determine the equation we choose two points which lie on the straight line. Do not use any of the experimental results from the table unless they happen to lie exactly on the line. Choose the points as far apart as is convenient because this will help the accuracy of your result.

The point $W=55$, $E=7{\cdot}5$ lies on the line. Hence

$$7{\cdot}5=55a+b \qquad \ldots(1)$$

The point $W=20$, $E=3{\cdot}3$ also lies on the line Hence

$$3{\cdot}3=20a+b \qquad \ldots(2)$$

Subtracting equation (2) from equation (1),

$$4{\cdot}2=35a$$
$$a=0{\cdot}12$$

Substituting for $a=0{\cdot}12$ in equation (2),

$$3{\cdot}3=20\times0{\cdot}12+b$$
$$b=0{\cdot}9$$

The required equation connecting E and W is therefore

$$E=0.12W+0.9$$

Exercise 63

Draw graphs of the following simple equations:

1) $y=x+2$ taking values of x between -3 and 2.

2) $y=2x+5$ taking values of x between -4 and 4.

3) $y=3x-4$ taking values of x between -4 and 3.

4) $y=5-4x$ taking values of x between -2 and 4.

The following equations represent straight lines. State in each case the gradient of the line and the intercept on the y-axis.

5) $y=x+3$

6) $y=-3x+4$

7) $y=-5x-2$

8) $y=4x-3$

9) Find the values of m and c if the straight line $y=mx+c$ passes through the point $(-2, 5)$ and has a gradient of 4.

10) Find the values of m and c if the straight line $y=mx+c$ passes through the point $(3, 4)$ and the intercept on the y-axis is -2.

In the following find the values of m and c if the straight line $y=mx+c$ passes through the given points:

11) $(-2, -3)$ and $(3, 7)$

12) $(1, 1)$ and $(2, 4)$

13) $(-2, 1)$ and $(3, -9)$

14) $(-3, 13)$ and $(1, 1)$

15) $(2, 17)$ and $(4, 27)$

16) The following table gives values of x and y which are connected by an equation of the type $y=ax+b$. Plot the graph and from if find the values of a and b.

x	2	4	6	8	10	12
y	10	16	22	28	34	40

17) The following observed values of P and Q are supposed to be related by the linear equation $P=aQ+b$, but there are experimental errors. Find by plotting the graph the most probable values of a and b.

Q	2·5	3·5	4·4	5·8
P	13·6	17·6	22·2	28·0

Q	7·5	9·6	12·0	15·1
P	35·5	47·4	56·1	74·6

18) In an experiment carried out with a machine the effort E and the load W were found to have the values given in the table below. The equation connecting E and W is thought to be of the type $E=aW+b$. By

plotting the graph check if this is so and hence find a and b.

W (kg)	10	30	50	60	80	100
E (kg)	8·9	19·1	29	33	45	54

19) A test on a metal filament lamp gave the following values of resistance (R ohms) at various voltages (V volts).

V	62	75	89	100	120
R	100	117	135	149	175

These results are expected to agree with an equation of the type $R=mV+c$ where m and c are constants. Test this by drawing the graph and find suitable values for m and c.

20) During an experiment to verify Ohm's Law the following results were obtained.

E (volts)	0	1·0	2·0	2·5	3·7
I (amperes)	0	0·24	0·5	0·63	0·92

E (volts)	4·1	5·9	6·8	8·0
I (amperes)	1·05	1·48	1·70	2·05

Plot these values with I horizontal and find the equation connecting E and I.

GRAPHS OF QUADRATIC FUNCTIONS

The expression ax^2+bx+c where a, b and c are constants is called a quadratic function of x. When plotted, quadratic functions always give a smooth curve known as a parabola.

Example

Plot the graph of $y=3x^2+10x-8$ between $x=-6$ and $x=4$.

A table may be drawn up as follows giving corresponding values of y for chosen values of x.

x	-6	-5	-4	-3	-2	-1
$3x^2$	108	75	48	27	12	3
$10x$	-60	-50	-40	-30	-20	-10
-8	-8	-8	-8	-8	-8	-8
y	40	17	0	-11	-16	-15

(cont.)

x	0	1	2	3	4
$3x^2$	0	3	12	27	48
$10x$	0	10	20	30	40
-8	-8	-8	-8	-8	-8
y	-8	5	24	49	80

The graph is shown in Fig. 18.13 and it is a smooth curve. Equations which are non-linear always give a graph which is a smooth curve.

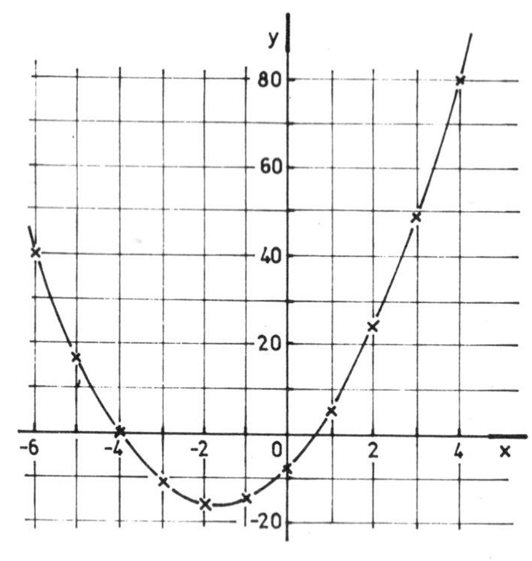

Fig. 18.13

Note that the gradient of curve is explained on page 148.

SOLUTION OF EQUATIONS

An equation may be solved by means of a graph. The following example shows the method.

Examples

1) Plot the graph of $y=6x^2-7x-5$ between $x=-2$ and $x=3$. Hence solve the equation $6x^2-7x-5=0$

A table is drawn up as follows.

x	-2	-1	0	1	2	3
y	33	8	-5	-6	5	28

The curve is shown in Fig. 18.14. To solve the equation $6x^2-7x-5=0$ we have to find the values of x when $y=0$. That is, we have

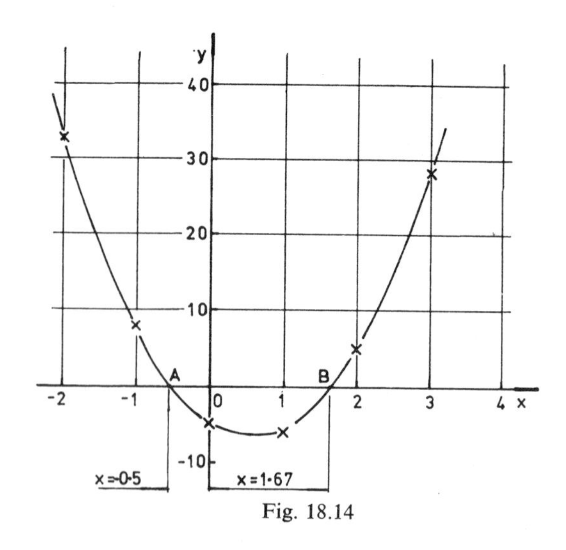

Fig. 18.14

to find the values of x where the graph cuts the x-axis. These are points A and B in Fig. 18.14 and hence the solutions are

$$x=-0{\cdot}5 \quad \text{or} \quad x=1{\cdot}67$$

2) Plot the graph of $y=2x^2-x-6$ and hence solve the equations a) $2x^2-x-6=0$ b) $2x^2-x-4=0$ c) $2x^2-x-9=0$. Take values of x between -4 and 6.

To plot $y=2x^2-x-6$ draw up a table values as shown below:

x	-4	-3	-2	-1	0
y	30	15	4	-3	-6

x	1	2	3	4	5	6
y	-5	0	9	22	39	60

a) The graph is plotted as shown in Fig. 18.15. The curve cuts the x-axis, i.e. where $y=0$, at the points where $x=-1{\cdot}5$ and $x=2$. Hence the solutions of the equation $2x^2-x-6=0$ are,

$$x=-1{\cdot}5 \quad \text{or} \quad x=2$$

b) The equation $2x^2-x-4$ may be written in the form

$$2x^2-x-6=-2$$

Hence if we find the values of x when $y=-2$ we shall obtain the solutions required. These are where the line $y=-2$

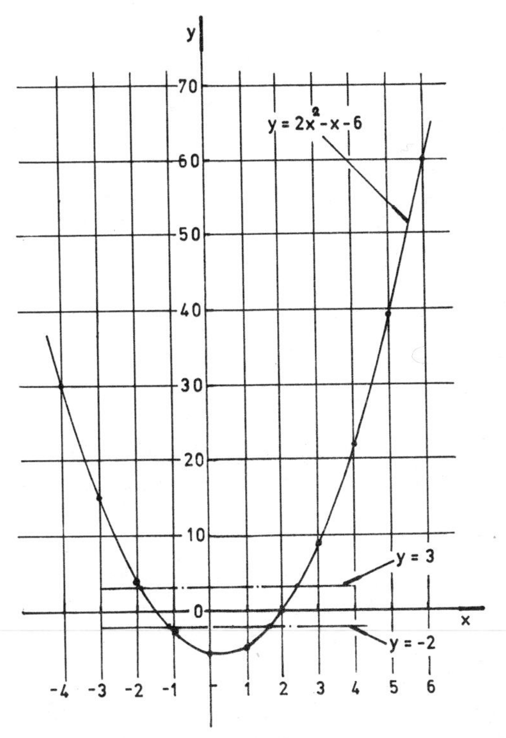

Fig. 18.15

cuts the curve (see Fig. 18.15). The solutions are therefore

$x = -1{\cdot}18$ or $1{\cdot}69$

c) The equation $2x^2 - x - 9 = 0$ may be written in the form

$2x^2 - x - 6 = 3$

Hence by drawing the line $y = 3$ and finding where it cuts the curve we shall obtain the solutions. They are

$x = -1{\cdot}89$ or $2{\cdot}36$

Exercise 64

Plot the graphs of the following equations:

1) $y = 2x^2 - 7x - 5$ between $x = -4$ and $x = 12$

2) $y = x^2 - 4x + 4$ between $x = -3$ and $x = 3$

3) $y = 6x^2 - 11x - 35$ between $x = -3$ and $x = 5$

4) $y = 3x^2 - 5$ between $x = -2$ and $x = 4$

5) $y = 1 + 3x - x^2$ between $x = -2$ and $x = 3$

By plotting suitable graphs solve the following equations:

6) $x^2 - 7x + 12 = 0$ (take values of x between 0 and 6)

7) $x^2 + 16 = 8x$ (take values of x between 1 and 7)

8) $x^2 - 9 = 0$ (take values of x between -4 and 4)

9) $3x^2 + 5x = 60$ (take values of x between 0 and 4)

10) Plot the graph of $y = x^2 + 7x + 3$ taking values of x between -12 and 2. Hence solve the equations:

a) $x^2 + 7x + 3 = 0$
b) $x^2 + 7x - 2 = 0$
c) $x^2 + 7x + 6 = 0$

11) Draw the graph of $y = 1 - 2x - 3x^2$ between $x = -4$ and $x = 4$. Hence solve the equations:

a) $1 - 2x - 3x^2 = 0$
b) $3 - 2x - 3x^2 = 0$
c) $9x^2 + 6x = 6$

12) Draw the graph of $y = x^2 - 9$ taking values of x between -5 and 5. Hence solve the equations:

a) $x^2 - 9 = 0$
b) $x^2 - 5 = 0$
c) $x^2 + 6 = 0$

INTERSECTING GRAPHS

Equations may also be solved graphically by using intersecting graphs. The method is shown in the following example.

Example
Plot the graph of $y = 2x^2$ and use it to solve the equation $2x^2 - 3x - 2 = 0$. Take values of x between -2 and 4.

The equation $2x^2-3x-2=0$ can be solved graphically by the method used in earlier examples, but the alternative method shown here is often preferable. The equation $2x^2-3x-2=0$ may be written in the form $2x^2=3x+2$. We now plot on the same axes and to the same scales the graphs

$y=2x^2$ and $y=3x+2$

x	-2	-1	0	1	2	3	4
$y=2x^2$	8	2	0	2	8	18	32
$y=3x+2$	-4		2				14

Note that to plot $y=3x+2$ we need only three points since this is a linear equation. The graphs are shown plotted in Fig. 18.16.

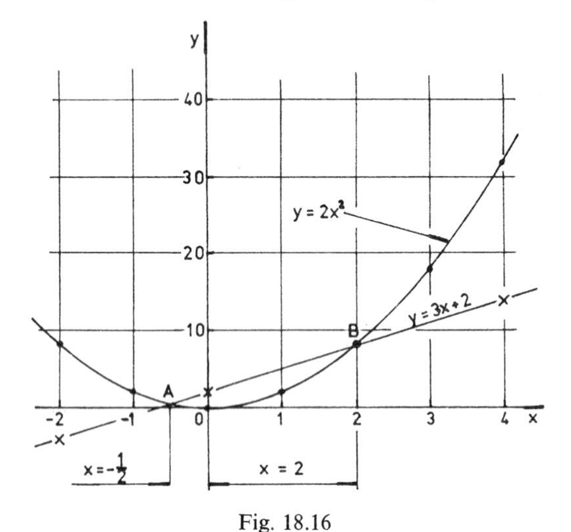

Fig. 18.16

At the points of intersection of the curve and the line (points A and B in Fig. 18.16) the y value of $2x^2$ is the same as the y value of $3x+2$. Therefore at these points the equation $2x^2=3x+2$ is satisfied. The required values of x may now be found by inspection of the graph. They are at A, where $x=-\frac{1}{2}$ and at B, where $x=2$. The required solutions are therefore $x=-\frac{1}{2}$ or $x=2$.

GRAPHICAL SOLUTIONS OF SIMULTANEOUS EQUATIONS

The method is shown in the following examples.

Examples

1) Solve graphically

$$y-2x=2 \quad \ldots(1)$$

$$3y+x=20 \quad \ldots(2)$$

Equation (1) may be written as:

$$y=2+2x$$

Equation (2) may be written as:

$$y=\frac{20-x}{3}$$

Drawing up the following table we can plot the two equations on the *same axes.*

x	-3	0	3
$y=2+2x$	-4	2	8
$y=\dfrac{20-x}{3}$	7·7	6·7	5·7

The solutions of the equations are the coordinates of the point where the two lines cross (that is, point P in Fig. 18.17). The co-ordinates of P are $x=2$ and $y=6$. Hence the solutions of the given equations are

$$x=2 \quad \text{and} \quad y=6$$

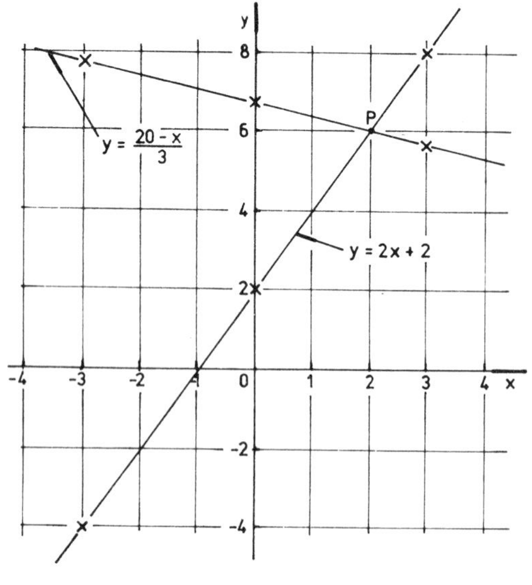

Fig. 18.17

2) Draw the graph of $y=(3+2x)(3-x)$ for values of x from $-1\frac{1}{2}$ to 3. On the same axes, and with the same scales, draw the graph of $3y=2x+14$. From your graphs determine the values of x for which $3(3+2x)(3-x)=2x+14$.
To plot the graph of $y=(3+2x)(3-x)$ we draw up the following table.

x	$-1\frac{1}{2}$	-1	$-\frac{1}{2}$	0
$y=(3+2x)(3-x)$	0	4	7	9

x	$\frac{1}{2}$	1	$1\frac{1}{2}$	2	$2\frac{1}{2}$	3
$y=(3+2x)(3-x)$	10	10	9	7	4	0

The equation $3y=2x+14$ may be rewritten as

$$y=\frac{2x+14}{3}$$

To draw this graph we need only take three points since it is a linear equation.

x	-1	1	3
$y=\dfrac{2x+14}{3}$	4	$5\frac{1}{3}$	$6\frac{2}{3}$

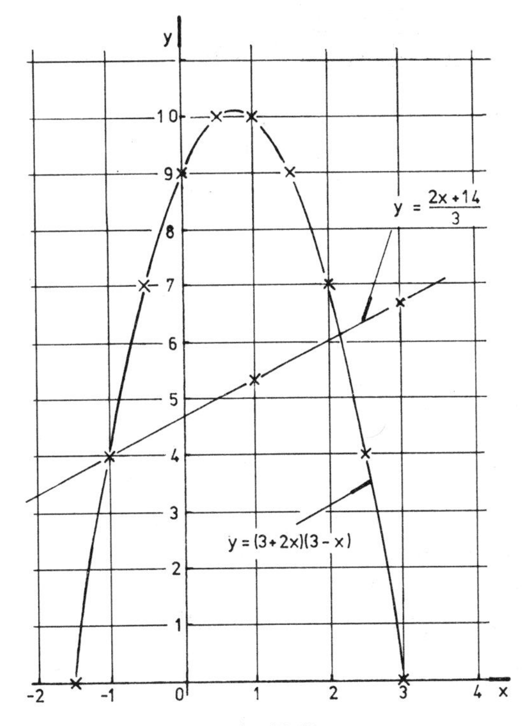

Fig. 18.18

The graphs are shown in Fig. 18.18. Since the equation $3(3+2x)(3-x)=2x+14$ may be rewritten to give

$$(3+2x)(3-x)=\frac{2x+14}{3}$$

the co-ordinates where the curve and the line intersect give the solutions which are:

$$x=-1 \quad \text{and} \quad x=2{\cdot}17$$

Exercise 65

1) Plot the graph of $y=3x^2$ taking values of x between -3 and 4. Hence solve the following equations:

a) $3x^2=4$
b) $3x^2-2x-3=0$
c) $3x^2-7x=0$

2) Plot the graph of $y=x^2+8x-2$ taking values of x between -12 and 2. On the same axes, and to the same scale, plot the graph of $y=2x-1$. Hence find the values of x which satisfy the equation $x^2+8x-2=2x-1$.

Solve graphically the following simultaneous equations:

3) $2x-3y=5$; $x-2y=2$

4) $7x-4y=37$; $6x+3y=51$

5) $\dfrac{x}{2}+\dfrac{y}{3}=\dfrac{13}{6}$; $\dfrac{2x}{7}-\dfrac{y}{4}=\dfrac{5}{14}$

6) If $y=x^2(15-2x)$ construct a table of values of y for values of x from -1 to $1\frac{1}{2}$ at half-unit intervals. Hence draw the graph of this function. Using the same axes and scales draw the straight line $y=10x+10$. Write down and simplify an equation which is satisfied by the values of x where the two graphs intersect. From your graph find the approximate value of the two roots of this equation.

7) Write down the three values missing from the following table which gives values of $2x^3+x+3$ for values of x from -2 to 2.

x	$-2{\cdot}0$	$-1{\cdot}5$	$-1{\cdot}0$	$-0{\cdot}5$
$2x^3+x+3$	$-15{\cdot}0$	$-5{\cdot}25$		$2{\cdot}25$

x	0	$0{\cdot}5$	$1{\cdot}0$	$1{\cdot}5$	$2{\cdot}0$
$2x^3+x+3$		$3{\cdot}75$	$6{\cdot}0$		$21{\cdot}0$

Using the same axes draw the graphs of $y=2x^3+x+3$ and $y=9x+3$. Use your graphs to write down

a) the range of values of x for which $2x^3+x+3$ is less than $9x+3$;

b) the solution of the equation $2x^3+x+3=5$. Write down and simplify the equation which is satisfied by the values of x at the points of intersection of the two graphs.

8) Write down the three values missing from the following table which gives values, correct to two decimal places, of $6-\dfrac{10}{2x+1}$ for values of x from 0·25 to 5.

x	$0{\cdot}25$	$0{\cdot}5$	1	$1{\cdot}5$
$6-\dfrac{10}{2x+1}$	$-0{\cdot}67$	$1{\cdot}00$	$2{\cdot}67$	$3{\cdot}50$

x	2	3	3.5	4	$4{\cdot}5$	5
$6-\dfrac{10}{2x+1}$		$4{\cdot}57$		$4{\cdot}89$		$5{\cdot}09$

Using the same axes draw the graphs of $y=6-\dfrac{10}{2x+1}$ and $y=x+1$. Use your graphs to solve the equation $2x^2-9x+5=0$.

9) If $y=\dfrac{x+10}{x+1}$ construct a table of values of y when $x=0$, 1, 2, 3, 4, 5. Draw the graph of this function and also using the same axes and scales draw the graph of $y=x-1$. Write down, and simplify, an equation which is satisfied by the value of x where the graphs intersect. From your graphs find the approximate value of the root of this equation.

10) Calculate the values of $\dfrac{x^2}{4}+\dfrac{24}{x}-12$ which are omitted from the table below.

x	2	$2{\cdot}5$	3	$3{\cdot}5$	4
$\dfrac{x^2}{4}+\dfrac{24}{x}-12$		$-0{\cdot}84$		$-2{\cdot}08$	

x	$4{\cdot}5$	5	$5{\cdot}5$	6
$\dfrac{x^2}{4}+\dfrac{24}{x}-12$	$-1{\cdot}60$	$-0{\cdot}95$	$-0{\cdot}07$	$1{\cdot}00$

Draw the graph of $y=\dfrac{x^2}{4}+\dfrac{24}{x}-12$ from $x=2$ to $x=6$. Using the same scales and axes draw the graph of $y=\dfrac{x}{3}-2$. Write down, but do not simplify, an equation which is satisfied by the values of x where the graphs intersect. From your graphs find approximate values for the two roots of this equation.

SELF TEST 18

In questions 1 to 20 the answer is either "true" or "false".

1) The intersection of the two axes of reference, used when plotting a graph, is called the origin. _ _ _ _ _

2) When a graph is a straight line it means that there is a definite law connecting the two quantities which are plotted. _ _ _ _ _

3) When a graph is a smooth curve it means that there is not a definite law connecting the two quantities which are plotted. _ _ _ _ _

4) Interpolation means using a graph to find values which are not given in the table from which the graph is drawn. _ _ _ _ _

5) In order to extrapolate the graph is extended just beyond the range of the values from which the graph was plotted. _ _ _ _ _

6) The co-ordinates of the point shown in Fig. 18.19 are (3, 5). - - - - -

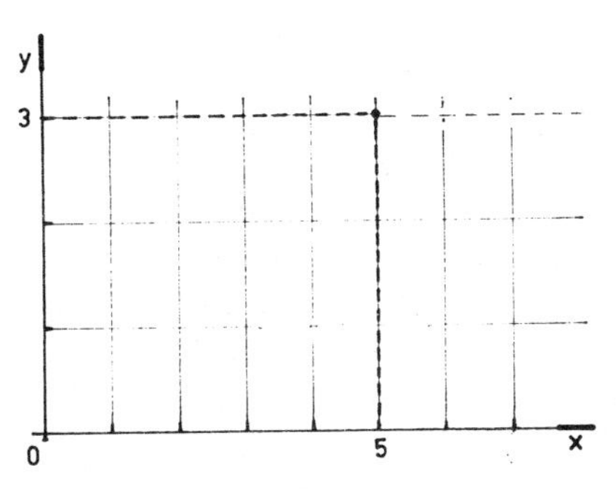

Fig. 18.19

7) The co-ordinates of the point shown in Fig. 18.20 are (2, 3). - - - - -

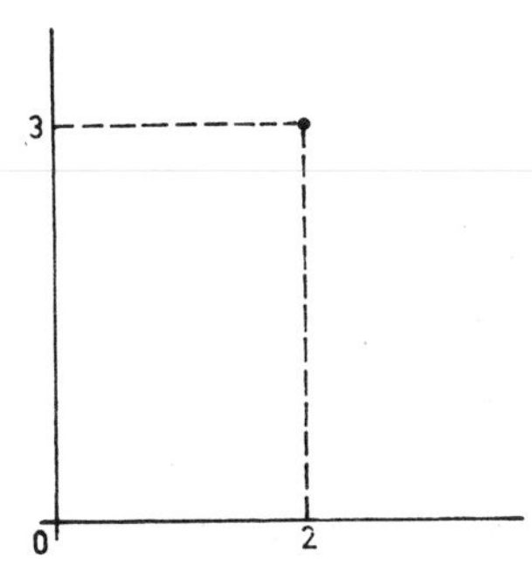

Fig. 18.20

8) When the co-ordinates of a point are stated as (3, 6) it means that $x=3$ and $y=6$. - - - - -

9) When the co-ordinates of a point are stated as $(-2, 4)$ it means that $y=-2$ when $x=4$. - - - - -

10) The equation $y=3x+7$ will give a graph which is a curve. - - - - -

11) The equation $y=3-5x$ will give a graph which is a straight line. - - - - -

12) The equation $p=\frac{5}{q}$ will give a graph which is a straight line. - - - - -

13) The equation $y=8-\frac{3}{x}$ will give a graph which is a straight line. - - - - -

14) The equation $y=3+x^2$ will give a graph which is a curve. - - - - -

15) The equation $y=3-2x^3$ will give a graph which is a curve. - - - - -

16) When drawing the graph of $y=5x^2+7x+8$ values of y are plotted on the vertical axis. - - - - -

17) When drawing the graph of $M=q^2+3$ values of q are plotted on the vertical axis. - - - - -

18) When $r=3s+7$, r is called the independent variable. - - - - -

19) When $q=7p-8$, p is called the independent variable. - - - - -

20) When $V=8r^3$, V is called the dependent variable. - - - - -

In questions 21 to 28 state the letter (or letters) which correspond to the correct answer (or answers).

21) The graph of $y=3+2x$ will look like one of the following diagrams (Fig. 18.21).

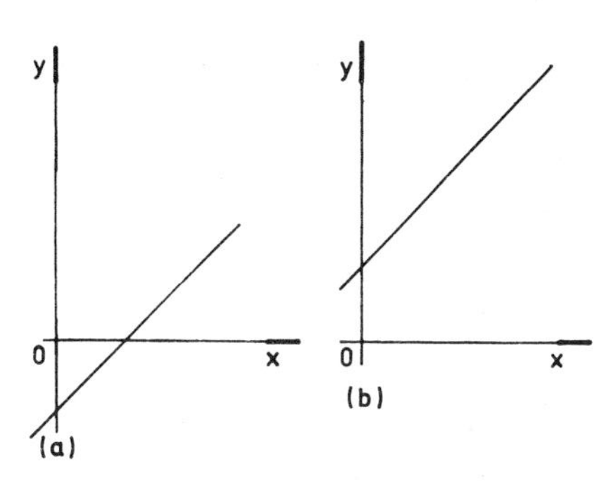

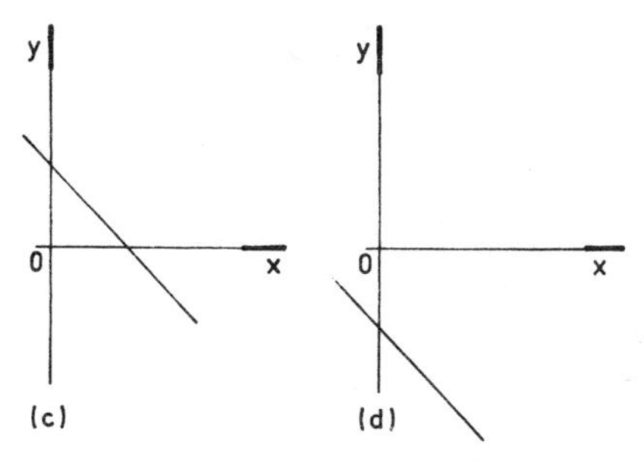

Fig. 18.21

22) The graph of $y=5-3x$ will look like one of the following diagrams (Fig. 18.22).

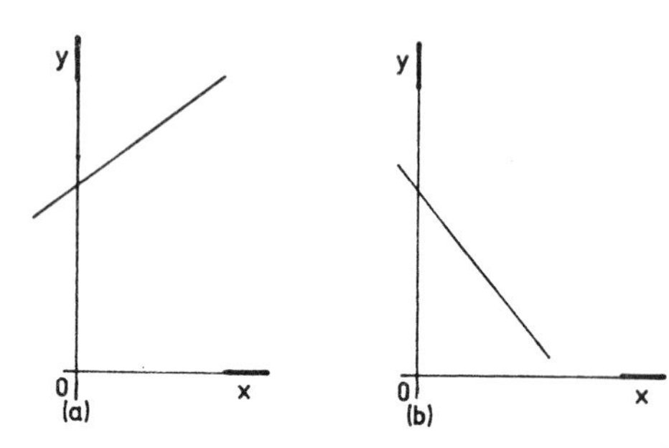

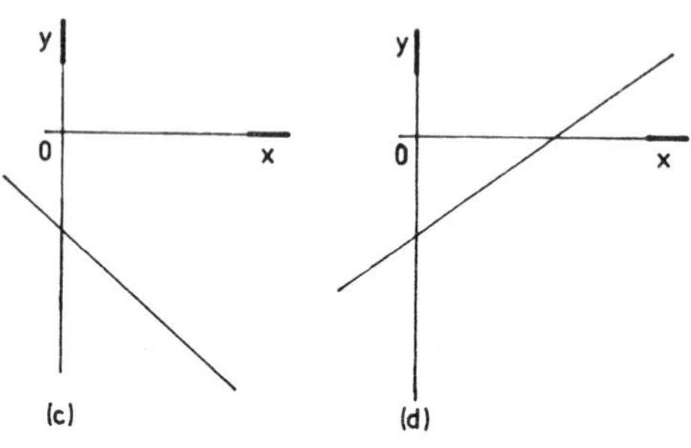

Fig. 18.22

23) A straight line passes through the points (0, 1) and (2, 7). The law of the line is therefore

a $y=3x+1$ b $y=3x-1$

c $y=\frac{3}{7}x+1$ d $y=\frac{3}{7}x-1$

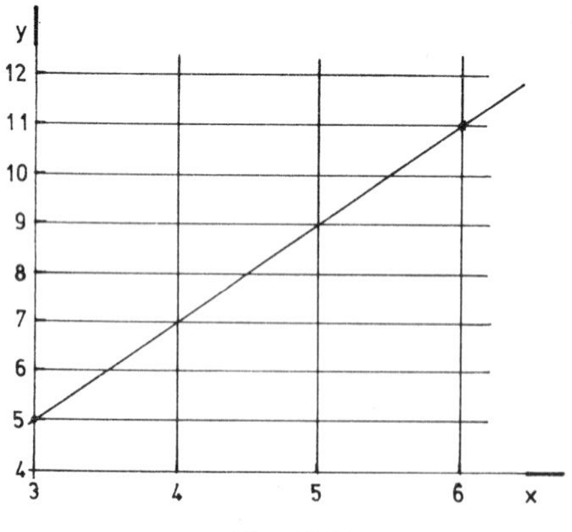

Fig. 18.23

24) The law of the line shown in Fig. 18.23 is

a $y=2x+5$ b $y=5-2x$

c $y=2x-1$ d $y=\frac{1}{2}x+\frac{1}{2}$

25) The graph showing the relationship between two quantities S and T is a straight line. Values of S are indicated on the horizontal axis. The gradient of the graph is 5 and the intercept on the vertical axis is 3. Hence the law of the line is

a $S=5T+3$ b $T=5S+3$
c $S=3T+5$ d $T=3S+5$

26) Figure 18.24 shows the graphs of $y=x^2-3x+2$ and $y=3x+6$, plotted on the same axes. The solutions of the equation $x^2-6x+8=0$

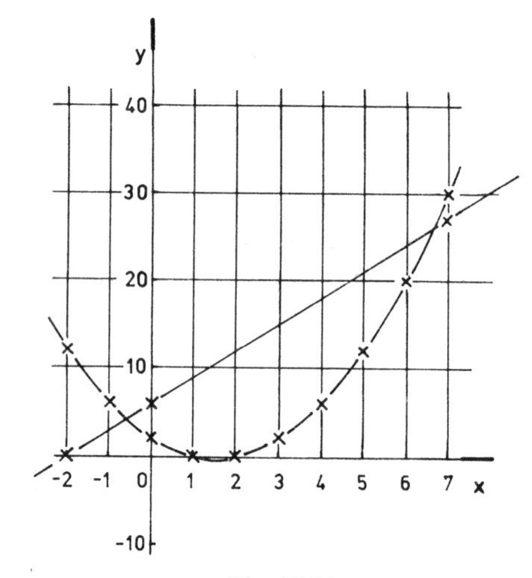

Fig. 18.24

a cannot be found from the graphs
b are $-0{\cdot}5$ and $6{\cdot}5$
c are 4 and 25
d are $-0{\cdot}5$ and 4 and $6{\cdot}5$ and 25

27) You are given the graph of $y=2x^2+x-15$. From the graph the solutions of the equation $2x^2-11x+15=0$ are required. Hence, on the same axes, you would plot

a $y=30-12x$ b $y=10x-30$
c $y=12x+30$ d $y=10x+30$

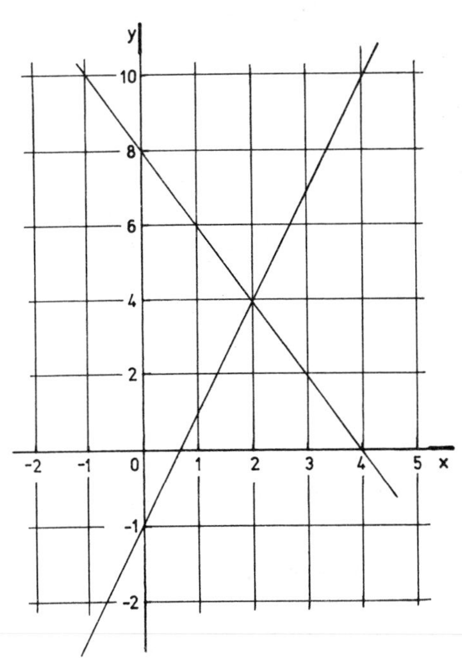

Fig. 18.25

28) Fig. 18.25 shows the graphs of $y=3x-2$ and $y=8-2x$, plotted on the same axes. The solutions of the simultaneous equations $3x-y=2$ and $2x+y=8$

a cannot be found from the graphs
b are $x=2$ and $y=4$
c are $x=4$ and $y=2$
d are $x=4$ and $y=-2$

Chapter 19 Variation

DIRECT VARIATION

The statement that y is proportional to x (often written $y \propto x$) means that the graph of y against x is a straight line passing through the origin (Fig. 19.1). If the gradient of this

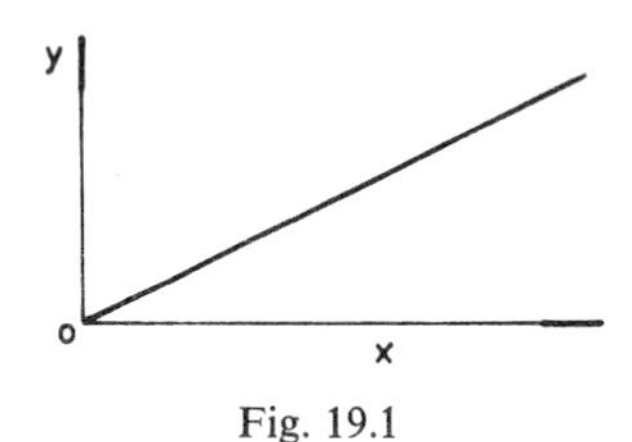

Fig. 19.1

line is k, then $y=kx$. The value of k is called the constant of proportionality. Thus, the ratio of y to x is equal to the constant of proportionality and y is said to vary directly as x. Hence direct variation means that if x is doubled then y is also doubled, if x is halved then y is halved and so on. Some examples of direct variation are as follows:

1) The circumference of a circle is directly proportional to its diameter.
2) The volume of a cone of given radius is directly proportional to its height.

Most problems on direct variation involve first finding the constant of proportionality from information given in the problem as shown in the following example.

Example

If y is directly proportional to x and $y=2$ when $x=5$, find the value of y when $x=6$.
Since $y \propto x$ then $y=kx$.
We are given that when $y=2$, $x=5$. Hence,

$$2=k \times 5 \quad \text{or} \quad k=\frac{2}{5}$$

$$\therefore \; y=\frac{2}{5}x$$

when $x=6$,

$$y=\frac{2}{5}\times 6=\frac{12}{5}$$

The volume of a sphere is given by the equation. $V=\frac{4}{3}\pi r^3$. From this equation we see that V varies directly as the cube of r (that is $V \propto r^3$) and the constant of proportionality is $\frac{4}{3}\pi$.
If y is proportional to x^2 ($y \propto x^2$) the graph of y against x^2 is a straight line passing through the origin. If the gradient of this line is k then $y=kx^2$.
Similarly if $y \propto \sqrt{x}$ then $y=k\sqrt{x}$.

Example

The surface area of a sphere, A square millimetres, varies directly as the square of its radius, r millimetres. If the surface area of a sphere 2 mm radius is $50{\cdot}24$ mm^2 find the surface area of a sphere whose radius is 4 mm.
Since $A \propto r^2$ then $A=kr^2$.
We are given that $A=50{\cdot}24$ when $r=2$. Hence

$$50{\cdot}24=k\times 2^2 \quad \text{or} \quad k=\frac{50{\cdot}24}{2^2}=12{\cdot}56$$

$$\therefore \; A=12{\cdot}56r^2$$

when $r=4$,

$$A=12{\cdot}56\times 4^2=200{\cdot}96 \text{ mm}^2$$

INVERSE VARIATION

If y is inversely proportional to x then the graph of y against $\frac{1}{x}$ is a straight line passing

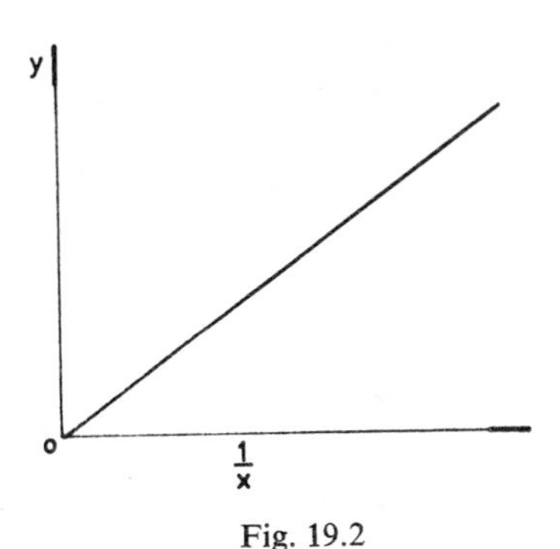

Fig. 19.2

through the origin (Fig. 19.2). If the gradient of this line is k, then

$$y = k \times \frac{1}{x} = \frac{k}{x}$$

Example
The electrical resistance, R ohms, of a wire of given length is inversely proportional to the square of the diameter of the wire, d mm. If R is 4·25 ohms when d is 2 mm find the value of R when $d=3$ mm.

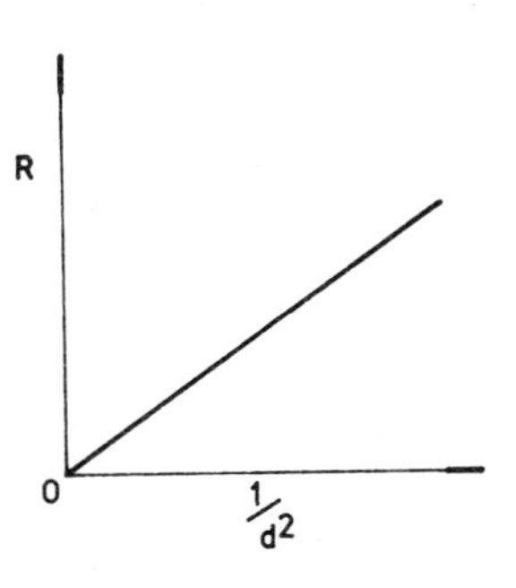

Fig. 19.3

Since R is inversely proportional to d^2 (Fig. 19.3)

$$R \propto \frac{1}{d^2}$$

$$R = \frac{k}{d^2}$$

when $R=4{\cdot}25$, $d=2$, hence

$$4{\cdot}25 = \frac{k}{2^2}$$

$$k = 4{\cdot}25 \times 2^2 = 17$$

$$\therefore \; R = \frac{17}{d^2}$$

when $d=3$,

$$R = \frac{17}{3^2} = 1{\cdot}9 \text{ ohms}$$

Exercise 66

1) Express the following with an equal sign and a constant:
  a) y varies directly as x^2.
  b) U varies directly as the square root of V.
  c) S varies inversely as T^3.
  d) h varies inversely as the cube root of m.

2) If $y=2$ when $x=4$ write down the value of y when $x=9$ for the following:
  a) y varies directly as the square of x.
  b) y varies inversely as the square root of x.
  c) y varies inversely as x.

3) If S varies inversely as T^3 and $S=54$ when $T=3$ find the value of T when $S=16$.

4) If U varies directly as $\sqrt{V}$ and $U=2$ when $V=9$ find the value of V when $U=4$.

5) The surface area of a sphere, V mm^2, varies directly as the square of its diameter, d mm. If the surface area is to be doubled by what ratio must the diameter be altered.

JOINT VARIATION

The volume of a cylinder varies directly as two dimensions: its height and the square of its radius. Hence the volume is directly proportional to the *product* of its height and the square of its radius. Written as an equation the statement becomes:

$$V = khr^2$$

Again, the constant k can be found from information given in the problem.

Example

A certain law in physics connecting three quantities p, v and t states that p varies directly as t and inversely as v. If it is known that $p=800$ when $t=300$ and $v=36$ calculate the value of v when $p=700$ and $t=350$.

We are given $p \propto t$ and $p \propto \frac{1}{v}$

$$\therefore \quad p=\frac{kt}{v}$$

Since $p=800$ when $t=300$ and $v=36$,

$$800=\frac{k \times 300}{36}$$

$$\therefore \quad k=\frac{800 \times 36}{300}=96$$

$$\therefore \quad p=\frac{96t}{v}$$

To find v when $p=700$ and $t=350$ substitute these values in the last equation. Thus,

$$700=\frac{96 \times 350}{v}$$

$$v=\frac{96 \times 350}{700}$$

$$\therefore \quad v=48$$

VARIATION AS THE SUM OF TWO PARTS

The function $(ax+bx^2)$ is the sum of two quantities:

ax which varies directly as x, and

bx^2 which varies directly as x^2.

An alternative description would describe the function $(ax+bx^2)$ as that function which varies partly as x and partly as x^2.
It is important to be able to recognise whether problems involve joint variation or variation as the sum of two parts.

Examples

1) A quantity p is the sum of two terms, one of which is constant, whilst the other varies inversely as the square of q. When $q=1$, $p=-1$ and when $q=2$, $p=2$. Find the positive value of q when $p=2\frac{3}{4}$.
If a and b are constants

$$p=a+\frac{b}{q^2} \text{ (see Fig. 19.4)}$$

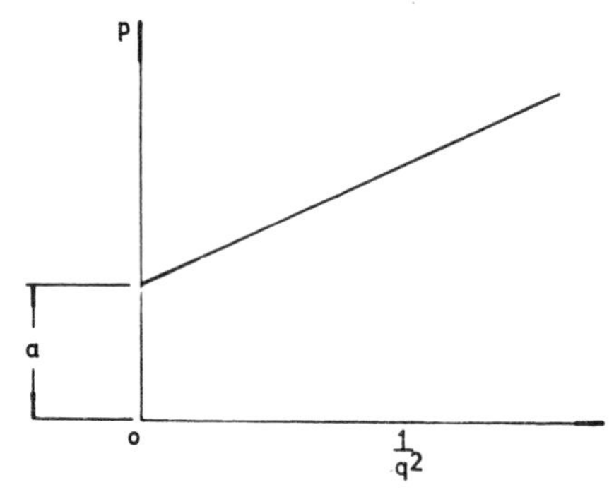

Fig. 19.4

In order to evaluate a and b we use the information given in the question. Thus, when $q=1$, $p=-1$

$$\therefore \quad -1=a+\frac{b}{1^2}$$

$$\text{or} \quad -1=a+b \qquad \ldots(1)$$

when $q=2$, $p=2$

$$\therefore \quad 2=a+\frac{b}{2^2}$$

$$\text{or} \quad 2=a+\frac{b}{4}$$

$$8=4a+b \qquad \ldots(2)$$

Subtracting equation (1) from (2),

$$9=3a$$

$$\therefore \quad a=3$$

Substituting for $a=3$ in equation (1),

$$-1=3+b$$

$$\therefore \quad b=-4$$

$$\therefore \quad p=3-\frac{4}{q^2}$$

when $p=2\frac{3}{4}$,

$$2\tfrac{3}{4}=3-\frac{4}{q^2}$$

$$\frac{4}{q^2}=\frac{1}{4}$$

$$q^2=16$$

$$q=\pm 4$$

Since only the positive value of q is required

$$q=4$$

Exercise 67

1) The mass of a solid cone varies jointly as the square of the radius of the base and as the height. If a cone has a mass of 15 g, a height of 5 mm and a radius of 3 mm find the radius of a second cone made from the same material which has a mass of 32 g and a height of 6 mm.

2) It is known that P varies inversely as the square of Q. Corresponding pairs of values are shown in the following table:

P	t	$t+1$
Q	5	4

Calculate the value of t.

3) If y varies as x^3 and $y=25$ when $x=10$, calculate the value of y when $x=6$.

4) Draw a neat sketch graph of the function $y=\dfrac{p}{x}$, giving p the value which will make y equal to 4 when $x=3$. State this value of p. You are not required to draw the graph accurately but simply to show its position and give a general idea of its shape.

5) A quantity C is the sum of two parts. The first part varies directly as the cube of t; the second part varies inversely as the square of t. Given that $C=74$ when $t=1$ and $C=34$ when $t=2$, find the value of C when $t=3$.

6) A quantity P is the difference between two parts. The first part is constant and the second varies inversely as the square of Q. If $P=1$ when $Q=2$ and $P=6$ when $Q=3$, find the positive value of Q when $P=7\frac{3}{4}$.

7) It is given that y is inversely proportional to the square of $(x+3)$. If $y=9$ when $x=1$, find the possible values of x when $y=1$.

8) The velocity, v metres per second, of a body moving in a straight line is at any given instant given by the sum of two terms, one of which is proportional to the time t seconds which has elapsed since the body started moving, and the other is proportional to the square of the time t. Given that $v=68$ when $t=1$ and $v=104$ when $t=2$, form an equation for v in terms of t. Use your equation to calculate the value of t when the body comes to rest.

9) For a certain series of experiments it is known that a quantity F is directly proportional to h and the square root of P, and inversely proportional to the square of d. If $d=8$, $h=40$ and $P=1000$ when $F=12$, calculate the value of P when $F=8$, $d=10$ and $h=30$.

10) Three quantities E, D and H are connected so that E varies directly as H and inversely as the square of root of D. If $E=5$ when $H=10$ and $D=16$ find the value of D when $E=20$ and $H=4$.

SELF TEST 19

In questions 1 to 15 the answer is either 'true' or 'false'.

1) The graph in Fig. 19.5 shows y plotted against $\sqrt{x}$. Hence y is proportional to $\sqrt{x}$. -----

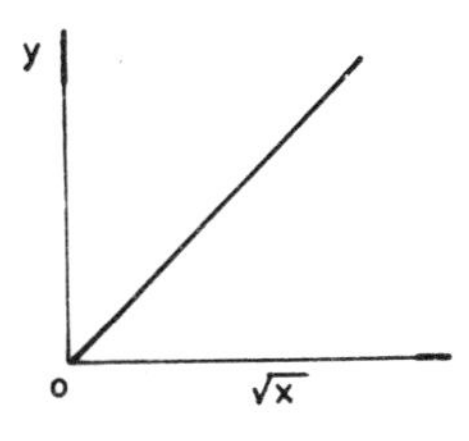

Fig. 19.5

2) The graph in Fig. 19.6 shows y plotted against x^2. Therefore $y=kx^2$ where k is a constant. _ _ _ _ _

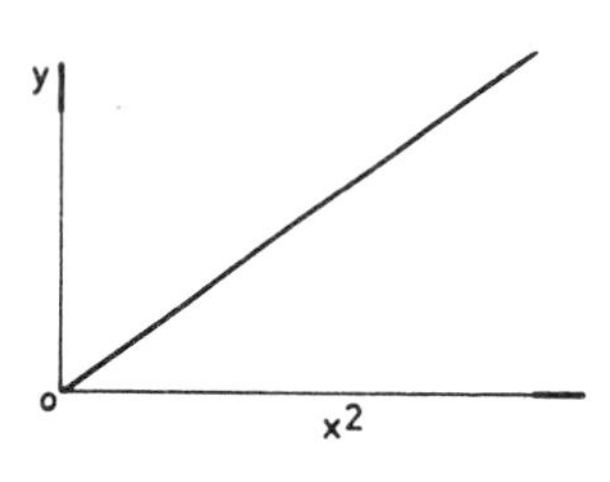

Fig. 19.6

3) The graph in Fig. 19.7 shows y plotted against $\frac{1}{x}$. Therefore $y=kx$. _ _ _ _

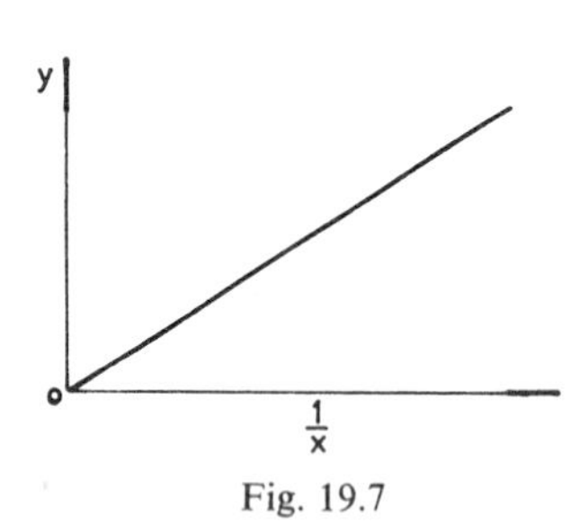

Fig. 19.7

4) If V is proportional to r^2 and V is 6 when r is 2 then $V=1{\cdot}5r^2$. _ _ _ _ _

5) If Q is inversely proportional to $\sqrt{x}$ then $Q=\dfrac{k}{\sqrt{x}}$ _ _ _ _ _

6) If y is inversely proportional to d^3 and y is $\frac{3}{8}$ when d is 2 then $y=3d^3$ _ _ _ _ _

7) If M varies directly as p and inversely as q^2 then $M=\dfrac{kp}{q^2}$ _ _ _ _ _

8) If y varies directly as the square root of v and inversely as u then $y=ku\sqrt{v}$ _ _ _ _ _

9) A quantity F varies directly as m and inversely as p. If $m=6$ and $p=3$ when $F=4$ then $F=2\,pm$ _ _ _ _ _

10) The graph shown in Fig. 19.8 means that y is the sum of two parts. _ _ _ _ _

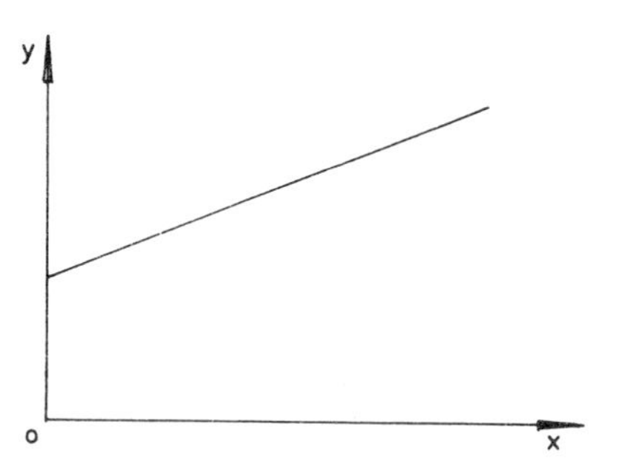

Fig. 19.8

11) In Fig. 19.8 $y=kx+c$ where k and c are constants. _ _ _ _ _

12) In Fig. 19.9, the graph indicates that the quantity Q is the difference between two parts, the first being a constant and the second varying directly with x^2. _ _ _ _ _

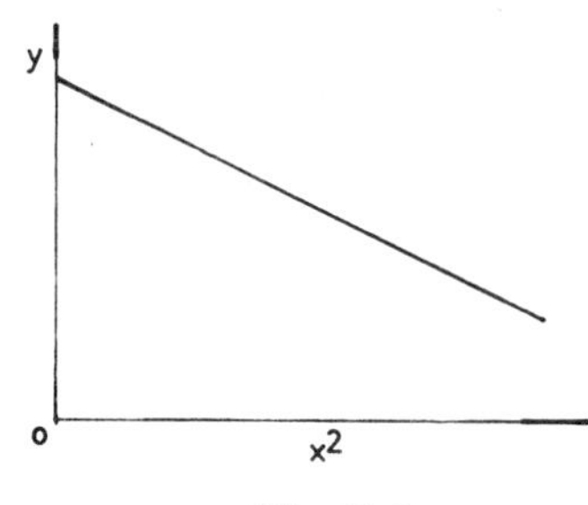

Fig. 19.9

13) A quantity M is the sum of two parts. The first part varies directly as the square of p and the second part as the cube of p. Hence $M=k_1p^2+k_2p^3$ where k_1 and k_2 are constants. _ _ _ _ _

14) The graph in Fig. 19.10 represents the relationship $y=\dfrac{k}{x}$. _ _ _ _ _

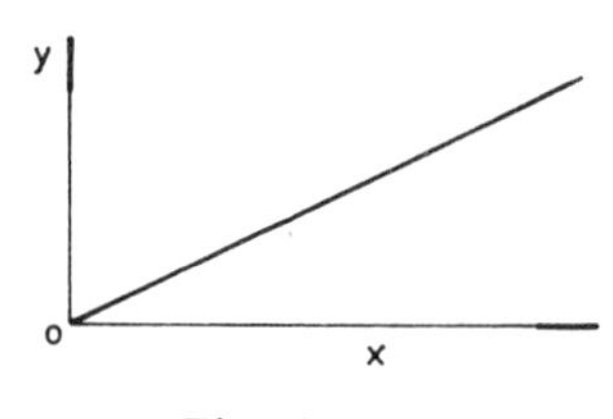

Fig. 19.10

15) A quantity C is the sum of two parts. The first part varies directly as the square root of m and the second part varies inversely as the square of m. Hence $C=\dfrac{k\sqrt{m}}{m^2}$ where k is a constant. - - - - -

In questions 16 to 24 state the letter (or letters) which correspond to the correct answer (or answers).

16) In Fig. 19.11,

a $y=kx^2$ b $y=\dfrac{k}{x^2}$

c $y=k_1+k^2x^2$ d $y=k_1+\dfrac{k_2}{x^2}$

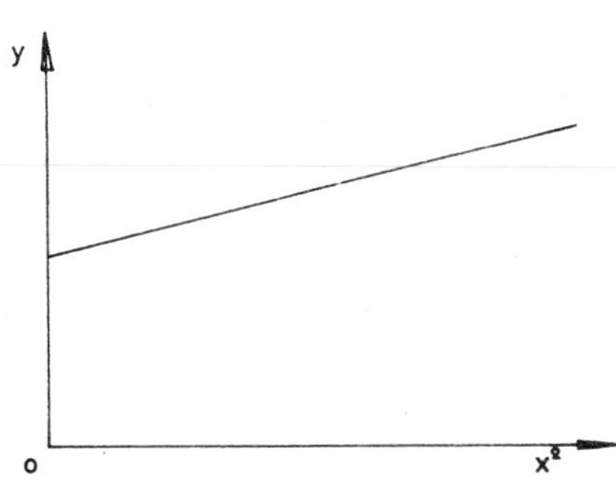

Fig. 19.11

17) In Fig. 19.12

a $y=k_1+k_2\sqrt{x}$ b $y=k_1-k_2\sqrt{x}$

c $y=k_1\sqrt{x}$ d $y=\dfrac{k_1}{\sqrt{x}}$

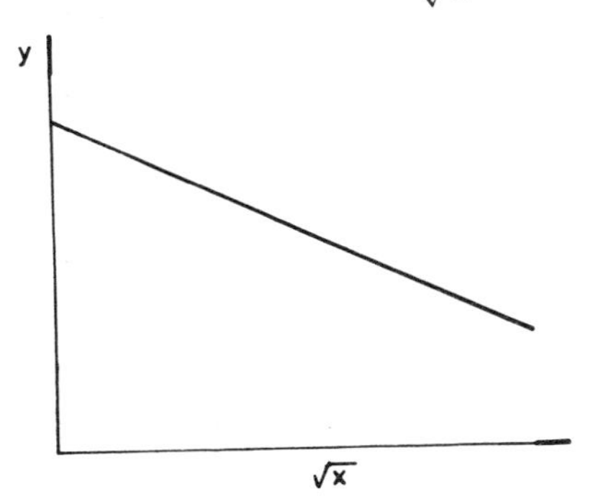

Fig. 19.12

18) If y varies as x^2 and $y=25$ when $x=10$ then the value of y when $x=2$ is

a 1 b 16 c 625 d 0·0625

19) If R varies inversely as $\sqrt{p}$ and $R=2$ when $p=4$, then the value of R when $p=16$ is

a $\dfrac{1}{4}$ b 4 c 1 d 8

20) A quantity M varies directly as d^2 and inversely as q. When $M=6$, $d=2$ and $q=2$. Thus when $d=4$ and $q=8$ the value of M is

a 96 b $\dfrac{1}{96}$ c $\dfrac{1}{6}$ d 6

21) A quantity P is the difference of two parts. The first part varies directly as r and the second part varies inversely as r^2. An expression for P is

a $P=k_1r+k_2r^2$ b $P=k_1r-k_2r^2$

c $P=k_1r+\dfrac{k_2}{r^2}$ d $P=k_1r-\dfrac{k_2}{r^2}$

22) A quantity y is the sum of two parts. The first part is a constant and the second part varies directly as $\sqrt{x}$. An expression for y is

a $y=C(1+\sqrt{x})$ b $y=C(1-\sqrt{x})$

c $y=C+k\sqrt{x}$ d $y=C-k\sqrt{x}$

23) A quantity m is the sum of two parts, one of which is constant and the other varies inversely as the square of P. When $p=2$, $m=5$ and when $p=1$, $m=14$. Hence an expression for m is

a $m=2+\dfrac{12}{p^2}$ b $m=2-\dfrac{12}{p^2}$

c $m=2+12p^2$ d $m=2-12p^2$

24) The electrical resistance, R, of a wire varies directly as the length, l, and inversely as the square of the diameter d. A formula giving d in terms of l, R and a constant of variation, k, is

a $d=\dfrac{kl}{R}$ b $d=\sqrt{\dfrac{kl}{R}}$ c $d=\dfrac{R}{kl}$

d $d=\sqrt{\dfrac{R}{kl}}$

Chapter 20 The Differential Calculus

THE GRADIENT OF A CURVE

In mathematics and science we often need to know the rate of change of one variable with respect to another. For instance, speed is the rate of change of distance with respect to time and acceleration is the rate of change of speed with respect to time.

Consider the graph $y=x^2$ part of which is shown in Fig. 20.1. As the values of x increase so do the values of y, but they do not increase at the same rate. A glance at the portion of the curve shown in Fig. 20.1 shows that the values of y increase faster when x is large because the gradient of the curve is increasing.

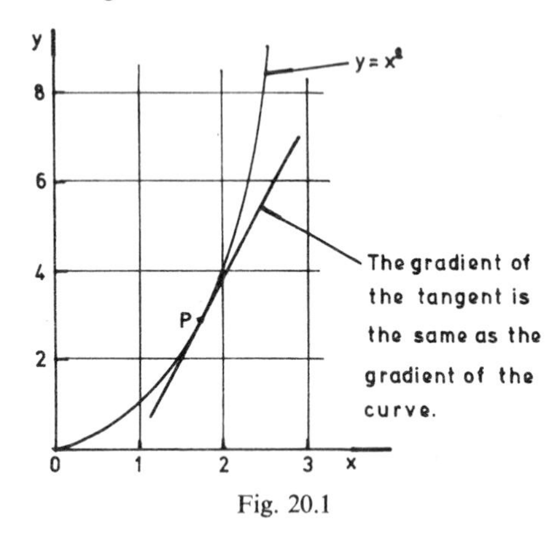

Fig. 20.1

To find the rate of change of y with respect to x at a particular point we need to find the gradient of the curve at that point. If we draw a tangent to the curve at the point, the gradient of the tangent will be the same as the gradient of the curve.

Examples

1) Draw the curve of $y=x^2$ and find the gradient of the curve at the points where $x=2$ and $x=-2$.

To draw the curve the table below is drawn up

x	-3	-2	-1	0	1	2	3
$y=x^2$	9	4	1	0	1	4	9

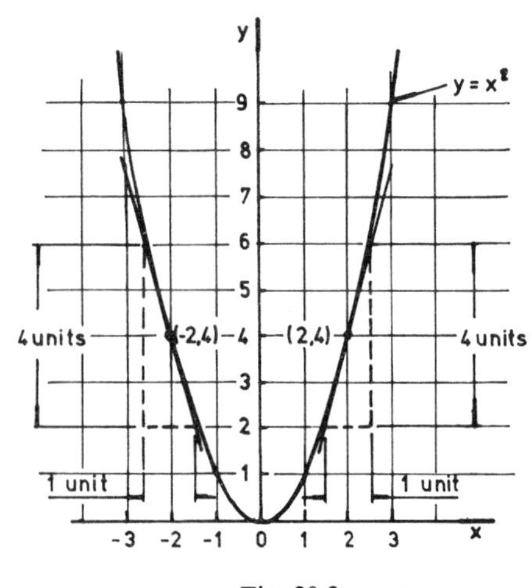

Fig. 20.2

The point where $x=2$ is the point (2, 4). We draw a tangent at this point as shown in Fig. 20.2. Then by constructing a right-angled triangle the gradient is found to be $\frac{4}{1}=4$. This gradient is positive since the tangent slopes upwards from left to right.

A positive value of the gradient indicates that y is increasing as x increases.

The point where $x=-2$ is the point $(-2, 4)$. By drawing the tangent at this point and constructing a right-angled triangle as shown in Fig. 20.2, the gradient is found to be

$$\frac{-4}{1}=-4.$$

The gradient is negative because the tangent slopes downwards from left to right.

A negative value of the gradient indicates that y is decreasing as x increases.

2) Draw the graph of $y=x^2-3x+7$ between $x=-4$ and $x=4$ and hence find the gradient of the curve at the points $x=-3$ and $x=2$. To plot the curve draw up the following table.

x	-4	-3	-2	-1	0	1	2	3	4
y	35	25	17	11	7	5	5	7	11

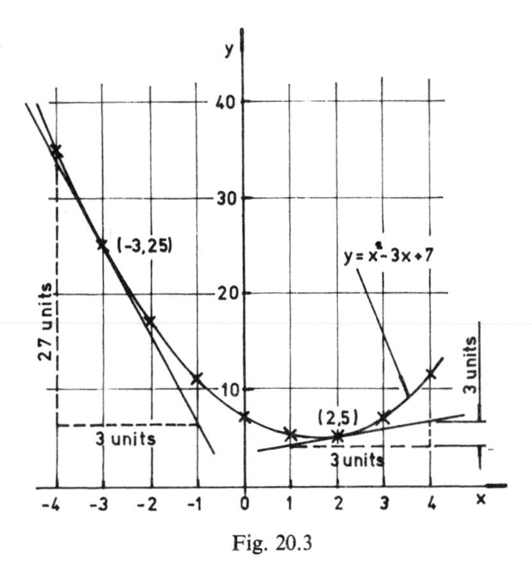

Fig. 20.3

At the point where $x=-3$, $y=25$. At the point $(-3, 25)$ draw a tangent to the curve as shown in Fig. 20.3. The gradient is found by drawing a right-angled triangle (which should be as large as possible for accuracy) as shown and measuring its height and base. Hence

$$\text{gradient at the point } (-3, 25) = -\frac{27}{3} = -9$$

at the point where $x=2$, $y=5$. Hence by drawing a tangent and a right-angled triangle at the point (2, 5),

$$\text{gradient at point } (2, 5) = \frac{3}{3} = 1$$

3) Draw the graph of $y=x^2+3x-2$ taking values of x between $x=-1$ and $x=4$. Hence find the value of x where the gradient of the curve is 7.

To plot the curve the following table is drawn up

x	-1	0	1	2	3	4
y	-4	-2	2	8	16	26

The curve is shown in Fig. 20.4.

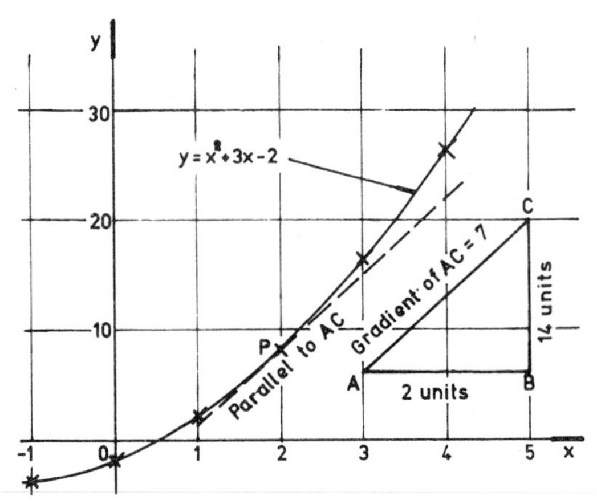

Fig. 20.4

To obtain a line whose gradient is 7 we draw the △ ABC making (for convenience) AB = 2 units to the scale on the x-axis and BC = 14 units to the scale on the y-axis. Hence

$$\text{gradient of AC} = \frac{BC}{AB} = \frac{14}{2} = 7$$

Using set-squares we draw a tangent to the curve so that the tangent is parallel to AC. As can be seen this tangent touches the curve at the point P where $x=2$. Hence the gradient of the curve is 7 at the point where $x=2$.

Exercise 68

1) Draw the graph of $y=3x^2+7x+3$ and find the gradient of the curve at the points where $x=-2$ and $x=2$.

2) Draw the graph of $y=2x^2-5$ for values of x between -2 and 3. Find the gradient of the curve at the points where $x=-1$ and $x=2$.

3) Draw the curve of $y=x^2-3x+2$ from $x=2{\cdot}5$ to $x=3{\cdot}5$ and find its gradient at the point where $x=3$.

4) For what values of x is the gradient of the curve $y=\frac{x^3}{3}+\frac{x^2}{2}-33x+7$ equal to 3?
In drawing the curve take values of x between -8 and 6.

5) If $y=(1+x)(5-2x)$ copy and complete the table below.

x	-2	$-1\frac{1}{2}$	-1	0	$\frac{1}{2}$	1	$1\frac{1}{2}$	2	3
y	-9		0	5		6	5	3	-4

Hence draw the graph of $y=(1+x)(5-2x)$.
Find the value of x at which the gradient of the curve is -2.

6) If $y=x^2-5x+4$ find by plotting the curve between $x=0$ and $x=5$ the value of x at which the gradient of the curve is 11.

DIFFERENTIATION

It is possible to find the gradient of a curve at any point by graphical means. However this method is often inconvenient and not very accurate. Hence the gradient of a curve is usually found by differentiation.
The gradient of a curve at any point on the curve is given by its derived function. Thus if

$$y=x^n$$

then it can be shown that the derived function is

$$\frac{dy}{dx}=nx^{n-1}$$

This formula is true for all values of n including fractional and negative indices.
The expression $\frac{dy}{dx}$, compares the rate of change of y with that of x and it must be realised that $\frac{dy}{dx}$ is not a fraction in the ordinary sense. The d in dy is not a multiple (compare with cos y or log y) and dy cannot be separated from its denominator dx.
The process of finding $\frac{dy}{dx}$ is called differentiation.

Examples

1) If $y=x^3$, $\frac{dy}{dx}=3x^2$

2) If $y=\frac{1}{x}$ then $y=x^{-1}$

and $\frac{dy}{dx}=-x^{-2}=-\frac{1}{x^2}$

3) If $y=\sqrt{x}$ then $y=x^{\frac{1}{2}}$

and $\frac{dy}{dx}=\frac{1}{2}x^{-\frac{1}{2}}=\frac{1}{2x^{\frac{1}{2}}}=\frac{1}{2\sqrt{x}}$

4) If $y=\sqrt[5]{x^2}$ then $y=x^{\frac{2}{5}}$

and $\frac{dy}{dx}=\frac{2}{5}x^{-\frac{3}{5}}=\frac{2}{5x^{\frac{3}{5}}}=\frac{2}{5\cdot\sqrt[5]{x^3}}$

When a power of x is multiplied by a constant, the constant remains unchanged by the process of differentiation. Hence if

$$y=ax^n$$

$$\frac{dy}{dx}=nax^{n-1}$$

Examples

1) If $y=3x^4$, $\frac{dy}{dx}=3\times 4x^3=12x^3$

2) If $y=2x^{1\cdot 3}$ $\frac{dy}{dx}=2\times 1\cdot 3x^{0\cdot 3}=2\cdot 6x^{0\cdot 3}$

3) If $y=\frac{3}{4}\sqrt[3]{x}=\frac{3}{4}x^{\frac{1}{3}}$

$$\frac{dy}{dx}=\frac{3}{4}\times\frac{1}{3}x^{-\frac{2}{3}}=\frac{1}{4}x^{-\frac{2}{3}}=\frac{1}{4\cdot\sqrt[3]{x^2}}$$

4) If $y=\frac{4}{x^2}=4x^{-2}$

$$\frac{dy}{dx}=4\times(-2)x^{-3}=-8x^{-3}=-\frac{8}{x^3}$$

When a numerical constant is differentiated the result is zero. Since $x^0=1$, we can write the numerical constant 4 as $4\times x^0$. Then differentiating with respect to x we get

$$4\times 0x^{-1}=0.$$

To differentiate an expression containing a sum of terms we differentiate each individual term separately.

Examples

1) If $y=3x^2+2x+3$

$$\frac{dy}{dx}=3\times 2x^1+2\times 1x^0+0=6x+2$$

2) If $y=ax^3+bx^2+cx+d$ where a, b, c and d are constants

$$\frac{dy}{dx}=3ax^2+2bx+c$$

3) If $y=\sqrt{x}+\frac{1}{\sqrt{x}}=x^{1/2}+x^{-1/2}$

$$\frac{dy}{dx}=\frac{1}{2}x^{-1/2}+\left(-\frac{1}{2}\right)x^{-3/2}$$

$$=\frac{1}{2\sqrt{x}}-\frac{1}{2\sqrt{x^3}}$$

4) If $s=\frac{t^3+t^2+2t}{t}$

then $s=\frac{t^3}{t}+\frac{t^2}{t}+\frac{2t}{t}=t^2+t+2$

$$\frac{ds}{dt}=2t+1$$

Exercise 69

Differentiate the following:

1) $y=x^2$

2) $y=x^7$

3) $y=4x^3$

4) $y=6x^5$

5) $s=0{\cdot}5t^3$

6) $A=\pi R^2$

7) $y=x^{1/2}$

8) $y=4x^{3/2}$

9) $y=2\sqrt{x}$

10) $y=3\sqrt[3]{x^2}$

11) $y=\frac{1}{x^2}$

12) $y=\frac{1}{x}$

13) $y=\frac{3}{5x}$

14) $y=\frac{2}{x^3}$

15) $y=\frac{1}{\sqrt{x}}$

16) $y=\frac{2}{\sqrt[3]{x}}$

17) $y=\frac{5}{x\sqrt{x}}$

18) $s=\frac{3\sqrt{t}}{5}$

19) $K=\frac{0{\cdot}01}{H}$

20) $y=\frac{5}{x}$

21) $y=4x^2-3x+2$

22) $s=3t^3-2t^2+5t-3$

23) $q=2u^2-u+7$

24) $y=5x^4-7x^3+3x^2+5$

25) $s=7t^3-3t^2+7$

26) $y=\frac{x+x^3}{\sqrt{x}}$

27) $y=\frac{3+x^2}{x}$

28) $y=\sqrt{x}+\frac{1}{\sqrt{x}}$

29) $y=x^3+\frac{3}{\sqrt{x}}$

30) $s=t^{1{\cdot}3}-\frac{1}{4t^{2{\cdot}3}}$

31) $y=\frac{3x^3}{5}-\frac{2x^2}{7}-\sqrt{x}$

32) $y=0{\cdot}08+\frac{0{\cdot}01}{x}$

33) $y=31x^{1{\cdot}5}-2{\cdot}4x^{0{\cdot}6}$

34) $y=\frac{x^3}{2}-\frac{5}{x}+3$

35) $s=10-6t+7t^2-2t^3$

THE GRADIENT OF A CURVE BY THE CALCULUS

It has been shown previously that $\frac{dy}{dx}$ represents the general expression for the gradient of a curve at any point.

Examples

1) If $y=x^2-5x+7$ find the gradient of the curve at the points where $x=-3$ and $x=2$.

Since

$$y=x^2-5x+7$$

$$\frac{dy}{dx}=2x-5$$

when $x=-3$, $\frac{dy}{dx}=2\times(-3)-5=-11$.

Hence the gradient of the curve when $x=-3$ is -11,

when $x=2$, $\frac{dy}{dx}=2\times 2-5=-1$.

Hence when $x=2$, the gradient of the curve is -1.

2) If $y=x^3-3x^2+7$, find the values of x at which the gradient of the curve is 24. Since

$$y=x^3-3x^2+7$$

$$\frac{dy}{dx}=3x^2-6x$$

When the gradient of the curve is 24, $\frac{dy}{dx}=24$

hence

$$3x^2-6x=24$$

$$3x^2-6x-24=0$$

$$x^2-2x-8=0$$

$$(x-4)(x+2)=0$$

$$x=4 \text{ or } -2$$

The gradient of the curve is 24 when $x=4$ and $x=-2$.

Exercise 70

1) Find the gradient of the curve $y=2x^2-5x+3$ at the point (1, 0).

2) Find the gradient of the curve $y=x^2+2x-3$ at the point where $x=2$.

3) Find the gradient of the curve $y=\frac{1}{x}+5$ at the point (2, 5·5).

4) Find the value of x where the gradient of the curve $y=x^2+2x+7$ is 8.

5) Find the co-ordinates of the points on the curve $y=x^3-2x^2+3x-5$ where its gradient is 2.

6) The curve $y=x^2+\frac{A}{x}$ has a gradient of 7 when $x=4$. Calculate the value of A.

7) Prove that the curve $y=4x-\frac{32}{x^2}$ crosses the x-axis at the point where $x=2$. Calculate the gradient of the curve at this point.

VELOCITY AND ACCELERATION

Suppose that a body travels a distance of s metres in a time of t seconds. Since velocity is the rate of change of distance with respect to time, $\frac{ds}{dt}$ gives an expression for the instantaneous velocity at any time t seconds. If the velocity is v metres per second then

$$v=\frac{ds}{dt}$$

Acceleration is the rate of change of velocity with respect to time. Hence $\frac{dv}{dt}$ gives an expression for the instantaneous acceleration at any time t seconds. If the acceleration is a metres per second per second then

$$a=\frac{dv}{dt}$$

Note that $\frac{ds}{dt}$ and $\frac{dv}{dt}$ represent the *instantaneous* velocity and acceleration as opposed to the *average* velocity and acceleration which were used previously (see Chapter 7).

Examples

1) A body moves a distance of s metres in a time of t seconds so that $s=2t^3-5t^2+4t+5$. Find a) its velocity after 3 seconds, b) its acceleration after 3 seconds, c) when its velocity is zero, d) when its acceleration is zero.

a) $s=2t^3-5t^2+4t+5$

$$\frac{ds}{dt}=6t^2-10t+4$$

Since $v=\frac{ds}{dt}$, when $t=3$

$$v=6\times 3^2-10\times 3+4=28$$

Hence the velocity after 3 seconds is 28 m/s.

b) $v=6t^2-10t+4$

$$\frac{dv}{dt}=12t-10$$

Since $a=\frac{dv}{dt}$, when $t=3$

$$a=12\times 3-10=26$$

Hence the acceleration after 3 seconds is 26 m/s^2.

c) When the velocity is zero, $\frac{ds}{dt}=0$

$$\therefore\ 6t^2-10t+4=0$$

or $3t^2-5t+2=0$

$$(3t-2)(t-1)=0$$

$$t=\frac{2}{3}\text{ or }1$$

Hence the velocity is zero when $t=\frac{2}{3}$ seconds or when $t=1$ second.

d) When the acceleration is zero, $\frac{dv}{dt}=0$

$$\therefore\ 12t-10=0$$

$$t=\frac{5}{6}$$

Hence the acceleration is zero when $t=\frac{5}{6}$ seconds.

2) A body moves so that the distance travelled, s metres, in a time t seconds is given by

$$s=8+12t-t^3$$

Calculate the distance at which the body will stop and reverse direction.
When the body stops its velocity is zero.
Since

$$s=8+12t-t^3$$

$$\frac{ds}{dt}=12-3t^2$$

when $\frac{ds}{dt}=0$,

$$12-3t^2=0$$

$$3t^2=12$$

$$t=\pm\sqrt{4}=\pm 2$$

It is possible to have negative time (e.g. -12 seconds to blast off) but in this type of question usually only the positive value is used. Hence the body stops and reverses direction when $t=2$ seconds. To find the distance travelled in this time we substitute for $t=2$ into the expressions $s=8+12t-t^3$. When $t=2$, $s=8+12\times 2-2^3=24$.
The distance at which the body will stop and reverse direction is 24 metres.

3) An electric train starts from A and travelled to its next stop 6 km from A. The following readings were taken of the time since leaving A (in seconds) and the distance from A (in km).

Time	30	60	90	120	150	180
Distance	0·10	0·34	0·80	1·46	2·46	3·50

Time	210	240	270	300	330	360
Distance	4·34	5·00	5·44	5·74	5·92	6·00

Draw a graph of these values taking time horizontally. From the graph estimate the speed of the train when it has travelled 5 km.
The graph is shown in Fig. 20.5. The point P on the graph coincides with a distance travelled of 5 km. The speed of the train at this instant is represented by the gradient of the curve at P. Drawing a tangent to the curve at P, we find its gradient by drawing the right-angled triangle ABC whose base is 150 and whose height is 2·6.

Hence the speed is $\frac{2 \cdot 6}{150} = 0 \cdot 017$ km per second

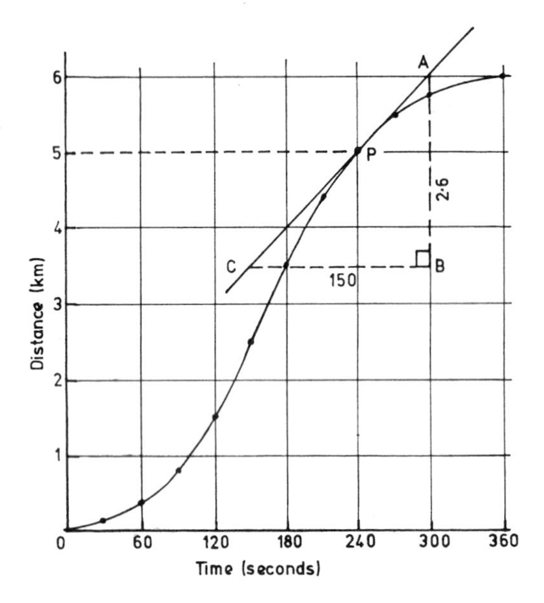

Fig. 20.5

Exercise 71

1) If $s = 10 + 50t - 2t^2$, where s metres is the distance travelled in t seconds by a body, what is the speed of the body after 2 seconds?

2) If $v = 5 + 24t - 3t^2$, where v metres per second is the speed of the body after t seconds, what is the acceleration of the body after 3 seconds?

3) A body moves s metres in a time t seconds so that $s = t^3 - 3t^2 + 8$. Find

a) its speed at the end of 3 seconds;
b) when its speed is zero;
c) its acceleration at the end of 2 seconds;
d) when its acceleration is zero.

4) A body moves s metres in t seconds where $s = \frac{1}{t^2}$. Find the speed and acceleration after 3 seconds.

5) The distance moved by a body in t seconds is given in metres by $s = 2t^2 + 5t - 3$. Find a) the initial velocity b) the velocity after 3 seconds c) the acceleration of the body.

6) A particle moves so that the distance travelled, s metres, in a time t seconds is $s = 5 + 6t - t^3$. Calculate the distance at which the body will stop and reverse direction.

7) A particle is moving in a straight line through 0, so that after t seconds its distance from 0 is given by $s = t^3 - 9t^2 + 30t$. Calculate the values of the acceleration when the velocity is 6 metres per second.

8) A body starts from A. The following readings were taken of the time (in seconds) since leaving A and the distance (in metres) from A.

Time	0	1	2	3	4	5
Distance	0	1	16	63	160	325

Find the speed in metres per second after 3 seconds.

9) The speed of a body at certain times is given in the table below. Draw the speed–time graph and find the acceleration after 5 seconds.

Time (seconds)	0	1	2	3	4	5	6
Speed (m/s)	0	3	6	11	18	27	38

10) A body starts from A and its distance from A after a time of t seconds is given by $s = 2t^3 - 5t^2 + 20t$. Find the acceleration when the velocity of the body is 24 m/s. What distance from A is the body when the velocity is 24 m/s.

TURNING POINTS

At the points P and Q (Fig. 20.6) the tangent to the curve is parallel to the x-axis. The points P and Q are called *turning points*. The turning point at P is called a *maximum* turning point and the turning point at Q is called a *minimum* turning point. It will be seen from Fig. 20.6 that the value of y at P is not the greatest value of y nor is the value of y at Q the least. The terms maximum and minimum values apply only to the values of y at the turning points and not to the values of y in general.

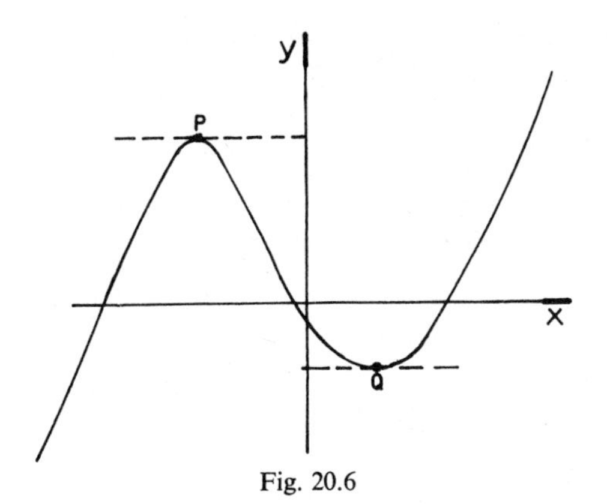

Fig. 20.6

Examples

1) Plot the graph of $y=x^3-5x^2+2x+8$ for values of x between -2 and 6. Hence find the maximum and minimum values of y.

To plot the graph we draw up a table in the usual way.

x	-2	-1	0	1
$y=x^3-5x^2+2x+8$	-24	0	8	6

x	2	3	4	5	6
$y=x^3-5x^2+2x+8$	0	-4	0	18	56

The graph is shown in Fig. 20.7. The maximum value occurs at the point P where

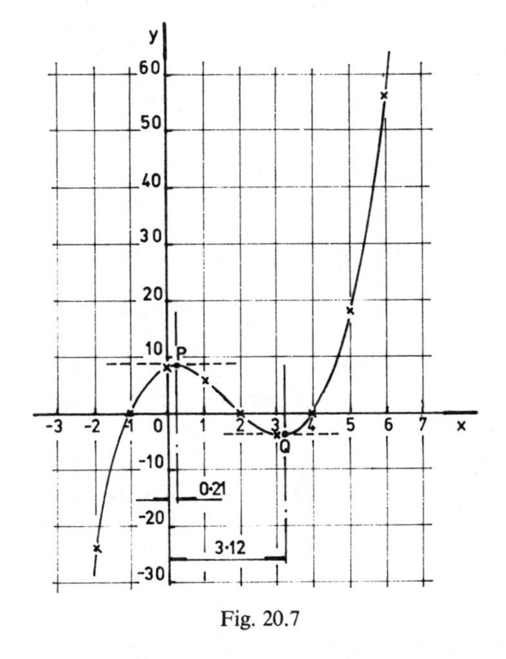

Fig. 20.7

the tangent to the curve is parallel to the x-axis. The minimum value occurs at the point Q where again the tangent to the curve is parallel to the x-axis. From the graph the maximum value of y is 8·21 and the minimum value of y is $-4{\cdot}06$.

Notice that the value of y at P is not the greatest value of y nor is the value of y at Q the least. However, the values of y at P and Q are called the maximum and minimum values of y respectively.

2) A small box is to be made from a rectangular sheet of metal 36 cm by 24 cm. Equal squares of side x cm are cut from each of the corners and the box is then made by folding up the sides. Prove that the volume V of the box is given by the expression $V=x(36-2x)(24-2x)$. Find the value of x so that the volume may be a maximum and find this maximum volume.

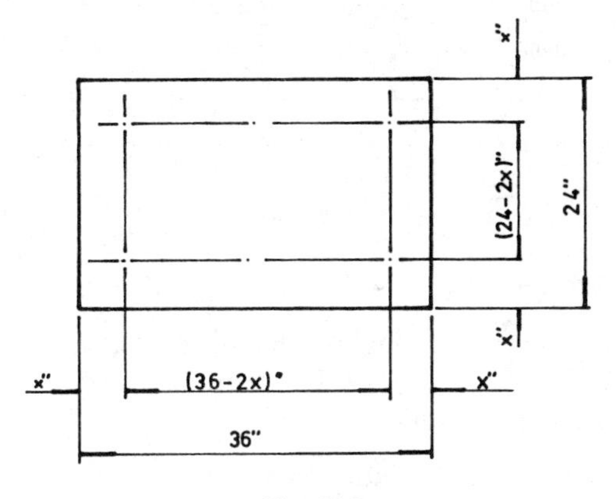

Fig. 20.8

Referring to Fig. 20.8 we see that after the box has been formed

its length $=36-2x$
its breadth $=24-2x$
its height $=x$

The volume of the box is

$$V=\text{length}\times\text{breadth}\times\text{height}$$
$$\therefore\ V=x(36-2x)(24-2x)$$

We now have to plot a graph of this equation and so we draw up the table below:

x	1	2	3	4
$36-2x$	34	32	30	28
$24-2x$	22	20	18	16
V	748	1280	1620	1792

x	5	6	7	8
$36-2x$	26	24	22	20
$24-2x$	14	12	10	8
V	1820	1728	1540	1280

The graph is shown in Fig. 20.9 and it can be seen that the maximum volume is 1825 cm^3 which occurs when $x=4{\cdot}71$ cm.

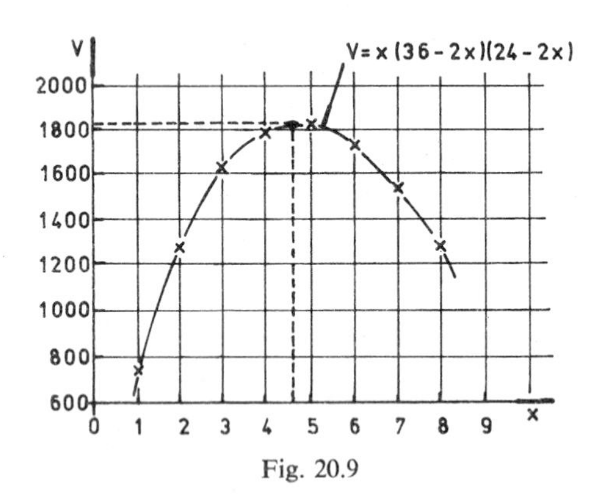

Fig. 20.9

Exercise 72

1) Find the minimum value of the curve $y=3x^2+2x-3$. Plot the graph for values of x between -2 and 3.

2) Find the maximum value of the curve $y=-x^2+5x+7$. Plot the graph for values of x between -2 and 4.

3) Plot the graph of $y=x^3-9x^2+15x+2$ taking values of x from 0 to 7. Hence find the maximum and minimum values of y.

4) Draw the graph of $y=x^2-3x$ from $x=-1$ to $x=4$ and use your graph to find

a) the least value of y;

b) the two solutions of the equation $x^2-3x=1$

c) the two solutions of the equation $x^2-2x-1=0$.

5) Draw the graph of $y=(x-1)(4-x)$ for values of x from 0 to 5. From your graph find the greatest value of $(x-1)(4-x)$.

6) Write down the three values missing from the following table which gives values of $\frac{1}{2}(3x^2-5x-1)$ for values of x from -2 to 3.

x	-2	$-1{\cdot}5$	-1	$-0{\cdot}5$	0	$0{\cdot}5$
$\frac{1}{2}(3x^2-5x-1)$	10·50	6·63		1·13		$-1{\cdot}38$

x	1	1·5	2	2·5	3·0
$\frac{1}{2}(3x^2-5x-1)$	$-1{\cdot}50$	$-0{\cdot}88$		2·63	5·50

Draw the graph of $y=\frac{1}{2}(3x^2-5x-1)$ and from it find the minimum value of $\frac{1}{2}(3x^2-5x-1)$ and the value of x at which it occurs.

7) A piece of sheet metal 20 cm × 12 cm is used to make an open box. To do this, squares of side x cm are cut from the corners and the sides and ends folded over. Show that the volume of the box is

$$V=x(20-2x)(12-2x)$$

By taking values of x from 1 cm to 5 cm in 0·5 cm steps, plot a graph of V against x and find the value of x which gives a maximum volume. What is the maximum volume of the box?

8) An open tank which has a square base of x metres has to hold 200 cubic metres of liquid when full. Show that the height of the tank is $\frac{200}{x^2}$ and hence prove that the surface area of the tank is given by $A=\left(x^2+\frac{800}{x}\right)$ square metres. By plotting a graph of A against x find the dimensions of the tank so that the surface area is a minimum. (Take values of x from 3 to 9).

9) A rectangular parcel of length x metres, width k metres and height k metres is to be sent through the post. The total length and girth (i.e. the distance round) of the parcel is to be exactly 2 metres. Show that the volume of the parcel is

$$V=\frac{x}{16}(2-x)^2$$

Draw a graph of V against x for values of x from 0·3 to 1 in steps of 0·1 and hence find the dimensions of the parcel which has the greatest possible volume.

10) A farmer uses 100 m of hurdles to make a rectangular cattle pen. If he makes a pen of length x metres show that the area enclosed is $(50x-x^2)$ square metres. Draw the graph of $y=50x-x^2$ for values of x between 0 and 50 and use your graph to find

a) the greatest possible area that can be enclosed

b) the dimensions of the pen when the area enclosed is 450 square metres.

MAXIMUM AND MINIMUM VALUES USING THE CALCULUS

In Fig. 20.10 the point P is a maximum turning point and the point Q is a minimum turning point. At both P and Q the tangent to the curve is parallel to the x-axis and hence at both points

$$\frac{dy}{dx}=0$$

By using the fact that $\frac{dy}{dx}=0$ at a turning point, we can find the turning points without drawing a graph.

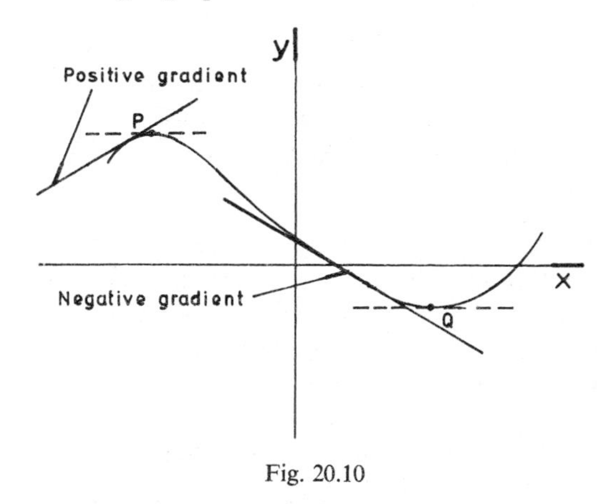

Fig. 20.10

Examples

1) Find the maximum and minimum values of $y=x^3-6x^2+9x+2$.

$$y=x^3-6x^2+9x+2$$

$$\frac{dy}{dx}=3x^2-12x+9$$

At a turning point $\frac{dy}{dx}=0$. Hence at a turning point

$$3x^2-12x+9=0$$

$$x^2-4x+3=0$$

$$(x-3)(x-1)=0$$

$$x=1 \text{ or } 3$$

Hence the turning points occur when $x=1$ and $x=3$. It now remains for us to determine which of these values of x makes y a maximum and which makes y a minimum. From Fig. 20.10 we see that the gradient of the curve is *negative* just before a *minimum* turning point and it is *positive* just before a *maximum* turning point. Therefore if we take a value of x just slightly less than the value of x at the turning point and substitute this value in the expression for $\frac{dy}{dx}$ we can discover which turning point is a maximum and which is a minimum.

When $x=1$: Take x slightly less than 1, say 0·9. Substituting this value in the expression for $\frac{dy}{dx}$:

$$\frac{dy}{dx}=3\times0{\cdot}9^2-12\times0{\cdot}9+9=0{\cdot}63$$

This value of $\frac{dy}{dx}$ is positive and hence when $x=1$ we have a maximum turning point. The maximum value of y is

$$y=1^3-6\times1^2+9\times1+2=6$$

When $x=3$: Take x as slightly less than 3, say 2·9. Substituting for $x=2{\cdot}9$ in the expression for $\frac{dy}{dx}$. We have

$$\frac{dy}{dx}=3\times 2{\cdot}9^2-12\times 2{\cdot}9+9=-0{\cdot}57$$

This value of $\frac{dy}{dx}$ is negative and hence when $x=3$ we have a minimum turning point. The minimum value of y is

$$y=3^5-6\times 3^2+9\times 3+2=2$$

2) The total area of the surface of a solid cylinder is 132 cm^2. If the height of the cylinder is h cm and its radius is r cm, show that $h=\frac{21}{r}-r$. Hence calculate the value of r for which the volume of the cylinder is a maximum. $\left(\text{Take } \pi=\frac{22}{7}\right)$.

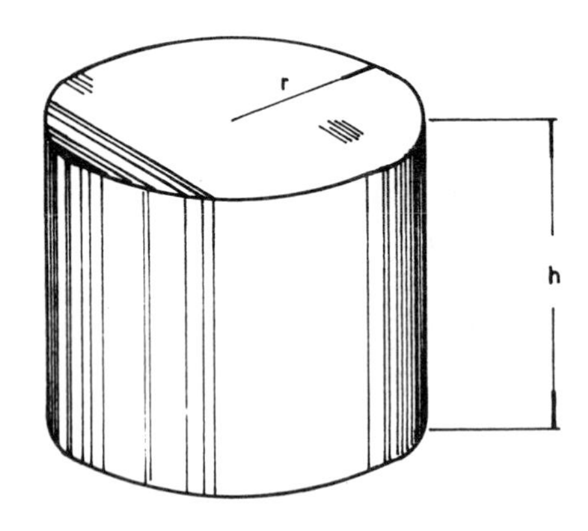

Fig. 20.11

From Fig. 20.11 the surface area of the solid cylinder is

$$A=2\pi r^2+2\pi rh$$

Since the surface area is 132 cm^2

$$2\pi r^2+2\pi rh=132$$

$$2\pi rh=132-2\pi r^2$$

$$h=\frac{132}{2\pi r}-\frac{2\pi r^2}{2\pi r}$$

$$h=\frac{21}{r}-r$$

The volume of the cylinder is

$$V=\pi r^2 h$$

Substituting for h,

$$V=\pi r^2\left(\frac{21}{r}-r\right)$$

$$V=21\pi r-\pi r^3$$

$$\frac{dV}{dr}=21\pi-3\pi r^2$$

For a maximum or a minimum $\frac{dV}{dr}=0$

$$\therefore\ 21\pi-3\pi r^2=0$$

$$3\pi r^2=21\pi$$

$$r^2=7$$

$$r=\pm\sqrt{7}$$

Only the positive value applies and hence $r=\sqrt{7}$. This must give a maximum volume since the minimum value of the volume must be zero. (i.e. a cylinder having a very large radius and practically no height or vice versa).

Exercise 73

1) Find the maximum and minimum values of
   a) $y=2x^3-3x^2-12x+4$
   b) $y=x^3-3x^2+4$
   c) $y=-6x^2+x^3$

2) Given that $y=60x+3x^2-4x^3$, calculate
   a) the gradient of the tangent to the curve of y at the point where $x=1$;
   b) the value of x for which y has its maximum value;
   c) the value of x for which y has its minimum value.

3) Calculate the co-ordinates of the points on the curve $y=x^3-3x^2-9x+12$ at each of which the tangent to the curve is parallel to the x-axis.

4) A curve has the equation $y=8+2x-x^2$. Find a) the value of x for which the gradient of the curve is 6; b) the value of x which gives the maximum value of y; c) the maximum value of y.

5) The curve $y=2x^2+\frac{k}{x}$ has a gradient of 5

when $x=2$. Calculate a) the value of k; b) the minimum value of y.

6) From a rectangular sheet of metal measuring 12 cm by 7·5 cm equal squares of side x are cut from each of the corners. The remaining flaps are then folded upwards to form an open box. Prove that the volume of the box is given by $V=90x-39x^2+4x^3$. Find the value of x such that the volume is a maximum.

7) An open rectangular tank of height h metres with a square base of side x metres is to be constructed so that it has a capacity of 500 cubic metres. Prove that the surface area of the four walls and the base will be $\left(\frac{2000}{x}+x^2\right)$ square metres. Find the value of x for this expression to be a minimum.

8) The volume of a cone is given by the formula $V=\frac{1}{3}\pi r^2 h$, where h is the height of the cone and r its radius. If $h=6-r$, calculate the value of r for which the volume is a maximum.

9) A box without a lid has a square base of side x cm and rectangular sides of height h cm. It is made from 108 cm^2 of sheet metal of negligible thickness. Prove that $h=\dfrac{108-x^2}{4x}$ and that the volume of the box is $(27x-\frac{1}{4}x^3)$. Hence calculate the maximum volume of the box.

10) A cylindrical tank, with an open top, is to be made to hold 300 cubic metres of liquid. Find the dimensions of the tank so that its surface area shall be a minimum.

SELF TEST 20

In questions 1 to 30 the answer is either 'true' or 'false'.

1) The gradient of the curve (Fig. 20.12) at the point where $x=1$ is 4. - - - - -

2) The gradient of the curve (Fig. 20.12) at the point where $x=-1$ is 4. - - - - -

3) The gradient of the curve (Fig. 20.12) at the point where $x=-1{\cdot}5$ is -6. - - - - -

4) In Fig. 20.12, the value of x is 0.5 when the gradient of the curve is 2. - - - - -

5) In Fig. 20.12, the value of $x=-1$ when the gradient of the curve is -4. - - - - -

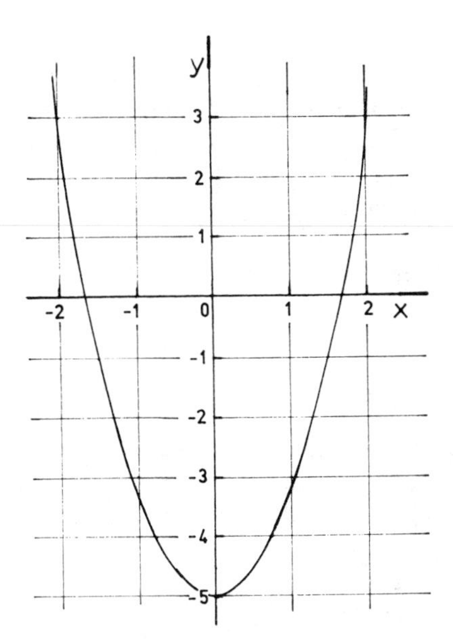

Fig. 20.12

6) If $y=x^4$ then $\dfrac{dy}{dx}=4x^3$. - - - - -

7) If $y=\dfrac{1}{x^2}$ then $\dfrac{dy}{dx}=2x$. - - - -

8) If $y=\dfrac{1}{x^3}$ then $\dfrac{dy}{dx}=-\dfrac{3}{x^4}$. - - - - -

9) If $y=\sqrt[3]{x^2}$ then $\dfrac{dy}{dx}=\dfrac{2}{3\sqrt[3]{x}}$. - - - - -

10) If $y=\sqrt[4]{x^3}$ then $\dfrac{dy}{dx}=-\dfrac{3}{4\sqrt[4]{x}}$. - - - - -

11) If $y=\frac{8}{x^4}$ then $\frac{dy}{dx}=-\frac{32}{x^5}$. - - - - -

12) If $y=7x^2-4x+3$ then $\frac{dy}{dx}=14x-4$. - - - - -

13) If $y=\frac{3+x^3}{x}$ then $\frac{dy}{dx}=-3$. - - - - -

14) If $y=\frac{5+x^4}{x^2}$ then $\frac{dy}{dx}=2x-\frac{10}{x^3}$. - - - - -

15) If $y=x^2-7x+5$, the gradient of the curve at the point where $x=2$ is -3. - - - - -

16) If $y=x^2-7x+5$, the point where the gradient is 10 is $x=8{\cdot}5$. - - - - -

17) If $y=2x-\frac{16}{x^2}$, the gradient of the curve at the point where $x=2$ is -2. - - - - -

18) If $y=2x-\frac{32}{x^2}$, the gradient of the curve at the point where $x=2$ is 10. - - - - -

19) A body moves a distance of s metres in t seconds so that $s=t^3-2t^2+3t+2$. The velocity of the body after 2 seconds is 7 metres per second. - - - - -

20) The acceleration of the body in question 19 after a time of 2 seconds is 8 m/s^2. - - - - -

21) A body moves a distance of s metres in t seconds so that $s=2t^3-15t^2+36t+2$. Its velocity will be zero when $t=3$ seconds and $t=2$ seconds. - - - - -

22) The body in question 21 will have zero acceleration when $t=2\frac{1}{2}$ seconds. - - - - -

23) A body moves a distance of s metres in t seconds so that $s=\frac{4+t^2}{t}$. The body will stop when $t=4$ seconds - - - - -

24) If $y=x^2+3x-2$, the turning point on the curve occurs when $x=-1{\cdot}5$. - - - - -

25) If $y=x^3-5x^2-8x+3$, then there will be two turning points on the curve. - - - - -

26) The turning points in question 25 occur when $x=-1{\cdot}5$ and $x=4$. - - - - -

27) In question 26, the turning point at $x=4$ is a minimum. - - - - -

28) The minimum value of the curve of $y=2x^2-8x+3$ is -5. - - - - -

29) The maximum value of the curve of $y=-x^2+8x+7$ is 23. - - - - -

30) The greatest value of $(x-2)(4-x)$ is 7. - - - - -

In questions 31 to 40 state the letter (or letters) corresponding to the correct answer (or answers).

31) In the curve (Fig. 20.13), the gradient of the curve at the point where $x=-2$ is

a -6 b 6 c $-\frac{1}{6}$ d $\frac{1}{6}$

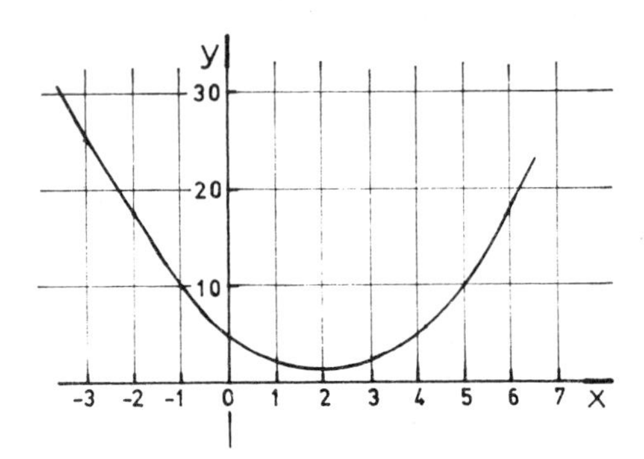

Fig. 20.13

32) In the curve (Fig. 20.13) the gradient of the curve at the point where $x=3$ is

a -2 b 2 c $-\frac{1}{2}$ d $\frac{1}{2}$

33) If $y=\frac{3x^3-2x^2}{x}$, then $\frac{dy}{dx}$ is equal to

a $9x^2-4x-\frac{1}{x^2}$ b $9x^2-4x+\frac{1}{x^2}$.

c $6x-2$ d $2-6x$

34) If $y=5\times\sqrt[3]{x^2}$, then $\frac{dy}{dx}$ is equal to

a $\frac{15\sqrt{x}}{2}$ b $\frac{15}{2\sqrt{x}}$

c $\frac{10}{3\times\sqrt[3]{x}}$ d $\frac{10\times\sqrt[3]{x}}{3}$

35) If $y=3x-\frac{30}{x}$ the gradient of the curve is 10·5 at the points where x is equal to

a 2 b -2 c ±2 d 5

36) The curve $y=x^2-4$ cuts the positive x-axis at a point P. The gradient of the curve at P is

a 2 b -2 c 0 d -8

37) The maximum value of the function $y=2x^3-21x^2+72x+5$ is

a 4 b 3 c 85 d 86

38) The maximum value of the function shown in Fig. 20.14 is

a -2 b 25 c 2 d 30

39) The minimum value of the function shown in Fig. 20.14 is

a -3 b 10 c -1 d 20

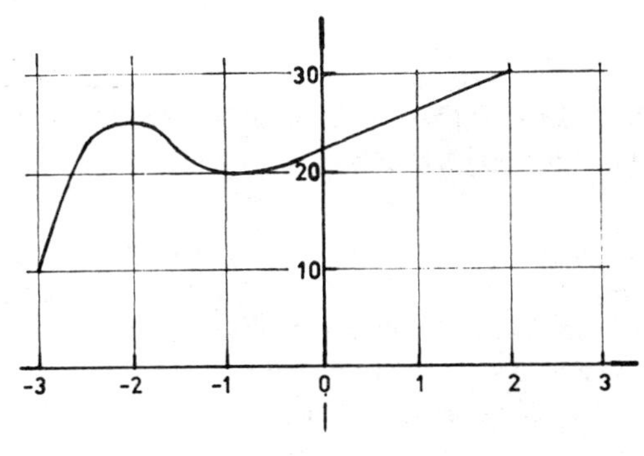

Fig. 20.14

40) A particle moves s metres in a time t seconds so that $s=2t^2-6t+5$. Hence the distance the particle travels before it comes to rest is

a 3 m b 3·5 m c 4 m
d 1·5 m

Chapter 21 Integration

INTEGRATION AS THE INVERSE OF DIFFERENTIATION

In Chapter 20 we discovered how to obtain the differential coefficients of various functions. Our objective in this chapter is to find out how to reverse the process. That is, being given the differential coefficient of a function we try to discover the original function.

If $y=\frac{x^4}{4}$ then $\frac{dy}{dx}=x^3$

We may write

$$dy=x^3\,dx$$

The expression $x^3\,dx$ is called the **differential** of $\frac{x^4}{4}$.

Reversing the process of differentiation is called integration. It is indicated by using the integration sign $\int$ in front of the differential. Thus if

$$dy=x^3\,dx$$

$$y=\int x^3\,dx=\frac{x^4}{4}$$

Similarly if $y=\frac{x^5}{5}$

$$\frac{dy}{dx}=x^4$$

$$dy=x^4\,dx$$

Reversing the process

$$y=\int x^4\,dx=\frac{x^5}{5}$$

If $y=\frac{x^{n+1}}{n+1}$, $\quad\frac{dy}{dx}=x^n$

$$dy=x^n\,dx$$

$$\mathbf{y=\int x^n\,dx=\frac{x^{n+1}}{n+1}}$$

$\frac{x^{n+1}}{n+1}$ is called the integral of $x^n\,dx$.

This rule applies to all indices, positive, negative and fractional except for

$$\int x^{-1}\,dx.$$

THE CONSTANT OF INTEGRATION

We know that the differential of $\frac{x^2}{2}$ is $x\,dx$.

Therefore if we are asked to integrate $x\,dx$, $\frac{x^2}{2}$ is one answer but it is not the only possible answer because $\frac{x^2}{2}+2$, $\frac{x^2}{2}+5$, $\frac{x^2}{2}+19$ etc. are all expressions whose differential is $x\,dx$. The general expression for $\int x\,dx$ is therefore $\frac{x^2}{2}+c$, where c is a constant known as the constant of integration. Every time we integrate the constant of integration must be added in.

$$\int x^n\,dx=\frac{x^{n+1}}{n+1}+c$$

Examples

1) $\int x^5\,dx=\frac{x^{5+1}}{5+1}+c=\frac{x^6}{6}+c$

2) $\int x\,dx=\dfrac{x^{1+1}}{1+1}+c=\dfrac{x^2}{2}+c$

3) $\int\sqrt{x}\,dx=\int x^{1/2}\,dx=\dfrac{x^{3/2}}{\frac{3}{2}}+c$

$=\dfrac{2x^{3/2}}{3}+c$

4) $\int\dfrac{dx}{x^3}=\int x^{-3}\,dx=\dfrac{x^{-2}}{-2}+c$

$=-\dfrac{2}{x^2}+c$

A constant coefficient may be taken outside the integral sign. Thus

$$\int 3x^2\,dx=3\int x^2\,dx=3\frac{x^3}{3}+c=x^3+c$$

The integral of a sum is the sum of their separate integrals.

Examples

1) $\int(x^2+x)\,dx$

Integrating each term separately

The integral of x^2 is $\dfrac{x^3}{3}$

The integral of x is $\dfrac{x^2}{2}$

$\therefore\ \int(x^2+x)\,dx=\dfrac{x^3}{3}+\dfrac{x^2}{2}+c$

2) $\int(5x^4+3x^2+2x-6)\,dx$

$=x^5+x^3+x^2-6x+c$

3) $\int(2x+5)^2\,dx=\int(4x^2+20x+25)\,dx$

$=\dfrac{4x^3}{3}+\dfrac{20x^2}{2}+25x+c$

$=\dfrac{4x^3}{3}+10x^2+25x+c$

Exercise 74

Integrate with respect to x

1) x^2 2) x^8

3) $\sqrt{x}$ 4) $\dfrac{1}{x^2}$

5) $\dfrac{1}{x^4}$ 6) $\dfrac{1}{\sqrt{x}}$

7) $3x^4$ 8) $5x^8$

9) x^2+x+3 10) $2x^3-7x-4$

11) $x^2-5x+\dfrac{1}{\sqrt{x}}+\dfrac{2}{x^2}$

12) $\dfrac{8}{x^3}-\dfrac{2}{x^2}+\sqrt{x}$

13) $(x-2)(x-1)$

14) $(x+3)^2$

15) $(2x-7)^2$

EVALUATING THE CONSTANT OF INTEGRATION

The value of the constant of integration may be found provided a corresponding pair of values of x and y are known.

Example

The gradient of the curve which passes through the point (2, 3) is given by x^2. Find the equation of the curve.

We are given $\dfrac{dy}{dx}=x^2$

$\therefore\ y=\int x^2\,dx=\dfrac{x^3}{3}+c$

We are also given that when $x=2$, $y=3$. Substituting these values in

$y=\dfrac{x^3}{3}+c$

$$3=\frac{2^3}{3}+c$$

$$3=2\tfrac{2}{3}+c$$

$$\therefore\ c=\frac{1}{3}$$

Hence the equation of the curve is

$$y=\frac{x^3}{3}+\frac{1}{3}=\frac{1}{3}(x^3+1)$$

Exercise 75

1) The gradient of the curve which passes through the point (2, 3) is given by x. Find the equation of the curve.

2) The gradient of the curve which passes through the point (3, 8) is given by (x^2+3). Find the value of y when $x=5$.

3) It is known that for a certain curve $\frac{dy}{dx}=3-2x$ and the curve cuts the x-axis where $x=5$. Express y in terms of x. State the length of the intercept on the y-axis and calculate the maximum value of y.

4) Find the equation of the curve which passes through the point (1, 4) and is such that $\frac{dy}{dx}=2x^2+3x+2$.

5) If $\frac{dp}{dt}=(3-t)^2$ find p in terms of t given that $p=3$ when $t=2$.

6) A curve passes through the point (1, −2) and is such that $\frac{dy}{dx}=3x^2-4x+1$. Prove that the equation of the curve may be written as $y=(x^2+1)(x-2)$.

7) The gradient of a curve is $ax+b$ at all points, where a and b are constants. Find the equation of the curve given that it passes through the points (0, 4) and (1, 3) and that the tangent at (1, 3) is parallel to the x-axis.

8) A curve passes through the point (0, 1) and is such that at every point of the curve $\frac{dy}{dx}=x^2$. Sketch the curve.

9) A curve passes through the point (2, 3) and is such that $\frac{dy}{dx}=2-x^2$. Find the value of y when $x=1$.

THE DEFINITE INTEGRAL

It has been shown that $\int x^n\,dx=\frac{x^{n+1}}{n+1}+c$. This expression is called an *indefinite integral* and it must contain an arbitrary constant. For many purposes we require *definite integrals* which are written $\int_b^a x^n\,dx$. The values of a and b are called the limits, a being the upper limit and b the lower limit. The method of evaluating a definite integral is shown in the following examples.

Examples

1) Find the value of $\int_2^3 x^2\,dx$

$$\int_2^3 x^2\,dx=\left[\frac{x^3}{3}\right]_2^3$$

$$=\left(\text{value of }\frac{x^3}{3}\text{ when }x\text{ is put equal to }3\right)$$

$$-\left(\text{value of }\frac{x^3}{3}\text{ when }x\text{ is put equal to }2\right)$$

$$=\frac{3^3}{3}-\frac{2^3}{3}=\frac{27}{3}-\frac{8}{3}=\frac{19}{3}=6\tfrac{1}{3}$$

2) Find the value of $\int_1^2 (3x^2-2x+5)\,dx$

$$\int_1^2 (3x^2-2x+5)=[x^3-x^2+5x]_1^2$$

$$=(2^3-2^2+5\times 2)$$

$$-(1^3-1^2+5\times 1)$$

$$=14-5=9$$

Exercise 76

Evaluate the following definite integrals:

1) $\int_1^2 x^2\,dx$

2) $\int_2^3 (2x+3)\,dx$

3) $\int_0^2 (x^2+3)\,dx$

4) $\int_1^2 (3x^2-4x+3)\,dx$

5) $\int_1^2 x(2x-1)\,dx$

6) $\int_0^2 \sqrt{x}\,dx$

7) $\int_1^3 \frac{1}{x^2}\,dx$

8) $\int_2^4 (x-1)(x-3)\,dx$

AREA UNDER A CURVE

The general expression for finding the area bounded by a curve, the x-axis and the lines $x=a$ and $x=b$ (Fig. 21.1) is

$$A=\int_a^b y\,dx$$

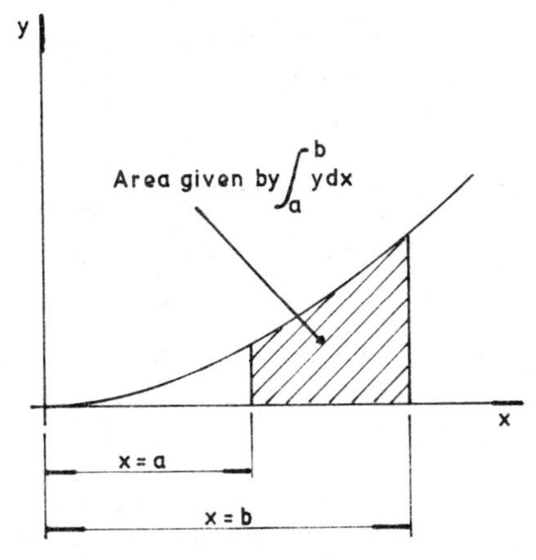

Fig. 21.1

Examples

1) Find the area bounded by the curve $y=x^3+3$, the x-axis and the lines $x=1$ and $x=3$

$$A=\int_1^3 (x^3+3)\,dx=\left[\frac{x^4}{4}+3x\right]_1^3$$

$$=\left[\frac{3^4}{4}+3\times 3\right]-\left[\frac{1^4}{4}+3\times 1\right]$$

$$=29\tfrac{1}{4}-3\tfrac{1}{4}=26 \text{ square units}$$

2) Find the area between the straight line $y=12+3x$ and the curve $y=2x^2+3$.

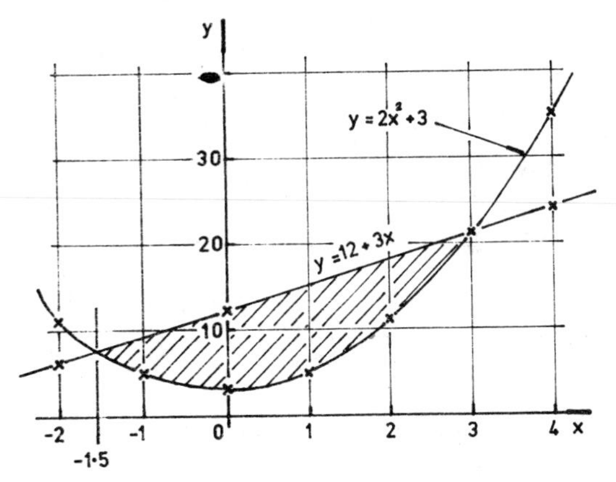

Fig. 21.2

Referring to Fig. 21.2 it will be seen that the shaded area is the one required. The first step is to find the points of the intersection of the line and curve. At the points of intersection:

$$2x^2+3=12+3x$$

$$2x^2-3x-9=0$$

$$(2x+3)(x-3)=0$$

$$x=-1{\cdot}5 \text{ and } 3$$

The shaded area = area under the line − area under the curve

$$=\int_{-1\cdot5}^{3} (12+3x)\,dx-\int_{-1\cdot5}^{3} (2x^2+3)\,dx$$

$$\int_{-1\cdot5}^{3}(12+3x)\,dx=\left[12x+\frac{3x^2}{2}\right]_{-1\cdot5}^{3}$$

$$=(36+13\cdot5)$$

$$-(-18+3\cdot375)$$

$$=64\cdot375 \text{ square units}$$

$$\int_{-1\cdot5}^{3}(2x^2+3)\,dx=\left[\frac{2x^3}{3}+3x\right]_{-1\cdot5}^{3}$$

$$=(18+9)-(-2\cdot25-4\cdot5)$$

$$=33\cdot75 \text{ square units}$$

$\therefore$ Shaded area $=64\cdot375-33\cdot75$

$=30\cdot625$ square units

Exercise 77

Find the areas under the following curves:

1) Between the curve $y=x^3$, the x-axis and the lines $x=5$ and $x=3$.

2) Between the curve $y=3+2x+3x^2$, the x-axis and the lines $x=1$ and $x=4$.

3) Between the curve $y=x^2(2x-1)$, the x-axis and the lines $x=1$ and $x=2$.

4) Between the curve $y=\frac{1}{x^2}$, the x-axis and the lines $x=1$ and $x=3$.

5) Between the curve $y=5x-x^3$, the x-axis and the lines $x=1$ and $x=2$.

6) A (2, 8) and B (4, 8) are two points on the curve $y=x(6-x)$. Calculate the area bounded by the straight line AB and the curve.

7) Calculate the area enclosed between the line $y=5$, the curve $y=12-4x+3x^2$ and the ordinates $x=-2$ and $x=3$.

8) Calculate the area enclosed between the x-axis and that part of the curve $y=10+3x-x^2$ which lies above it.

9) Prove that the curve $y=4x-\frac{32}{x^2}$ crosses the x-axis at the point where $x=2$. Calculate the area between the curve, the x-axis and the ordinate $x=4$.

10) Find the area between the straight line $y=3x+2$ and the curve $y=x^2+4$.

DISTANCE, VELOCITY AND ACCELERATION

It was shown in Chapter 20 that if a body travels s metres in time of t seconds then the velocity of the body is

$$v=\frac{ds}{dt}$$

Hence $s=\int v\,dt+c$

Therefore to find the distance travelled we integrate the expression for the velocity.
The acceleration of the body is

$$a=\frac{dv}{dt}$$

or $v=\int a\,dt+c$

Hence to find the velocity we integrate the expression for the acceleration.
The area under a velocity-time graph represents the distance travelled, whilst the area under an acceleration time graph represents the velocity.

Examples

1) The velocity of a body, v metres per second, after a time of t seconds is given by

$$v=t^2+1$$

Find the distance travelled at the end of 2 seconds.
When $t=0$ the distance travelled will be 0 metres. Hence the distance travelled at the end of 2 seconds is found by integrating the

expression for v between the limits of 2 and 0.

$$s=\int_0^2 (t^2+1)\,dt$$

$$=\left[\frac{t^3}{3}+t\right]_0^2=\frac{2^3}{3}+2=4\tfrac{2}{3} \text{ metres}$$

2) The acceleration of a moving body at the end of t seconds from the commencement of motion is (9 − t) metres per second. Find the velocity and the distance travelled at the end of 2 seconds if the initial velocity is 5 metres per second.

$$v=\int a\,dt+c=\int (9-t)\,dt+c$$

$$=9t-\frac{t^2}{2}+c$$

The initial velocity is the velocity when $t=0$. Hence when $t=0$, $v=5$

$$\therefore\ 5=9\times 0-0+c \quad \text{or} \quad c=5$$

$$\therefore\ v=9t-\frac{t^2}{2}+5$$

When

$$t=2, v=9\times 2-\frac{2^2}{2}+5=21 \text{ metres per second.}$$

$$\text{Now } s=\int v\,dt=\int\left(9t-\frac{t^2}{2}+5\right)dt+d$$

$$=\frac{9t^2}{2}-\frac{t^3}{6}+5t+d$$

Unless information is given to the contrary it is always assumed that $s=0$ when $t=0$
Hence $d=0$

$$\text{and } s=\frac{9t^2}{2}-\frac{t^3}{6}+5t$$

When $t=2$,

$$s=\frac{9\times 2^2}{2}-\frac{2^3}{6}+5\times 2=26\tfrac{2}{3} \text{ metres}$$

Exercise 78

1) The velocity of a body is $(t+1)$ metres per second after a time of t seconds. Find the distance travelled at the end of 3 seconds.

2) The acceleration of a moving body at the end of t seconds from the commencement of motion is $(5-t)$ m/s^2. If the initial velocity is 10 m/s find the velocity and distance travelled at the end of 3 seconds.

3) The acceleration of a moving body is constant at 15 m/s^2. If the initial velocity is 10 m/s, derive an expression for the distance moved in t seconds. Hence find the distance travelled at the end of 4 seconds.

4) The motion of a body is given by

$$\frac{ds}{dt}=3t^2-2t+c$$

If the displacement of the particle is 5 m when $t=1$ second and 2 m when $t=\frac{1}{2}$ second, find the displacement when $t=2$.

5) A body moving along a straight path passes a fixed point O with a velocity of 12 m/s and t seconds later, when it is s metres from O its acceleration is $6t$. Find the velocity of the body when $t=4$ and its distance from O at that instant.

6) The acceleration of a body is $(3t^2+5)$ m/s^2 after a time of t seconds. If its initial velocity is 8 m/s, find the distance travelled at the end of 4 seconds.

SOLID OF REVOLUTION

If the area under a curve is rotated about the x-axis, the solid which results is called a solid of revolution. Any section of this solid by a plane perpendicular to the x-axis is a circle.

It can be shown that the volume of a solid of revolution is (Fig. 21.3):

$$V=\int_a^b \pi y^2\,dx$$

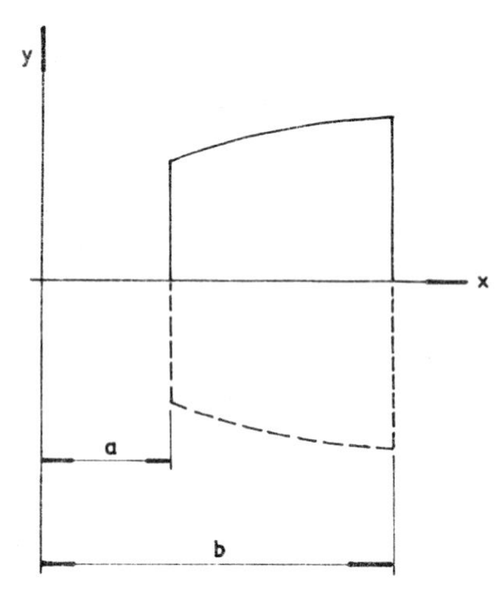

Fig. 21.3. When a curve is rotated about the x-axis the volume of the solid of revolution so produced is $\int_a^b \pi y^2\,dx$

Examples

1) The area between the curve $y=x^2$, the x-axis and the ordinates at $x=1$ and $x=3$ is rotated about the x-axis. Calculate the volume of the solid generated.

$$V=\int_1^3 \pi y^2\,dx=\pi\int_1^3 (x^2)^2\,dx=\pi\int_1^3 x^4\,dx$$

$$=\pi\left[\frac{x^5}{5}\right]_1^3=\pi\left[\frac{3^5}{5}-\frac{1^5}{5}\right]=\pi\left[\frac{243}{5}-\frac{1}{5}\right]$$

$$=48{\cdot}4\,\pi \text{ cubic units.}$$

2) The area between the curve $y^2=3x-1$, the x-axis and the ordinates at $x=2$ and $x=5$ is rotated about the x-axis. Calculate the volume of the solid generated. (You may leave π as a factor in your answer.)

$$V=\int_2^5 \pi y^2\,dx=\pi\int_2^5 (3x-1)\,dx$$

$$=\pi\left[\frac{3x^2}{2}-x\right]_2^5$$

$$=\pi\left\{\left[\frac{3\times5^2}{2}-5\right]-\left[\frac{3\times2^2}{2}-2\right]\right\}$$

$$=\pi(37{\cdot}5-5)-(6-2)$$

$$=28{\cdot}5\,\pi \text{ cubic units.}$$

Exercise 79

Find the volume of solid of revolution when the areas under the following curves are rotated about the x-axis. Leave the answers as multiples of π.

1) $y=2x^2$ between $x=0$ and $x=2$.

2) $y=\sqrt{x}$ between $x=2$ and $x=4$.

3) $y=\dfrac{1}{x^2}$ between $x=1$ and $x=3$.

4) $y=x+2$ between $x=0$ and $x=4$.

5) $y=2x+5$ between $x=1$ and $x=3$.

6) The part of the curve $y=x(x+2)$ between the ordinates $x=1$ and $x=2$ is rotated about the x-axis. Calculate the volume of the solid of revolution so formed.

7) The area bounded by the x-axis, the curve $y=3x-\dfrac{2}{x}$ and the ordinates $x=1$ and $x=4$ is rotated about the x-axis. Calculate the volume of the solid generated.

8) The area between the curve $y^2=2x+5$, the x-axis and the ordinates at $x=1$ and $x=3$ is rotated about the x-axis. Calculate the volume of the solid generated.

9) The curve represented by the equation $y=x^2-3x$ cuts the x-axis at two points A and B. The area bounded by the x-axis and the arc of the curve AB is rotated through a complete revolution about the x-axis. Calculate the volume generated.

10) The area bounded by the curve $y=x^3-6x^2$, the line $x=1$ and the x-axis from $x=0$ to $x=1$ is rotated about the x-axis. Calculate the volume generated.

SELF TEST 21

In the following questions state the letter (or letters) corresponding to the correct answer (or answers).

1) $\int x^2\,dx$ is equal to

a $\frac{x^3}{3}$ b $\frac{x^3}{3}+c$ c $2x$ d $2x+c$

2) $2\int x^2\,dx$ is equal to

a $\frac{2x^3}{3}$ b $\frac{2x^3}{3}+c$ c $4x$

d x^3+c

3) $\int \frac{dx}{x^2}$ is equal to

a $-\frac{2}{x}$ b $-\frac{2}{x}+c$ c $-\frac{1}{x}$

d $-\frac{1}{x}+c$

4) $\int \frac{1}{\sqrt{x}}\,dx$ is equal to

a $\frac{1}{2\sqrt{x}}+c$ b $\frac{1}{2\sqrt{x}}$ c $2\sqrt{x}+c$

d $2\sqrt{x}$

5) $\int (3x^2+8x+5)\,dx$ is equal to

a x^3+4x^2+5x+c b x^3+4x^2+5x
c $6x+8+c$ d $6x+8$

6) $\int (x^2+2)\,dx$ is equal to

a $\frac{x^3}{3}+2x$ b $\frac{x^3}{3}+2x+c$

c $2x+c$ d $2x$

7) $\int (x+4)^2\,dx$ is equal to

a $2(x+4)+c$ b $\frac{(x+4)^3}{3}+c$

c $\frac{x^3}{3}+4x^2+16x+c$ d $\frac{x^3}{3}+16x+c$

8) $\int (2x-3)^2\,dx$ is equal to

a $2(2x-3)+c$ b $\frac{(2x-3)^3}{3}+c$

c $\frac{4x^3}{3}-9x+c$ d $\frac{4x^3}{3}-6x^2+9x+c$

9) The equation of a curve is $y=x^2+3x+c$. The curve passes through the point (2, 13). Hence c is equal to

a 3 b -3 c -206 d 206

10) For a certain curve $\frac{dy}{dx}=2x$ and the curve passes through the point (4, 9). Hence the equation of the curve is

a $y=x^2$ b $y=x^2-7$
c $y=x^2-65$ d $y=x^2+7$

11) In Fig. 21.4, the graph of $y=x^2-2$ is drawn. The shaded area is given by the expression

a $\int_0^1 y\,dx$ b $\int_0^{-1} y\,dx$ c $\int_{-1}^1 y\,dx$

d $\int_{-1}^1 (x^2-2)\,dx$

12) The shaded area in Fig. 21.4 is

a $-3\frac{1}{3}$ b $3\frac{1}{3}$ c $\frac{2}{3}$ d $-4\frac{2}{3}$

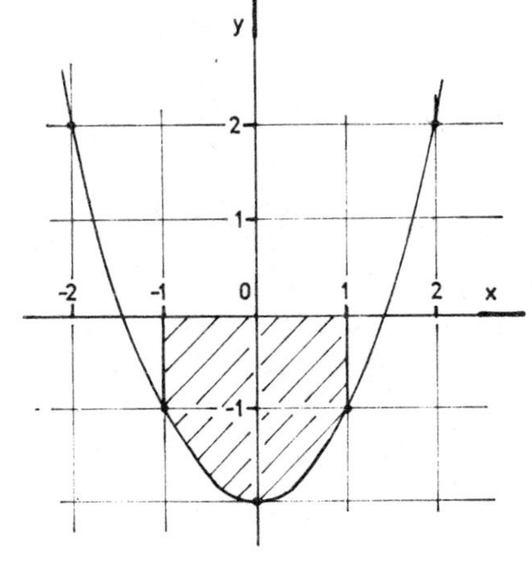

Fig. 21.4

13) The area represented by the integral $\int_{-2}^{2} y\,dx$ is the shaded area in Fig. 21.5.

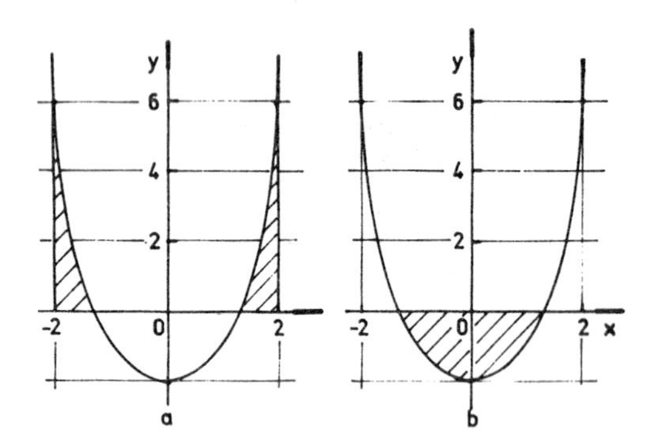

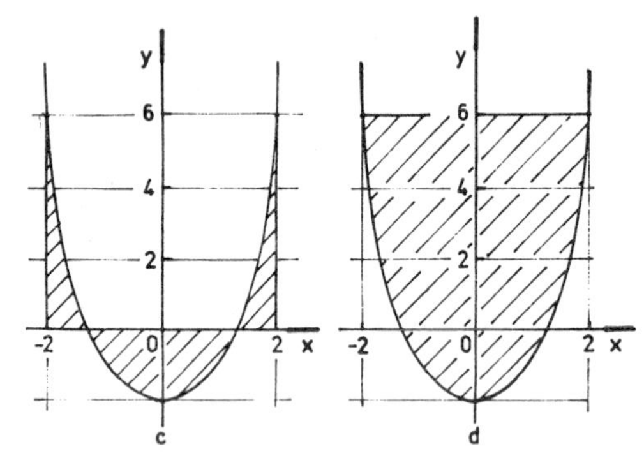

Fig. 21.5

14) The integral $\int_{-1}^{2} (3x^3-3)\,dx$ is equal to

a $2\frac{1}{4}$ b $9\frac{3}{4}$ c $5\frac{3}{4}$ d $14\frac{1}{4}$

15) The area between the straight line $y=13-2x$ and the curve $y=2x^2+1$ is given by

a $\int_{-3}^{2} (13-2x)\,dx - \int_{-3}^{2} (2x^2+1)\,dx$

b $\int_{-2}^{3} (13-2x)\,dx - \int_{-2}^{3} (2x^2+1)\,dx$

c $\int_{-2}^{3} (2x^2+1)\,dx - \int_{-2}^{3} (13-2x)\,dx$

d $\int_{-3}^{2} (2x^2+1)\,dx - \int_{-3}^{2} (13-2x)\,dx$

16) The acceleration of a moving body at the end of t seconds is $(9-t)$ m/s^2. The velocity at the end of 3 seconds if its initial velocity is 3 m/s is

a 6 m/s b 3 m/s c 25·5 m/s
c 8 m/s

17) The acceleration of a body is $(3t^2+2)$ m/s^2 after a time t seconds. Its initial velocity is 5 m/s. Hence at the end of 2 seconds the body will have travelled a distance of

a 17 m b 6 m c 18 m
d 8 m

18) The shaded area shown in Fig. 21.6 is rotated about the x-axis, The volume of the solid revolution so formed is

a $\pi \int_{0}^{2} y\,dx$ b $\pi \int_{-2}^{2} y\,dx$

c $\pi \int_{0}^{2} y^2\,dx$ d $\pi \int_{-2}^{2} y^2\,dx$

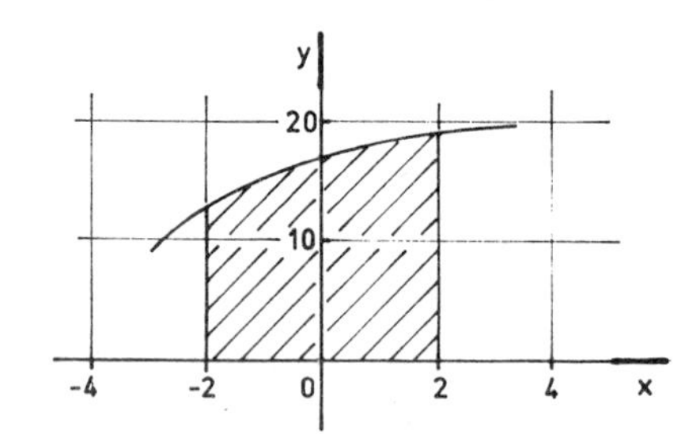

Fig. 21.6

Miscellaneous Exercise

Exercise 80

These questions are of the type found in O Level papers.

1) Calculate the distance between the points $(-2, 5)$ and $(-4, 1)$.

2) Given that $\frac{dy}{dx}=3x^2-1$ and $y=6$ when $x=1$, find y in terms of x.

3) The curve $y=2x-\frac{6}{x}$ cuts the positive x-axis at P. At the points Q and R on the curve $x=1$ and 3 respectively. Find the gradient of the curve at P and prove that the tangent at P is parallel to QR.

4) Find the gradient of the curve $(x-1)(2x-3)$ at the point where the curve cuts the y-axis.

5) The volume V cubic metres of a pyramid varies as the product of the area of the base A square metres and the height h metres. If $V=2400$ when $A=150$ and $h=48$, find the value of h when $V=6000$ and the base of the pyramid is a square of side 24 metres.

6) A rectangular tank has a horizontal square base of side 7 m and a height of 10 m.
A solid iron cylinder of radius 3 m and height 7 m rests with its curved surface in contact with the bottom of the tank. Water is poured into the tank until the cylinder is just covered. If the cylinder is now taken out of the tank, calculate the depth of water in the tank assuming none is lost in removing the cylinder.
The cylinder is now placed with one circular face resting on the bottom of the tank and with its axis vertical Find the new depth of water in the tank. $\left(\text{Take } \pi=\frac{22}{7}.\right)$

7) The table below gives the values of y corresponding to certain values of x for the function $y=9-2x-\frac{12}{x^2}$

x	1	1·5	2	2·5	3	4	5	6
y	-5	0·67		2·08		0·25		

Copy and complete this table and use it to draw the graph of this function for values of x between $x=1$ and $x=6$. From your graph estimate two roots of each of the equations
a) $2x^3-9x^2+12=0$
b) $2x^3-11x^2+12=0$

8) Prove that the function $y=3+3x^2-x^3$ has a maximum value and find this value.

9) If $\frac{dy}{dx}=2+3x$ and $y=12$ when $x=2$, find y in terms of x.

10) A quantity y is the sum of two parts, one being proportional to x and the other proportional to x^2. If $y=0$ when $x=3$, find the value of x at which $\frac{dy}{dx}=0$.

11) Draw the graph of $y=2x(x-1)(x-3)$ from $x=0$ to $x=3$. Estimate, from your graph, two values of x which satisfy the equation $2x^3-8x^2+6x+1=0$. Calculate the smaller of the two areas enclosed by the curve and the x axis.

12) A particle moves in a straight line so that its distance s metres from a fixed point in the line t seconds after being set in motion is given by the equation $s=t^3-2t^2+6t$. Find the instant when the particle first attains a velocity of 5 m/s and find the further time that elapses before the velocity is once again 5 m/s.

13) A hollow metal container consists of two parts. One of them is a hollow cylindrical pipe of internal radius $5r$ cm, external radius $6r$ cm and length $4r$ cm. The other is a hollow hemisphere of internal radius $5r$ cm and external radius $6r$ cm. The hemisphere is joined to one end of the cylinder so that the flat surfaces coincide. Find the capacity of the container in cm^3 giving your answer in terms of r and π. Prove that the volume of the metal in the container is $\dfrac{314\pi r^3}{3}$ cm^3.
Find the volume of metal in the container when its capacity is $27\frac{1}{2}$ cm^3.
(The volume of a sphere of radius r cm is $\frac{4}{3}\pi r^3$ cm^3.)

14) Prove that the function $y=x+\dfrac{4}{x}$ has a minimum value and find this value.
Prove that the graph of this function meets line $y=5$ at two points P and Q and find the co-ordinates of these points. The area bounded by the arc PQ, the x-axis and the lines through P and Q parallel to the y-axis is rotated completely about the x-axis. Find the volume generated giving your answer as a multiple of π.

15) Differentiate with respect to x, $3x-5+6x^3$.

16) Find the gradient of the line joining the points (11, 8) and (7, −2).

17) Find the length of an arc of a circle whose radius is 140 m and which subtends 54° at the centre.

18) A rectangular block of metal 38 cm long, 29 cm wide and 14 cm thick is melted and cast into a cube. Calculate the length of an edge of the cube, in cm, correct to 1 decimal place.

19) Draw up a table showing values of $y=27x-x^3-34$ from $x=0$ to $x=5$ at unit intervals of x. Draw the graph of y in this range. From the graph estimate the two roots of the equation $x^3=27x-34$. Calculate the area between the curve and the lines $x=2$, $x=4$ and $y=0$.

20) A rectangular water tank has a horizontal base of area 800 m^2. It is being filled so that the volume of water in it at the end of t minutes is V m^3 where $V=10t^2+100t$.
If the tank is full when $t=30$, calculate
a) the depth of water when the tank is full;
b) the value of t when the tank is half full;
c) the average rate of filling the tank in m^3/min in the first 7 min;
d) the rate of filling the tank when $t=7$.

21) A particle is moving in a straight line and t seconds after the commencement of motion its acceleration is $2t$ m/s^2. Given that the initial velocity of the particle was 10 m/s when the particle was at a point O in the line, calculate the distance of the particle from O after 3 seconds.

22) A flask of capacity 10 cm^3 is 2 cm high. Calculate the height of a flask of the same shape holding 5 litres.

23) Prove that the curve represented by the equation $y=x^2-3x$ passes through two points O and A on the x-axis. Write down the co-ordinates of each of these two points. Calculate the y-co-ordinate of the point P on the curve where $x=1{\cdot}5$ and calculate the gradient of the curve at this point.
Sketch the arc of the curve between O and A. The area bounded by the x-axis and the arc OA is rotated through a complete revolution about the x-axis. Calculate the volume generated, giving the answer as a multiple of π.

24) Find the co-ordinates of the points on the curve $y=x^3-6x^2$ at which y has a maximum or minimum value, distinguishing between the two. The area bounded by the curve, the line $x=1$ and the x-axis between $x=0$ to $x=1$, is rotated about the x-axis. Calculate the volume generated, giving the answer as a multiple of π.

25) Evaluate $\int_1^3 2x^2\,dx$.

26) A cone has a vertical height of 3·75 m and a base radius of 2·58 m. Use tables to calculate the volume of the cone. (Take $\pi = 3{\cdot}142$).

27) The co-ordinates of the points A and B are $(2a, 5)$ and $(a, -3)$ respectively. If AB is 10 units, calculate the possible values of a.

28) The gradient of the line joining the points (5, 6) and $(7, 3k)$ is $\frac{3}{4}$. Find k.

29) Calculate the co-ordinates of the point on the curve $y=3+14x-5x^2$ at which the tangent is parallel to the x-axis. Show that the value of y at this point is a maximum.

30) Given that $y=\sqrt{2500-9x^2}$, calculate the value of y when $x=16$. Calculate the value of x, correct to three significant figures, when $x=y$. Calculate the value of $\int_1^2 y^2\,dx$.

31) Integrate $x^3+2x-\dfrac{1}{x^3}$ with respect to x.

32) The area of the curved surface of a cone of base radius 3·52 cm is 100 cm^2. Find the slant height of the cone.

33) On the same axes draw the graphs of

$$y=\frac{x}{2}+\frac{6}{x} \quad \text{and} \quad y=7-x$$

for values of x from 1 to 6. Write down, and express in its simplest form, the equation in x whose roots are the x-co-ordinates of the points of intersection of these graphs. Write down an approximate value of the larger root of this equation.

34) A curve is represented by the equation $y=2x^3+3x^2-12x+2$. Prove that the tangent at the point P on this curve where $x=-2$ is parallel to the x-axis. Find the value of y at this point. Find also the y co-ordinates of the point Q where the curve crosses the y-axis.
Calculate the area bounded by the curve, the x and y co-ordinates and the line through P parallel to the y-axis. Hence calculate the area bounded by the curve PQ, the y-axis and the tangent at P.

35) Given that y is proportional to x^3 and that $y=24$ when $x=2$, find the value of y when $x=5$.

36) Calculate the area bounded by the x-axis and the portion of the curve $y=4x-3-x^2$ from $x=1$ to $x=3$.

37) Draw the graph of $y=10x^2-2x^3$ from $x=0$ to $x=5$ plotting points at unit intervals of x. From your graph estimate two roots of the equation $5x^2-x^3=14$. With the same axes and to the same scales draw the line $y=20-x$. Estimate two roots of the equation

$$2x^3-10x^2-x+20=0.$$

38) The temperature of a piece of metal is S degrees at time t minutes where $S=10t^2-18t+5$. Calculate
 a) the times when the temperature is zero;
 b) the minimum temperature.

39) If a graph has the equation $y=a+bx$ and $y=1$ when $x=-3$ and $y=-7$ when $x=5$, find the value of y when $x=1$.

40) Using the same axes and scales, draw the graphs of

$$y=5-x^2 \quad \text{and} \quad y=\frac{6}{2x+5}$$

for values of x from -2 to 2·5. Use these graphs to obtain approximate values for two roots of the equation

$$2x^3+5x^2-10x+19=0.$$

Chapter 22 Angles and Straight Lines

MEASUREMENT OF ANGLES

An angle may be considered to be an amount of rotation or turning. In Fig. 22.1 the line OA has been rotated about O, in an *anti-clockwise* direction, until it takes up the position OB. The angle through which the line has turned is the amount of opening between the lines OA and OB.

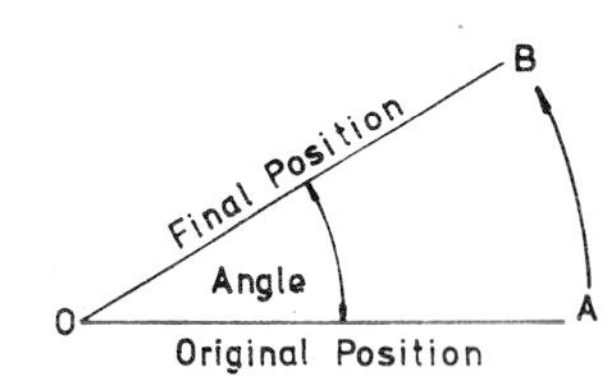

Fig. 22.1

If the line OA is rotated until it returns to its original position it will have described one revolution. Angles are often measured in degrees, minutes and seconds as follows:

60 seconds = 1 minute
60 minutes = 1 degree
360 degrees = 1 revolution

An angle of 25 degrees 7 minutes 30 seconds is written $25°7'30''$. A right angle is $\frac{1}{4}$ of a revolution and hence it contains 90°.

Examples

1) Add together 22°35′ and 49°42′.

$$\begin{array}{r} 22°35' \\ 49°42' \\ \hline 72°17' \end{array}$$

The minutes 35 and 42 add up to 77 minutes which is 1°17′. The 17 is written in the minutes column and 1° carried over to the degrees column. The degrees 22, 49 and 1 add up to 72 degrees.

2) Subtract 17°49′ from 39°27′.

$$\begin{array}{r} 39°27' \\ 17°49' \\ \hline 21°38' \end{array}$$

We cannot subtract 49′ from 27′ so we borrow 1 from the 39° making it 38°. The 27′ now becomes $27'+60'=87'$. Subtracting 49′ from 87′ gives 38′ which is written in the minutes column. The degree column is now $38°-17°=21°$.

RADIAN MEASURE

We have seen that an angle is measured in degrees. There is however a second way of measuring an angle. In this second system the unit is known as the *radian*. Referring to Fig. 22.1A,

$$\text{angle in radians} = \frac{\text{length of arc}}{\text{radius of circle}}$$

$$\theta \text{ radians} = \frac{l}{r}$$

$$l = r\theta$$

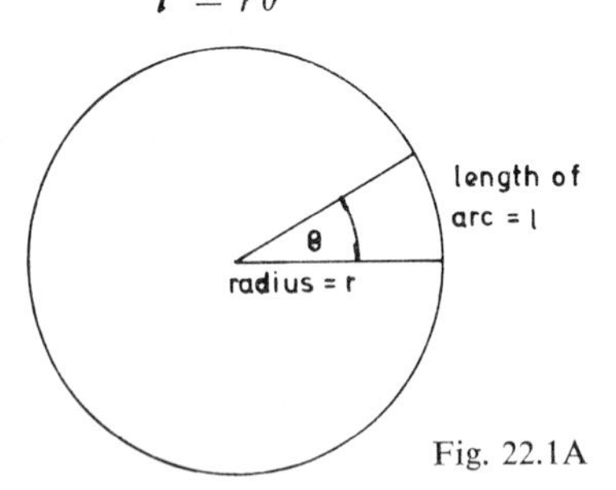

Fig. 22.1A

RELATION BETWEEN RADIANS AND DEGREES

If we make the arc AB (Fig. 22.1A) equal to a semi-circle then,

$$\text{length of arc} = \pi r$$
$$\text{angle in radians} = \frac{\pi r}{r} = \pi$$

But the angle at the centre subtended by a semi-circle is 180° and hence

$$\pi \text{ radians} = 180°$$
$$1 \text{ radian} = \frac{180°}{\pi} = 57{\cdot}3°$$

It is worth remembering that

$\theta° = \frac{\pi\theta}{180}$ radians	$90° = \frac{\pi}{2}$ radians
$60° = \frac{\pi}{3}$ radians	$30° = \frac{\pi}{6}$ radians
$45° = \frac{\pi}{4}$ radians	

Examples

1) Find the angle in radians subtended by an arc 12·9 cm long whose radius is 4·6 cm.

$$\text{Angle in radians} = \frac{\text{length of arc}}{\text{radius of circle}} = \frac{12{\cdot}9}{4{\cdot}6} = 2{\cdot}804 \text{ radians}$$

2) Express an angle of 1·26 radians in degrees and minutes.

$$\text{Angle in degrees} = \frac{180 \times \text{angle in radians}}{\pi} = \frac{180 \times 1{\cdot}26}{\pi} = 72{\cdot}18°$$

Now $0{\cdot}18° = 0{\cdot}18 \times 60$ minutes $= 11$ minutes
Angle $= 72°11'$.

3) Express an angle of 104° in radians.

$$\text{Angle in radians} = \frac{\pi \times \text{angle in degrees}}{180} = \frac{\pi \times 104}{180} = 1{\cdot}815 \text{ radians}$$

Exercise 81

Add together the following angles:

1) 11°8′ and 17°29′

2) 25°38′ and 43°45′

3) 8°38′49″ and 5°43′45″

4) 27°4′52″ and 35°43′19″

5) 72°15′4″, 89°27′38″ and 17°28′43″

Subtract the following angles:

6) 8°2′ from 29°5′

7) 17°28′ from 40°16′

8) 12°34′16″ from 20°18′12″

9) 0°7′15″ from 6°2′5″

10) 48°19′21″ from 85°17′32″

11) Find the angle in radians subtended by the following arcs:
a) arc = 10·9 cm, radius = 3·4 cm
b) arc = 7·2 m, radius = 2·3 m

12) Express the following angles in degrees and minutes:
a 5 radians b 1·73 radians c 0·159 radians

13) Express the following angles in radians:
a 83° b 189° c 295° d 5·21°.

TYPES OF ANGLES

An *acute angle* (Fig. 22.2) is less than 90°.
An *obtuse angle* (Fig. 22.3) lies between 90° and 180°.

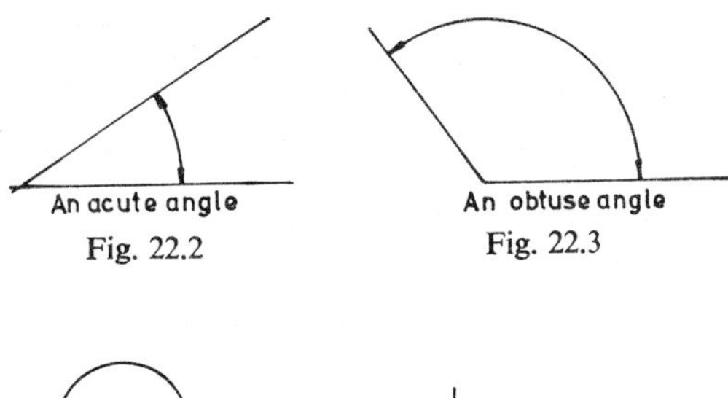

Fig. 22.2 Fig. 22.3

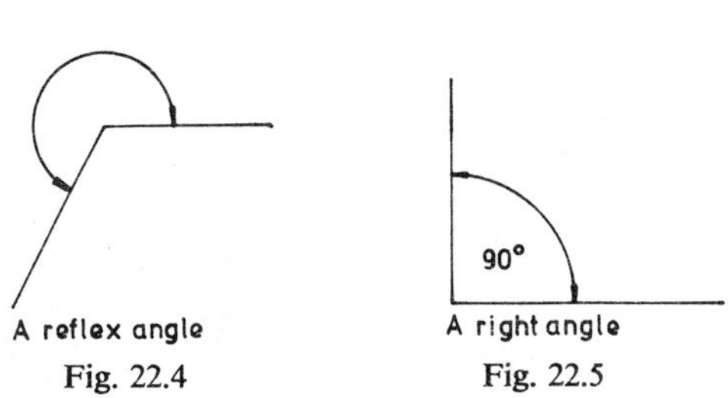

Fig. 22.4 Fig. 22.5

A *reflex angle* (Fig. 22.4) is greater than 180°.
A *right angle* (Fig. 22.5) is equal to 90°.
Complementary angles are angles whose sum is 90°.
Supplementary angles are angles whose sum is 180°.

PROPERTIES OF ANGLES AND STRAIGHT LINES

1) *The total angle on a straight line is 180°* (Fig. 22.6). The angles A and B are called adjacent angles. They are also supplementary.

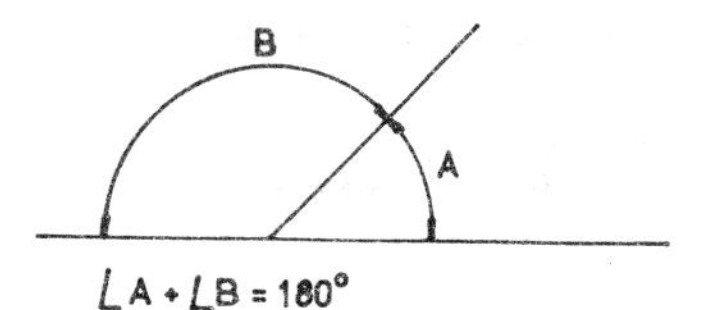

Fig. 22.6

2) *When two straight lines intersect the opposite angles are equal* (Fig. 22.7). The angles A and C are called vertically opposite angles. Similarly the angles B and D are also vertically opposite angles.

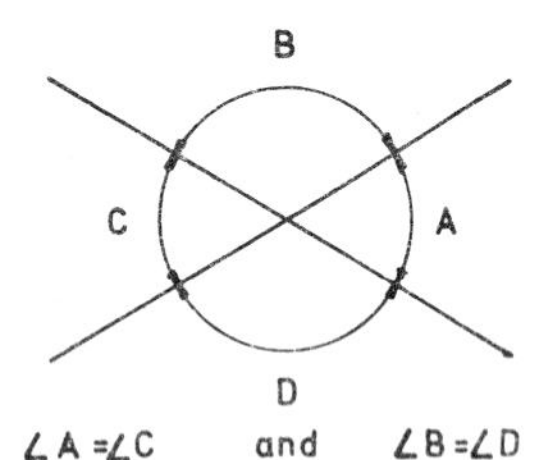

Fig. 22.7

3) *When two parallel lines are cut by a transversal* (Fig. 22.8)

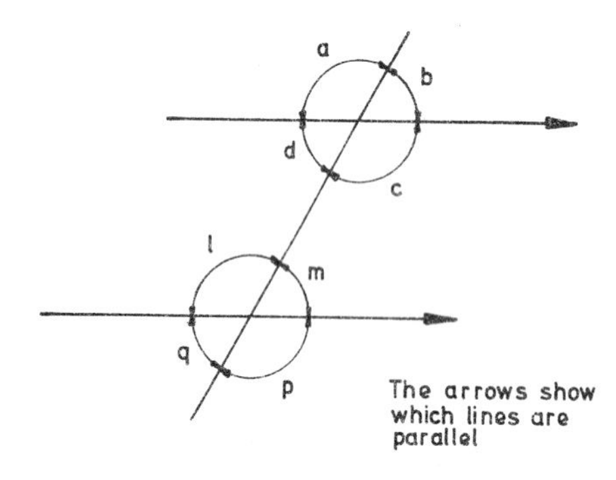

Fig. 22.8

a) The corresponding angles are equal $a=l$; $b=m$; $c=p$; $d=q$.

b) the alternate angles are equal $d=m$; $c=l$.

c) the interior angles are supplementary $d+l=180°$; $c+m=180°$.

Conversely if two straight lines are cut by a transversal the lines are parallel if any *one* of the following is true:

i) Two corresponding angles are equal.
ii) Two alternate angles are equal.
iii) Two interior angles are supplementary.

Examples

1) Find the angle A shown in Fig. 22.9.

$\angle B = 180° - 138° = 42°$
$\angle B = \angle A$ (corresponding angles)
$\angle A = 42°$

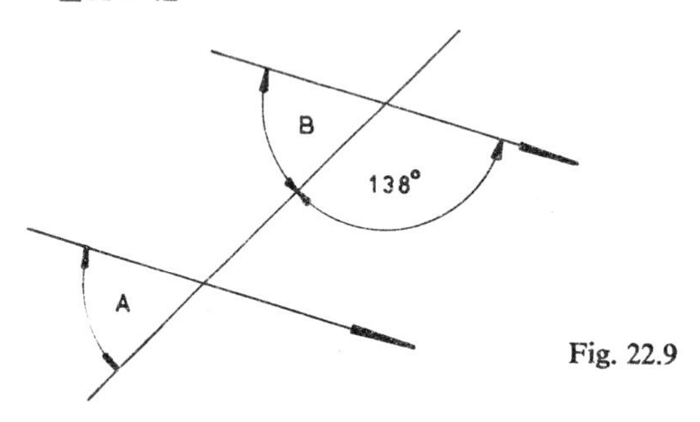

Fig. 22.9

2) In Fig. 22.10 the line BF bisects $\angle ABC$. Find the value of the angle α.

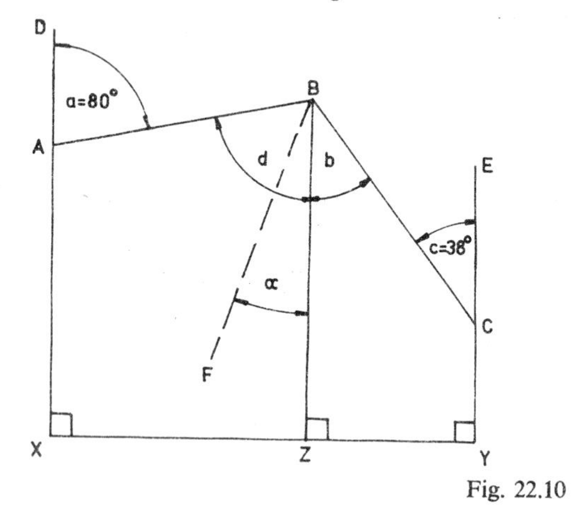

Fig. 22.10

The lines AX, BZ and EY are all parallel because they lie at right-angles to the line XY.

$\therefore$ $c=b$ (alternate angles: BZ∥EY)
$\therefore$ $b=38°$ (since $c=38°$)
$a=d$ (alternate angles: XD∥BZ)
$\therefore$ $d=80°$ (since $a=80°$)

$\angle ABC = b+d = 80° + 38° = 118°$

$\angle FBC = 118° \div 2 = 59°$ (since BF bisects $\angle ABC$)

$\therefore \; b + \alpha = 59°$

$38° + \alpha = 59°$

$\therefore \; \alpha = 59° - 38° = 21°$

Exercise 82

1) Find x in Fig. 22.11.

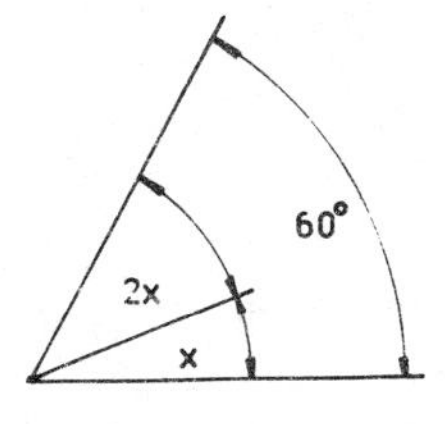

Fig. 22.11

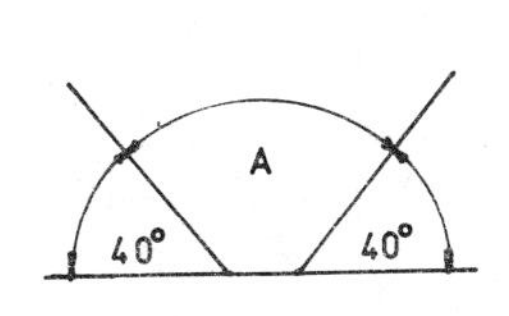

Fig. 22.12

2) Find A in Fig. 22.12.

3) Find x in Fig. 22.13.

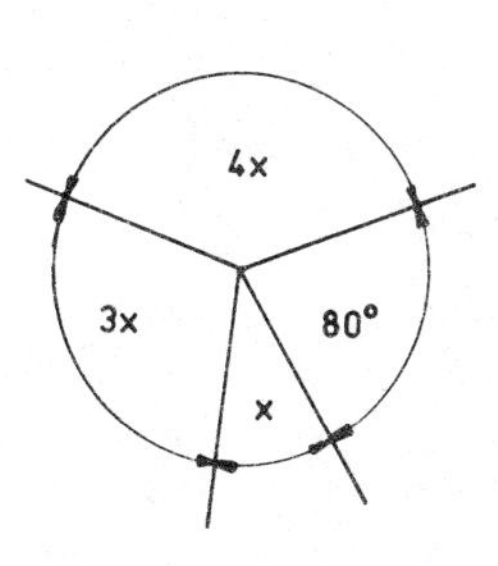

Fig. 22.13

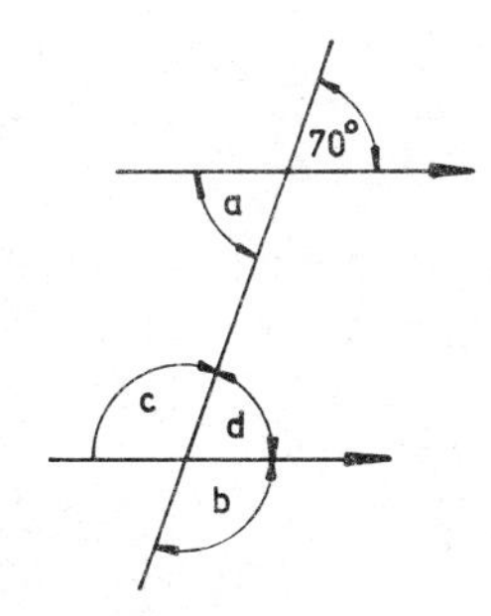

Fig. 22.14

4) In Fig. 22.14 find a, b, c and d.

5) In Fig. 22.15 find A.

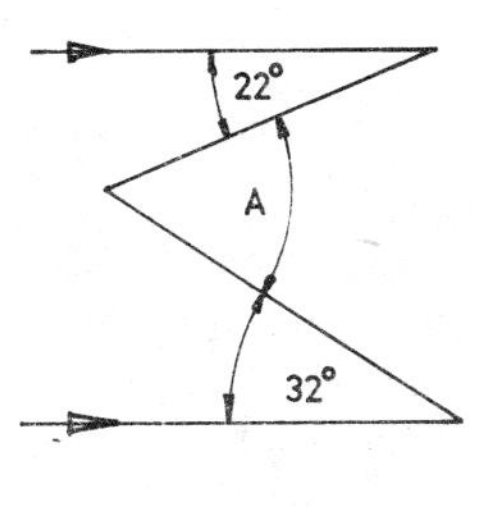

Fig. 22.15

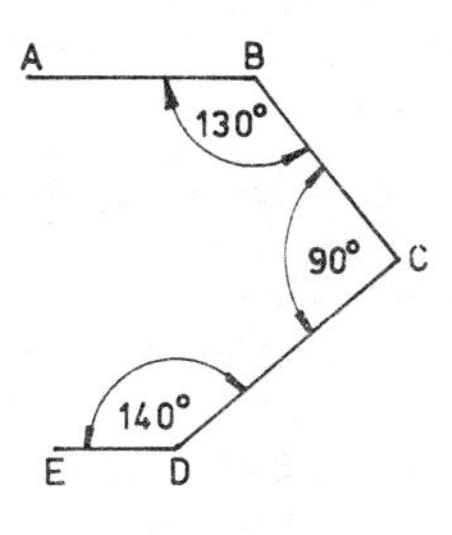

Fig. 22.16

6) In Fig. 22.16 prove that AB is parallel to ED.

7) Find A in Fig. 22.17.

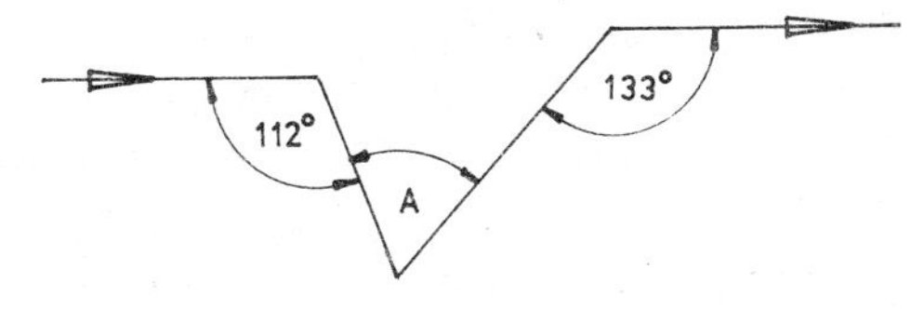

Fig. 22.17

8) In Fig. 22.18 the lines AB, CD and EF are parallel. Find the values of x and y.

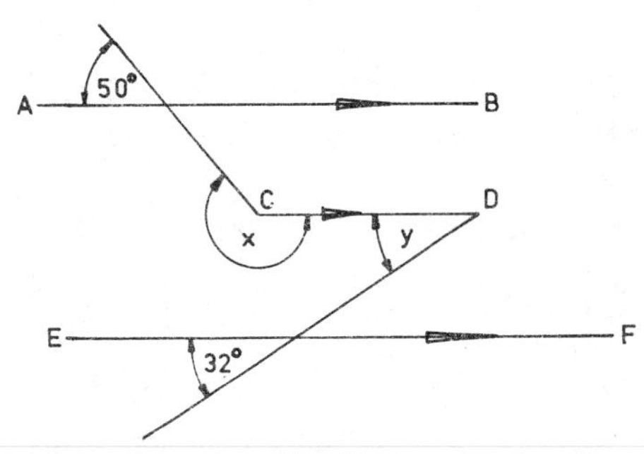

Fig. 22.18

9) Find the angle x in Fig. 22.19.

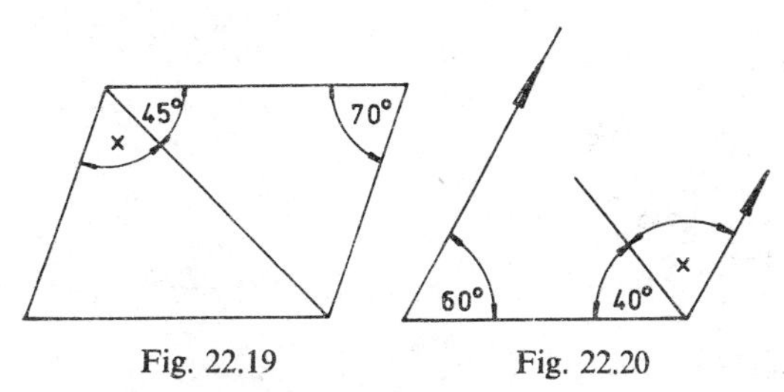

Fig. 22.19 Fig. 22.20

10) Find x in Fig. 22.20.

11) In Fig. 22.21 prove that AB, CD and EF are parallel to each other.

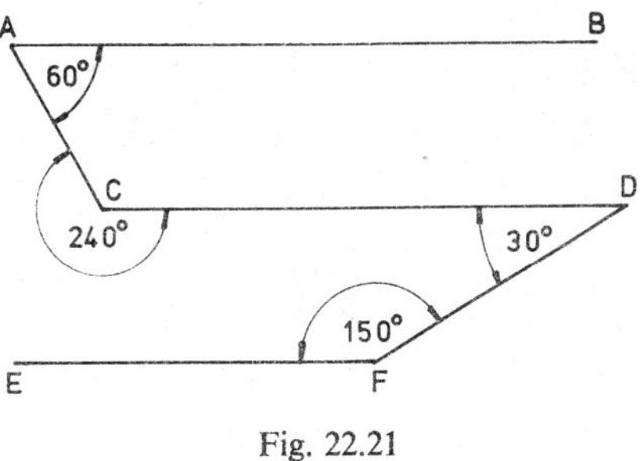

Fig. 22.21

12) In Fig. 22.22 prove that AB and CD are parallel.

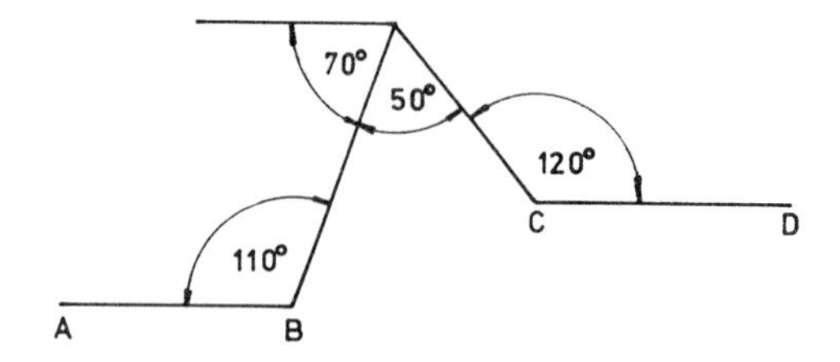

Fig. 22.22

SELF TEST 22

In the following state the letter (or letters) corresponding to the correct answer (or answers).

1) The angle shown in Fig. 22.23 is

a acute b right c reflex
d obtuse

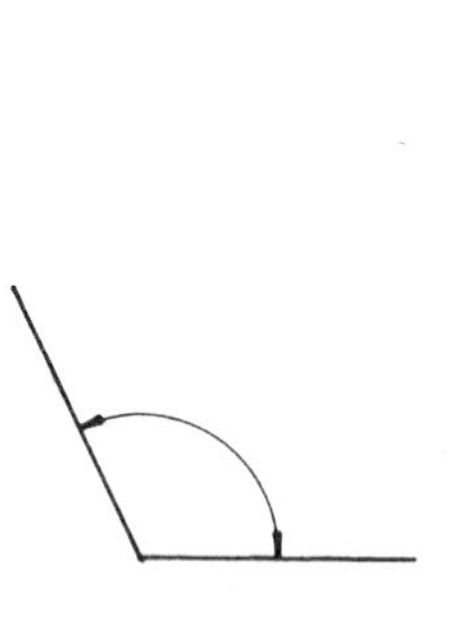
Fig. 22.23

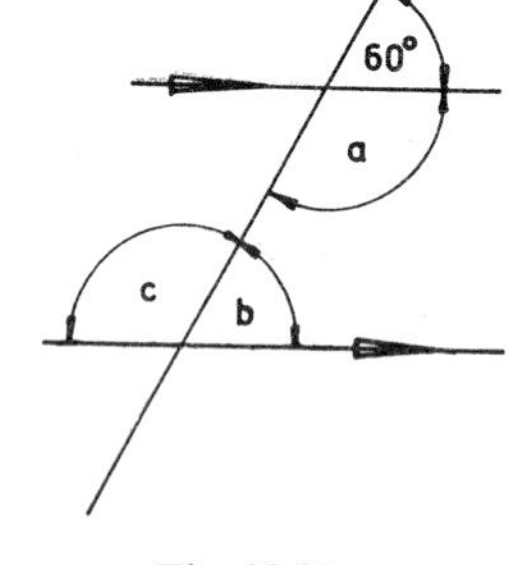

Fig. 22.24

2) The angle a shown in Fig. 22.24 is equal to

a 120° b 60° c neither of these

3) The angle b shown in Fig. 22.24 is equal to

a 120° b 60° c neither of these

4) The angle c shown in Fig. 22.24 is equal to

a 120° b 60° c neither of these

5) In Fig. 22.25

a $a=d$ b $a=e$ c $e=b$
d $a=c$

6) In Fig. 22.26

a $q=p+r$ b $p+q+r=360°$
c $q=r-p$ d $q=360-p-r$

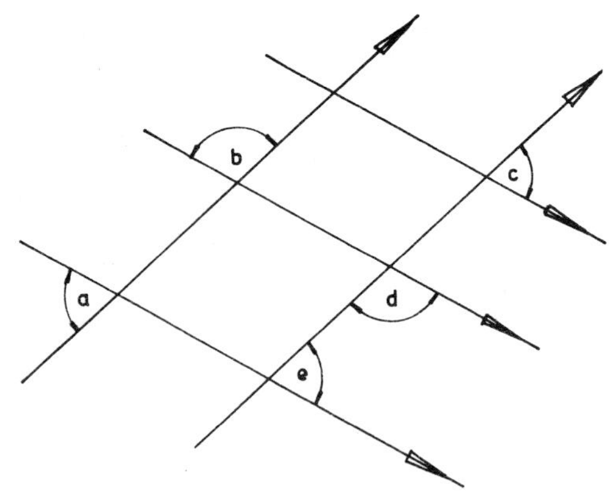

Fig. 22.25

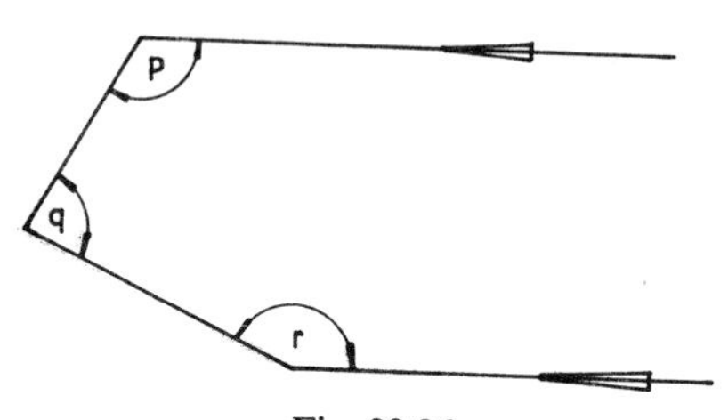

Fig. 22.26

7) In Fig. 22.27

a $x=y$ b $x=180°-y$
c $x=y-180°$ d $x+y=180°$

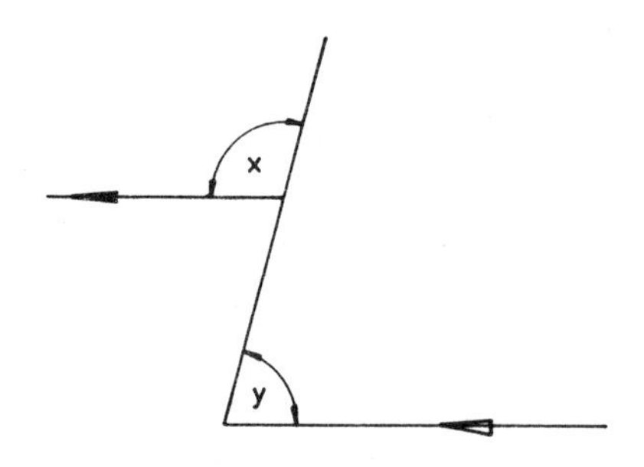

Fig. 22.27

8) A reflex angle is

a less than 90° b greater than 90°
c greater than 180° d equal to 180°

9) Angles whose sum is 180° are called

a complementary angles
b alternate angles
c supplementary angles
d corresponding angles

Chapter 23 Triangles

TYPES OF TRIANGLES

1) An *acute-angled* triangle has all its angles less than 90° (Fig. 23.1).

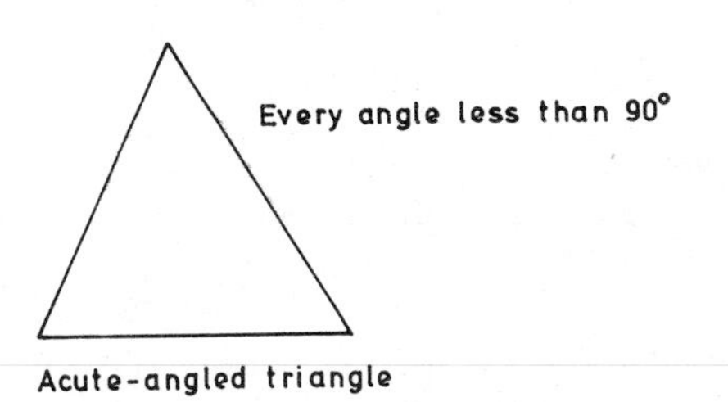

Fig. 23.1

2) A *right-angled* triangle has one of its angles equal to 90°. The side opposite to the right-angle is the longest side and it is called the hypotenuse (Fig. 23.2).

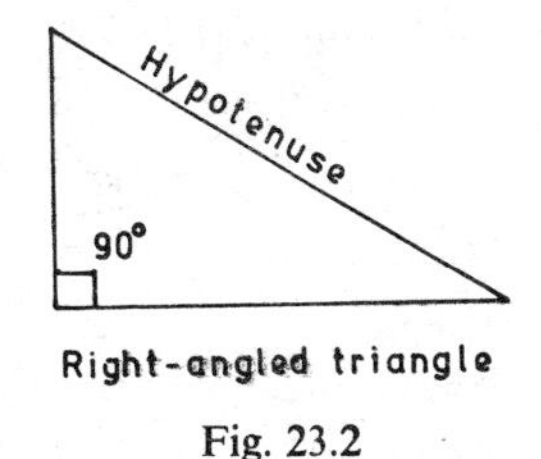

Fig. 23.2

3) An *obtuse-angled* triangle has one angle greater than 90° (Fig. 23.3).

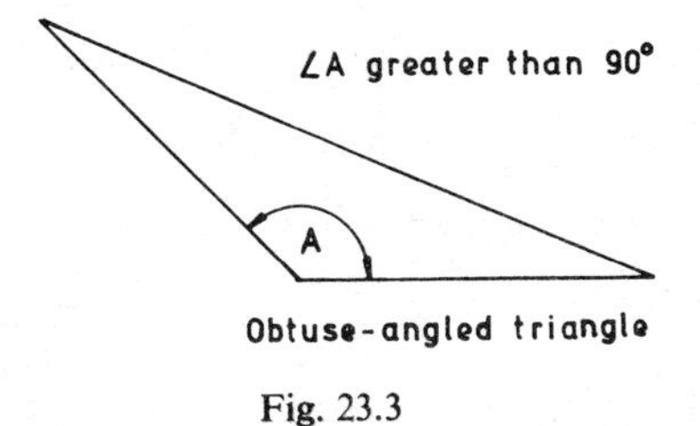

Fig. 23.3

4) A *scalene* triangle has all three sides of different length.

5) An *isosceles* triangle has two sides and two angles equal. The equal angles lie opposite to the equal sides (Fig. 23.4).

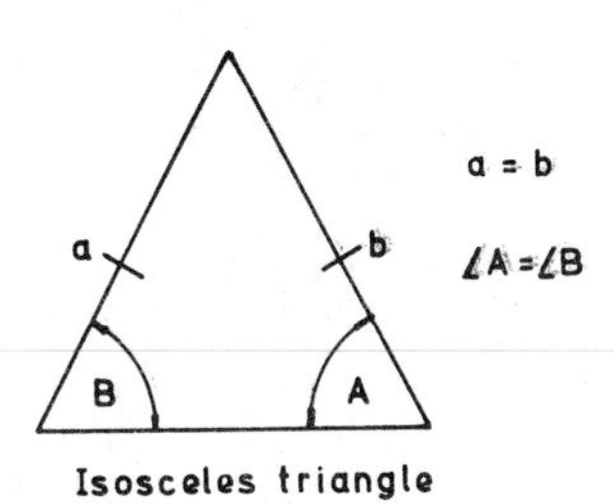

Fig. 23.4

6) An *equilateral* triangle has all its sides and angles equal. Each angle of the triangle is 60° (Fig. 23.5).

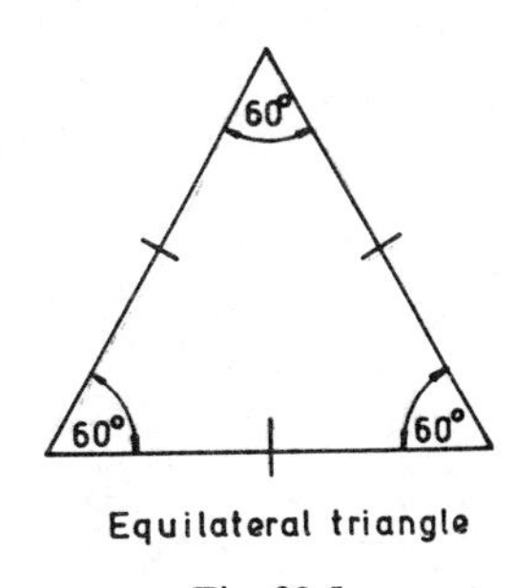

Fig. 23.5

ANGLE PROPERTIES OF TRIANGLES

1) *The sum of the angles of a triangle are equal to 180°* (Fig. 23.6).

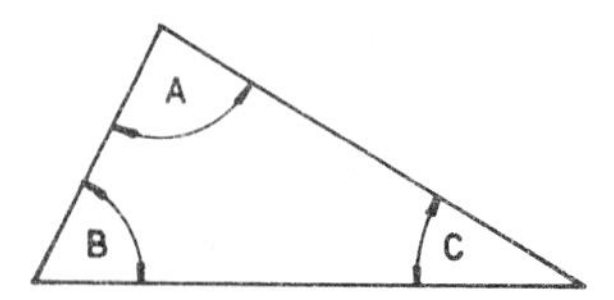

Fig. 23.6 $A+B+C=180°$

2) *In every triangle the greatest angle is opposite to the longest side. The smallest angle is opposite to the shortest side.* In every triangle the sum of the lengths of any two sides is always greater than the length of the third side (Fig. 23.7).

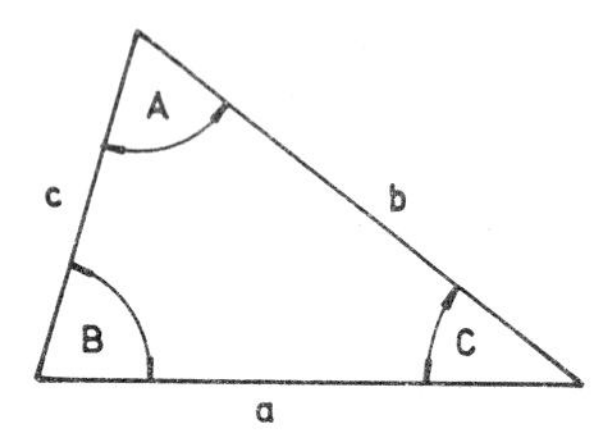

a is the longest side since it lies opposite to the greatest angle A. c is the shortest side since it lies opposite to the smallest angle C. $a+b$ is greater than c, $a+c$ is greater than b and $b+c$ is greater than a.

Fig. 23.7

3) *When the side of a triangle is produced the exterior angle so formed is equal to the sum of the opposite interior angles* (Fig. 23.8). See Theorem 1.

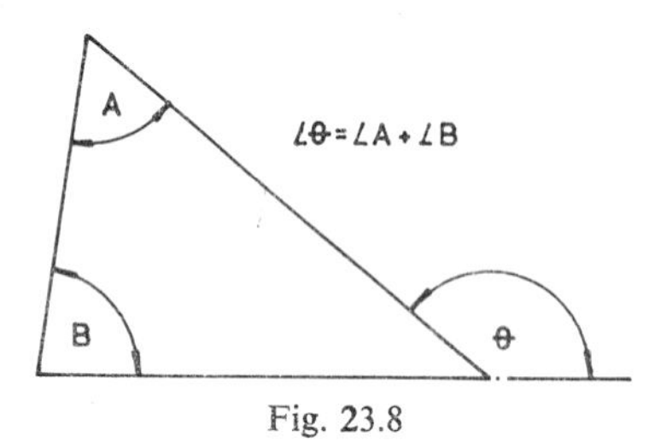

Fig. 23.8

4) *In an isosceles triangle the perpendicular (drawn from point where the two equal sides meet) to the base bisects the angle between the two equal sides. It also bisects the base of the triangle* (Fig. 23.9).

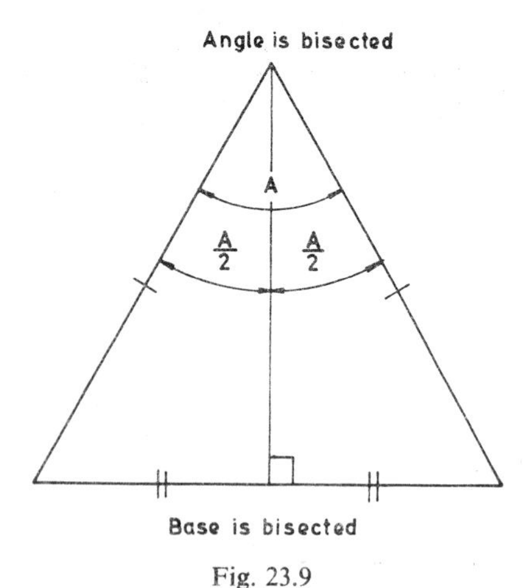

Fig. 23.9

Example

In Fig. 23.10, find the angle θ.

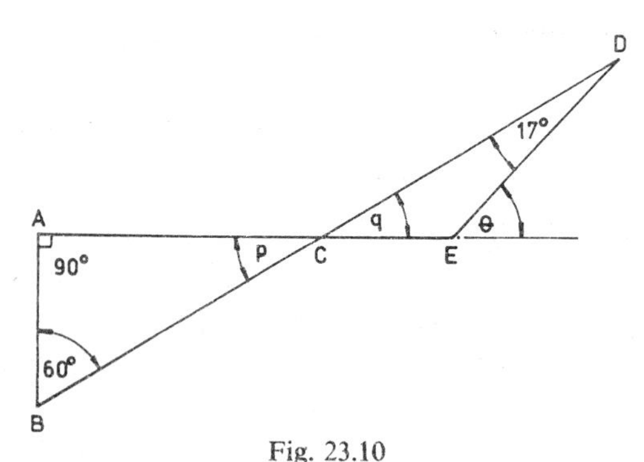

Fig. 23.10

In $\triangle ABC$,

$$p+90°+60°=180°$$
$$\therefore\ p=30°$$

but $p=q$ (vertically opposite angles)

$$\therefore\ q=30°$$

$\theta=q+17°$ (ext. angle = sum of the opp. interior angles)

$$\therefore\ \theta=30°+17°=47°$$

Exercise 83

1) Find the angles a and b in Fig. 23.11.

2) Find the angle x in Fig. 23.12.

3) Find the angles p, q and r in Fig. 23.13.

4) The triangle ABC (Fig. 23.14) is isosceles. Find the angle y.

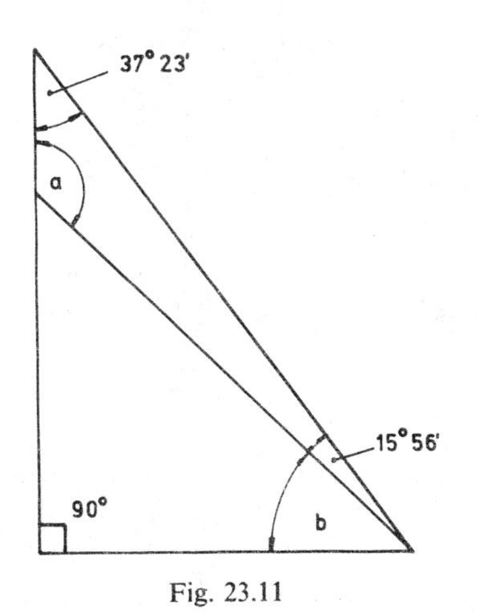

Fig. 23.11

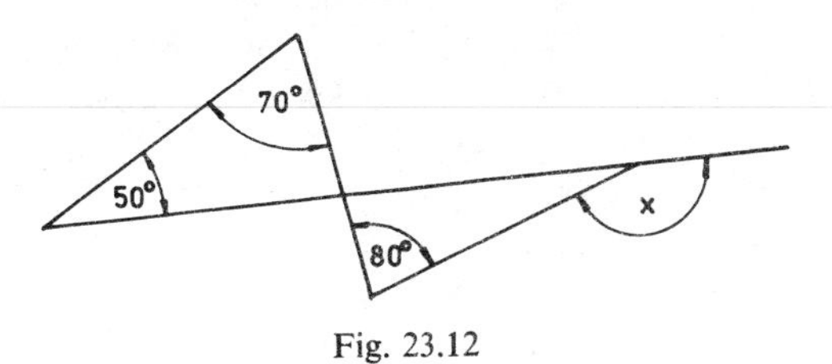

Fig. 23.12

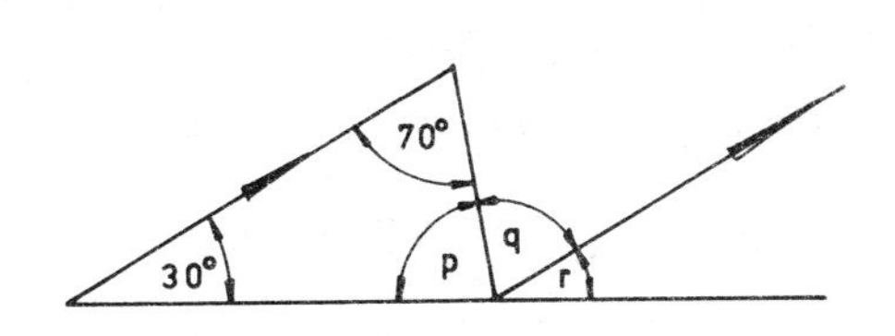

Fig. 23.13

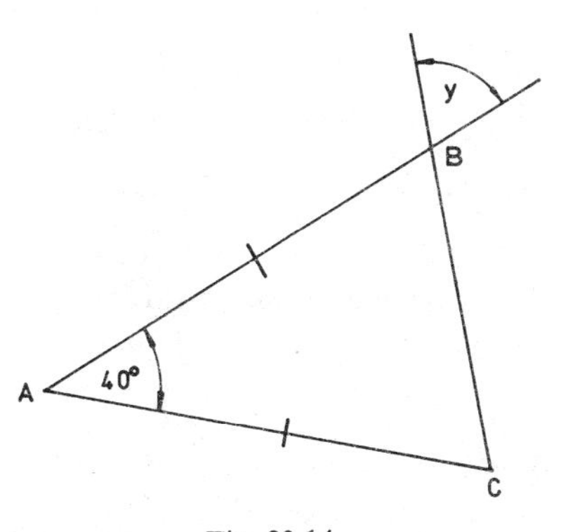

Fig. 23.14

5) Two sides of a triangle are each 6 cm long. The angle between these sides is 48°24′. What are the other two angles?

6) In △ABC, the angle A is 72° and the angle B is 63°. A perpendicular is drawn from B to AC cutting AC at P. Show that BP=CP.

7) In △ABC, ∠A=37° and ∠B=69°. The bisector of ∠C meets AB in N. Show that CN=AN.

CONGRUENT TRIANGLES

Two triangles are said to be congruent if they are equal in every respect. Thus in Fig. 23.15 the triangles ABC and XYZ are congruent because

AC=XZ		∠B=∠Y
AB=XY	and	∠C=∠Z
BC=ZY		∠A=∠X

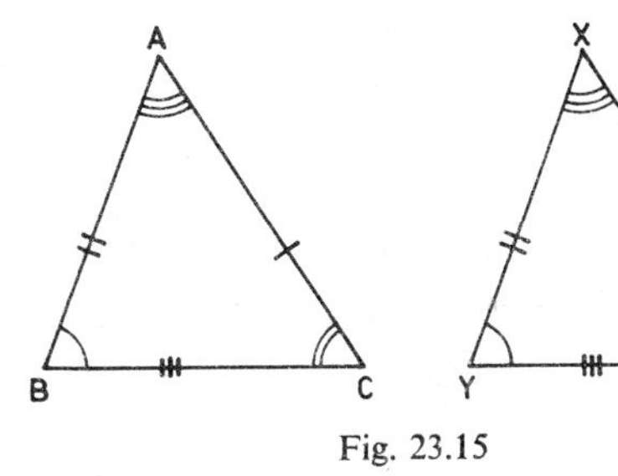

Fig. 23.15

Note that the angles which are equal lie opposite to the corresponding sides.

If two triangles are congruent they will also be equal in area. The notation used to express the fact that △ABC is congruent to △XYZ is △ABC≡△XYZ.

For two triangles to be congruent the six elements of one triangle (three sides and three angles) must be equal to the six elements of the second triangle. However to prove that two triangles are congruent it is not necessary to prove all six equalities. Any of the following are sufficient to prove that two triangles are congruent:

1) *One side and two angles in one triangle equal to one side and two similarly located angles in the second triangle* (Fig. 23.16).

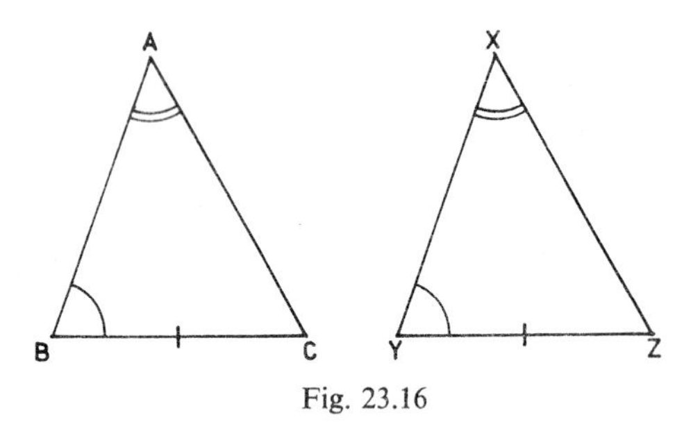

Fig. 23.16

2) *Two sides and the angle between them in one triangle equal to two sides and the angle between them in the second triangle* (Fig. 23.17).

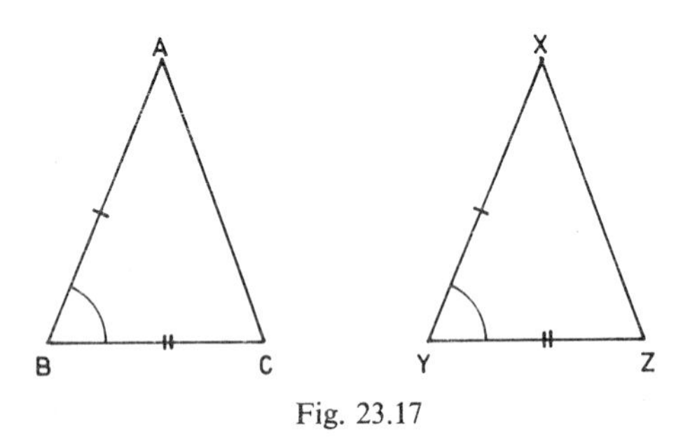

Fig. 23.17

3) *Three sides of one triangle equal to three sides of the other triangle* (Fig. 23.18).

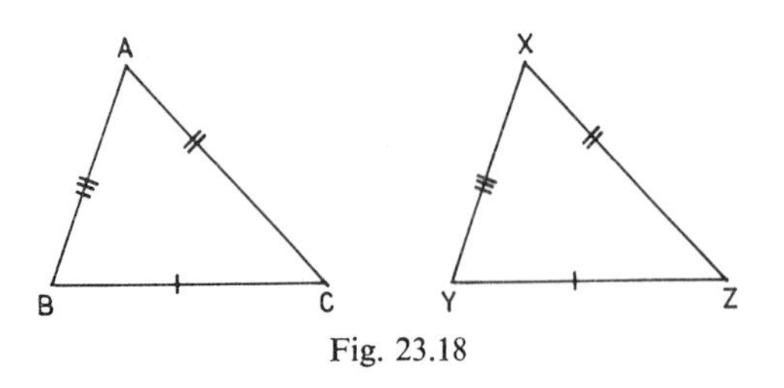

Fig. 23.18

4) *In right-angled triangles if the hypotenuses are equal and one other side in each triangle are also equal* (Fig. 23.19).

Note that three equal angles are not sufficient to prove congruency and neither are two sides

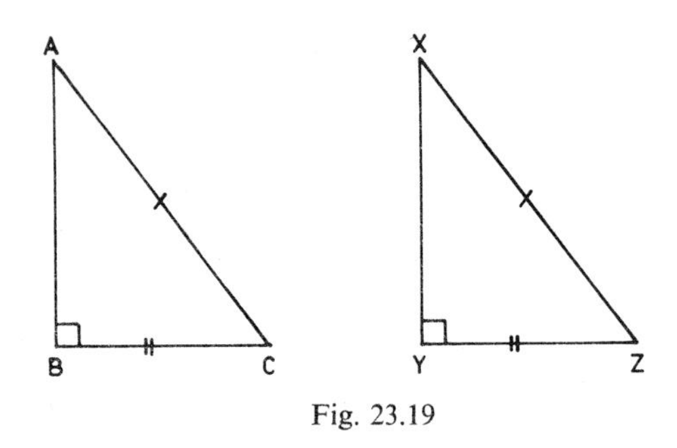

Fig. 23.19

and a non-included angle. An included angle is an angle between the two equal sides of the triangles (e.g. $\angle ABC$ and $\angle XYZ$ in Fig. 23.17 and $\angle ACB$ and $\angle XZY$ in Fig. 23.19).

Examples

1) The mid-points of the sides MP and ST of $\triangle LMP$ and $\triangle RST$ are X and Y respectively. If $LM=RS$, $MP=ST$ and $LX=RY$ prove that $\triangle LMP \equiv \triangle RST$.
Referring to Fig. 23.20

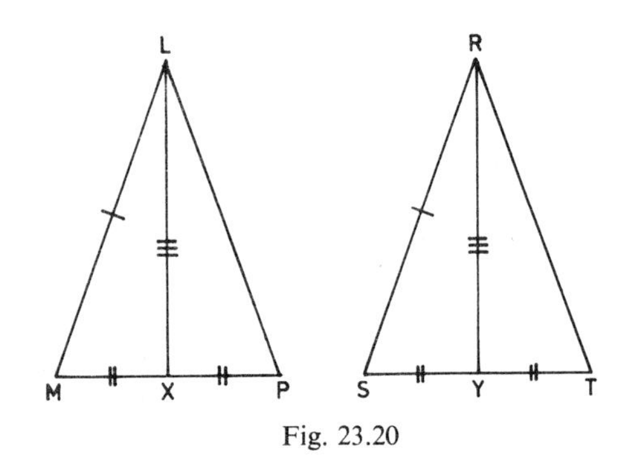

Fig. 23.20

$\triangle LMX \equiv \triangle RSY$ (condition (3) above)
$\therefore\ \angle M = \angle S$

In $\triangle$s LMP and RST

$LM=RS$; $MP=ST$; $\angle M=\angle S$.

That is, two sides and the included angle in $\triangle LMP$ equal the two sides and the included angle in $\triangle RST$. Hence $\triangle LMP \equiv \triangle RST$.

2) The diagonals of the quadrilateral XYZW intersect at O. Given that $OX=OW$ and $OY=OZ$ prove that $XY=ZW$.

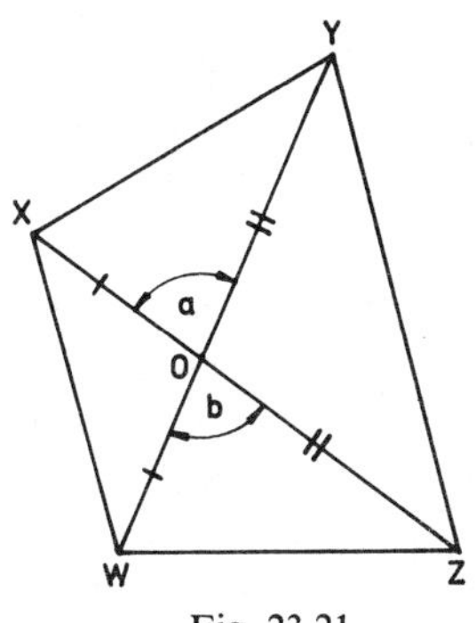

Fig. 23.21

Referring to Fig. 23.21
In △s XOY and WOZ

OX=OW and OY=OZ (given)
$a=b$ (vertically opposite angles)

Hence the two sides and the included angle in △XOY equal two sides and the included angle in △WOZ. Hence △XOY≡△WOZ.

∴ XY=ZW

Exercise 84

1) Two straight lines PQ and RS cut at X. If PX=RX and ∠SPX=∠QRX prove that △SPX≡△QRX.

2) If two straight lines PQ and XY bisect each other, prove that PX=YQ.

3) State which of the following must be congruent triangles and which need not necessarily be congruent triangles:
a) △ABC and △DEF in which AB=DE, BC=EF and ∠C=∠F=90°.
b) △KLM and △PQR in which ∠K=∠P, ∠L=∠Q and ∠M=∠R.
c) △STU and △WXZ in which ∠S=∠W, SU=XZ and ∠U=∠Z.

4) D is a point on the base BC of an isosceles triangle ABC in which AB=AC. The triangle ADE is drawn so that AD=AE, ∠DAE=∠BAC and D and E are on opposite sides of AC. Prove that
a) ∠BAD=∠CAE;
b) △s BAD and CAE are congruent.

5) PQRST is a pentagon (a five sided figure) in which PQ=PT, QR=TS and ∠PQR=∠PTS. Prove that a) PR=PS; b) ∠QRS=∠TSR.

6) In the equilateral triangle ABC a line through A meets BC at R and a line through B meets CA at S such that BR=CS. Prove △ABR≡△BCS. If AR and BS intersect at T prove that ∠RTB=60°.

7) D is a point on the hypotenuse BC of a right-angled triangle ABC, such that DB=DA. Prove that D is the mid-point of BC.

8) In Fig. 23.22, AE=BE and ∠C=∠D=∠AEB=90°. Prove △ADE≡△ECB.

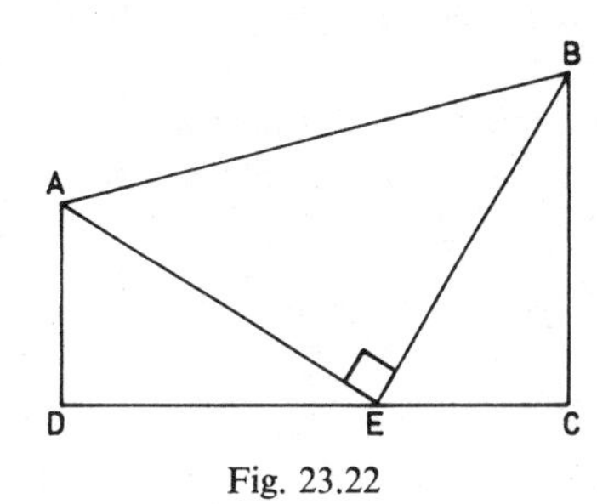

Fig. 23.22

9) In Fig. 23.23, SP=PT and PQ=QS=SR. Prove that

a) ∠TPQ=∠PSR; b) TQ=PR.

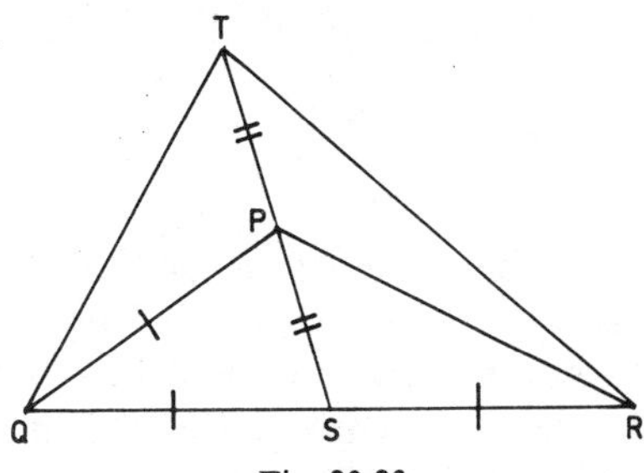

Fig. 23.23

10) In Fig. 23.24, $\triangle$s ABC and CBE are both equilateral. Prove that $\triangle CAD \equiv \triangle CDE$.

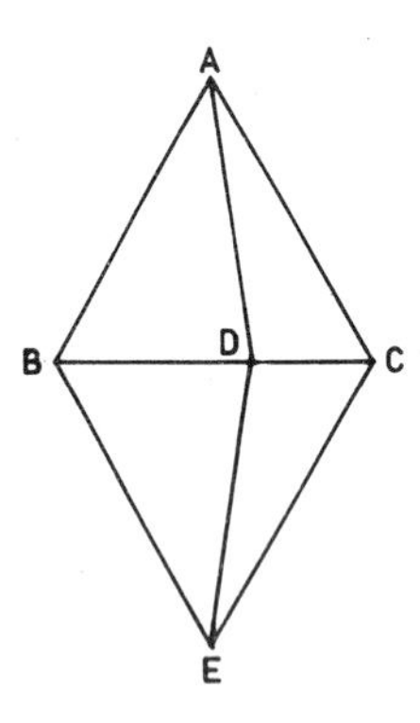

Fig. 23.24

SIMILAR TRIANGLES

Two triangles are said to be similar if they are equi-angular.

i) *If two triangles are equi-angular their corresponding sides are proportional.*

Thus in Fig. 23.25:

$$\frac{AB}{XY} = \frac{AC}{XZ} = \frac{BC}{YZ}$$

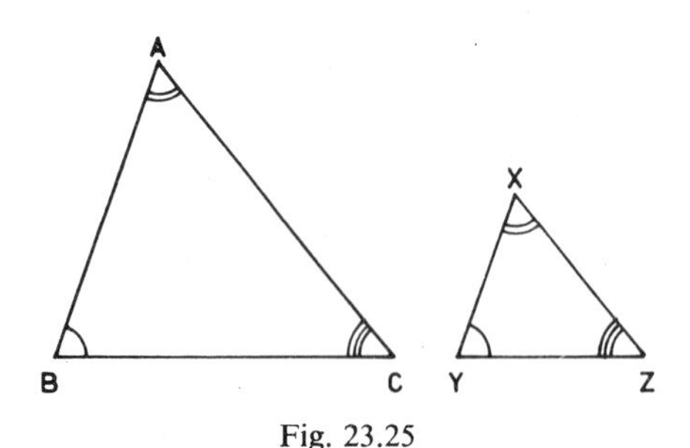

Fig. 23.25

ii) *If two triangles have their corresponding sides proportional then they are equi-angular.* Note that by corresponding sides we mean the sides opposite to the equal angles. It helps in solving problems on similar triangles if we write the two triangles with the equal angles under each other. Thus in $\triangle$s ABC and XYZ if $\angle A = \angle X$, $\angle B = \angle Y$ and $\angle C = \angle Z$ we write $\frac{ABC}{XYZ}$.

The equations connecting the sides of the triangles are then easily obtained by writing any two letters in the first triangle over any two corresponding letters in the second triangle. Thus

$$\frac{AB}{XY} = \frac{AC}{XZ} = \frac{BC}{YZ}$$

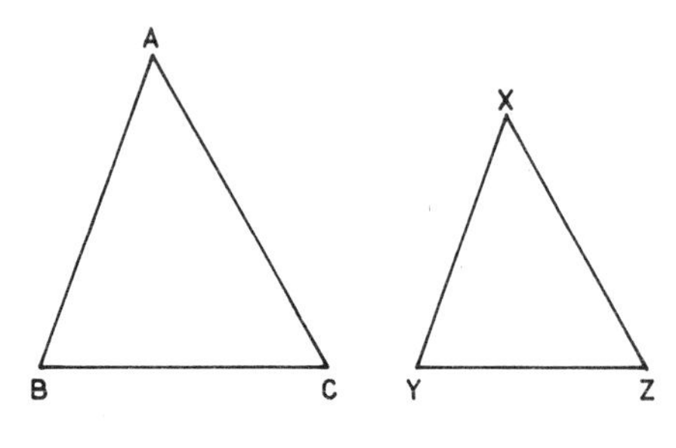

Fig. 23.26

In Fig. 23.26, to prove $\triangle ABC$ is similar to $\triangle XYZ$ it is sufficient to prove any one of the following:

i) *Two angles in $\triangle ABC$ equal to two angles in $\triangle XYZ$.* For instance, the triangles are similar if $\angle A = \angle X$ and $\angle B = \angle Y$, since it follows that $\angle C = \angle Z$.

ii) *The three sides of $\triangle ABC$ are proportional to the corresponding sides of $\triangle XYZ$.* Thus $\triangle ABC$ is similar to $\triangle XYZ$ if

$$\frac{AB}{XY} = \frac{AC}{XZ} = \frac{BC}{YZ}.$$

iii) *Two sides in $\triangle ABC$ are proportional to two sides in $\triangle XYZ$ and the angles included between these sides in each triangle are equal.* Thus $\triangle ABC$ is similar to $\triangle XYZ$ if

$$\frac{AB}{AC} = \frac{XY}{XZ} \quad \text{and} \quad \angle A = \angle X.$$

Examples

1) In Fig. 23.27 prove that △s PTS and PQR are similar and calculate the length of TS.

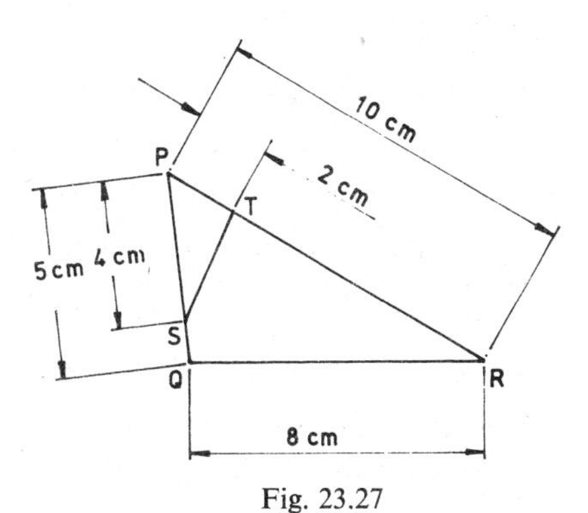

Fig. 23.27

In △s PTS and PQR

$$\frac{PS}{PT} = \frac{4}{2} = 2$$

$$\frac{PR}{PQ} = \frac{10}{5} = 2$$

$$\therefore \quad \frac{PS}{PT} = \frac{PR}{PQ}$$

also ∠P is common to both triangles and it is the included angle between PS and PT in △PTS and PR and PQ in △PQR. Hence △s PTS and PQR are similar. Writing $\frac{\triangle PTS}{\triangle PQR}$ we see that

$$\frac{TS}{QR} = \frac{PT}{PQ}$$

$$\frac{TS}{8} = \frac{2}{5}$$

$$TS = \frac{2 \times 8}{5} = 3{\cdot}2 \text{ cm}$$

2) In the triangle PQR the line ST is drawn parallel to QR such that PS = 3SQ (Fig. 23.28). Calculate the value of the ratio $\frac{ST}{QR}$

In △s PQR and PST

$$\angle Q = \angle S \quad \angle T = \angle R \quad \text{(since ST} \parallel \text{QR)}$$

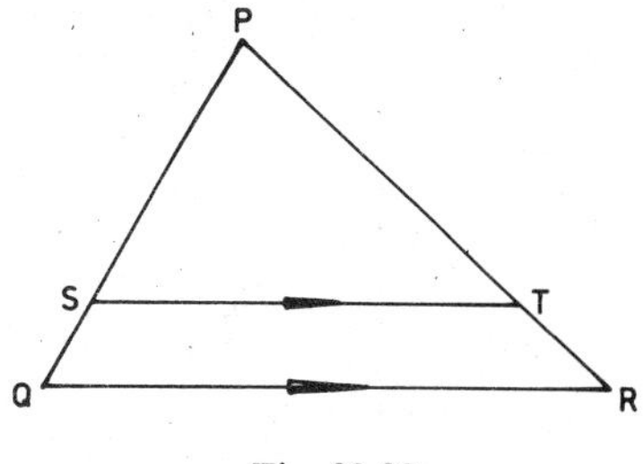

Fig. 23.28

∴ △s PQR and PST are similar. Writing

$\frac{\triangle PST}{\triangle PQR}$ we see that $\frac{ST}{QR} = \frac{PS}{PQ}$

now $\quad PS = 3SQ$

and $\quad SQ = PQ - PS$

$\therefore \quad PS = 3(PQ - PS) = 3PQ - 3PS$

$\quad 4PS = 3PQ$

and $\quad \frac{PS}{PQ} = \frac{3}{4}$

$\therefore \quad \frac{ST}{QR} = \frac{3}{4}$

Exercise 85

1) In Fig. 23.29 find BC.

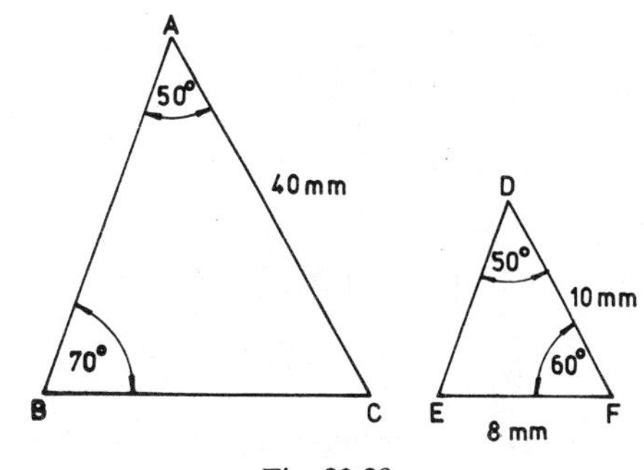

Fig. 23.29

2) In △s ABC and PQR, ∠A = ∠R and ∠C = ∠P. State the ratios between the sides of the two triangles.

3) In Fig. 23.30 calculate AE and EH.

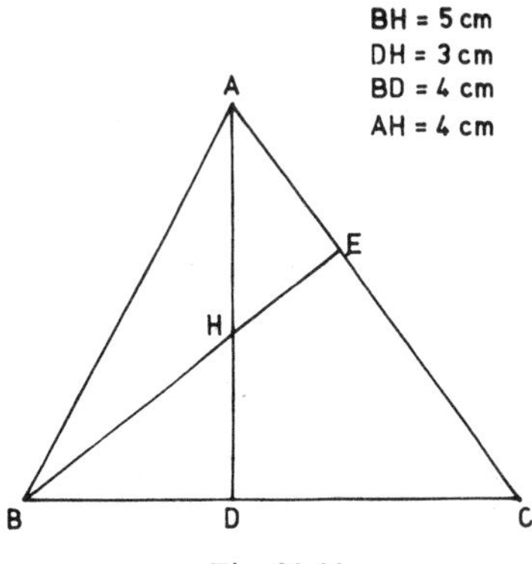

Fig. 23.30

4) A triangle ABC has sides AB = 5 cm and BC = 2 cm. Calculate the length of the side corresponding to BC in a similar triangle if that corresponding to AB = 4 cm.

5) In the parallelogram ABCD, the point P is taken on the side AB so that AP = 2PB. The lines BD and PC intersect at O. Prove that △s PBO and CDO are similar and hence calculate $\frac{PO}{OC}$.

6) WXYZ is a parallelogram. A line through W meets ZY at T and XY produced at U. Prove that △s WZT and UYT are similar. If ZT = 3 cm, TY = 2 cm and UY = 1 cm find WZ.

7) ABCD is a trapezium in which AB is parallel to DC. AB = 3 cm, DC = 6 cm and the diagonal BD = 7·8 cm. If BD and AC meet at K, calculate KB. If X is the mid-point of BD and the parallel to DC through X meets AC at Y, calculate XY.

8) The line MN is drawn parallel to the side BC of △ABC to meet AB at M and AC at N so that $\frac{MN}{BC} = \frac{1}{4}$. Calculate $\frac{AM}{MB}$.

9) In △ABC the point P is the mid-point of AB. A line through P parallel to BC meets AC at Q. Prove that $PQ = \frac{1}{2}BC$.

AREAS OF TRIANGLES

In Fig. 23.31, AP, BQ and CR are perpendiculars drawn from the vertices of the triangle ABC to the opposite sides. AP, BQ and CR are called the altitudes of the triangle ABC. AP is the altitude corresponding to the base BC, BQ is the altitude corresponding to the base AC and CR is the altitude corresponding to the base AB.

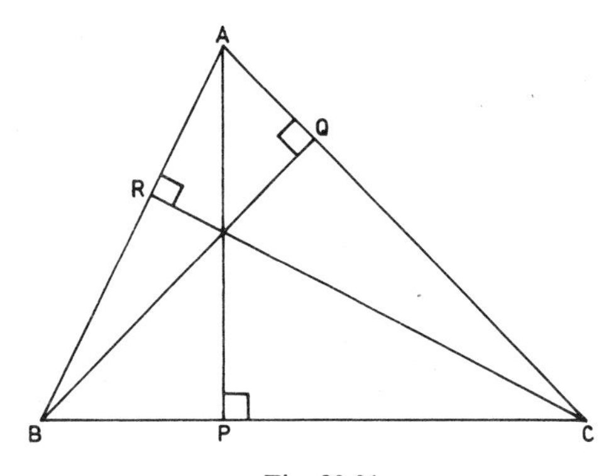

Fig. 23.31

The area for any triangle is:

$$\text{area} = \frac{1}{2} \times \text{base} \times \text{altitude}$$

Thus in Fig. 23.31,

$$\text{area of } \triangle ABC = \frac{1}{2} \times AP \times BC$$
$$= \frac{1}{2} \times BQ \times AC$$
$$= \frac{1}{2} \times RC \times AB$$

Triangles having equal bases and equal heights are equal in area. Thus in Fig. 23.32, △s ABC and XYZ have the same area since they have the same height *h* and the same base *b*.

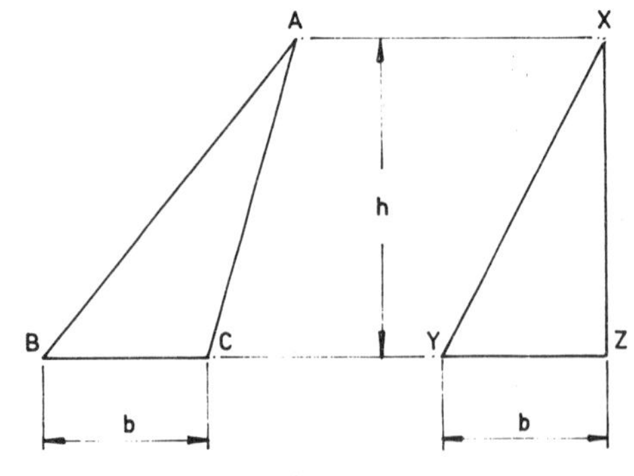

Fig. 23.32

Triangles which are congruent are equal in every respect and hence the areas of congruent triangles are equal.

Examples

1) The $\triangle ABC$ is right-angled at A. $AB=6$ cm, $AC=8$ cm and $BC=10$ cm. Calculate the altitude from A to BC.

Referring to Fig. 23.33,

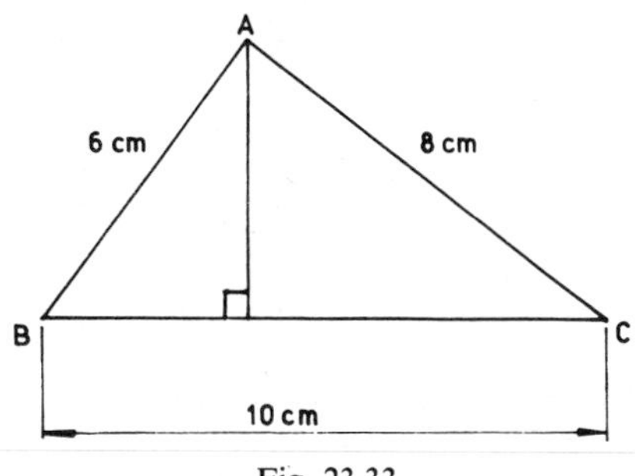

Fig. 23.33

$$\text{area of } \triangle ABC=\frac{1}{2}\times 6\times 8=24 \text{ cm}^2$$

$$=\frac{1}{2}\times 10\times AD=5\times AD$$

$$\therefore\ 5\times AD=24$$

$$AD=4{\cdot}8 \text{ cm}$$

2) In $\triangle ABC$, the side AC is greater than AB and D is the mid-point of BC. The perpendicular from B to AD meets AD at X and the perpendicular from C to AD meets AD produced at Y. Prove that the area of $\triangle ABC=AD.BX$

Referring to Fig. 23.34,

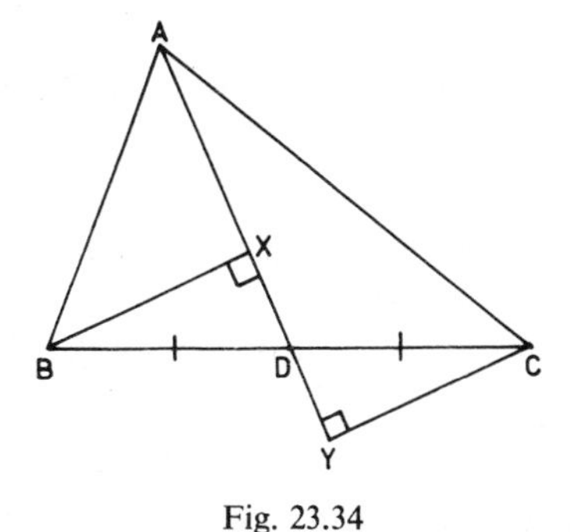

Fig. 23.34

$$\text{area of } \triangle ABC=\text{area of } \triangle ABD+\text{area of } \triangle ADC$$

$$=\frac{1}{2}AD.BX+\frac{1}{2}.AD.CY$$

In $\triangle$s BXD and DYC,

$\angle BXD=\angle DYC=90°$

$\angle BDX=\angle YDC$ (vertically opposite angles)

$BD=CD$ (given)

$\therefore\ \triangle BXD\equiv\triangle DYC$

$\therefore\ CY=BX$

Hence, area of

$$\triangle ABC=\frac{1}{2}AD.BX+\frac{1}{2}AD.BX$$

$$\therefore\ \text{area of } \triangle ABC=AD.BX$$

AREAS OF SIMILAR TRIANGLES

The ratio of the areas of similar triangles is equal to the ratio of the squares on corresponding sides.

If in Fig. 23.35 $\triangle$s ABC and XYZ are similar then

$$\frac{\text{area of } \triangle ABC}{\text{area of } \triangle XYZ}=\frac{AB^2}{XY^2}=\frac{AC^2}{XZ^2}=\frac{BC^2}{YZ^2}$$

It follows that if two triangles are similar then the ratio of their areas is equal to the ratios

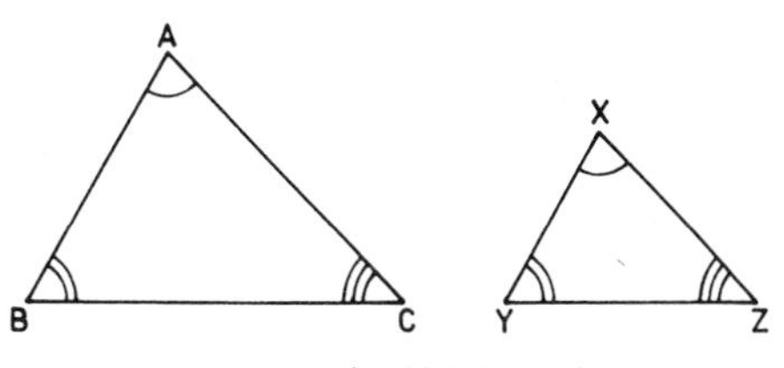

Fig. 23.35

of the squares of corresponding linear dimensions. Thus in Fig. 23.36:

$$\frac{\text{area of } \triangle ABC}{\text{area of } \triangle XYZ} = \frac{AD^2}{WX^2} = \frac{BE^2}{VY^2} \text{ (see theorem 8).}$$

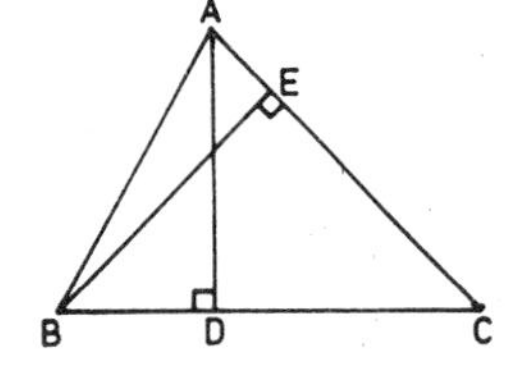

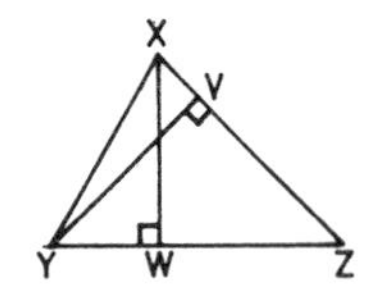

Fig. 23.36

Example

In $\triangle$s ABC and XYZ, $\angle A = \angle X$ and $\angle B = \angle Y$.

$\triangle ABC$ has an area of 9 cm^2 and $\triangle XYZ$ has an area of 4 cm^2.

a) If AB = 3 cm find XY.

b) The perpendicular from A to BC meets BC at D and the perpendicular from X to YZ meets YZ at W. If XW = 1·5 cm find AD.

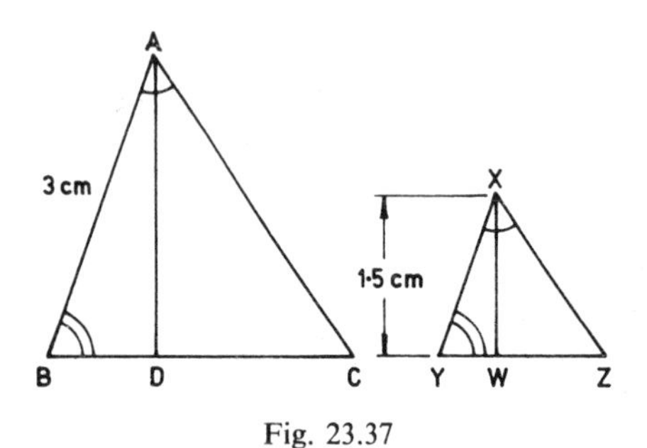

Fig. 23.37

Referring to Fig. 23.37, $\triangle$s ABC and XYZ are equi-angular and hence the two triangles are similar.

a) $$\frac{\text{area } \triangle XYZ}{\text{area } \triangle ABC} = \frac{XY^2}{AB^2}$$

$$\frac{4}{9} = \frac{XY^2}{3^2}$$

$$XY^2 = \frac{3^2 \times 4}{9} = 4$$

$$\therefore \; XY = 2 \text{ cm}$$

b) $$\frac{\text{area } \triangle ABC}{\text{area } \triangle XYZ} = \frac{AD^2}{XW^2}$$

$$\frac{9}{4} = \frac{AD^2}{1{\cdot}5^2}$$

$$AD^2 = \frac{9 \times 1{\cdot}5^2}{4}$$

$$AD = \frac{3 \times 1{\cdot}5}{2} = 2{\cdot}25 \text{ cm}$$

Exercise 86

1) In $\triangle ABC$, BC = 8 cm. The perpendicular from A to BC meets BC at P. If AP = 5 cm, calculate the area of $\triangle ABC$.

2) In $\triangle ABC$, BC = 10 cm and AC = 6 cm. The perpendicular from A to BC meets BC at X and the perpendicular from B to AC meets AC at Y. If AX = 4 cm find BY.

3) The sides of a triangle are 3 cm, 5 cm and 7 cm long. The shortest side of a similar triangle is 2 cm long. What is the ratio of the areas of the two triangles? What are the lengths of the other two sides of the smaller triangle?

4) In $\triangle ABC$ the line XY is drawn parallel to BC and meets AB at X and AC at Y. If the ratio between the areas of $\triangle$s ABC and AXY is 9 : 4, prove that AX = 2BX.

5) PQRS is a parallelogram. PS is produced to L so that SL = SR and LR produced meets PQ produced at M. Prove that QM = QR. If the area of the parallelogram is 20 cm^2 and PQ = 2PS, find the areas of LSR and LPM.

6) In a $\triangle ABC$ the side AB is divided at X so that $\frac{AX}{XB} = \frac{3}{2}$. A line through X parallel to BC, meets AC at Y. Express as a fraction of the area of $\triangle ABC$:

i) area of $\triangle BCY$; (ii) area of $\triangle AXY$;
iii) area of $\triangle BXY$.

7) In the parallelogram ABCD, a line through C cuts the diagonal BD at Q and the side AB

at P. Prove that triangles PBQ and CDQ are similar. Given that AP = 3PB, calculate

a) $\dfrac{PQ}{QC}$; b) $\dfrac{\text{area PBQ}}{\text{area CDQ}}$.

8) The sides AB and BC of $\triangle ABC$ are 5 cm and 6 cm respectively. Points H and K on AB and AC respectively are such that HK and BC are parallel. If the areas of $\triangle$s AHK and ABC are in the ratio of 4 : 9, calculate HK and HB.

ANGLE BISECTOR THEOREMS

1) *The internal bisector of an angle of a triangle divides the opposite side in the ratio of the sides containing the angle.* Thus in Fig. 23.38, if AD bisects the angle A then

$$\frac{AB}{AC} = \frac{BD}{DC} \text{ (see Theorem 9).}$$

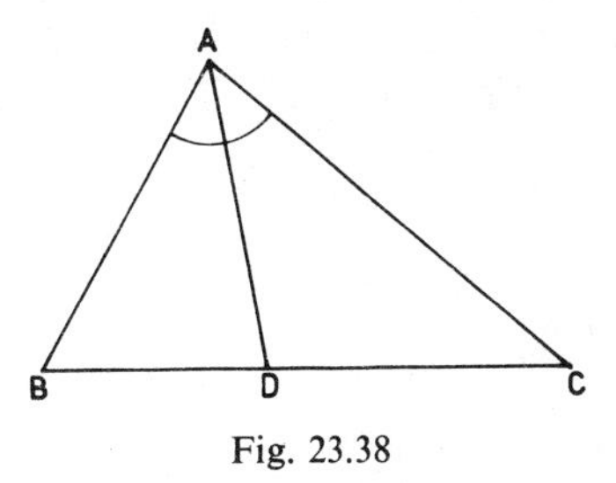

Fig. 23.38

2) *The external bisector of an angle of a triangle divides the opposite side externally in the ratio of the sides containing the angle.* Thus in Fig. 23.39, if AD bisects the angle $\angle CAX$ then

$$\frac{AB}{AC} = \frac{BD}{DC}$$

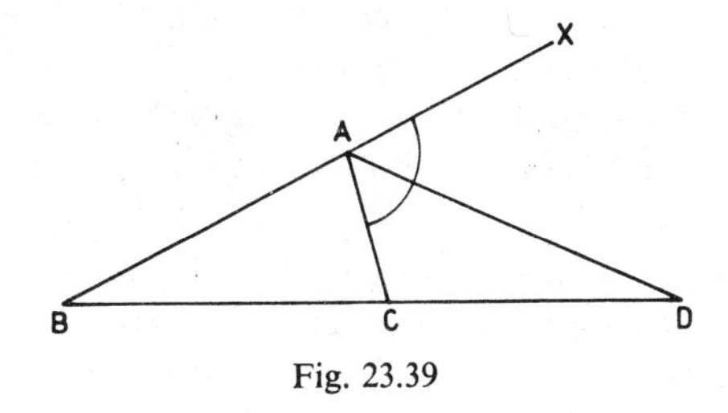

Fig. 23.39

3) *The converse of the above theorems is also true.* Thus if in Fig. 23.40

$$\frac{BD}{DC} = \frac{AB}{AC} \quad \text{then} \quad \angle BAD = \angle DAC$$

(that is the angle A is bisected).

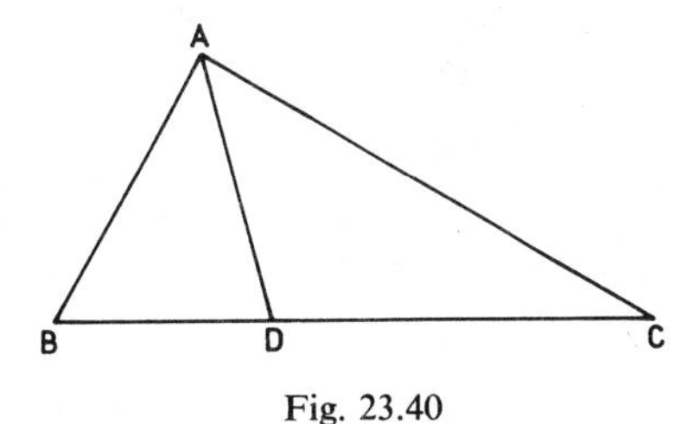

Fig. 23.40

Example
The area of a triangle ABC is 15 square centimetres and $\dfrac{AB}{AC} = \dfrac{3}{2}$. The line AD bisects angle A and meets BC internally at D. State the value of the ratio $\dfrac{BD}{BC}$ and calculate the area of the triangle ABD.

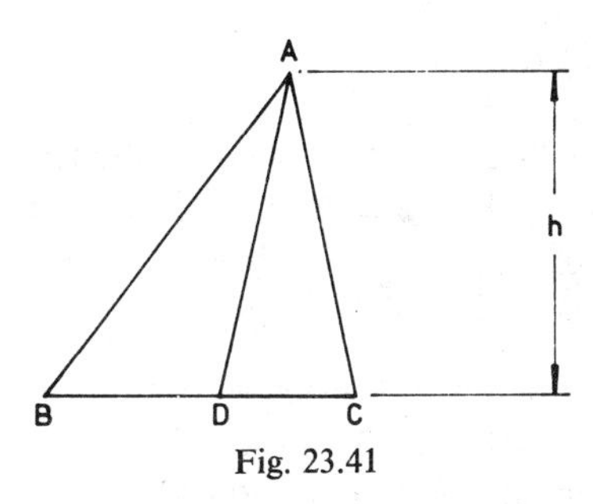

Fig. 23.41

Referring to Fig. 23.41, since $\angle A$ is bisected by AD, then

$$\frac{AB}{AC} = \frac{BD}{DC} = \frac{3}{2}$$

$$\therefore \ 2BD = 3DC$$

But DC = BC − BD

$$\therefore \ 2BD = 3(BC - BD)$$

$$2BD = 3BC - 3BD$$

$$5BD = 3BC$$

or $\dfrac{BD}{BC}=\dfrac{3}{5}$

$$\text{Area } \triangle ABC=\frac{1}{2}.BC.h$$

$$15=\frac{1}{2}.BC.h$$

$$h=\frac{30}{BC}$$

$$\text{Area } \triangle ABD=\frac{1}{2}.BD.h$$

$$=\frac{1}{2}.BD.\frac{30}{BC}$$

$$=15.\frac{BD}{BC}$$

but $\dfrac{BD}{BC}=\dfrac{3}{5}$

$\therefore$ area $\triangle ABD=15\times\dfrac{3}{5}=9\text{ cm}^2$

Exercise 87

1) In $\triangle ABC$, $\angle A$ is bisected by the line AD which meets BC at D. If AB = 8 cm and AC = 10 cm find the ratio $\dfrac{BD}{DC}$.

2) In $\triangle ABC$, $\angle A$ is bisected by the line AD which meets BC at D. If AB = 5 cm, AC = 4 cm and BC = 6 cm, find BD.

3) In $\triangle XYZ$ the exterior angle at X is bisected by the line XW which meets YZ produced at W. If $\dfrac{XY}{XZ}=\dfrac{3}{2}$, find $\dfrac{WY}{YZ}$.

4) In $\triangle ABC$, AB = 8 cm, AC = 12 cm and BC = 10 cm. D is a point on BC such that BD = 4 cm. Prove that AD bisects $\angle BAC$.

5) The area of a triangle ABC is 20 cm^2 and $\dfrac{AB}{AC}=\dfrac{3}{5}$. Calculate the area of triangle ABD if the line AD bisects $\angle A$ and meets BC at D.

6) ABC is a triangle in which AB = 8 cm, AC = 6 cm and BC = 7 cm. The internal bisector of $\angle BAC$ meets BC at X and the external bisector meets BC produced at Y.
a) Calculate CX; b) Calculate XY.

7) The area of a triangle HKL is 16 cm^2 and HK : HL = 5 : 3. The line HX bisects $\angle KHL$ and meets KL internally at X. State the value of the ratio KX : KL and calculate the area of $\triangle HKX$.

8) The internal and external bisectors of $\angle BAC$ of a triangle ABC meet BC and BC produced at P and Q respectively. If AB = 5 cm, AC = 3 cm and BC = 4 cm, calculate the lengths of BP and PQ.

THEOREM OF PYTHAGORAS

In a right-angled triangle the square described on the hypotenuse is equal to the sum of the squares described on the other two sides (see Theorem 4). Thus in Fig. 23.42

$$AC^2=AB^2+BC^2$$

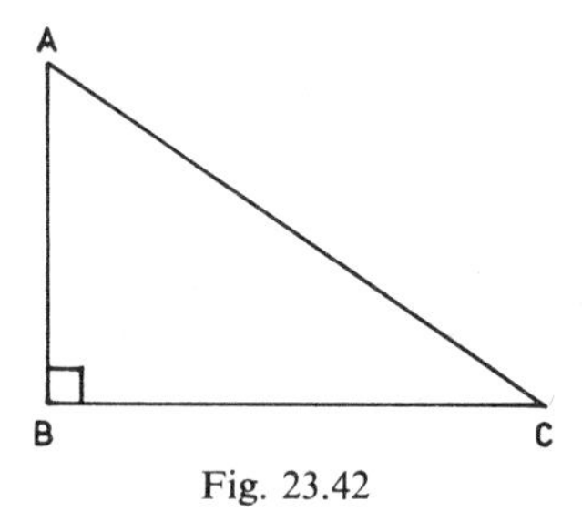

Fig. 23.42

The converse of this theorem is also true: *If the square on one side of a triangle is equal to the sum of the squares on the other two sides, the angle included between these sides is a right-angle.* Thus in Fig. 23.43 if $BC^2=AB^2+AC^2$ then $\angle BAC=90°$.
It is worth remembering that triangles with sides of 3, 4, 5; 5, 12, 13; 7, 24, 25; and any multiples of these are right-angled triangles. These triangles often occur in geometrical problems.

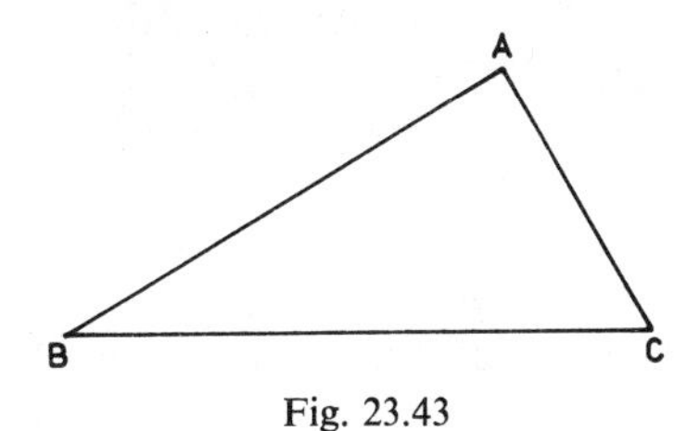

Fig. 23.43

Examples

1) In $\triangle ABC$, $\angle B = 90°$, $BC = 4{\cdot}2$ cm and $AB = 3{\cdot}7$ cm.
Find AC.
In Fig. 23.44, since $\angle B = 90°$, by Pythagoras

$$b^2 = a^2 + c^2 = 3{\cdot}7^2 + 4{\cdot}2^2 = 31{\cdot}33$$

$$b = \sqrt{31{\cdot}33} = 5{\cdot}598 \text{ cm}$$

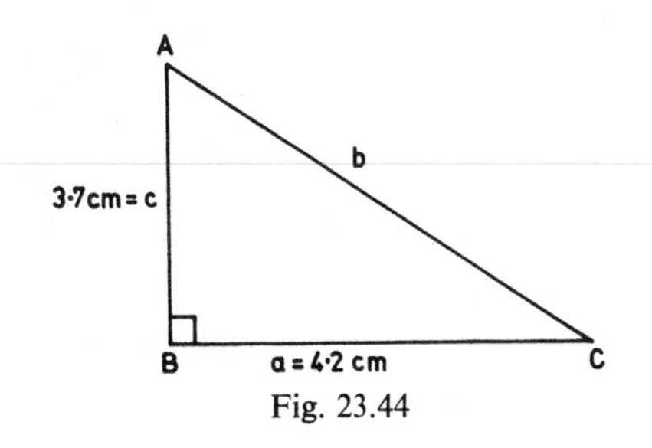

Fig. 23.44

2) In $\triangle ABC$, $\angle A = 90°$, $BC = 6{\cdot}4$ cm and $AC = 5{\cdot}2$ cm. Find AB.
In Fig. 23.45, since $\angle A = 90°$, by Pythagoras

$$a^2 = b^2 + c^2$$

or $c^2 = a^2 - b^2 = 6{\cdot}4^2 - 5{\cdot}2^2 = 13{\cdot}92$

$$c = \sqrt{13{\cdot}92} = 3{\cdot}731 \text{ cm}$$

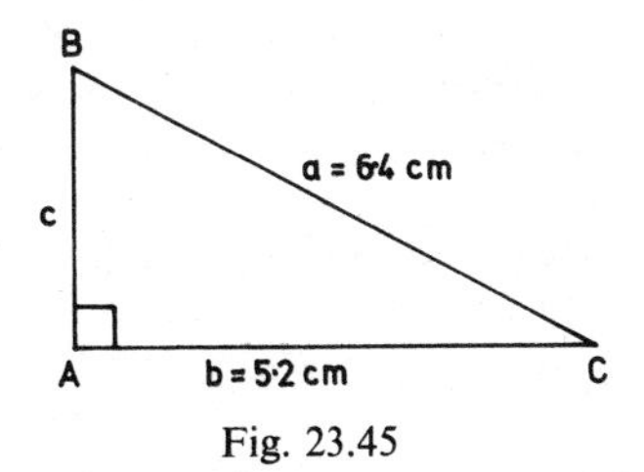

Fig. 23.45

(The difference of two squares may be used. Thus

$$c^2 = a^2 - b^2 = (a+b)(a-b) = (6{\cdot}4 + 5{\cdot}2) \times (6{\cdot}4 - 5{\cdot}2)$$

$$= 11{\cdot}6 \times 1{\cdot}2 = 13{\cdot}92).$$

3) In $\triangle ABC$, $AB = 9$ cm, $BC = 12$ cm and $AC = 15$ cm. Show that $\angle B = 90°$.
In Fig. 23.46 if $\angle B = 90°$ then

$$b^2 = a^2 + c^2$$

$$b^2 = 15^2 = 225$$

$$a^2 + c^2 = 12^2 + 9^2 = 225$$

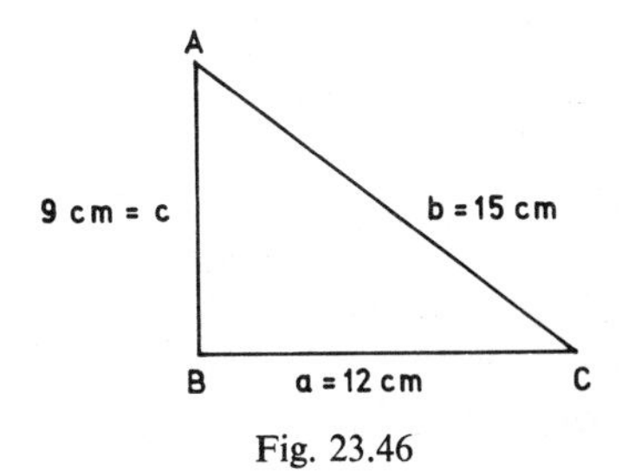

Fig. 23.46

Hence the square on one side of $\triangle ABC$ is equal to the sum of the squares on the other two sides and therefore the angle included by these sides is a right-angle. $\therefore \angle B = 90°$.

Exercise 88

1) In $\triangle ABC$, $\angle B = 90°$, $BC = 6$ cm and $AC = 7$ cm. Find AB.

2) In $\triangle ABC$, $\angle C = 90°$, $BC = 6{\cdot}1$ cm and $AC = 3{\cdot}4$ cm. Find AB.

3) In $\triangle ABC$, $\angle A = 90°$, $BC = 5{\cdot}3$ cm and $AC = 4{\cdot}8$ cm. Find AB.

4) In Fig. 23·47 find x.

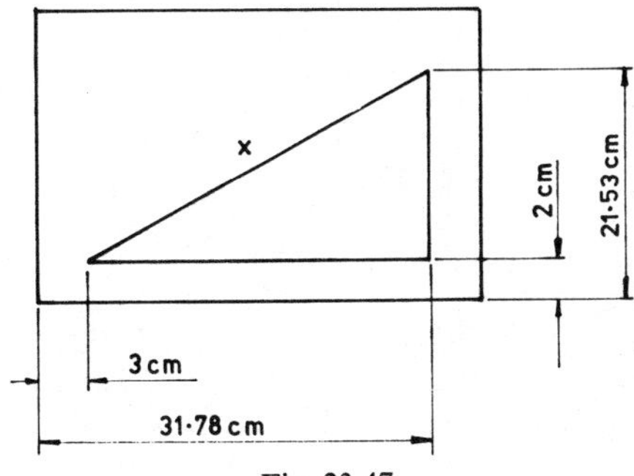

Fig. 23.47

5) In Fig. 23.48, find the lengths x and y.

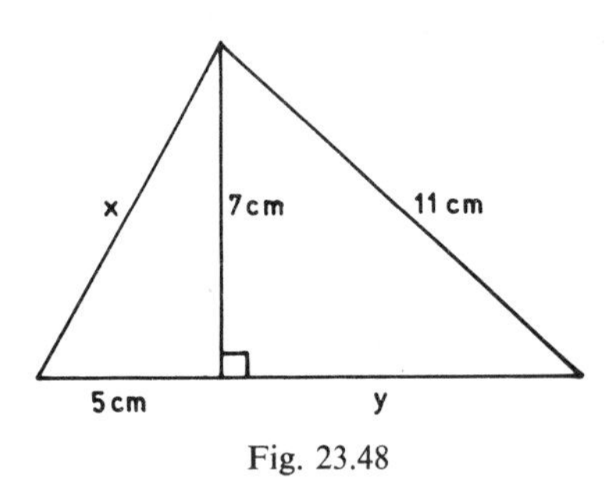

Fig. 23.48

6) $\triangle ABC$ is isosceles and $AB = AC = 18$ cm. If BC is 12 cm, find the altitude of the triangle drawn from A to BC.

7) In $\triangle PQR$, $PR = 12$ cm, $RQ = 5$ cm and $PQ = 13$ cm. Show that $\angle R = 90°$.

8) The altitude AD of a triangle ABC is 6 cm long. If $BD = 4$ cm and $DC = 9$ cm, prove that $\angle BAC = 90°$.

9) The side of an equilateral triangle is b. The length of an altitude is c. Prove that $3b^2 = 4c^2$.

10) In an isosceles triangle, the two equal sides are each a in length. The third side is $\frac{a}{2}$ in length. Show that, if the altitude drawn from the intersection of the equal sides is h, then $16h^2 = 15a^2$.

Exercise 89 (Miscellaneous)

All the questions in this exercise are of the type found in examination papers at O level.

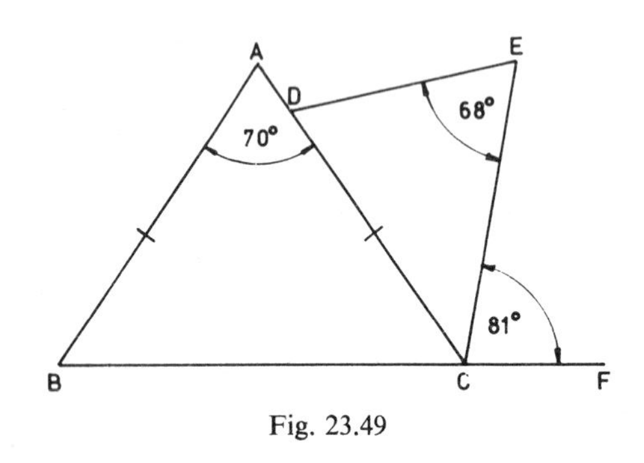

Fig. 23.49

1) In Fig. 23.49, $AB = AC$ and BCF is a straight line. $\angle BAC = 70°$, $\angle CED = 68°$ and $\angle ECF = 81°$. Prove that two of the sides of $\triangle CDE$ are equal.

2) In $\triangle ABC$, $\angle A$ is obtuse and $\angle C = 45°$. Name the shortest side of the triangle.

3) In Fig. 23.50, UWR is a straight line. $RS = RW$, $ST = SW$, $WT = WU$ and $\angle R = \angle TSW = x°$. Prove that WS bisects $\angle RWT$ and $\angle TWU = x°$.

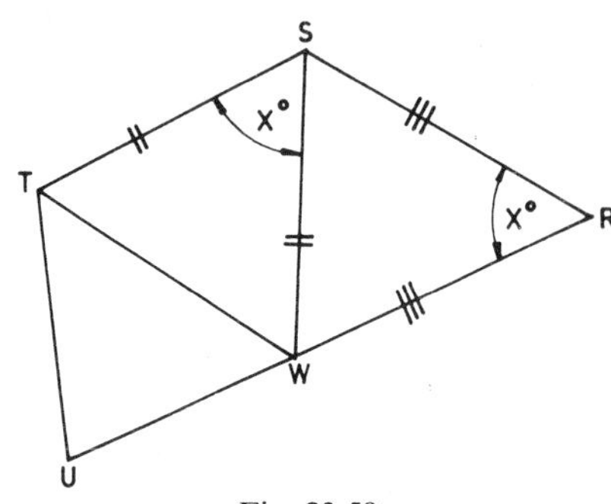

Fig. 23.50

4) In $\triangle ABC$, the sides AB and AC are equal. The side CA is produced to D and $\angle BAD = 148°$. Calculate $\angle ABC$.

5) The mid-points of the sides MP and ST of $\triangle LMP$ and $\triangle RST$ are X and Y respectively. If $LM = RS$, $MP = ST$ and $LX = RY$ prove that $\triangle LMP \equiv \triangle RST$.

6) Two similar triangles have areas of 27 cm^2 and 48 cm^2. Find the ratio of the lengths of a pair of corresponding sides of the two triangles.

7) In $\triangle ABC$, AN is the perpendicular from A to BC. If $BN = 9$ cm, $CN = 16$ cm and $AN = 12$ cm, prove $\angle BAC = 90°$.

8) In Fig. 23.51, ABCD is a quadrilateral in which $\frac{AB}{AD} = \frac{5}{3}$. AX bisects $\angle BAD$ and XY is parallel to BC. Calculate the ratios

$$\frac{BX}{XD} \text{ and } \frac{XY}{BC}.$$

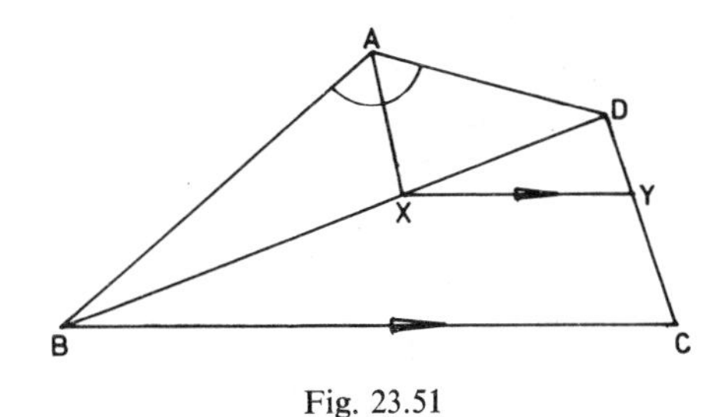

Fig. 23.51

Also find

$$\frac{\text{area } \triangle AXD}{\text{area } \triangle ABD} \text{ and } \frac{\text{area } \triangle DXY}{\text{area } \triangle DBC}.$$

9) In Fig. 23.52, F is the mid-point of the side AB of $\triangle ABC$ and FE is parallel to BC. If AC = 12 cm, BD = 8 cm and DC = 2 cm, calculate

a) the lengths FE and CX;
b) the ratio of the areas of $\triangle$s XCD and XEF.

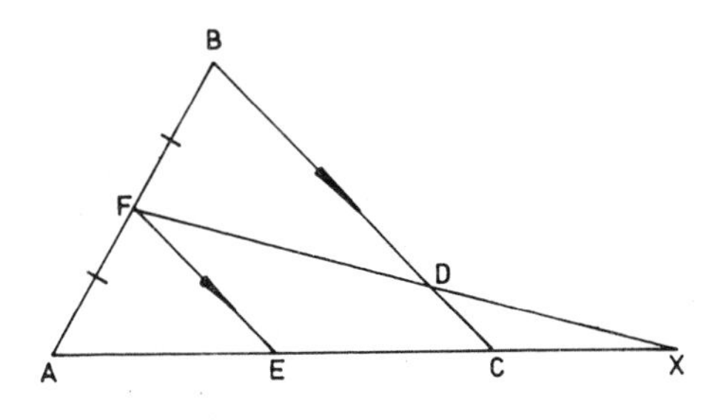

Fig. 23.52

10) BM and CN are altitudes of $\triangle ABC$ and BN = CM. Prove that

a) $\triangle BCN \equiv \triangle BCM$; b) AB = AC.

11) PQR is a triangle in which PQ = 5 cm, QR = 8 cm and RP = 10 cm. S and T are points on PQ and PR respectively such that PS = 4 cm and PT = 2 cm. Prove that $\triangle$s PTS and PQR are similar and calculate the length of ST.

12) PQR is a triangle in which $\angle PRQ = 62°$. S is a point on QR between Q and R such that SP = SR and $\angle QPS = 27°$. Calculate $\angle PQR$ and hence prove that SR is greater than QS.

13) In $\triangle ABC$, the perpendicular from A to BC meets BC at D and the perpendicular from D to AB meets AB at E. Given that BD = 4 cm, DC = 6 cm and the area of $\triangle ABC$ is 15 cm^2, prove that AD = 3 cm and calculate AE.

14) In $\triangle ABC$, D is the mid-point of BC and E is the mid-point of CA. The lines AD and BE meet at G. Prove that

a) $\triangle$s ABG and DEG are similar;
b) $\triangle$s AGE and BGD are equal in area.

15) In $\triangle PQR$, the line ST is drawn parallel to QR meeting PQ at S and PR at T such that PS = 3SQ. Calculate $\frac{ST}{QR}$. Given that the area of $\triangle TQS$ is 3 cm^2, calculate

a) the area of $\triangle PST$; b) the area of $\triangle TQR$.

16) A line cuts three parallel lines at A, B and C such that AB = BC. Another line cuts the parallel lines at P, Q and R. Draw lines through P and Q parallel to AC and use congruent triangles to prove that PQ = QR.

SELF TEST 23

State the letter (or letters) corresponding to the correct answer (or answers).

1) The triangle shown in Fig. 23.53 is

a acute-angled b obtuse-angled
c scalene d isosceles

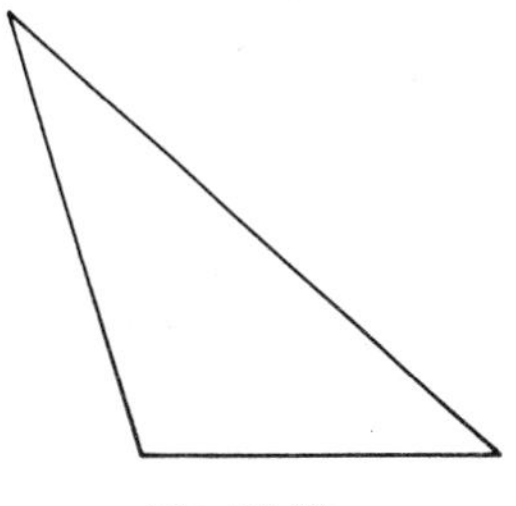
Fig. 23.53

2) In Fig. 23.54, ∠B is equal to
a 80° b 40° c 50° d 90°

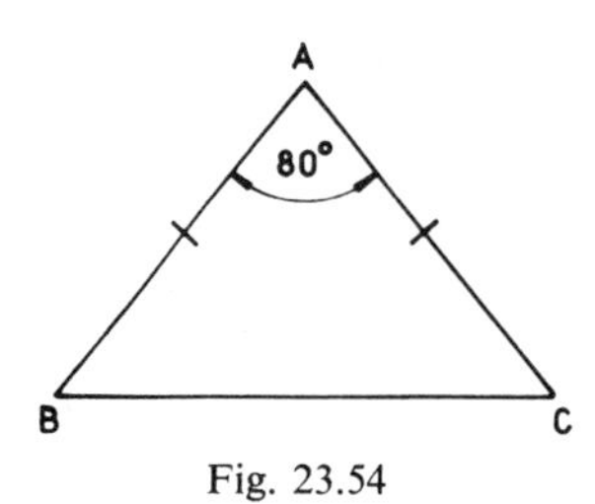

Fig. 23.54

3) In Fig. 23.55, ∠A is equal to
a 20° b 40° c 60° d 80°

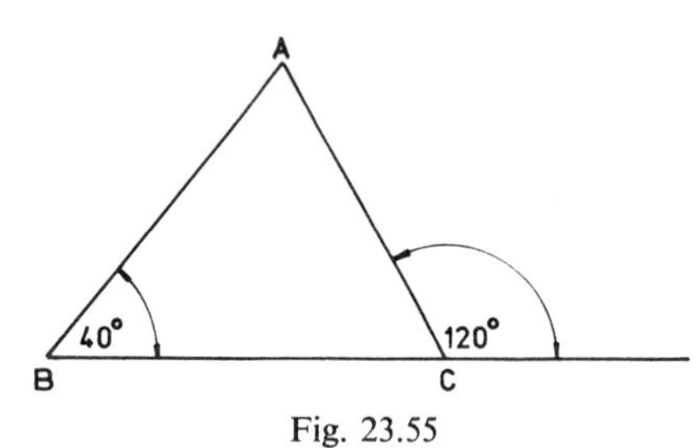

Fig. 23.55

4) In Fig. 23.56, x is equal to
a 140° b 70° c 80° d 60°

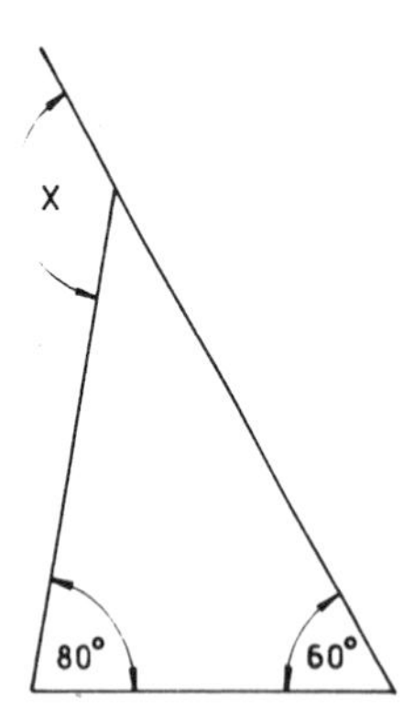

Fig. 23.56

5) In Fig. 23.57 the largest angle of the triangle is
a ∠A b ∠B c ∠C

6) A triangle is stated to have sides whose lengths are 5 cm, 8 cm and 14 cm.
a It is possible to draw the triangle.
b It is impossible to draw the triangle.

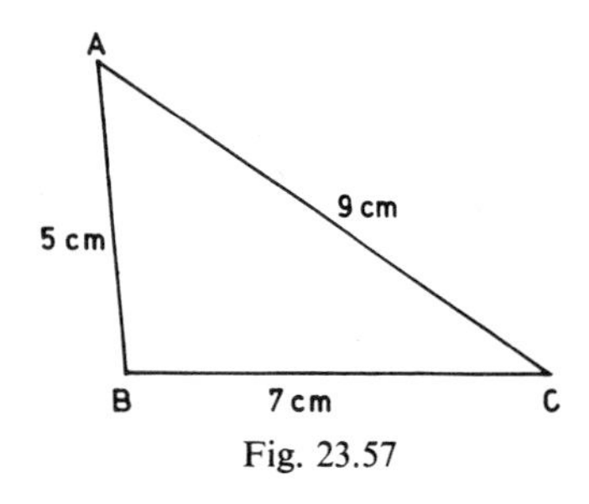

Fig. 23.57

7) In Fig. 23.58
a ∠BAD = ∠DAC
b ∠ABD = ∠ACD
c BD = DC d AD = BD

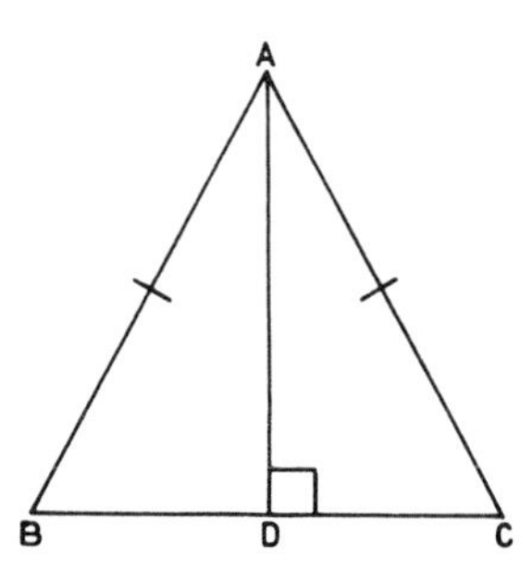

Fig. 23.58

8) In Fig. 23.59, x is equal to
a 17° b 48° c 8°30′ d 25°

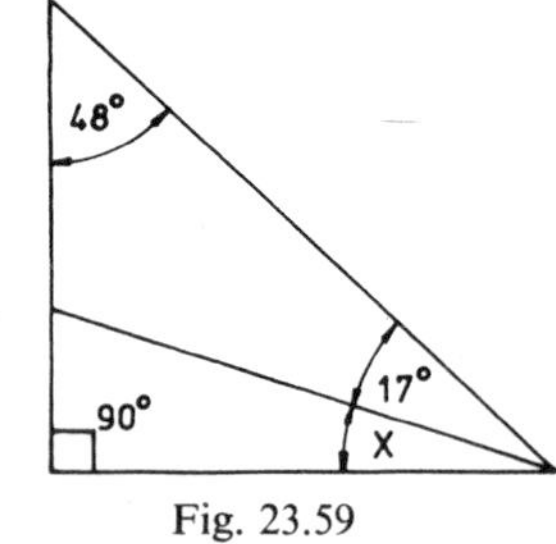

Fig. 23.59

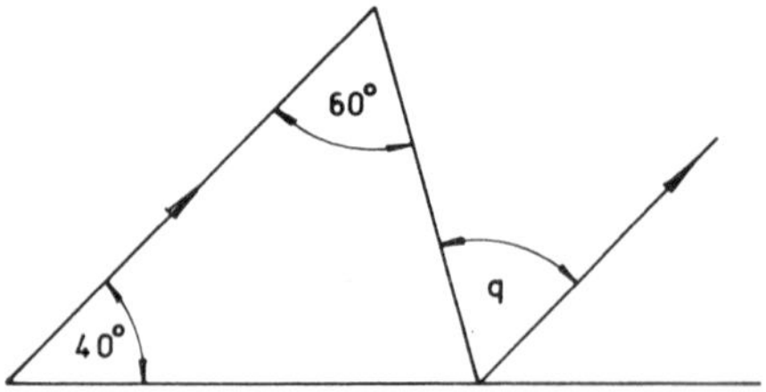

Fig. 23.60

9) In Fig. 23.60, q is equal to
a 80° b 60° c 40° d 100°

10) Two angles of a triangle are $(2x-40)°$ and $(3x+10)°$. The third angle is therefore
a $(210-5x)°$ b $(230+x)°$
c $(220-5x)°$

11) The three angles of a triangle are $(2x+20)°$, $(3x+20)°$ and $(x+20)°$. The value of x is
a 60° b 40° c 20° d 10°

12) In Fig. 23.61 state the letter which corresponds to those triangles which are definitely congruent.

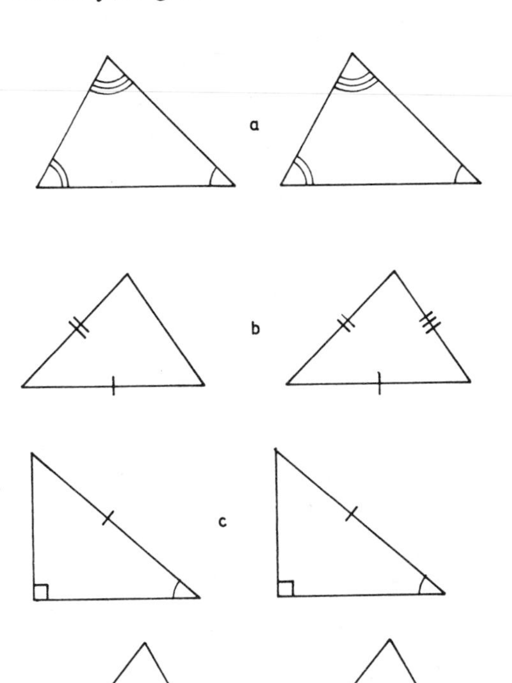

Fig. 23.61

13) In Fig. 23.62 ring the letter which corresponds to those triangles which are definitely congruent.

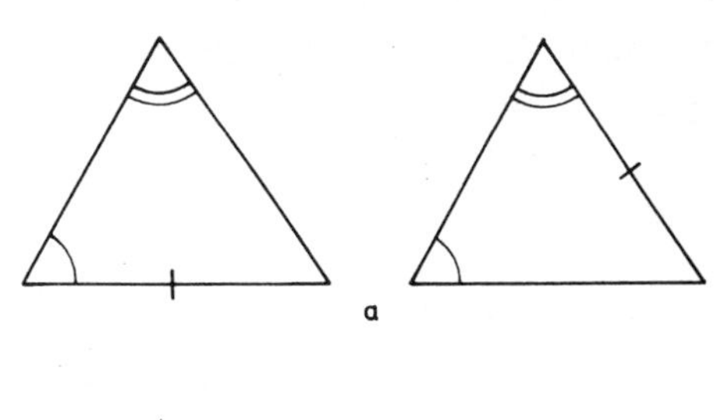

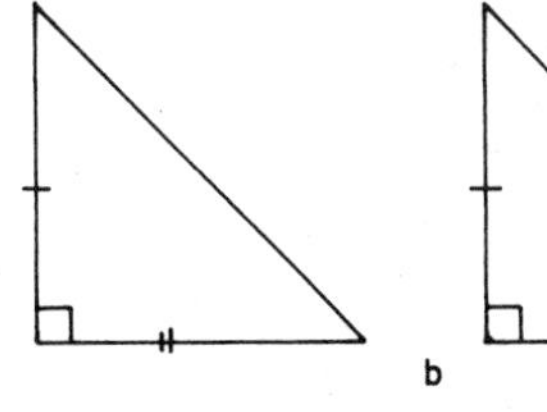
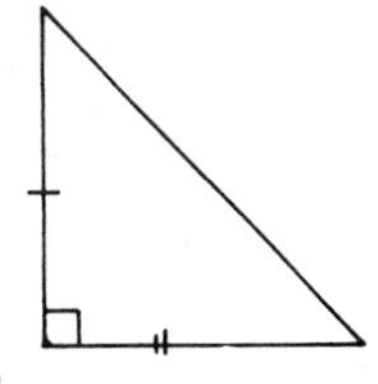

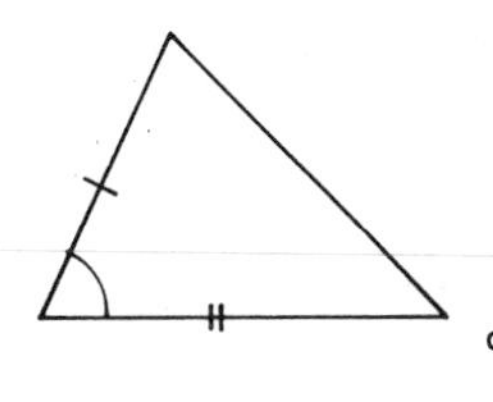
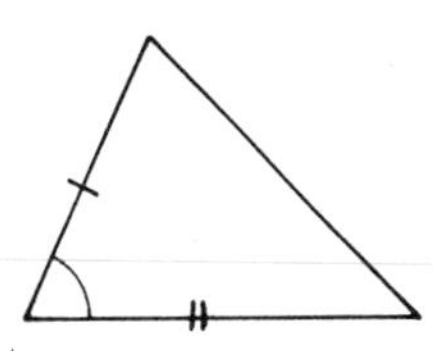

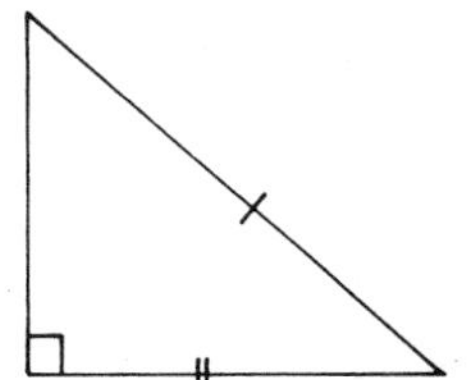
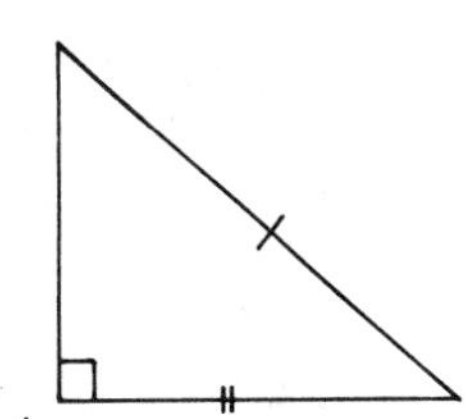

Fig. 23.62

14) In Fig. 23.63, AD = BC and AC = DB. Hence
a $\angle DAC = \angle DBC$
b $\angle ADB = \angle DBC$
c $\angle ADC = \angle BCD$
d $\angle ADC = \angle BDC$

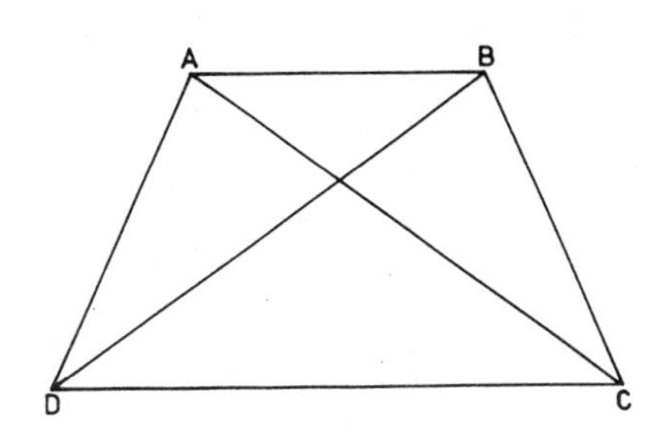

Fig. 23.63

15) In Fig. 23.64 $\triangle DEC$ is equilateral and ABCD is a square. $\angle DEA$ is therefore

a 30° b 15° c 45° d 20°

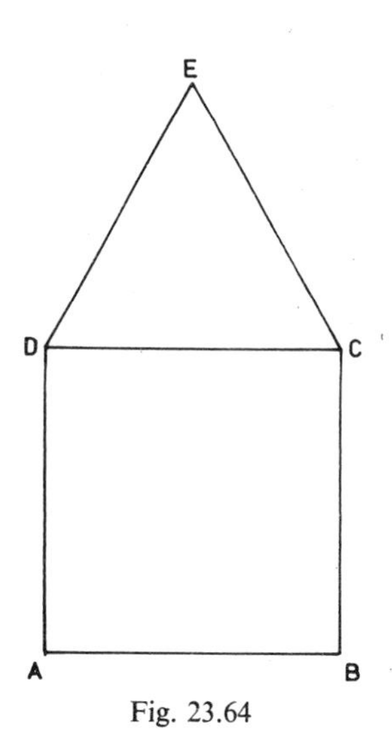

Fig. 23.64

16) In Fig. 23.65 two straight lines bisect each other at X. Therefore

a $\triangle AXC \equiv \triangle DXB$
b $\triangle ABD \equiv \triangle ABC$
c $\triangle ADX \equiv \triangle CXB$
d $\angle CAX = \angle XDB$

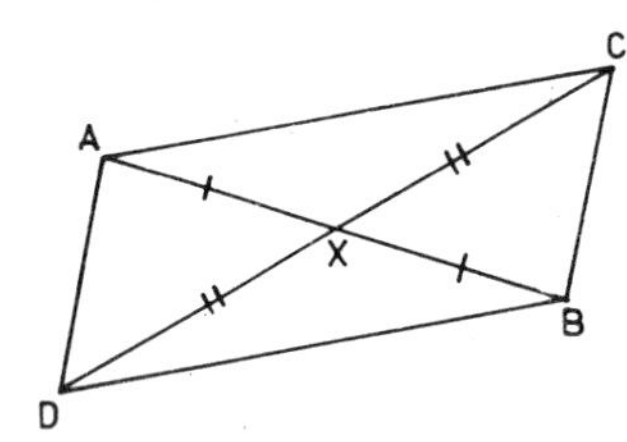

Fig. 23.65

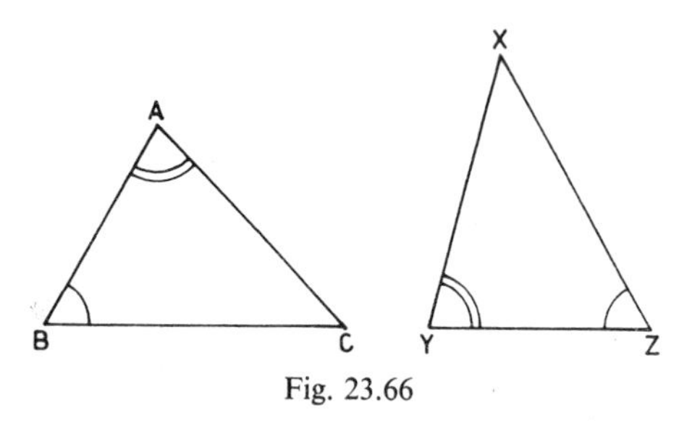

Fig. 23.66

17) The triangles shown in Fig. 23.66 are

a congruent b similar
c neither of these

18) If the triangles ABC and XYZ shown in Fig. 23.66 are similar then

a $\frac{AC}{XY} = \frac{XZ}{BC}$ b $\frac{AC}{XY} = \frac{BC}{XZ}$

c $\frac{BC}{AB} = \frac{YZ}{XZ}$ d $\frac{BC}{AB} = \frac{XZ}{YZ}$

19) In Fig. 23.67 if $\frac{AB}{XY} = \frac{AC}{XZ}$ and $\angle B = \angle Y$ then

a $\frac{AB}{XY} = \frac{BC}{YZ}$ b $\angle A = \angle X$

c $\angle C = \angle Z$

d none of the foregoing are necessarily true.

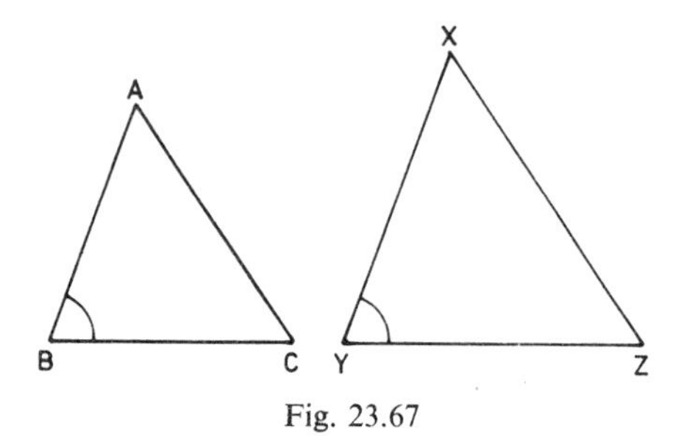

Fig. 23.67

20) In Fig. 23.68, $\angle A = \angle X$ and $\angle B = \angle Y$. Hence

a $XY = 6\frac{7}{8}$ cm b $XY = 17\frac{3}{5}$ cm
c $YZ = 19\frac{1}{5}$ cm d $YZ = 7\frac{1}{2}$ cm

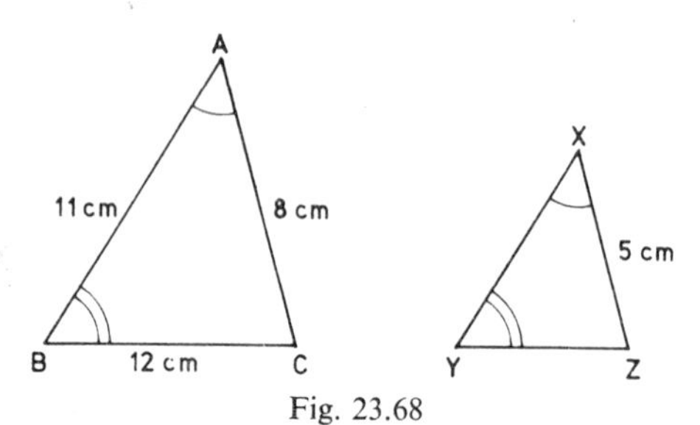

Fig. 23.68

21) In Fig. 23.69, PS=4SQ. Hence $\frac{ST}{QR}$ is equal to

a $\frac{1}{4}$ b $\frac{4}{1}$ c $\frac{4}{5}$ d $\frac{5}{4}$

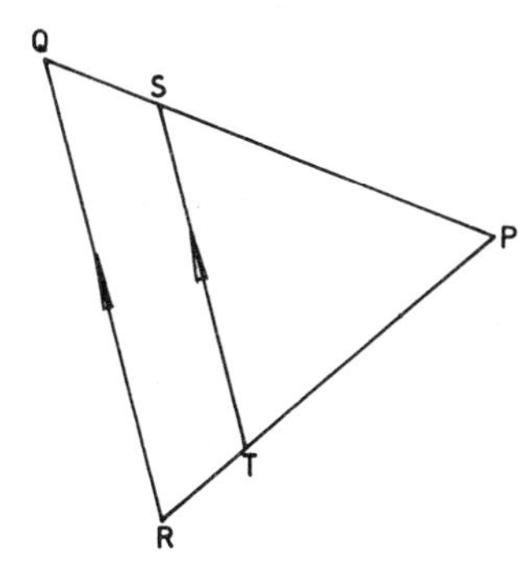

Fig. 23.69

22) In Fig. 23.70, XY is parallel to BC and AB is parallel to YZ. Hence

a $\angle B = \angle Z$

b $\triangle$s ABC and YZC are similar

c $\frac{YZ}{ZC} = \frac{AC}{BC}$ d $\frac{ZC}{AC} = \frac{YZ}{AB}$

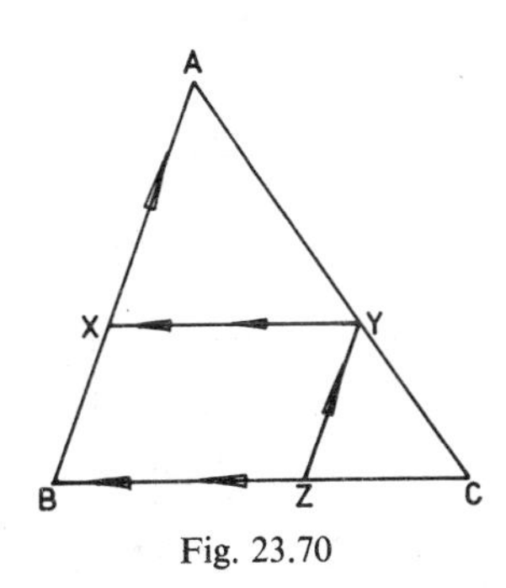

Fig. 23.70

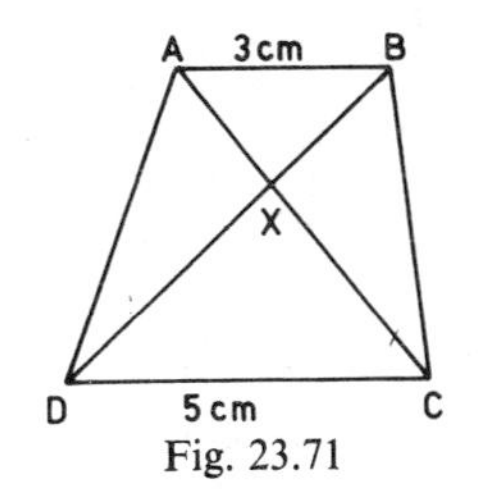

Fig. 23.71

23) In Fig. 23.71, AB is parallel to DC and AB=3 cm and DC=5 cm. Hence $\frac{XD}{XB}$ is equal to

a $\frac{3}{5}$ b $\frac{5}{3}$ c neither of these

d $\frac{5}{8}$

24) In Fig. 23.72, AD=DE=EB. Hence

a $\triangle$s ADC and CEB are equal in area.

b $\triangle$BCD has twice the area of $\triangle$ADC.

c $\triangle$ABC has three times the area of $\triangle$ACD.

d none of the foregoing is correct.

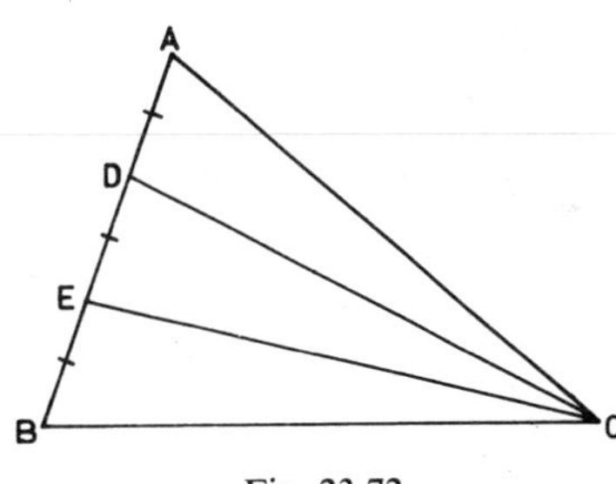

Fig. 23.72

25) In Fig. 23.73, $\triangle$s ABC and DEF are similar triangles. If the area of $\triangle$ABC is 20 cm^2 then the area of $\triangle$DEF is

a 10 cm^2 b 5 cm^2 c 8 cm^2

d none of these

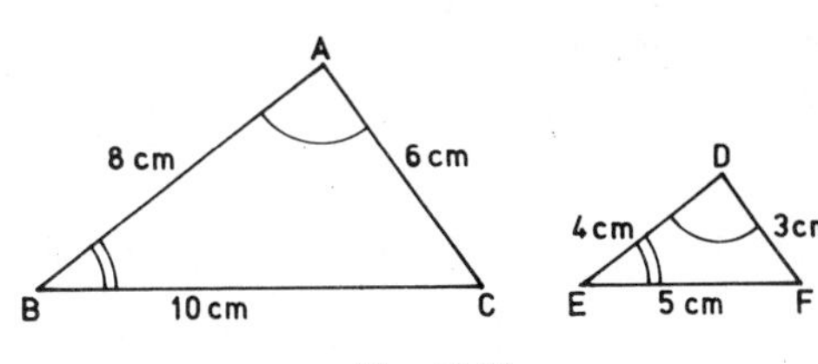

Fig. 23.73

26) In Fig. 23.74, if the area of $\triangle$XYZ is 10 cm^2 then the area of $\triangle$ABC is

a impossible to find from the given information

b 40 cm^2 c 80 cm^2 d 160 cm^2

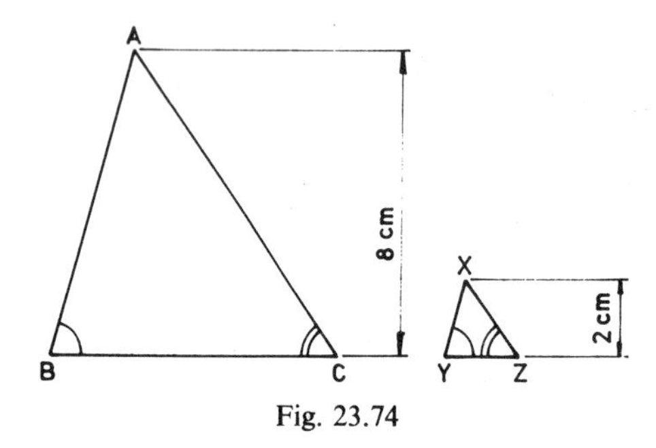

Fig. 23.74

27) In Fig. 23.75 $\angle A = \angle X$ and $\angle B = \angle Y$. $\triangle ABC$ has an area of 36 cm^2 and $\triangle XYZ$ has an area of 4 cm^2. If AB = 4 cm then XY is equal to

a $\frac{3}{4}$ cm b $\frac{4}{3}$ cm c $\frac{4}{9}$ cm

d $\frac{9}{4}$ cm

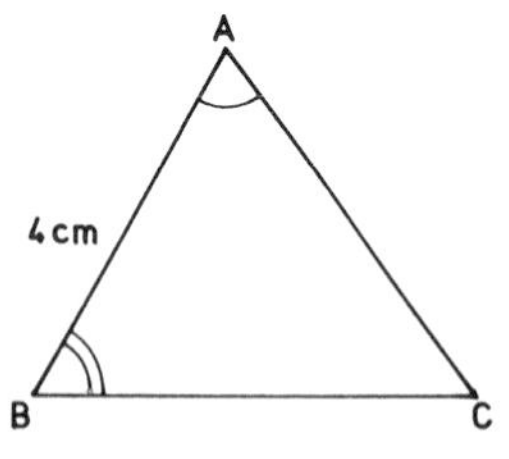

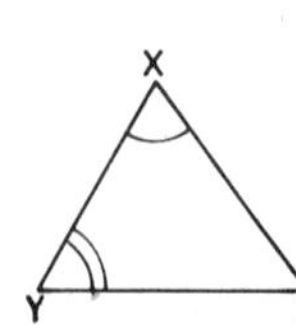

Fig. 23.75

28) In Fig. 23.76, $\frac{AX}{XB} = \frac{2}{1}$. The area of $\triangle ABC$ is 36 cm^2. Hence the area of XYBC is

a 12 cm^2 b 18 cm^2 c 20 cm^2

d 16 cm^2

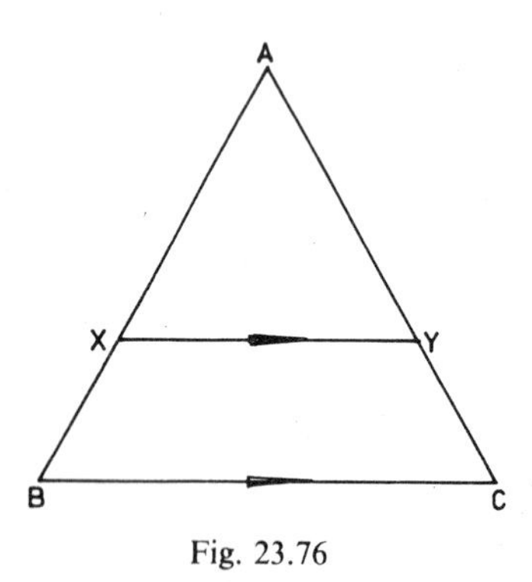

Fig. 23.76

29) In Fig. 23.77 the line AD bisects $\angle A$. Therefore

a $\frac{AB}{AC} = \frac{BD}{DC}$ b $\frac{AB}{AC} = \frac{DC}{BD}$

c neither of the foregoing is true.

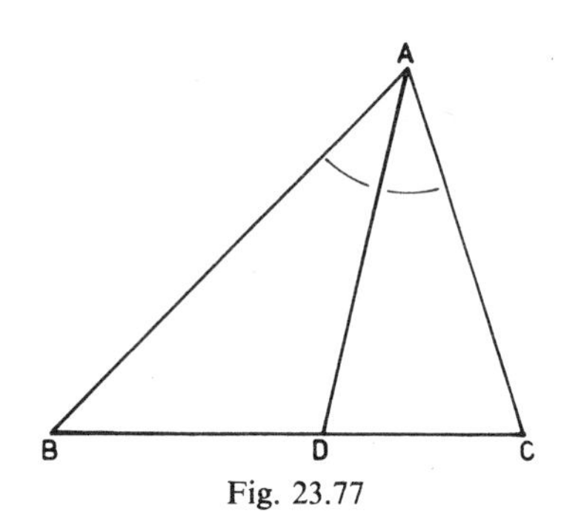

Fig. 23.77

30) In Fig. 23.78, the line AD bisects $\angle CAX$. Hence

a $\frac{BD}{DC} = \frac{AC}{AB}$ b $\frac{CD}{BC} = \frac{AC}{AB}$

c $\frac{BD}{DC} = \frac{AB}{AC}$ d $\frac{CD}{BC} = \frac{AB}{AC}$

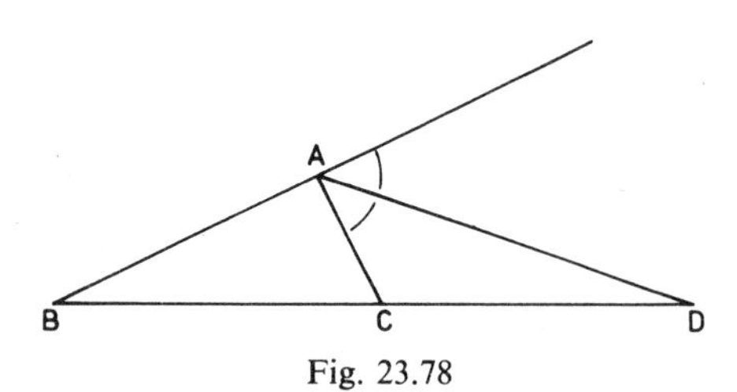

Fig. 23.78

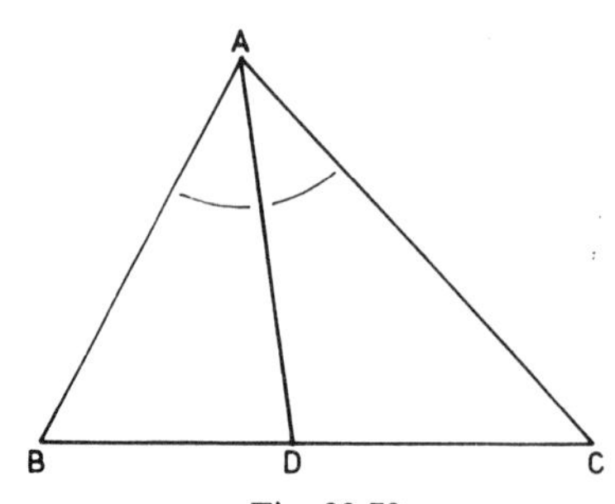

Fig. 23.79

31) In Fig. 23.79, the area of $\triangle ABC$ is 30 cm^2. The line AD bisects $\angle A$. If $\frac{AB}{AC}=\frac{3}{2}$ then the area of $\triangle ABD$ is

a impossible to find from the given data
b 9 cm^2 c 18 cm^2 d 60 cm^2

32) In $\triangle ABC$ (Fig. 23.79), $\angle A$ is bisected by the line AD and AB=4 cm and AC=5 cm.

The ratio $\frac{BD}{DC}$ is equal to

a $\frac{4}{5}$ b $\frac{5}{4}$ c neither of these

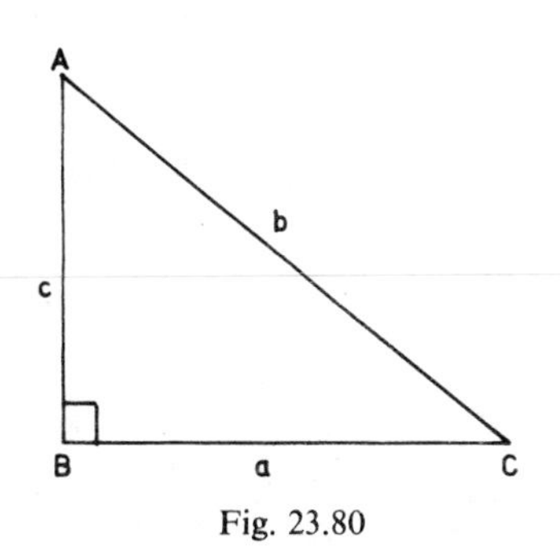

Fig. 23.80

33) In $\triangle ABC$ (Fig. 23.80)

a $a^2=b^2+c^2$ b $b^2=a^2+c^2$
c $c^2=a^2+b^2$
d none of the foregoing is true

34) In a triangle ABC, AB=3 cm, BC=4 cm and AC=5 cm. Hence

a $\angle A=90°$ b $\angle B=90°$
c $\angle C=90°$
d none of the angles is 90°.

35) In $\triangle ABC$, $\angle B=90°$, BC=5 cm, AB=12 cm. Hence AC is equal to

a 10·91 cm b 34·50 cm
c 41·11 cm d 13·00 cm

36) In $\triangle ABC$, $\angle A=90°$, BC=7·8 cm, AC=6·3 cm. Hence AB is equal to

a 14·53 cm b 4·60 cm
c 10·26 cm d 3·25 cm

37) If P is a point inside a rectangle ABCD then

a $AP^2+BP^2=CP^2+DP^2$
b $AP^2+DP^2=BP^2+CP^2$
c $AP^2+CP^2=CB^2+CD^2$
d None of these is correct.

Chapter 24 Quadrilaterals and Polygons

QUADRILATERALS

1) A *quadrilateral* is any four sided figure. Since a quadrilateral can be split up into two triangles, the sum of its angles is 360° (Fig. 24.1).

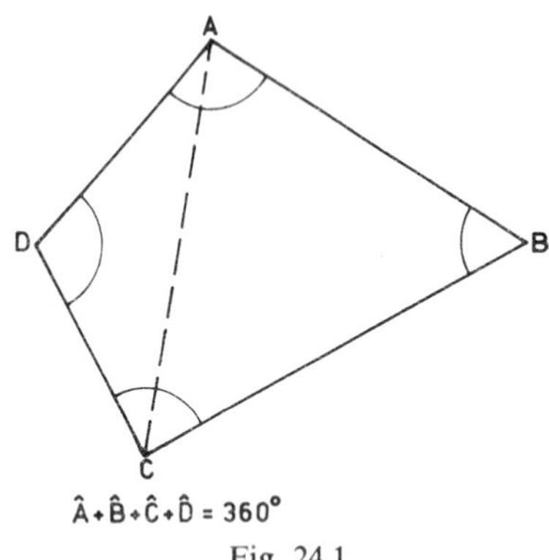

Fig. 24.1

2) A *parallelogram* has both pairs of opposite sides parallel. A parallelogram has the following properties:

i) The sides which are opposite to each other are equal in length.
ii) The angles which are opposite to each other are equal.
iii) The diagonals bisect each other.
iv) The diagonals each bisect the parallelogram.

3) A *rectangle* is a parallelogram with all its angles equal to 90°. A rectangle has all the properties of a parallelogram, but in addition the diagonals are equal in length.

4) A *rhombus* is a parallelogram with all its sides equal in length. It has all the properties of a parallelogram, but in addition it has the following properties:

i) The diagonals bisect at right-angles.
ii) The diagonal bisects the angle through which it passes.

A *square* is a rectangle with all its sides equal in length. A square has all the properties of a parallelogram, rectangle and rhombus.

Examples

1) X, P, Q, Y are points in order on a straight line and XP = QY. The parallelogram PQRS is drawn such that ∠PQR = 130° and QR = QY. The lines XS and YR are produced to meet at Z. Calculate ∠XZY.

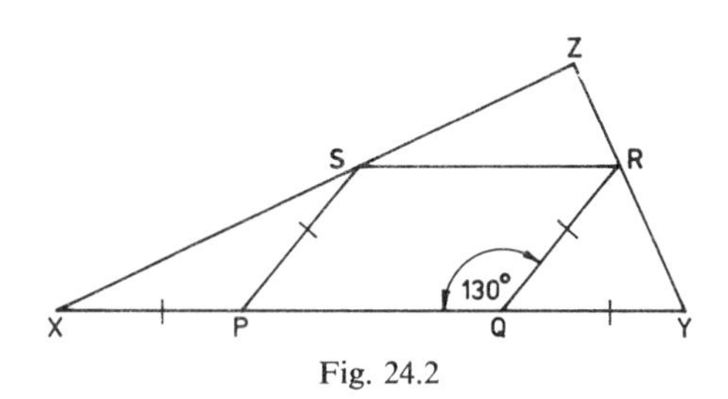

Fig. 24.2

Referring to Fig. 24.2

Since ∠PQR = 130°

then ∠PSR = 130° (PQRS is a parallelogram)

and ∠SPQ = ∠SRQ = 50°

Hence ∠XPS = 130° and ∠RQY = 50°

△RQY is isosceles (since QR = QY)

∴ ∠QRY = ∠RYQ = 65°

△SXP is isosceles (since XP = SP)

∴ ∠SXP = ∠PSX = 25°

In △ZXY

∠XZY = 180° − ∠SXP − ∠RYQ
= 180° − 65° − 25° = 90°

2) A rhombus ABCD and an equilateral triangle ABX lie on opposite sides of AB. If ∠BCD = 82°, calculate ∠ADX and ∠BDX

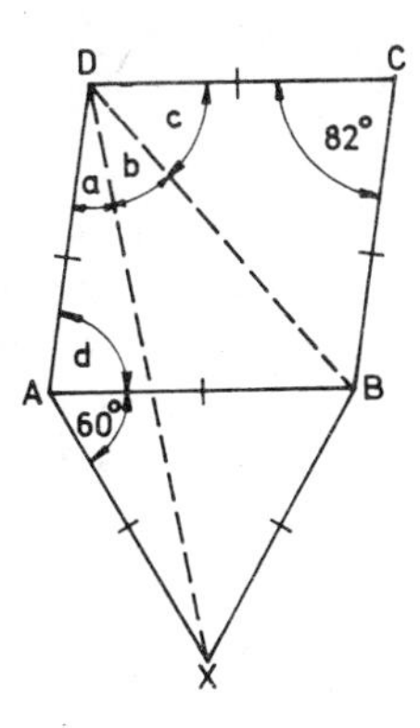

Fig. 24.3

Referring to Fig. 24.3

$d=82°$ (opp. angles of a rhombus are equal)

$\angle XAB=60°$ (angle of an equilateral triangle ABX)

$\triangle DAX$ is isosceles since $AD=AX$

$\angle DAX=82°+60°=142°$

$\therefore\ a=\frac{1}{2}(180°-142°)=19°$

$\triangle CDB$ is isosceles since $CD=CB$

$\therefore\ c=\frac{1}{2}(180°-82°)=49°$

In the rhombus ABCD

$\angle D=180°-82°=98°$

$a+b+c=98°$

$b=98°-a-c=98°-19°-49°=30°$

$\therefore\ a=\angle ADX=19°$ and $b=\angle XDB=30°$

Exercise 90

1) Calculate the angle x in Fig. 24.4.

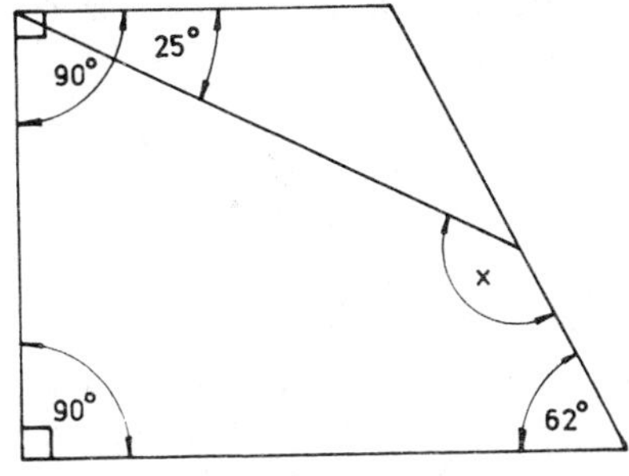

Fig. 24.4

2) Calculate the angle x in Fig. 24.5.

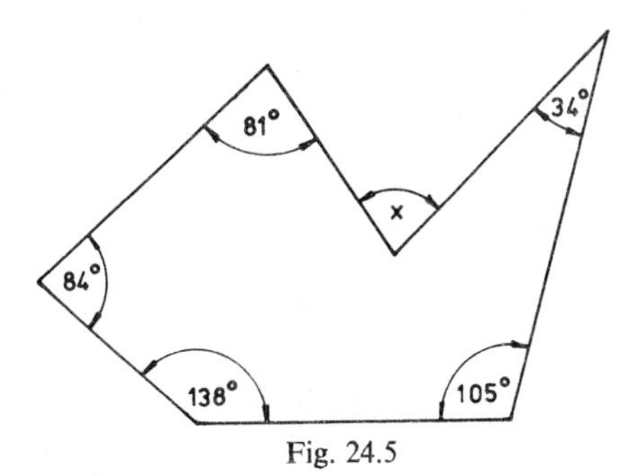

Fig. 24.5

3) Find the angle x in Fig. 24.6.

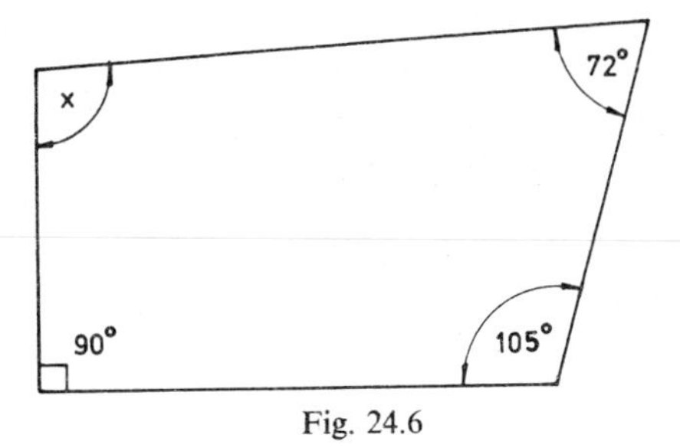

Fig. 24.6

4) In Fig. 24.7, ABCD is a parallelogram. Calculate the angles x and y.

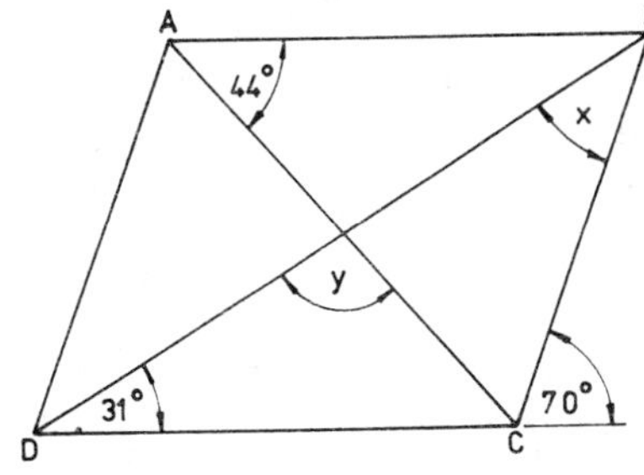

Fig. 24.7

5) ABCD is a quadrilateral in which $\angle A=86°$, $\angle C=110°$ and $\angle D=40°$. The angle ABC is bisected to cut the side AD at E. Find $\angle AEB$.

6) In Fig. 24.8, ABCD is a quadrilateral with its sides AB and DC produced to meet at E. The angles EAC and AEC are 40° and 30° respectively. If FD is parallel to AC find $\angle EDF$.

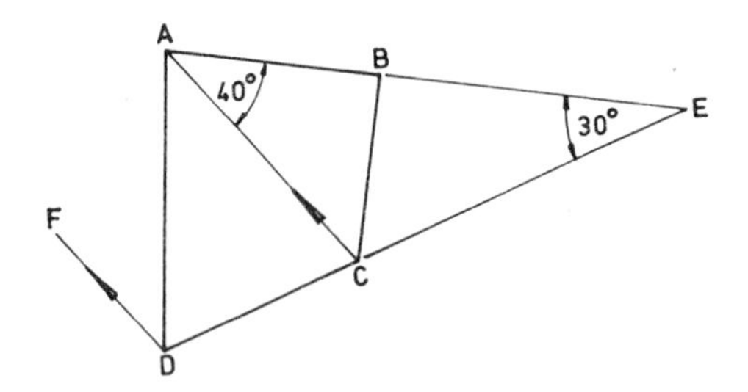

Fig. 24.8

7) In the quadrilateral ABCD, $\angle DAB = 60°$ and the other three angles are equal. The line CE is drawn parallel to BA to meet AD at E. Calculate the angles ABC and ECD.

8) PQRS is a square. T is a point on the diagonal PR such that PT = PQ. The line through T, perpendicular to PR cuts QR at X. Prove that QX = XT = TR.

9) The diagonals of a rhombus are 12 cm and 9 cm long. Calculate the length of the sides of the rhombus.

10) The diagonals of a quadrilateral XYZW intersect at O. Given that OX = OW and OY = OZ prove that
a) XY = ZW; b) YZ is parallel to XW.

11) ABCD is a parallelogram. Parallel lines BE and DF meet the diagonal AC at E and F respectively. Prove that
a) AE = FC; b) BEDF is a parallelogram.

POLYGONS

Any plane closed figure bounded by straight lines is called a polygon.

1) A *convex* polygon (Fig. 24.9) has no interior angle greater than 180°.

2) A *re-entrant* polygon (Fig. 24.10) has at least one angle greater than 180°.

3) A *regular* polygon has all of its sides and all of its angles equal.

4) A *pentagon* is a polygon with 5 sides.

5) A *hexagon* is a polygon with 6 sides.

6) An *octagon* is a polygon with 8 sides.

In a convex polygon having n sides the sum of the interior angles is $(2n-4)$ *right-angles. The sum of the exterior angles is* 360°, *no matter how many sides the polygon has.* Note that these statements apply to all polygons not just regular polygons.

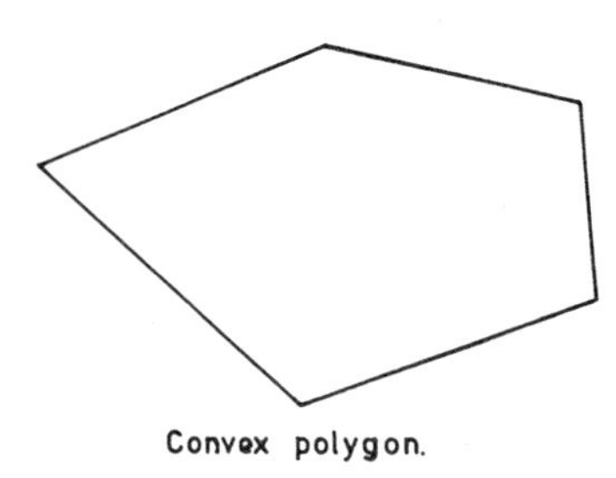

Convex polygon.

Fig. 24.9

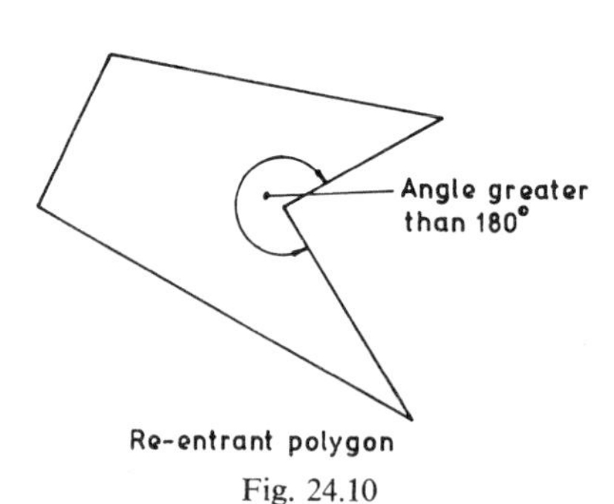

Re-entrant polygon

Fig. 24.10

Examples

1) Each interior angle of a regular polygon is 140°. How many sides has it?

Let the polygon have n sides.

The sum of the interior angles is then $140n$ degrees. But the sum of the interior angles is also $(2n-4)$ right-angles or $90(2n-4)$ degrees.

$$\therefore \quad 90(2n-4) = 140n$$
$$180n - 360 = 140n$$
$$40n = 360$$
$$n = 9$$

Hence the polygon has 9 sides.

2) In a regular polygon, each interior angle is greater by 140° than each exterior angle. How many sides has the polygon?

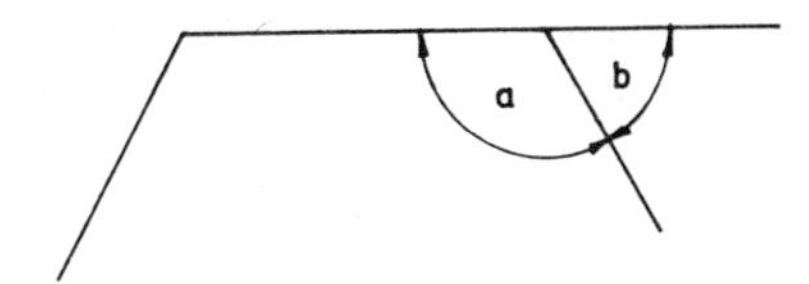

Fig. 24.11

In Fig. 24.11 let a be the interior angle and b the exterior angle of a polygon having n sides. Then

$$a - b = 140° \qquad ...(1)$$

Also since the sum of the exterior angles is 360°

$$nb = 360 \qquad ...(2)$$

and since the sum of the interior angles is $90(2n-4)°$

$$na = 90(2n-4) \qquad ...(3)$$

From equation (2)

$$b = \frac{360}{n}$$

From equation (3)

$$a = \frac{90(2n-4)}{n}$$

From equation (1)

$$\frac{90(2n-4)}{n} - \frac{360}{n} = 140$$

$$180n - 360 - 360 = 140n$$
$$40n = 720$$
$$n = 18$$

Hence the polygon has 18 sides.

Exercise 91

1) Find the sum of the interior angles of a convex polygon with
a) 5 b) 8 c) 10 d) 12 sides.

2) If the polygons in question 1 are all regular find the size of the interior angle of each.

3) A hexagon has interior angles of 100°, 110°, 120° and 128°. If the remaining two angles are equal, what is their size?

4) Each interior angle of a regular polygon is 150°. How many sides has it?

5) ABCDE is a regular pentagon and ABX is an equilateral triangle drawn outside the pentagon. Calculate ∠AEX.

6) In a regular polygon each interior angle is greater by 150° than each exterior angle. Calculate the number of sides of the polygon.

7) A polygon has n sides. Two of its angles are right-angles and each of the remaining angles is 144°. Calculate n.

8) In a pentagon ABCDE, ∠A = 120°, ∠B = 138° and ∠D = ∠E. The sides AB, DC when produced meet at right-angles. Calculate ∠BCD and ∠E.

9) In a regular pentagon ABCDE, the lines AD and BE intersect at P. Calculate the angles ∠BAD and ∠APE.

10) Calculate the exterior angle of a regular polygon in which the interior angle is four times the exterior angle. Hence find the number of sides in the polygon.

11) Each exterior angle of a regular polygon of n sides exceeds by 6° each exterior angle of a regular polygon of $2n$ sides. Find an equation for n and solve it.

12) Calculate the number of sides of a regular polygon in which the exterior angle is one-fifth of the interior angle.

AREA OF A PARALLELOGRAM

1) *The area of a parallelogram is the product of the base and altitude.* Thus in Fig. 24.12:

$$\text{Area of parallelogram ABCD} = \text{CD} \times \text{EF} = \text{BC} \times \text{GH}$$

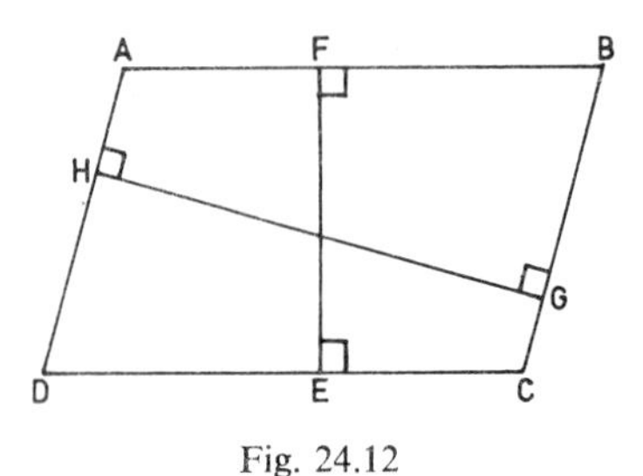

Fig. 24.12

2) *Parallelograms having equal bases and equal altitudes are equal in area.* Thus in Fig. 24.13:

Area parallelogram ABCD
= area parallelogram CDEF

(see Theorem 3.)

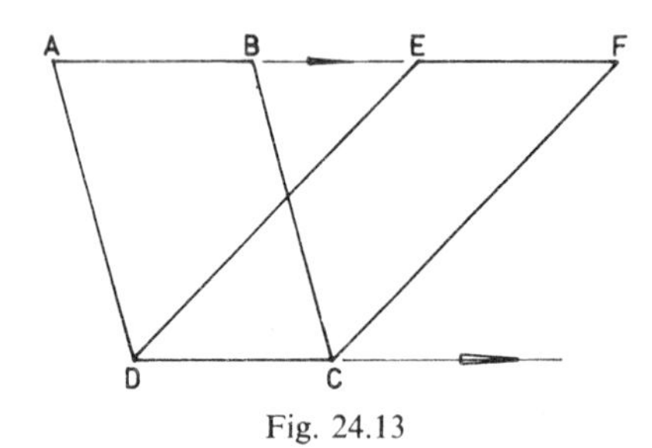

Fig. 24.13

3) It follows that:
   i) Parallelograms which have equal areas and equal bases must have equal alitudes.
   ii) Parallelograms which have equal areas and equal altitudes must have equal bases.

4) *The area of a triangle is half the area of a parallelogram drawn on the same base and between the same parallels.* Thus in Fig. 24.14:

Area $\triangle$ABC
$= \frac{1}{2}$ area of parallelogram ABDE

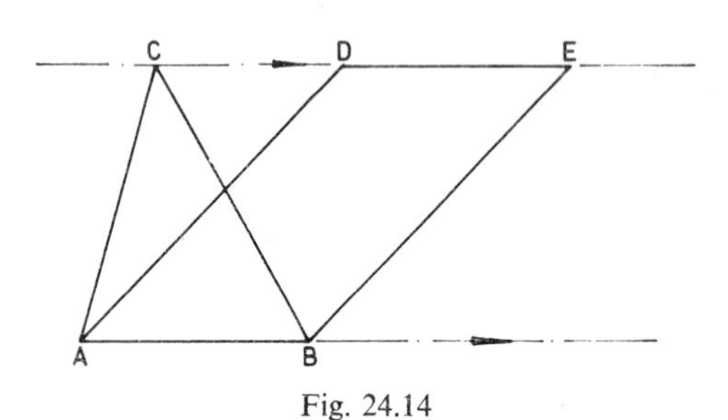

Fig. 24.14

AREA OF A TRAPEZIUM

A *trapezium* is a quadrilateral which has one pair of sides parallel (Fig. 24.15). Its area is easily found by dividing it up into two triangles as shown in the diagram.

Area $\triangle ABC = \frac{1}{2}ah$
Area $\triangle ACD = \frac{1}{2}bh$
Area of trapezium $ABCD = \frac{1}{2}ah + \frac{1}{2}bh$
$= \frac{1}{2}h(a+b)$

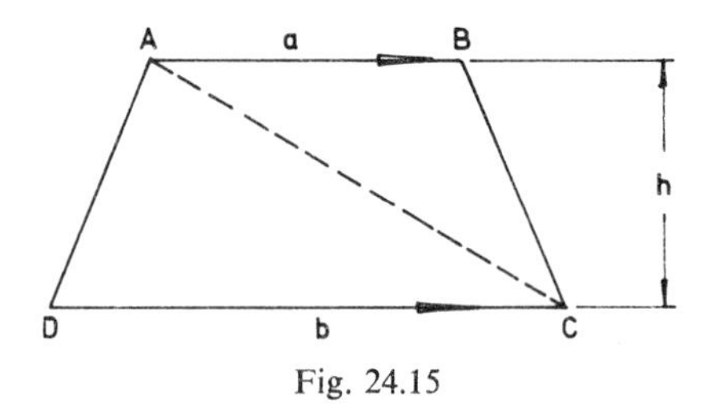

Fig. 24.15

Hence the area of a trapezium is half the product of the sum of the parallel sides and the distance between them.

Examples

1) Find the area of the parallelogram ABCD (Fig. 24.16).

Area of parallelogram ABCD
= base × altitude
$= 8 \times 10 = 80 \text{ cm}^2$

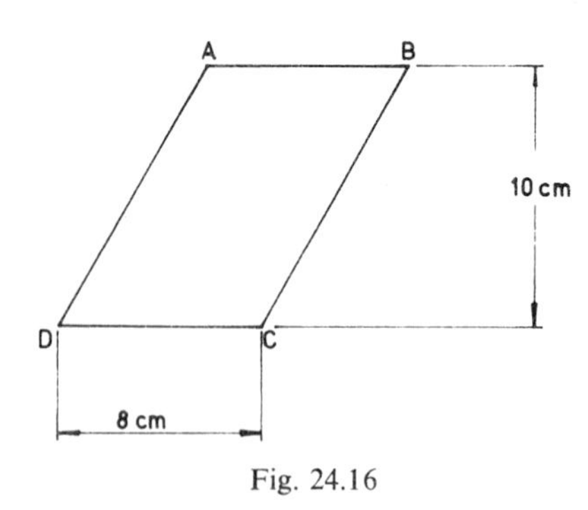

Fig. 24.16

2) In the trapezium PQRS (Fig. 24.17) the parallel sides PQ and SR are both perpendicular to QR. If PQ = 16 cm, PS = 17 cm and RS = 8 cm, calculate the area of the trapezium.

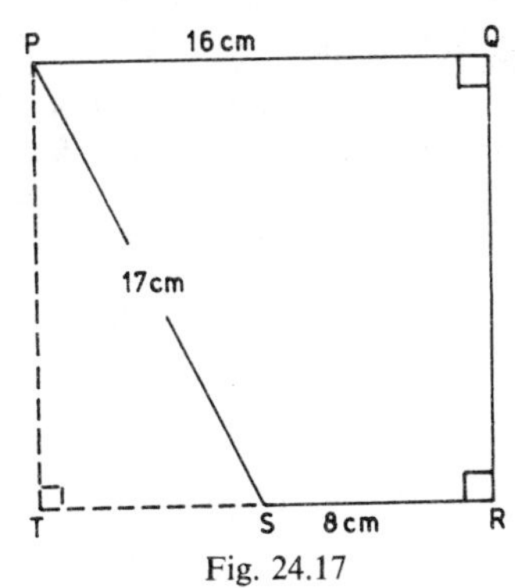

Fig. 24.17

In Fig. 24.17 draw the lines PT and TS as shown.
In $\triangle PST$, $TS = 8$ cm and $PS = 17$ cm.
Using Pythagoras,

$$PT^2 = PS^2 - TS^2 = 17^2 - 8^2 = 225$$

$$PT = \sqrt{225} = 15 \text{ cm}$$

Area of trapezium

$$PQRS = \frac{1}{2} \times 15 \times (16+8)$$

$$= \frac{1}{2} \times 15 \times 24$$

$$= 180 \text{ cm}^2$$

3) Two parallelograms ABCD and ABEF are as shown in Fig. 24.18. Prove that:
i) DCEF is a parallelogram;
ii) area ABCD = area ABEF − area DCEF

i) Since ABCD and ABEF are both parallelograms

$$AB = CD = EF$$

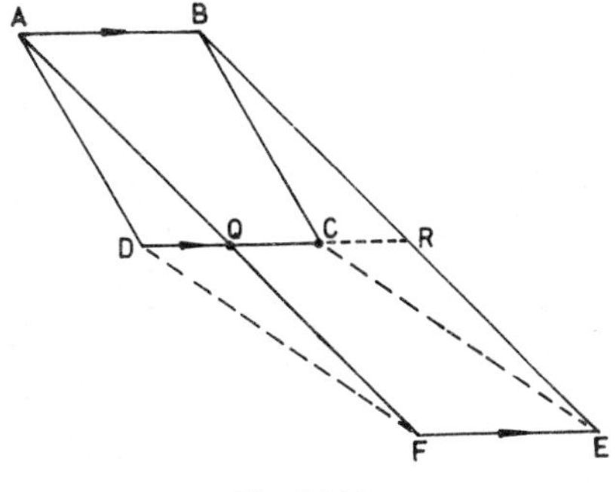

Fig. 24.18

and AB, CD and EF are all parallel to each other. Hence DCEF is a parallelogram because the two opposite sides CD and EF are equal and parallel.

ii) Draw CR as shown in Fig. 24.18. Parallelograms ABCD and ABRQ are equal in area since they have the same base AB and the same altitude. Similarly DCEF and QRFE are equal in area.

$$\text{Area ABEF} = \text{area ABRQ} + \text{area QREF}$$
$$= \text{area ABCD} + \text{area DCEF}$$
$$\text{Area ABCD} = \text{area ABEF} - \text{area DCEF}$$

Exercise 92

1) What is the area of a parallelogram whose base is 7 cm long and whose vertical height is 4 cm.

2) A parallelogram ABCD has $AB = 12$ cm and $BC = 10$ cm. $\angle ABC = 45°$. Calculate the area of ABCD.

3) The area of a parallelogram is 80 cm^2. If the base is 12 cm long what is its altitude?

4) Fig. 24.19 shows a trapezium. Find x.

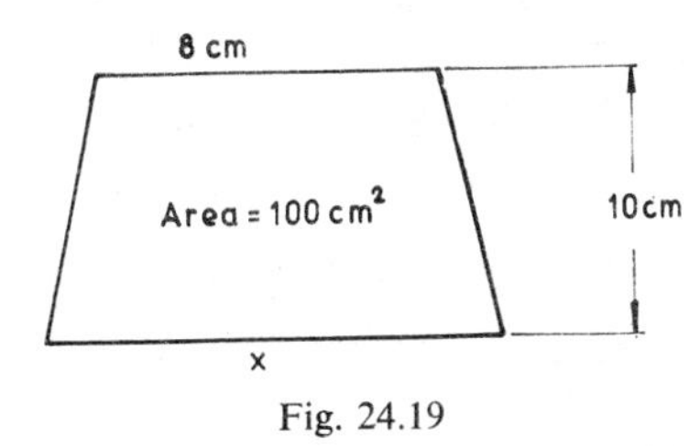

Fig. 24.19

5) E is the mid-point of the side AB of the parallelogram ABCD whose area is 80 cm^2. Find the area of $\triangle DEC$.

6) In the rhombus PQRS the side $PQ = 17$ cm and the diagonal $PR = 16$ cm. Calculate the area of the rhombus.

7) WXYZ is a parallelogram. A line through W meets ZY at T and XY produced at U.

Prove that $\triangle$s WZT and UYT are similar. If $\frac{ZT}{TY} = \frac{3}{2}$ and the area of WXYZ is 20 cm^2, calculate:

i) the area of the trapezium WXYT;
ii) the area of $\triangle$UYT.

8) The area of a rhombus is 16 cm^2 and the length of one of its diagonals is 6 cm. Calculate the length of the other diagonal.

9) A point P is taken on the side CD of the parallelogram ABCD and CD is produced to Q making DQ=CP. A line through Q parallel to AD meets BP produced at S. AD is produced to meet BS at R. Prove that ARSQ is a parallelogram and that its area is equal to the area of ABCD.

10) In the parallelogram ABCD the side AB is produced to X so that BX=AB. The line DX cuts BC at E. Prove that:

i) DBXC is a parallelogram;
ii) Area AED=twice area CEX.

SELF TEST 24

State the letter (or letters) corresponding to the correct answer (or answers).

1) A quadrilateral has one pair of sides parallel. It is therefore a

a rhombus b parallelogram
c rectangle d trapezium

2) A quadrilateral has diagonals which bisect at right-angles. It is therefore a

a rhombus b square c rectangle
d parallelogram

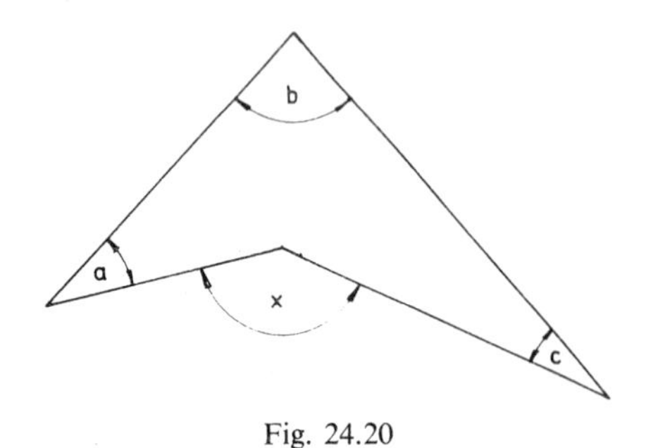

Fig. 24.20

3) In Fig. 24.20, x is equal to

a $a+b+c$ b $360°-(a+b+c)$
c $a+b+c+180°$ d $360°-a+b+c$

4) In Fig. 24.21, y is equal to

a 80° b 70° c 40° d 100°

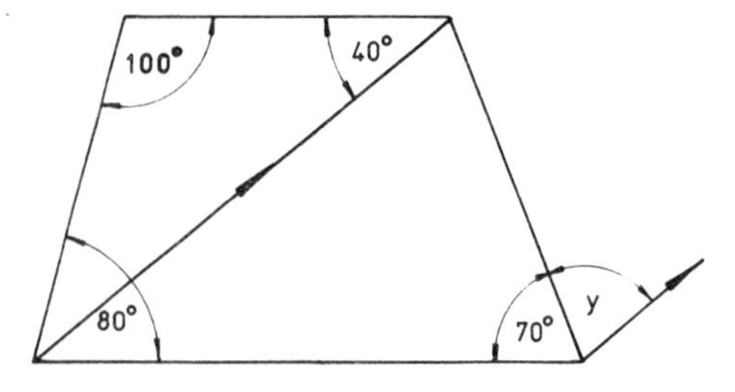

Fig. 24.21

5) In Fig. 24.22, x is equal to

a 190° b 110° c 70° d 60°

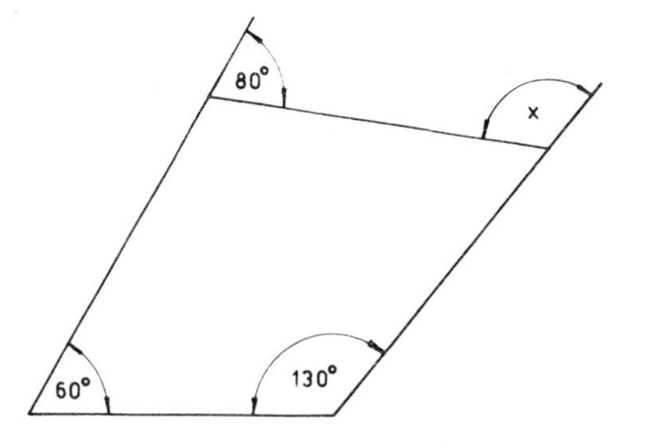

Fig. 24.22

6) In Fig. 24.23, p is equal to

a 120° b 115° c 60° d 65°

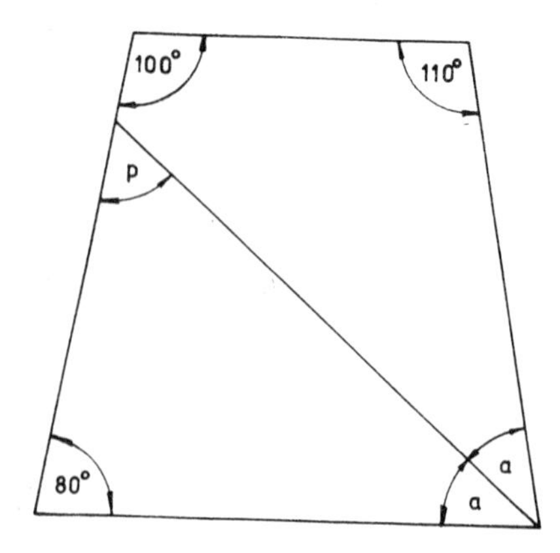

Fig. 24.23

7) A polygon has all its interior angles less than 180°. Hence it is definitely a

a convex polygon b regular polygon
c re-entrant polygon d quadrilateral

8) A regular polygon has each interior angle equal to 108°. It therefore has

a 4 sides b 5 sides c 6 sides
d 7 sides

9) A regular polygon has each exterior angle equal to 40°. It therefore has

a 7 sides b 8 sides c 9 sides
d 10 sides

10) A regular polygon has each interior angle greater by 60° than each exterior angle. It therefore has

a 4 sides b 6 sides c 7 sides
d 8 sides

11) Fig. 24.24 shows a trapezium. The side marked x is equal to

a 10 cm b 15 cm c 5 cm
d 0·75 cm

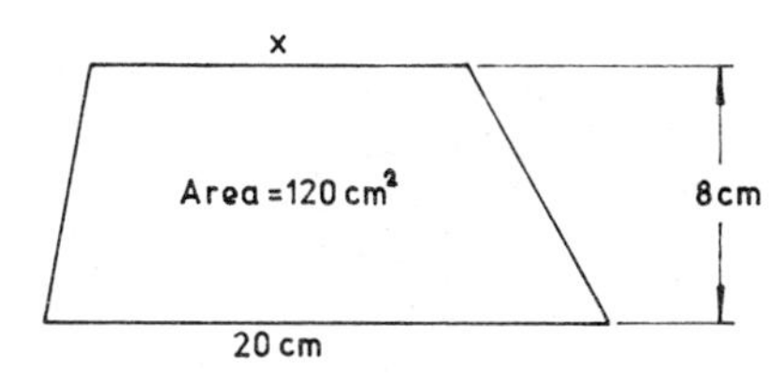

Fig. 24.24

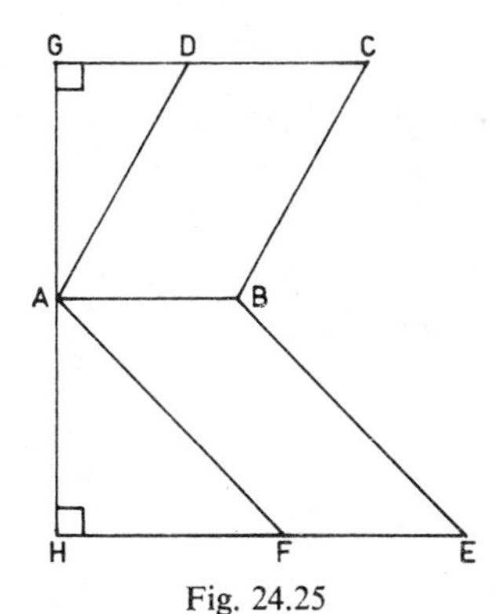

Fig. 24.25

12) In Fig. 24.25, ABCD and ABEF are parallelograms which have equal areas. It *must* be true that

a AF=AD b ∠D=∠F
c AG=AH d ∠C=∠E

13) In Fig. 24.26, ABCD is a parallelogram and CF=EF. Hence

a BC=BE b AD=AE
c Area ABCD=area AFE
d Area ABCD=2×area ABE

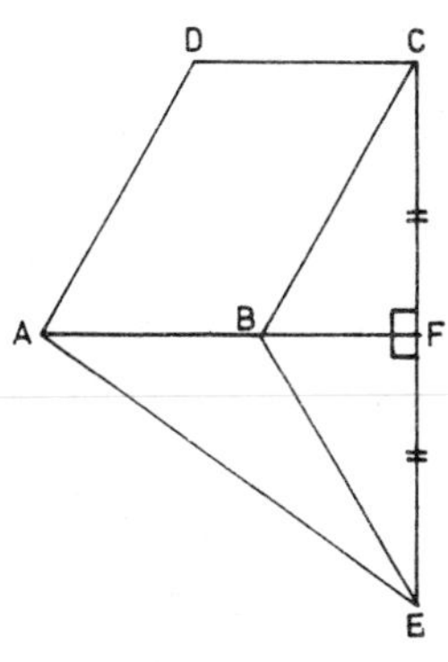

Fig. 24.26

14) In Fig. 24.27, ABCD is a trapezium. Hence

a ∠ADB=40° b ∠ADB=70°
c ∠ADC=90° d ∠ADC=120°

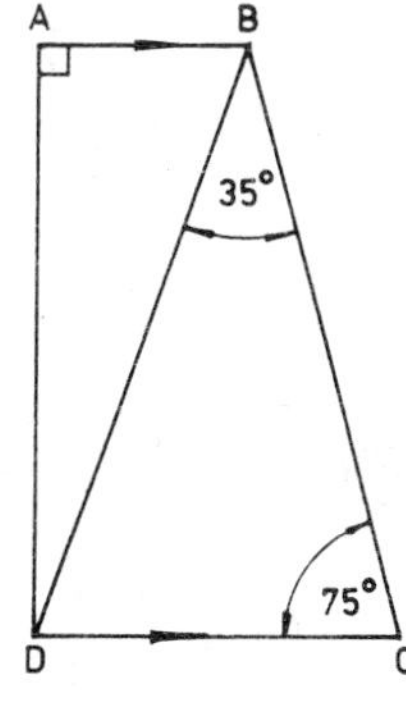

Fig. 24.27

Chapter 25 The Circle

CHORDS

A *chord* is a straight line which joins two points on the circumference of a circle.
A *diameter* is a chord drawn through the centre of the circle. The *radius* of a circle is half the diameter. (see Fig. 25.1)

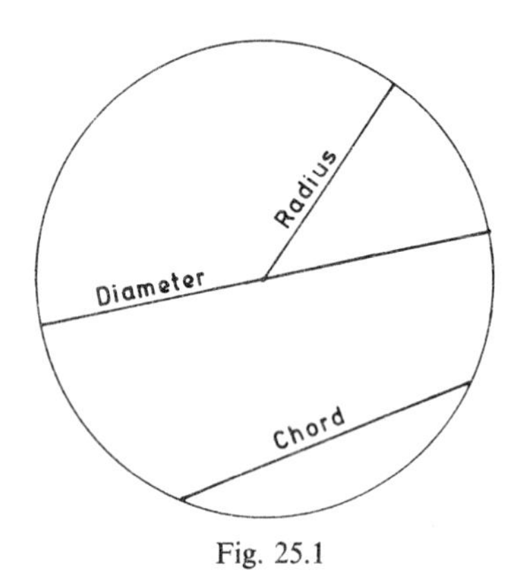

Fig. 25.1

1) *If a diameter of a circle is at right-angles to a chord then it divides the chord into two equal parts* (Fig. 25.2). The converse is also true.

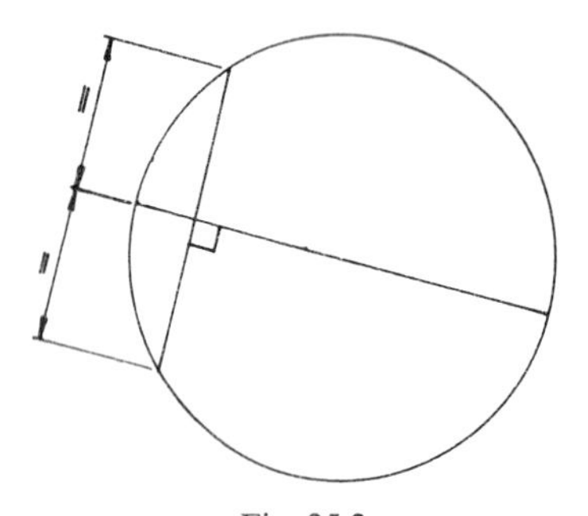

Fig. 25.2

2) *Chords which are equal in length are equi-distant from the centre of the circle.* Thus, in Fig. 25.3 if the chords AB and CD are equal in length, then the distances OX and OY are also equal. The converse is also true, i.e. chords which are equi-distant from the centre of the circle are equal in length.

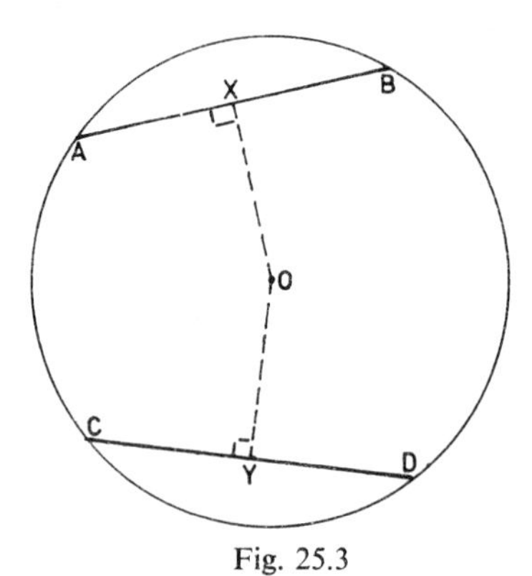

Fig. 25.3

3) *If two chords intersect inside or outside a circle the product of the segments of one chord is equal to the product of the segments of the other chord.* (see Theorem 6). Thus in Fig. 25.4, $AE \times EB = CE \times ED$
The converse is also true, i.e. if $AE \times EB = CE \times ED$ then the points ABCD are concyclic.

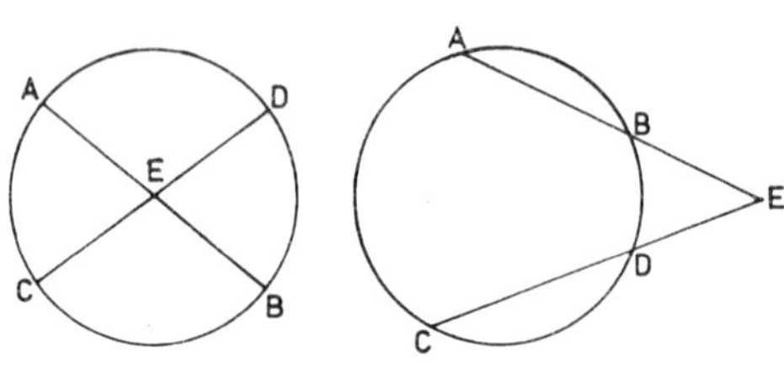

Fig. 25.4

4) *If a triangle is inscribed in a semi-circle the angle opposite the diameter is a right-angle* (Fig. 25.5). The converse is also true. Thus if a triangle is right-angled the hypotenuse is the diameter of its circumscribing circle.

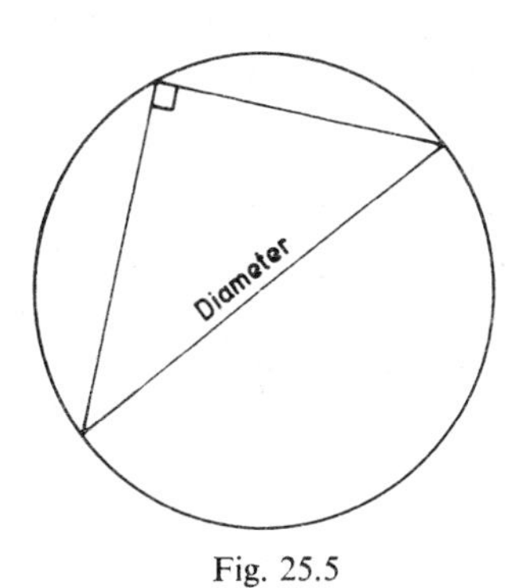

Fig. 25.5

Examples

1) Fig. 25.6 shows the segment of a circle. Find the diameter of the circle of which the segment is part.

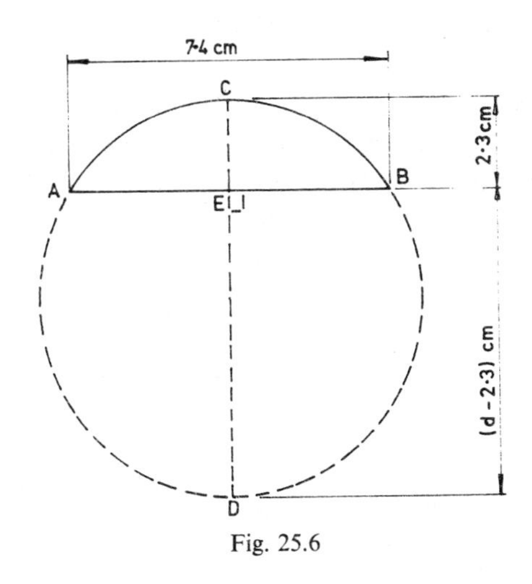

Fig. 25.6

In Fig. 25.6 draw a diameter CD at right-angles to the given chord AB. Then

$$AE = EB = 3{\cdot}7 \text{ cm}$$

also,

$$AE \times EB = CE \times ED.$$

If d be the diameter of the circle then,

$$3{\cdot}7^2 = 2{\cdot}3(d - 2{\cdot}3)$$

$$13{\cdot}69 = 2{\cdot}3d - 2{\cdot}3^2$$

$$2{\cdot}3d = 13{\cdot}69 + 2{\cdot}3^2 = 13{\cdot}69 + 5{\cdot}29 = 18{\cdot}98$$

$$d = \frac{18{\cdot}98}{2{\cdot}3} = 8{\cdot}25 \text{ cm}$$

2) A chord AB, 6 cm long, is drawn in a circle whose radius is 5 cm. A second chord DC is drawn in the same circle so that it lies at a distance of 4 cm from the centre of the circle. Show that CD = AB.

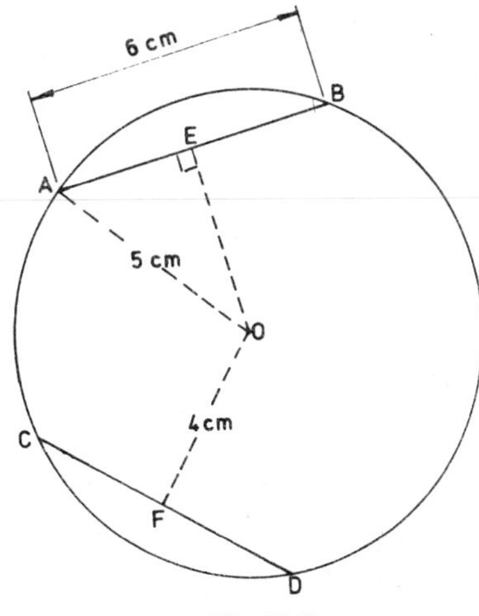

Fig. 25.7

In Fig. 25.7, let O be the centre of the circle. Draw OE perpendicular to AB. Then AE = EB = 3 cm. In $\triangle$AOE, by Pythagoras,

$$OE^2 = AO^2 - AE^2 = 5^2 - 3^2 = 16$$

$$\therefore \ OE = 4 \text{ cm}$$

$$\therefore \ OE = OF = 4 \text{ cm}$$

Hence since the chords AB and CD are equidistant from O, they are equal to each other.

3) Three points A, B and C are marked on the circumference of a circle as shown in Fig. 25.8. Find the diameter of the circle. In Fig. 25.8, join BC. Since $\angle$CAB is a right-angle, this is the angle in a semi-circle and hence BC is a diameter.

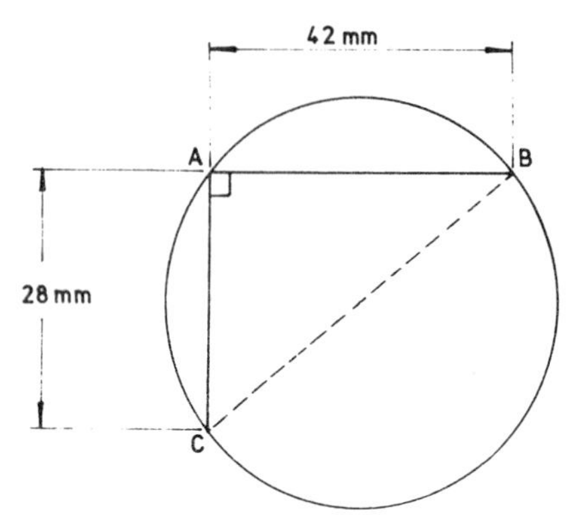

Fig. 25.8

In $\triangle ABC$, by Pythagoras,

$$BC^2 = AB^2 + AC^2 = 42^2 + 28^2$$

$$= 1764 + 784 = 2548$$

$$BC = \sqrt{2548} = 50{\cdot}48 \text{ mm}$$

The diameter of the circle is therefore 50·48 mm.

Exercise 93

1) A chord AB, 8 cm long, is drawn in a circle whose radius is 5 cm. Calculate its distance from the centre of the circle.

2) A chord XY is drawn in a circle whose radius is 10 cm. XY is 8 cm from the centre of the circle. What is its length?

3) An equilateral triangle of side 8 cm is drawn in a circle. Calculate the diameter of the circle.

4) An isosceles triangle whose sides are 5 cm, 5 cm and 6 cm long is inscribed in a circle. Find the diameter of the circle.

5) A chord AB, 8 cm long, is drawn in a circle so that the height of the minor arc is 3 cm. What is the diameter of the circle?

6) A chord AB, 8 cm long, is drawn in a circle whose radius is 5 cm. A second chord BC is drawn so that it lies at a distance of 3 cm from the centre of the circle. Show that AB = BC.

7) Two chords PQ and RS are parallel to each other and they lie on opposite sides of the centre of the circle. If PQ = 10 cm and RS = 8 cm and the circle is 12 cm radius, find the distance between the chords.

8) Three points X, Y and Z are marked on the circumference of a circle, so that YZ is a diameter. If YZ = 62 mm and XZ = 41 mm find XY.

9) Two chords AB and CD intersect at E. If CE = 3 cm, ED = 2 cm and AE = 4 cm find BE.

10) In Fig. 25.9, find W.

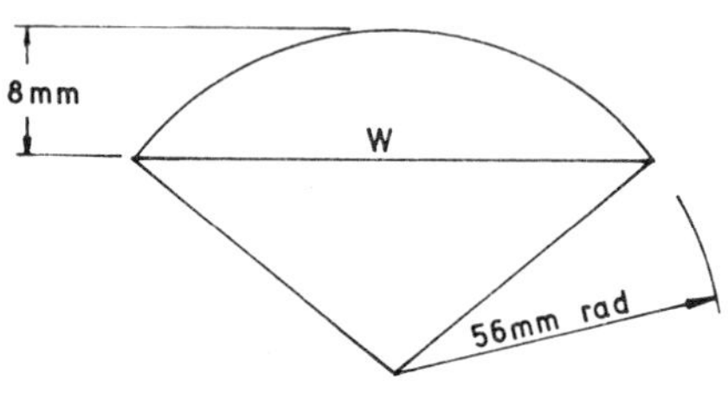

Fig. 25.9

ANGLES IN A CIRCLE

The chord AB (Fig. 25.10) divides the circle into two arcs. ABP is called the *major* arc and ABQ the *minor* arc. The areas ABP and ABQ are called *segments*.

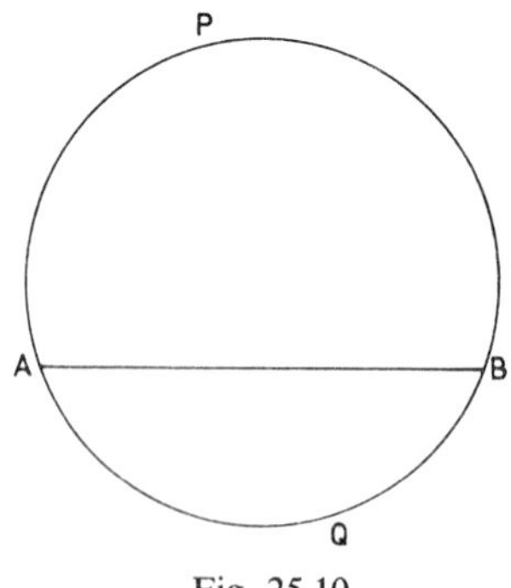

Fig. 25.10

The angles ARB and ASB (Fig. 25.11) are called angles in the segment APB. The angle ATB is called an angle in the segment ABQ.

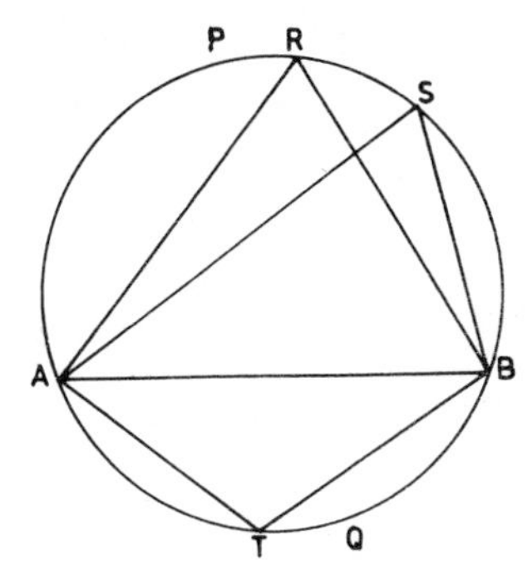

Fig. 25.11

1) *Angles in the same segment of a circle are equal*. Thus in Fig. 25.12, $\angle APB = \angle AQB$ since they are angles in the same segment ABQP. The converse is useful when proving that 4 points are concyclic. Thus in Fig. 25.12(a) if $\angle Z = \angle Y$, then the points W, X, Y and Z are concyclic.

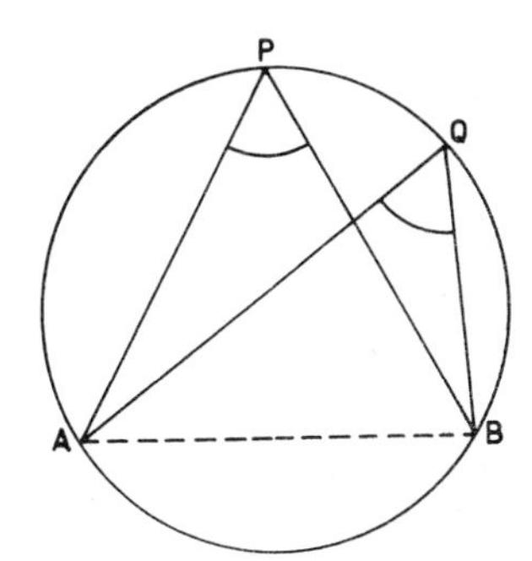

Fig. 25.12

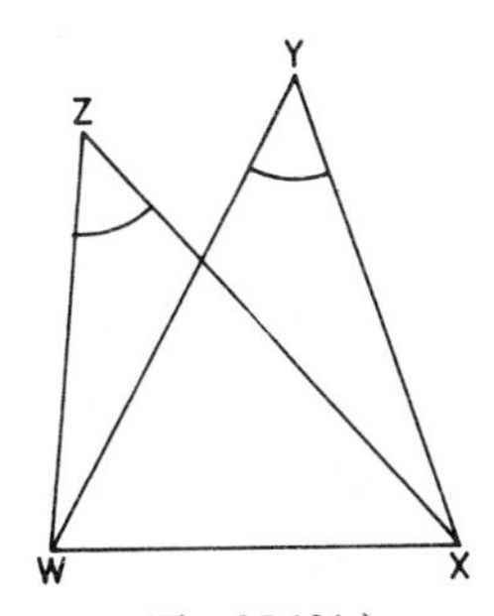

Fig. 25.12(a)

2) *The angle which an arc of a circle subtends at the centre is twice the angle which the arc subtends at the circumference* (see Theorem 5). Thus in Fig. 25.13, $\angle AOB = 2 \times \angle APB$.

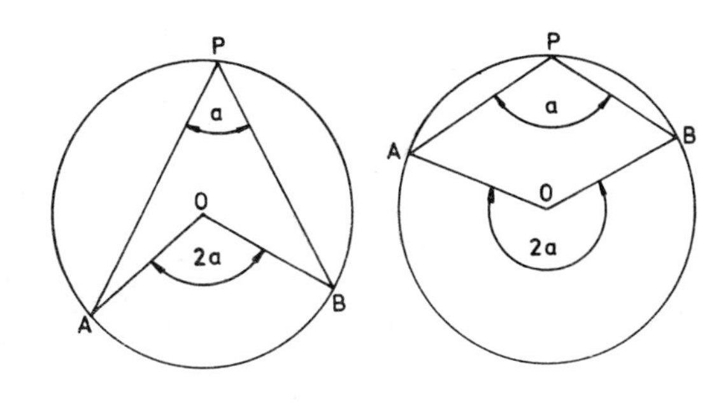

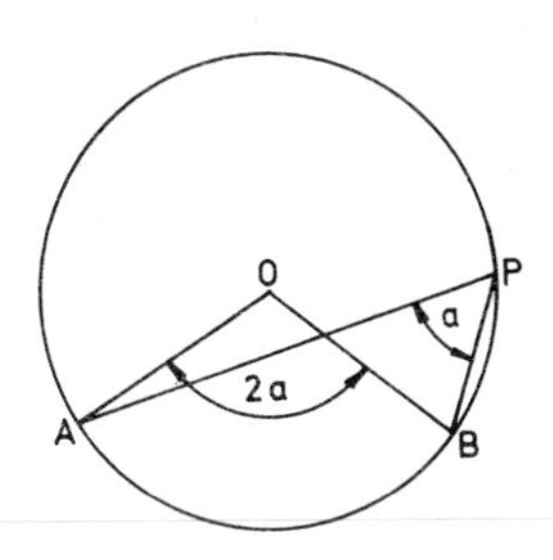

Fig. 25.13

3) *The opposite angles of any quadrilateral inscribed in a circle are supplementary* (i.e. equal to 180°). It follows that *the exterior angle is equal to the interior opposite angle.* A quadrilateral inscribed in a circle is called a *cyclic quadrilateral.* Thus the cyclic quadrilateral ABCD (Fig. 25.14) has $\angle A + \angle C = 180°$ and $\angle D + \angle B = 180°$. Also $\angle CDX = \angle B$, $\angle BCY = \angle A$ etc. The converse is also true i.e. a quadrilateral with two opposite angles that are supplementary is a cyclic quadrilateral.

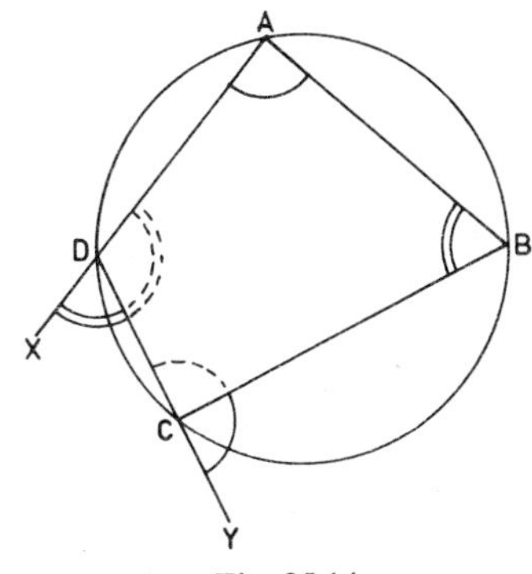

Fig. 25.14

Examples

1) In Fig. 25.15, X is the centre of the circle drawn on AB as diameter, BC = CD and $\angle XDA = 50^\circ$. Calculate the angles of the quadrilateral ABCD.

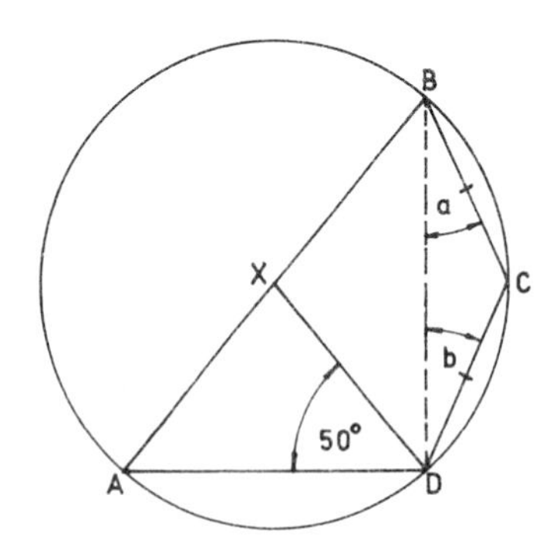

Fig. 25.15

In $\triangle AXD$,

AX = XD (radii)

Hence $\triangle AXD$ is isosceles

$\therefore \angle A = \angle XDA = 50^\circ$

Since ABCD is a cyclic quadrilateral

$\angle A + \angle C = 180^\circ$

$\angle C = 180^\circ - \angle A = 180^\circ - 50^\circ = 130^\circ$

In $\triangle BCD$,

BC = CD (given)

Hence $\triangle BCD$ is isosceles

$\therefore \angle a = \angle b$

also,

$\angle a + \angle b = 180^\circ - \angle C = 180^\circ - 130^\circ = 50^\circ$

$\therefore \angle a = \angle b = 25^\circ$

$\angle BDA = 90^\circ$ (angle in a semi-circle)

$\angle ADC = 90^\circ + \angle b = 90^\circ + 25^\circ = 115^\circ$

$\angle ABC + \angle ADC = 180^\circ$

$\angle ABC = 180^\circ - \angle ADC = 180^\circ - 115^\circ = 65^\circ$

Hence the four angles of the quadrilateral ABCD are $\angle A = 50^\circ$, $\angle B = 65^\circ$, $\angle C = 130^\circ$ and $\angle D = 115^\circ$.

2) In Fig. 25.16, AB = BE. Prove that CD = CE

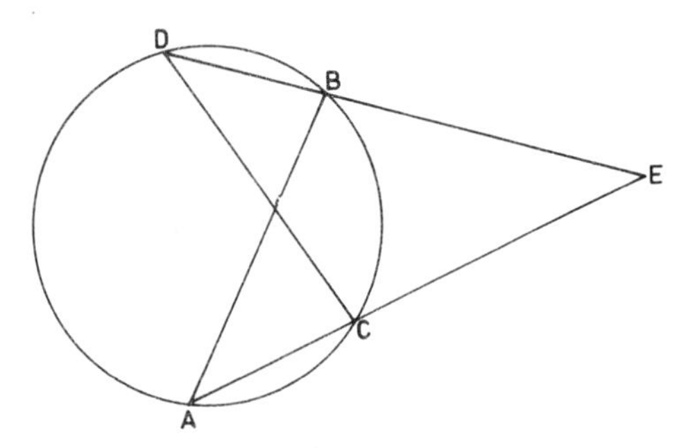

Fig. 25.16

$\triangle ABE$ is isosceles since AB = BE (given)

$\therefore \angle E = \angle A$

$\angle A = \angle D$ (angles in the same segment)

$\therefore \angle D = \angle E$

Hence $\triangle CDE$ is isosceles and hence CD = CE.

Exercise 94

1) In Fig. 25.17, O is the centre of the circle. If $\angle AOB = 60^\circ$, find $\angle ACB$.

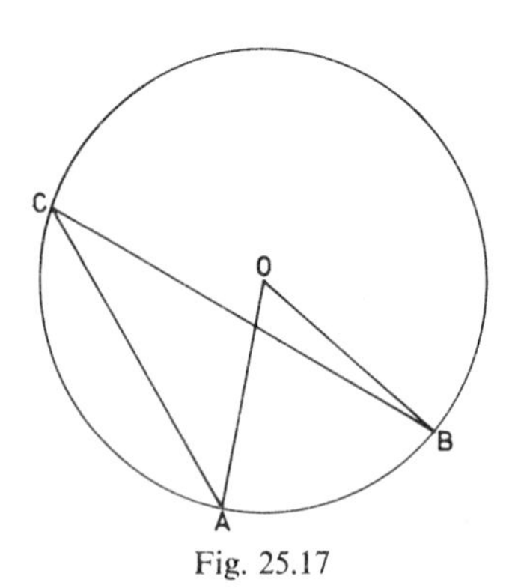

Fig. 25.17

2) The angles A and B of a cyclic quadrilateral are 80° and 120° respectively. Find the angles C and D.

3) In Fig. 25.18, prove that AE.CD=CE.AB

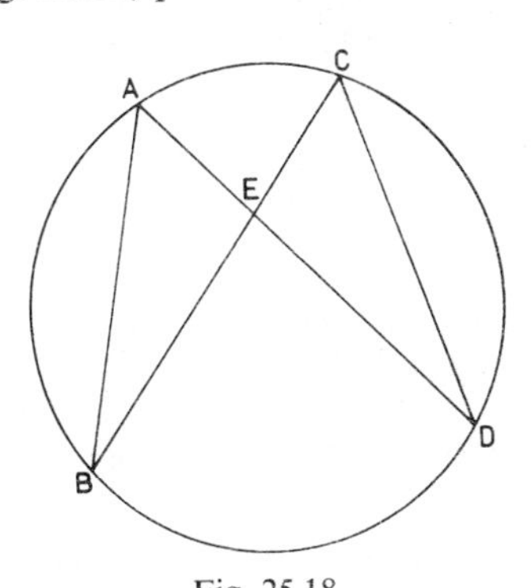

Fig. 25.18

4) In Fig. 25.19 show that WY = WX, WXYZ being a cyclic quadrilateral.

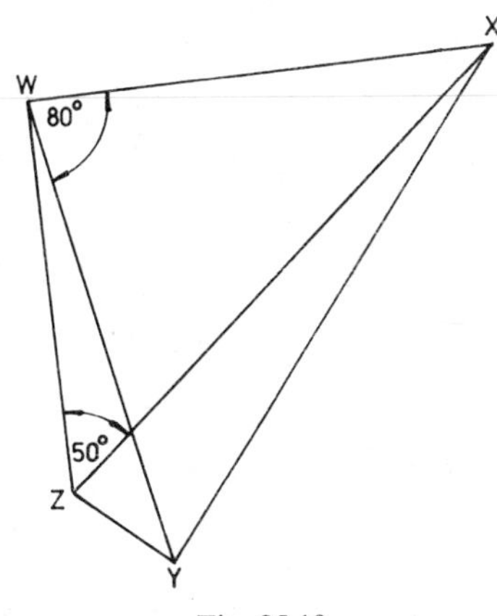

Fig. 25.19

5) In Fig. 25.20, O is the centre of the circle and AB is a diameter. BC=CD and ∠AOD=70°. Calculate the angles of the quadrilateral ABCD if BC=CD.

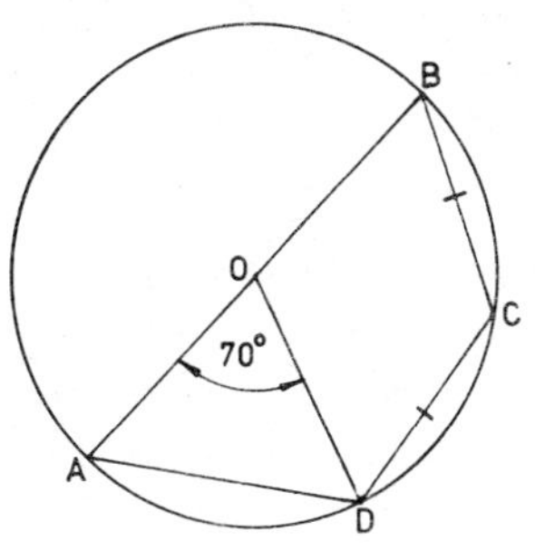

Fig. 25.20

6) In Fig. 25.21, OB=OC. Prove that ABCD is a trapezium and that OA=OD.

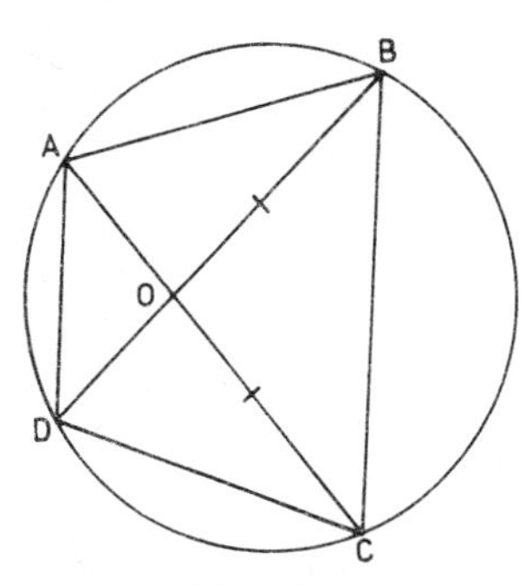

Fig. 25.21

7) ABCD is a quadrilateral inscribed in a circle whose centre if O. AB is a diameter of the circle. If BC=CD, prove that

i) ∠BDC=∠CAD;
ii) ∠BOD=4∠CAD;
iii) ∠ABD+2∠DBC=90°.

8) In Fig. 25.22, PL is perpendicular to QS and QM is perpendicular to PR. Prove that

i) the points P, Q, M and L lie on a circle;
ii) OR.OL=OS.OM.

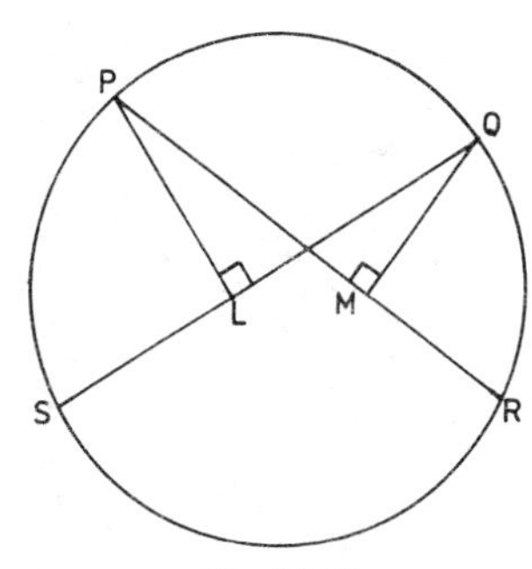

Fig. 25.22

9) In the circle drawn on PQ as diameter, a chord SR is parallel to PQ. T is a point on the minor arc SP. If ∠PQR=58°, calculate ∠RPQ and ∠STP.

10) The triangle ABC in which A is obtuse is inscribed in a circle. ∠ABC=22° and ∠ACB=40°. M is the mid-point of the major arc BC. Calculate ∠BMC and ∠BAM.

TANGENT PROPERTIES OF A CIRCLE

A tangent is a line which just touches a circle at one point only (Fig. 25.23). This point is called the point of tangency.

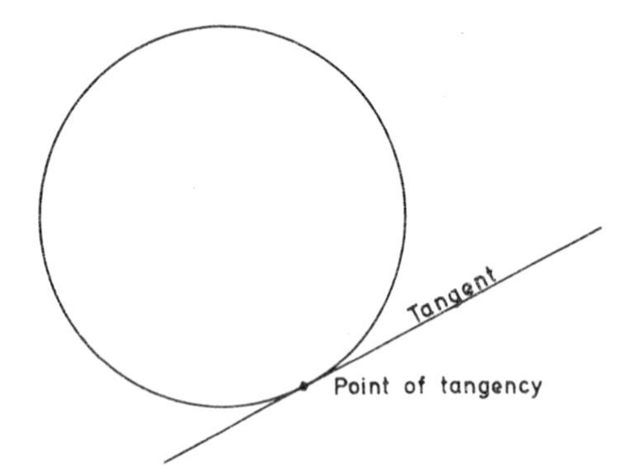

Fig. 25.23

1) *A tangent to a circle is at right-angles to a radius drawn from the point of tangency* (Fig. 25.24).

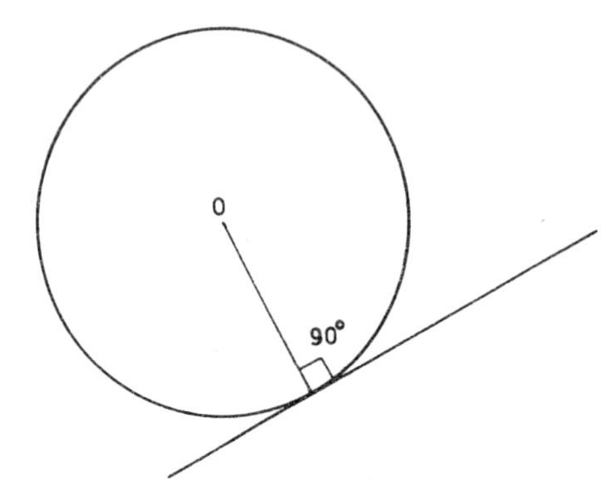

Fig. 25.24

2) *If from a point outside a circle, tangents are drawn to the circle, then the two tangents are equal in length.* They also make equal angles with the chord joining the points of tangency (Fig. 25.25). It follows that the line drawn from the point where the tangents meet to the centre of the circle bisects the angle between the two tangents.

3) *The angle between a tangent and a chord drawn from the point of tangency equals*

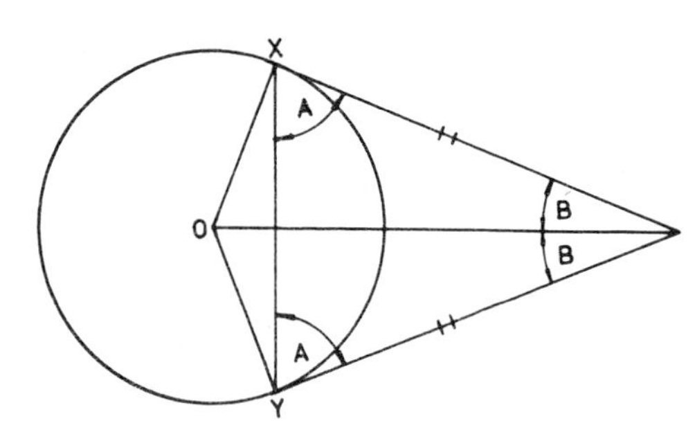

Fig. 25.25

one-half of the angle at the centre subtended by the chord.
Thus in Fig. 25.26, $\angle B = \frac{1}{2}\angle A$.

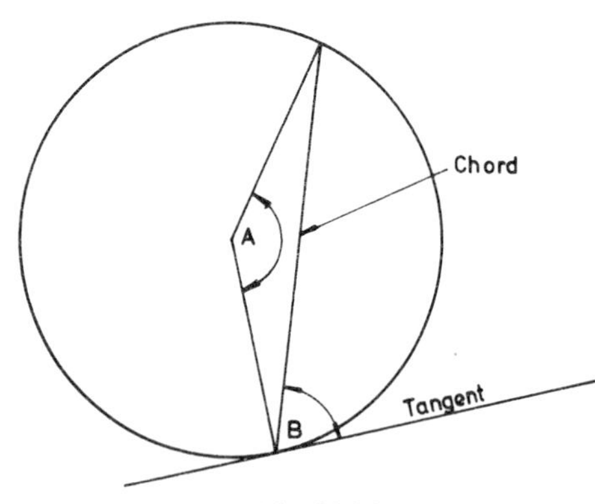

Fig. 25.26

4) *The angle between a tangent and a chord drawn from the point of tangency equals the angle at the circumference subtended by the chord.* The angle at the circumference must be in the alternate segment (see Figs. 25.27 (a) and (b)). Thus in Fig. 25.27 (c), $\angle B$, the angle between the chord XY and the tangent YT equals, $\angle A$, the angle subtended by the chord at the circumference. Note that $\angle A$ is in the alternate segment to $\angle B$.
In Fig. 25.27 (d), $\angle C = \angle D$.

5) A line (Fig. 25.28) which cuts a circle at two points is called a *secant.*
If from a point outside a circle two lines are drawn, one a secant and the other a tangent to the circle, then the square on the tangent is equal to the rectangle contained by the whole

secant and that part of it which lies outside the circle (see Theorem 7).
Thus in Fig. 25.29, $CT^2 = AC.BC$

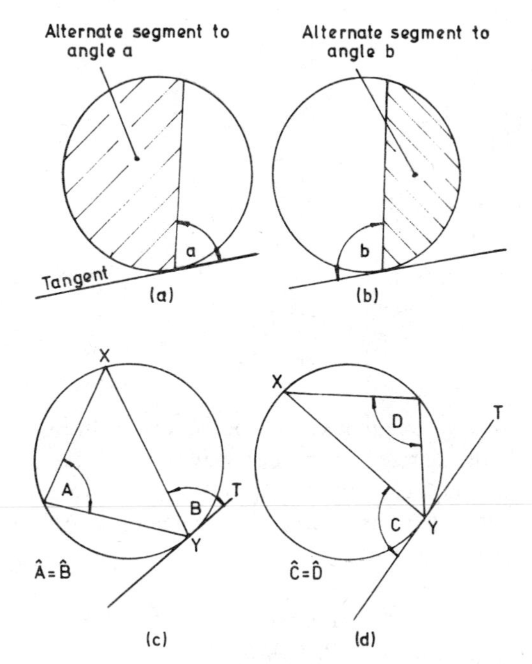

Fig. 25.27

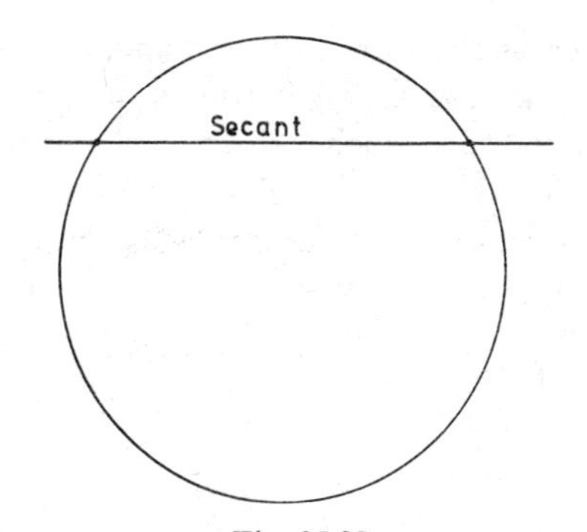

Fig. 25.28

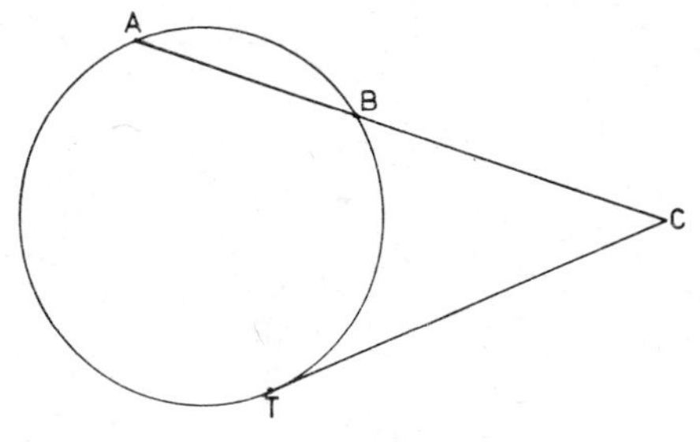

Fig. 25.29

6) *If two circles touch internally or externally then the line which passes through their centres, also passes through the point of tangency* (Fig. 25.30).

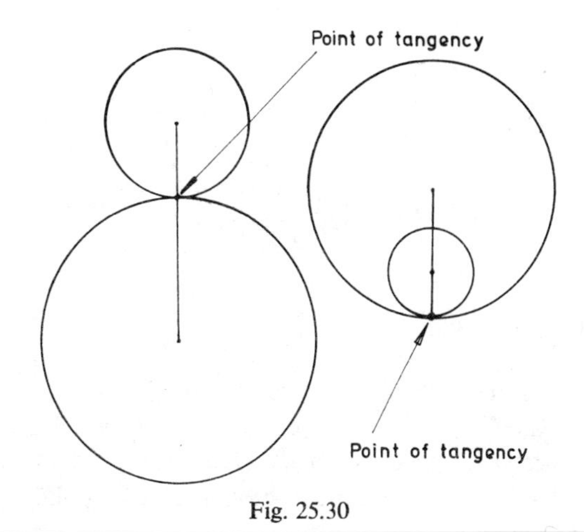

Fig. 25.30

Example
The tangent at the point C on a circle meets the diameter AB produced at T. If $\angle BCT = 27°$ calculate $\angle CTA$. If $CT = t$ and $BT = x$, prove that the radius of the circle is $\frac{t^2 - x^2}{2x}$.
In Fig. 25.31, join AC.
Then $\angle BCT = \angle A = 27°$
also $\angle BCA = 90°$ (angle in a semi-circle)

$\therefore$ $\angle B = 90° + 27° = 117°$
$\angle T = 180° - 117° - 27° = 36°$

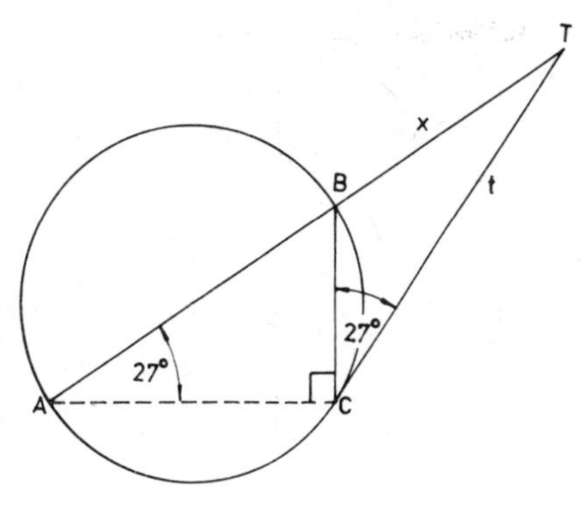

Fig. 25.31

H

Let r be the radius of the circle, then

$$AT = 2r + x$$

But $AT.BT = CT^2$

or $(2r+x)x = t^2$

$$2rx + x^2 = t^2$$

$$2rx = t^2 - x^2$$

$$r = \frac{t^2 - x^2}{2x}.$$

Exercise 95

1) In Fig. 25.32, O is the centre of the circle. AC = CB and AE is the tangent at A which meets BD produced at E. Given that ∠EAD = 32°, calculate ∠BOC and ∠AED.

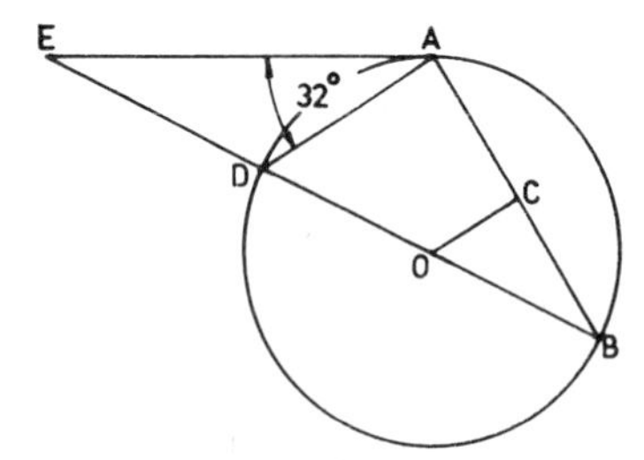

Fig. 25.32

2) In Fig. 25.33, X is the mid-point of the chord AB and XY is parallel to AT, the tangent at A. Prove that

i) ∠AYX = ∠ABC;
ii) BXYC is a cyclic quadrilateral;
iii) $AB^2 = 2AY.AC$

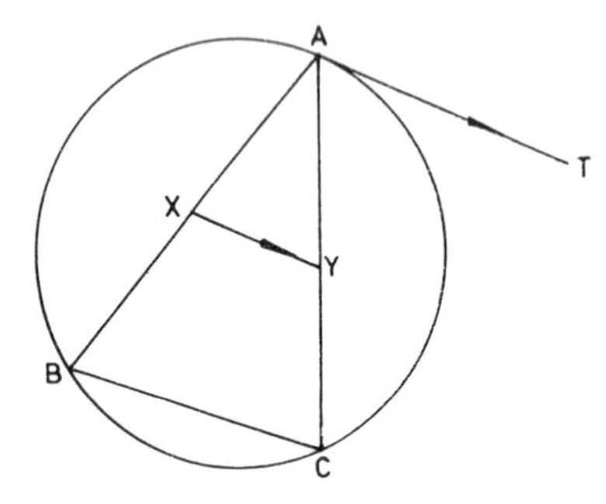

Fig. 25.33

3) In Fig. 25.34, XC is a tangent and Y is the mid-point of the arc BC. If ∠X = 28° and ∠BCA = 2∠ACX, calculate ∠CBA and ∠CBY.

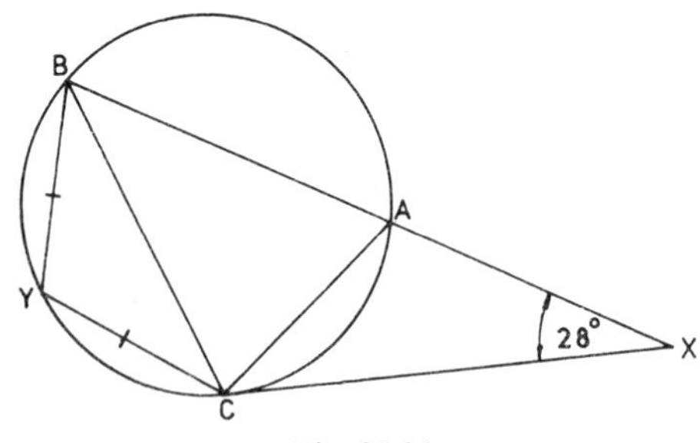

Fig. 25.34

4) Two unequal circles intersect at P and Q with their centres on opposite sides of the common chord PQ. Through P the diameters PA and PB are drawn. The tangents at A and B meet at C. Prove that

i) AQB is a straight line;
ii) a circle can be drawn through the points A, P, B and C;
iii) ∠APQ = ∠BPC.

5) Two circles of radii 16 cm and 9 cm touch each other externally. A common tangent to the two circles touches them at R and S. Calculate the length of RS.

6) In Fig. 25.35, BP = 8 cm, DC = 7 cm and CP = 9 cm. Calculate the lengths of the chord AB and the tangent PT.

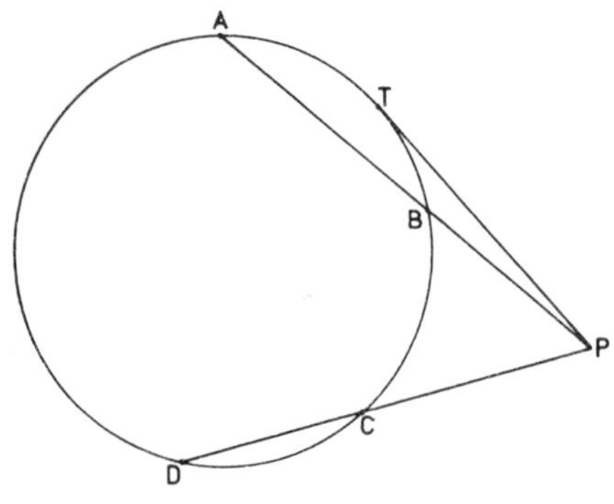

Fig. 25.35

7) In Fig. 25.36, QRX is parallel to the tangent PT. Prove that

i) $\angle PSQ = \angle RSX$;
ii) $PQ.RS = PS.RX$.

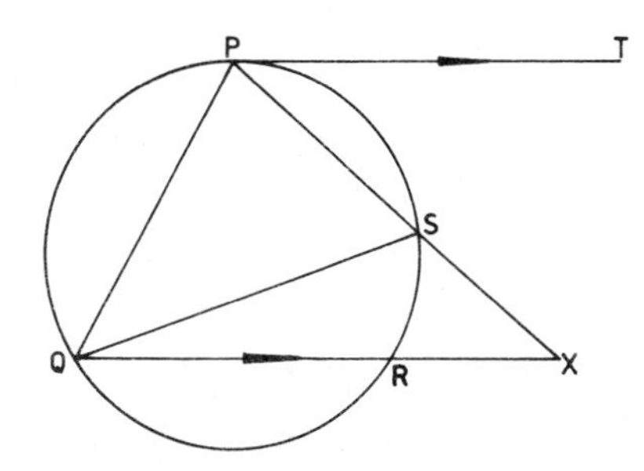

Fig. 25.36

8) In Fig. 25.37, AC is parallel to the tangent DE. Prove that

i) $\triangle ADC$ is isosceles;
ii) $\angle ABC = 2\angle DAC$.

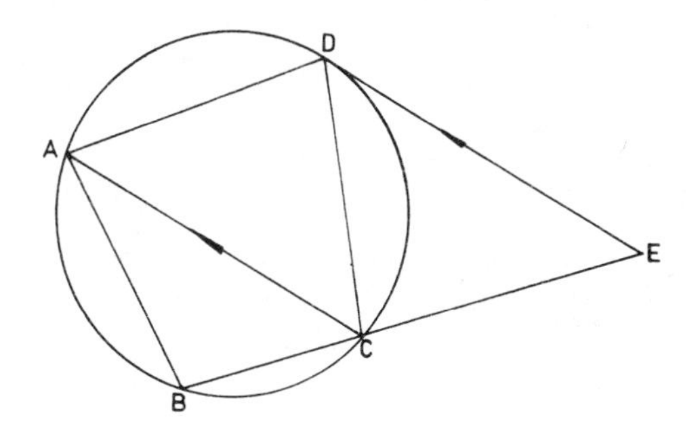

Fig. 25.37

9) The line TCB cuts a circle at C and B and the line TA touches the circle at A. Given that AB = AT prove that CA = CT. Given also that BC is a diameter of the circle, calculate $\angle ATC$.

10) AC is a diameter of a circle centre O and CD is a chord. M is the mid-point of CD. The tangent at A meets MO produced at T. Prove that

i) $\triangle CMO$ is similar to $\triangle TAO$;
ii) $TA.MO = AO.MC$.

11) In Fig. 25.38, if $AB = x$, $BC = y$ and $CT = t$ show that $x = \dfrac{t^2 - y^2}{y}$.

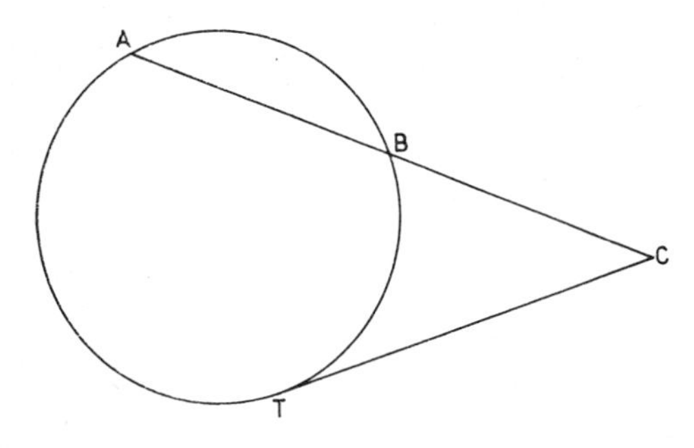

Fig. 25.38

12) The diameter of a circle, XY, is produced to T. From T a tangent is drawn which touches the circle at W. If the radius of the circle is r and $XT = x$ find CT in terms of r and x.

SELF TEST 25

In the following questions state the letter (or letters) corresponding to the correct answer (or answers).

1) In Fig. 25.39 the line AB is called a

a secant b chord c diameter
d tangent.

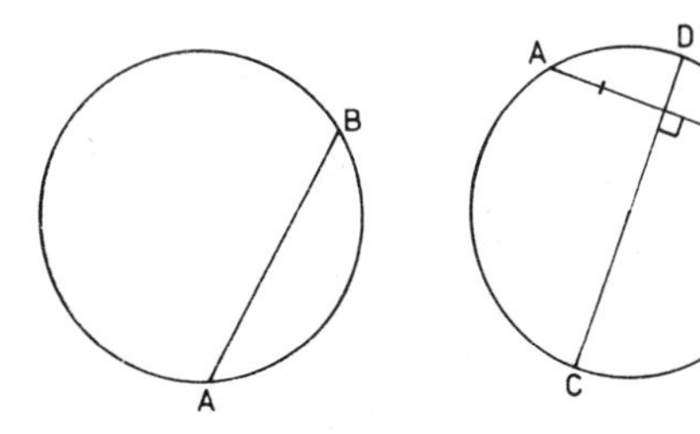

Fig. 25.39 Fig. 25.40

2) In Fig. 25.40 the line AB is bisected at right-angles by the line CD. Hence CD is a

a secant b chord c tangent
d diameter

3) In Fig. 25.41, O is the centre of the circle and AB = CD. It is necessarily true that

a ABCD is a parallelogram
b ABCD is a rhombus

c ABCD is a rectangle
d ABCD is a square

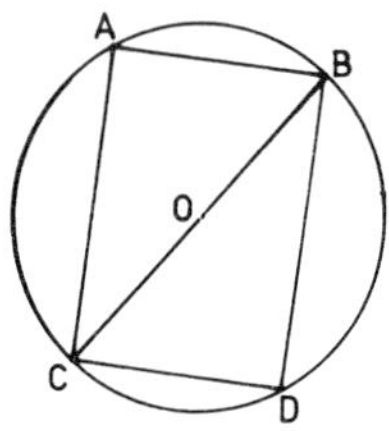

Fig. 25.41

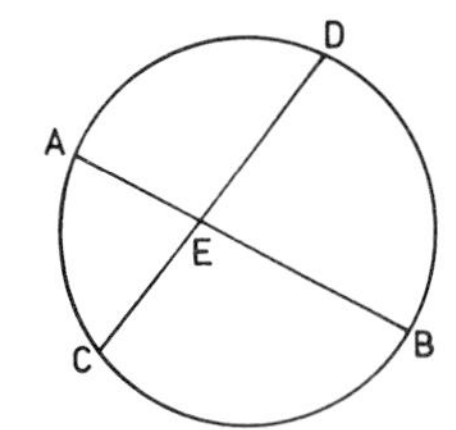

Fig. 25.42

4) In Fig. 25.42, AB and CD are two chords intersecting at E.
a AE.EB = CE.ED
b AE.ED = CE.ED
c AE.CE = ED.EB
d $AE^2 = CE^2$

5) In Fig. 25.43 it is true that
a $\angle A = \angle D$ b $\angle A = \angle B$
c $\angle B = \angle D$ d $\angle D = \angle F$

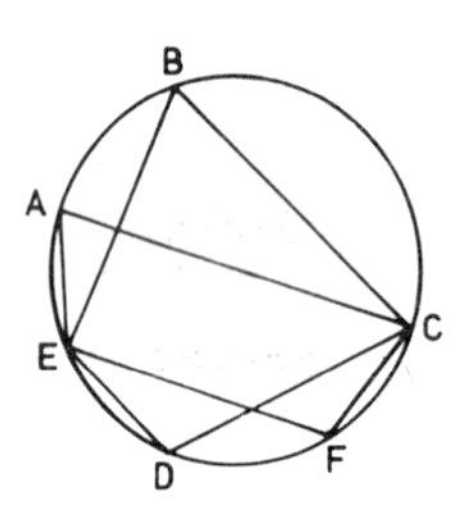

Fig. 25.43

6) A regular five sided figure is inscribed in a circle. The angle subtended at the circumference by the figure is
a 72° b 36° c 54° d 108°

7) In Fig. 25.44 it is true that
a AB = BC b AB = CD
c AB = AD d BC = AD

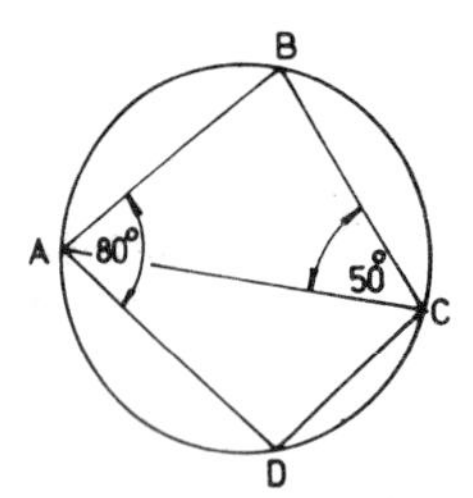

Fig. 25.44

8) In Fig. 25.45, it is true that
a AC.OD = AO.DB
b AC.OD = DB.OC
c OB.BD = OC.AC
d AO.AC = OB.DB

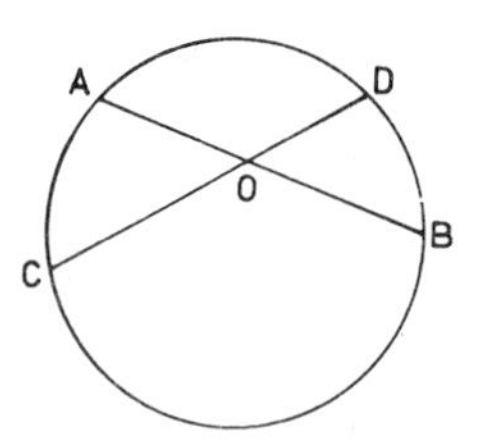

Fig. 25.45

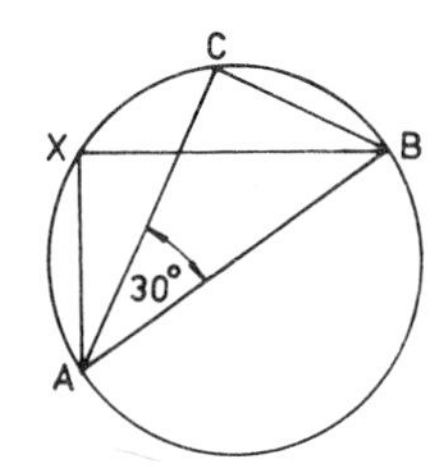

Fig. 25.46

9) In Fig. 25.46, AB is a diameter of the circle and $\angle CBX = \angle ABX$. Hence $\angle CAX$ is equal to
a 15° b 45° c 60° d 30°

10) In Fig. 25.47, X is the centre of the circle. The angle B is equal to
a 120° b 60° c 90° d 80°

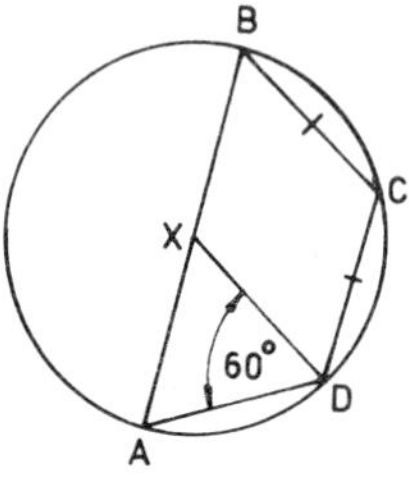

Fig. 25.47

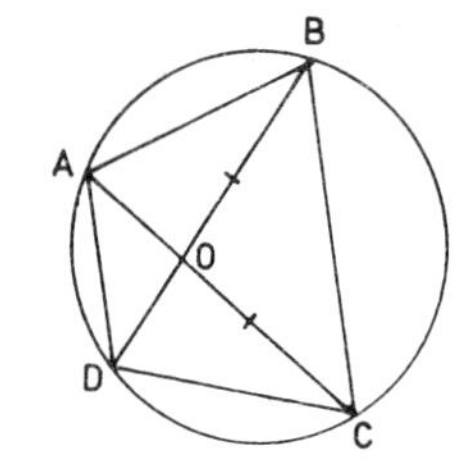

Fig. 25.48

11) In Fig. 25.48, OB = OC. Hence it is necessarily true that
a ABCD is a parallelogram
b ABCD is a rectangle
c ABCD is a trapezium
d ABCD is a rhombus

12) In Fig. 25.49, AD and BD are tangents to the circle whose centre is O. If $\angle ADB = 40°$ then $\angle ACB$ is
a 140° b 70° c 35° d 55°

13) In Fig. 25.50, $\angle AEB = \angle BED$. Hence $\angle BCA$ is equal to
a 94° b 61° c 53° d 47°

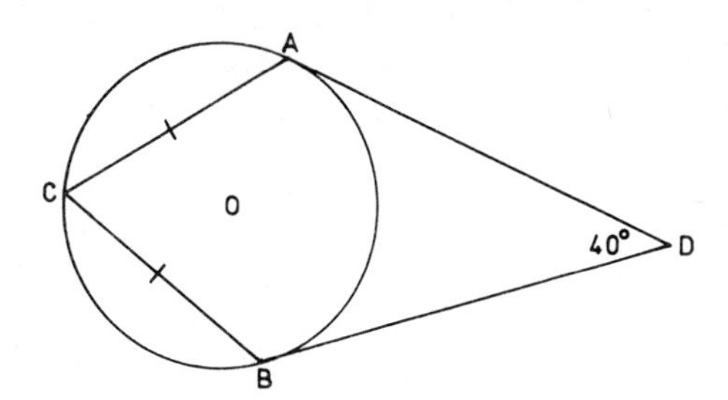

Fig. 25.49

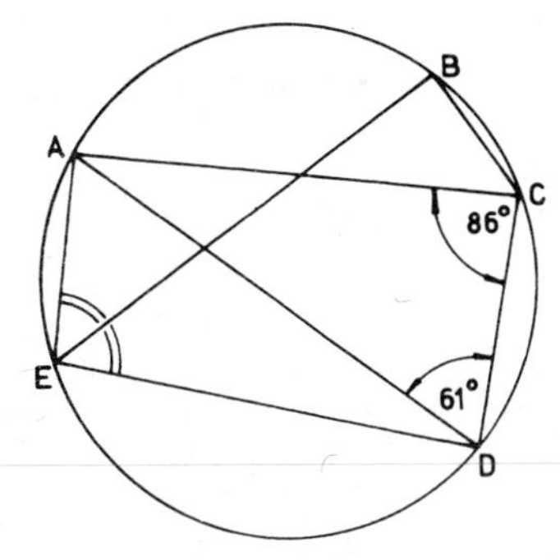

Fig. 25.50

14) In Fig. 25.51, the length of BC is
a 18 cm b $4\frac{1}{2}$ cm c $5\frac{1}{3}$ cm
d 14 cm

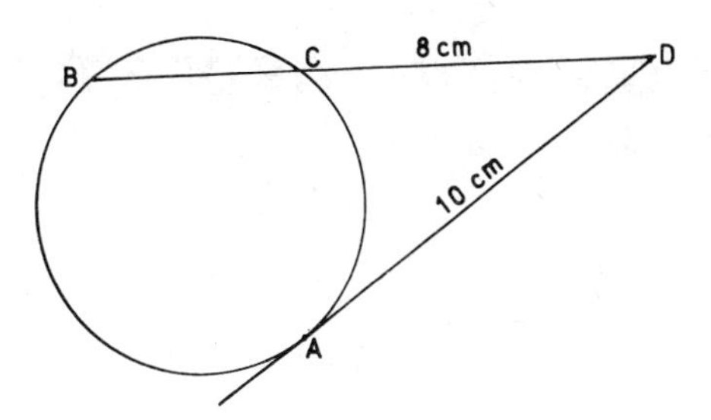

Fig. 25.51

15) In Fig. 25.52, AP is a tangent to the circle. ∠ADC is equal to

a 102° b 78° c 70° d 110°

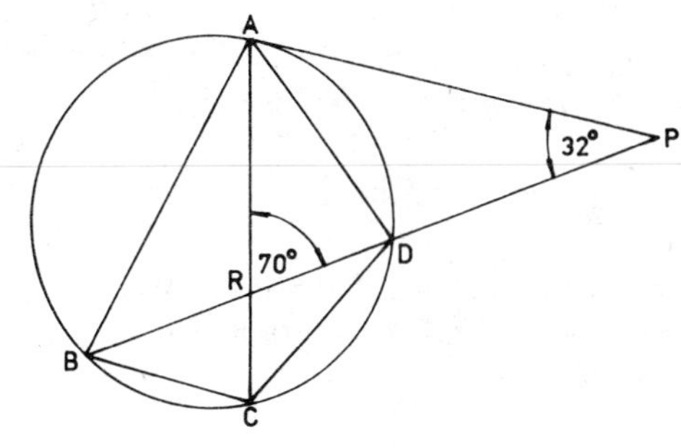

Fig. 25.52

Proofs of Theorems

Any of the following proofs may be required. If the diagrams and constructions are remembered the proof can usually be reproduced.

Theorem 1

When the side of a triangle is produced the exterior angle so formed is equal to the sum of the opposite interior angles.

Given: △ABC with the side BC produced to D.
To prove: ∠ACD = ∠CBA + ∠BAC.
Construction: Draw CY parallel to AB.
Proof: In Fig. 1,

$x=a$ (alternate angles, AB ∥ CY)
$y=b$ (corresponding angles, AB ∥ CY)

By addition, $x+y=a+b$
i.e. ∠BAC + ∠CBA = ∠ACD

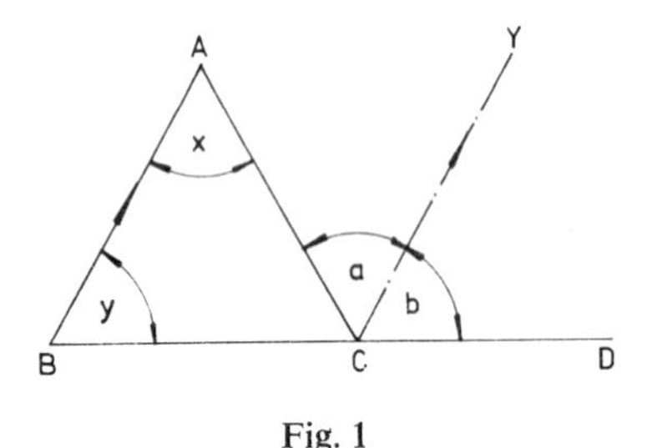

Fig. 1

Theorem 2

The straight line joining the mid-points of two sides of a triangle is parallel to the third side and equal to half the length of the third side.

Given: AD = BD and AE = EC
To prove: DE is parallel to BC and DE = $\frac{1}{2}$BC
Construction: From B draw a line parallel to EC to meet ED produced at X (Fig. 2).

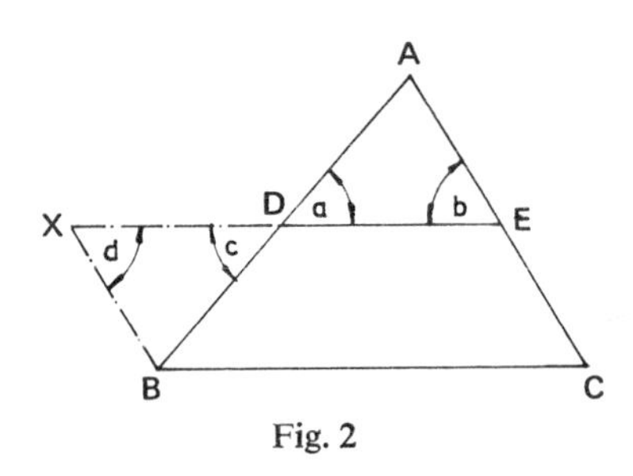

Fig. 2

Proof: In △s ADE and XDB
AD = BD (given)
$a=c$ (vertically opp. angles)
$b=d$ (alternate angles: AE ∥ BX)

∴ △ADE = △XDB

Hence DX = DE and BX = AE
But AE = EC (given)

∴ BX = EC

That is, BX is equal and parallel to EC.
Hence XECB is a parallelogram

∴ XE = BC
∴ DE = $\frac{1}{2}$BC

Also since XE is parallel to BC, DE is parallel to BC.

Theorem 3

Parallelograms on the same base and between the same parallels are equal in area.

Given: DCEF is a straight line and ABCD and ABFE are parallelograms (Fig. 3).

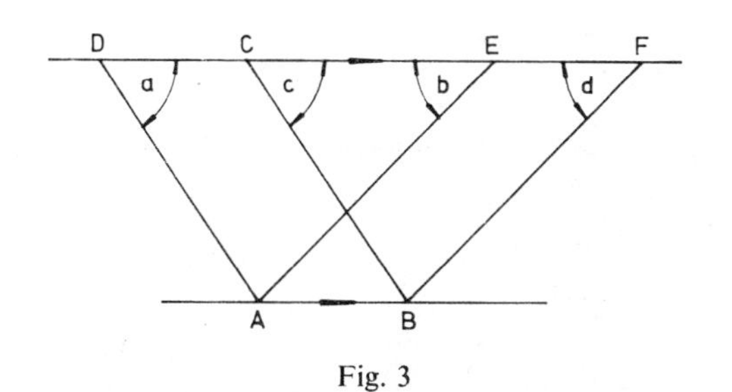

Fig. 3

To prove: ABCD and ABFE are equal in area.
Proof: In △s ADE and BCF
AD = BC (opp. sides of a parallelogram)
$a = c$ (corresponding angles: AD ∥ BC)
$b = d$ (corresponding angles: AE ∥ BF)

∴ △ADE ≡ △BCF

Congruent triangles are equal in every respect. Hence △s ADE and BCF are equal in area.

area ABCD = area ABFD − area BCF
area ABFE = area ABFD − area ADE

∴ area ABCD = area ABFE

Theorem 4

Theorem of Pythagoras

Given: △ABC with ∠B a right-angle (Fig 4).

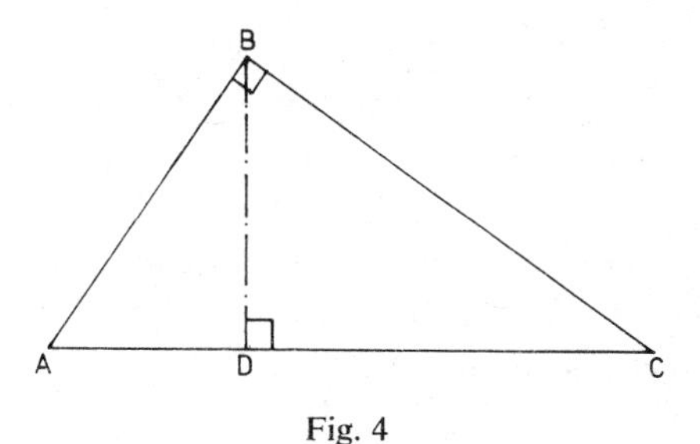

Fig. 4

To prove: $AC^2 = AB^2 + BC^2$.
Construction: Draw BD perpendicular to AC.

Proof: In △s DBA and BCA
∠ADB = ∠ABC = 90°
∠BAC is common
Hence △s BCA and DCB are similar

In △s BCA and DCB
∠ABC = ∠ BDC
∠BCD is common
Hence △s BCA and DCB are similar
Therefore △s DBA, BCA and DCB are all similar.

In △s DBA and BCA,

$$\frac{AB}{AC} = \frac{AD}{AB} \quad \text{or} \quad AB^2 = AC.AD$$

In △s DCB and BCA,

$$\frac{BC}{CD} = \frac{AC}{BC} \quad \text{or} \quad BC^2 = AC.CD$$

$$\begin{aligned} AB^2 + BC^2 &= AC.AD + AC.CD \\ &= AC(AD + CD) \\ &= AC.AC \end{aligned}$$

$$\therefore AB^2 + BC^2 = AC^2$$

Theorem 5

The angle at the centre of a circle is twice any angle at the circumference, standing on the same arc.

Given: ACD, an arc of a circle centre O.
P is any point on the remaining arc (Fig. 5).

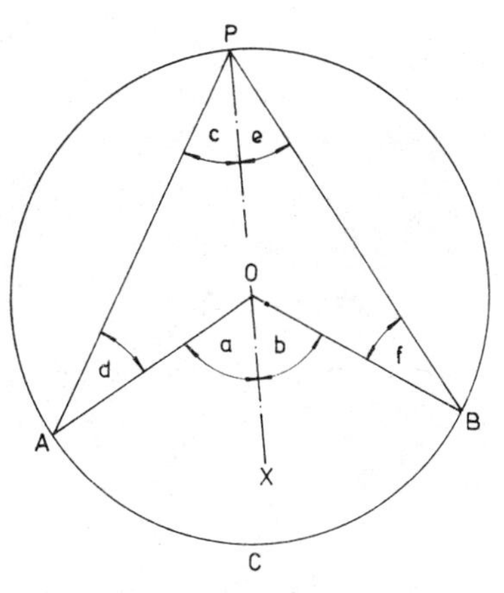

Fig. 5

To prove: $\angle AOB = 2\angle APB$.
Construction: Join PO and produce it to X.
Proof: Since

$$OA = OP \quad \text{(radii)}$$
$$c = d$$
$$a = c + d = 2c \quad \text{(exterior angle of } \triangle APO\text{)}$$

Similarly

$$b = e + f = 2e \quad \text{(exterior angle of } \triangle BPO\text{)}$$

Hence $a + b = 2c + 2e = 2(c + e)$
i.e. $\angle AOB = 2\angle APB$.

Theorem 6

If two chords of a circle intersect inside or outside a circle the product of the segments of one chord is equal to the product of the segments of the other chord.

Given: AB and CD are two chords of a circle intersecting at E (Fig. 6).

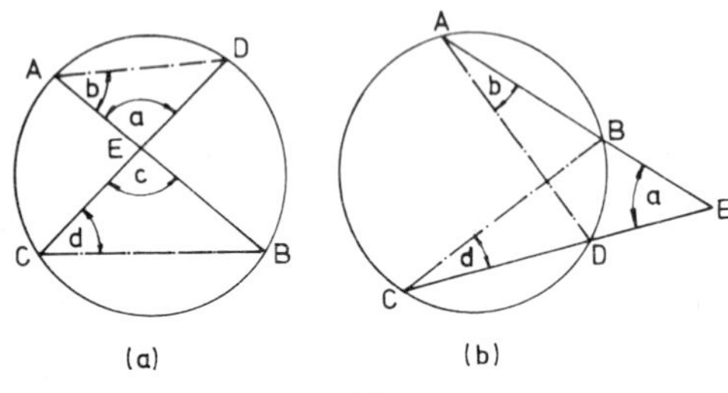

Fig. 6

To prove: AE . EB = CE . ED
Construction: Join AD and BC
Proof: In $\triangle$s AED and CEB,

$b = d$ (angles on arc BD)
$a = c$ (vertically opp. angles in Fig. (6a))

(or a is common in Fig. (6b))

$\therefore$ In both diagrams AED and CEB are similar triangles.

$$\text{Hence } \frac{AE}{CE} = \frac{ED}{EB}$$

or AE . EB = CE . ED

Theorem 7

If from a point outside a circle two lines are drawn, one a secant and the other a tangent to the circle, then the square on the tangent is equal to the rectangle contained by the whole secant and that part of it outside the circle.

Given: secant ABC and tangent CT (Fig. 7).

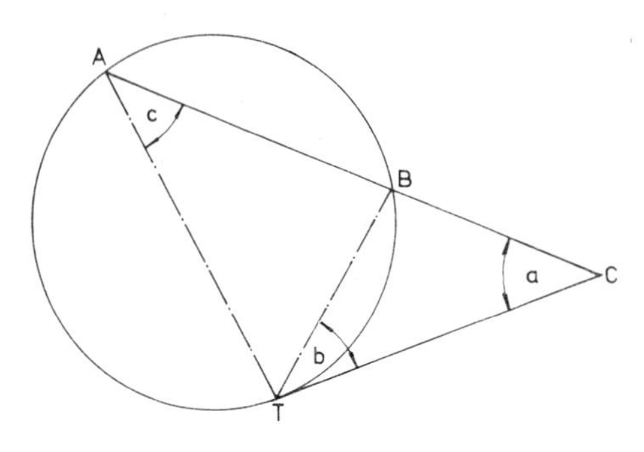

Fig. 7

To prove: $CT^2 = AC . CB$
Construction: Join AT and BT
Proof: In $\triangle$s CBT and CTA,

a is common
$b = c$ (alternate segment)

$\therefore$ $\triangle$s CBT and CTA are similar.

$$\text{Hence } \frac{CB}{CT} = \frac{CT}{AC}$$

or $AC . CB = CT^2$

Theorem 8

The ratio of the areas of similar triangles is equal to the ratio of the squares on corresponding sides.
Given: Two similar triangles ABC and XYZ.

$$\textit{To prove: } \frac{\text{area of } \triangle ABC}{\text{area of } \triangle XYZ} = \frac{BC^2}{YZ^2}.$$

Construction: Draw the altitudes AD and XN of the two triangles.
Proof: area of $\triangle ABC = \frac{1}{2}BC.AD$
area of $\triangle XYZ = \frac{1}{2}YZ.XN$

In $\triangle$s ABD and XYN

$\angle B = \angle Y$ (given)
$\angle D = \angle N = 90°$

$\therefore$ $\triangle$s ABD and XYN are similar.

Hence $\frac{AD}{XN} = \frac{AD}{XY} = \frac{BC}{YZ}$

$$\frac{\text{area } \triangle ABC}{\text{area } \triangle XYZ} = \frac{\frac{1}{2}BC.AD}{\frac{1}{2}YZ.XN} = \frac{BC.AD}{YZ.XN}$$

$$= \frac{BC.BC}{YZ.YZ} = \frac{BC^2}{YZ^2}$$

$\left(\text{since } \frac{AD}{XN} = \frac{BC}{YZ}\right)$.

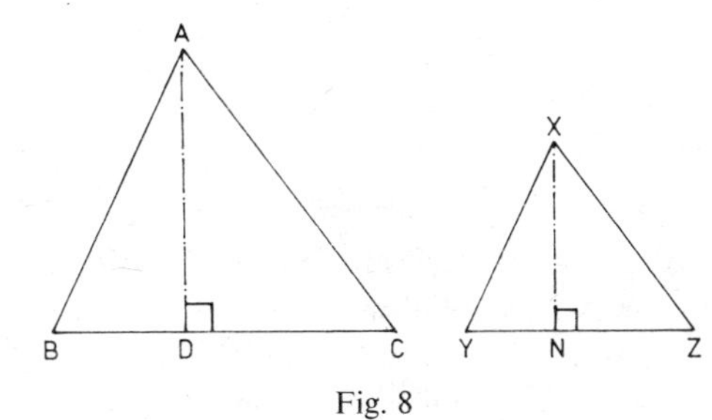

Fig. 8

Theorem 9

The bisector of any angle of a triangle divides the opposite side in the ratio of the sides containing the angle.

Given: $\triangle ABC$ with AD bisecting $\angle CAB$ (Fig. 9).

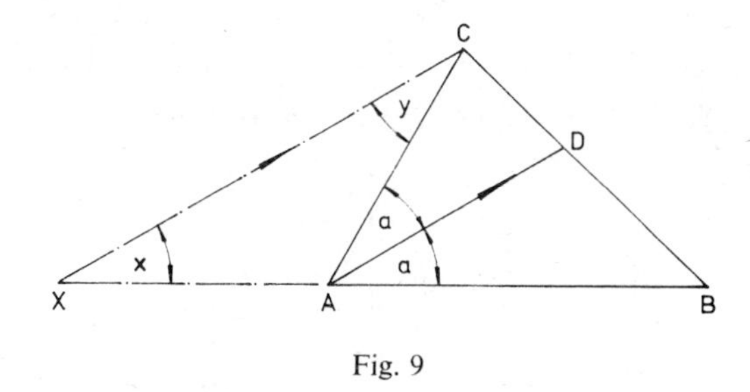

Fig. 9

To prove: $\frac{BD}{DC} = \frac{BA}{AC}$

Construction: From C draw a line parallel to AD to meet BA produced at X.
Proof:

$a = x$ (corresponding angles: AD ||l CX)
$a_1 = y$ (alternate angles: AD ||l CX)

but $a = a_1$ (given)

$\therefore x = y$.

Hence $\triangle CXA$ is isosceles and AC=AX also, since AD is parallel to CX

$$\frac{BD}{DC} = \frac{BA}{AX}$$

$$\therefore \frac{BD}{DC} = \frac{BA}{AC}.$$

Chapter 26 Geometrical Constructions

1) *To divide a line AB into two equal parts*

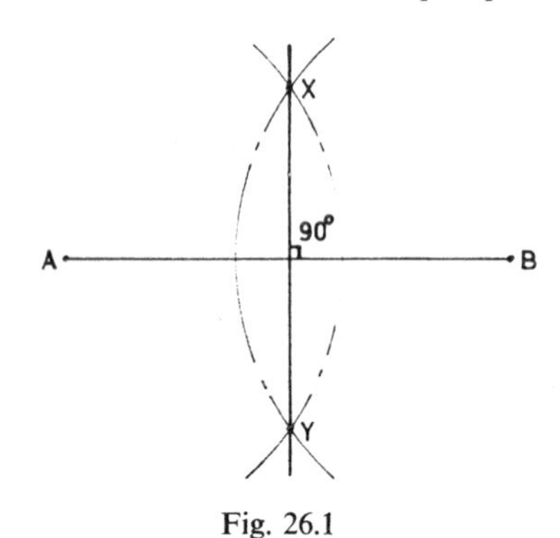

Fig. 26.1

Construction: With A and B as centres and a radius greater than $\frac{1}{2}$AB, draw circular arcs which intersect at X and Y (Fig. 26.1). Join XY. The line XY divides AB into two equal parts and it is also perpendicular to AB.

2) *To draw a perpendicular from a given point A on a straight line.*

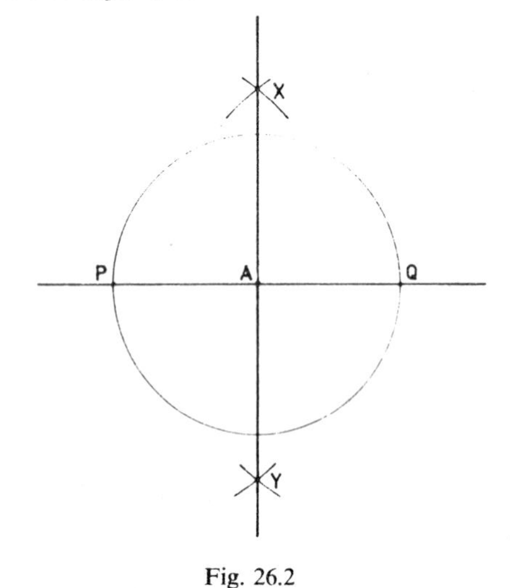

Fig. 26.2

Construction: With centre A and any radius draw a circle to cut the straight line at points P and Q (Fig. 26.2). With centres P and Q and a radius greater than AP (or AQ) draw circular arcs to intersect at X and Y. Join XY. This line will pass through A and it is perpendicular to the given line.

3) *To draw a perpendicular from a point A at the end of a line* (Fig. 26.3)

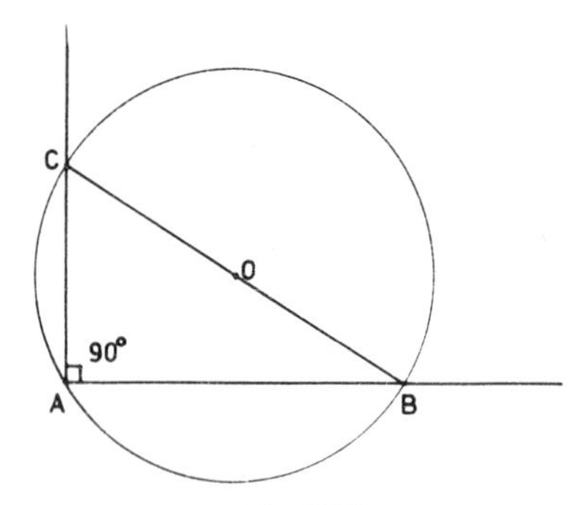

Fig. 26.3

Construction: From any point O outside the line and radius OA draw a circle to cut the line at B. Draw the diameter BC and join AC. AC is perpendicular to the straight line (because the angle in a semi-circle is 90°.)

4) *To draw the perpendicular to a line AB from a given point P which is not on the line.*

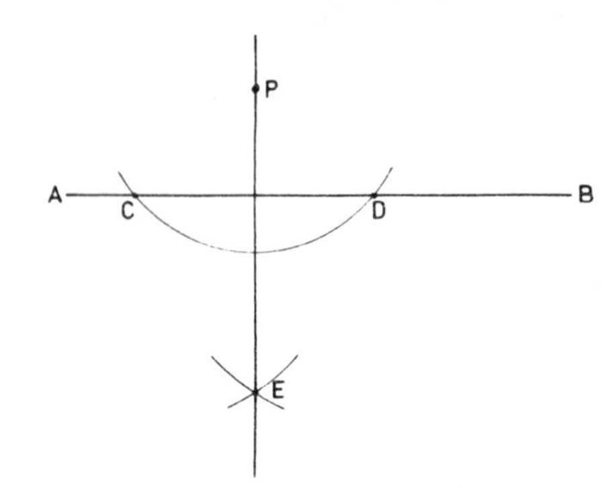

Fig. 26.4

Construction: With P as centre draw a circular arc to cut AB at points C and D. With C and D as centres and a radius greater than $\frac{1}{2}$CD, draw circular arcs to intersect at E. Join PE. The line PE is the required perpendicular (Fig. 26.4).

5) *To construct an angle of* 60°

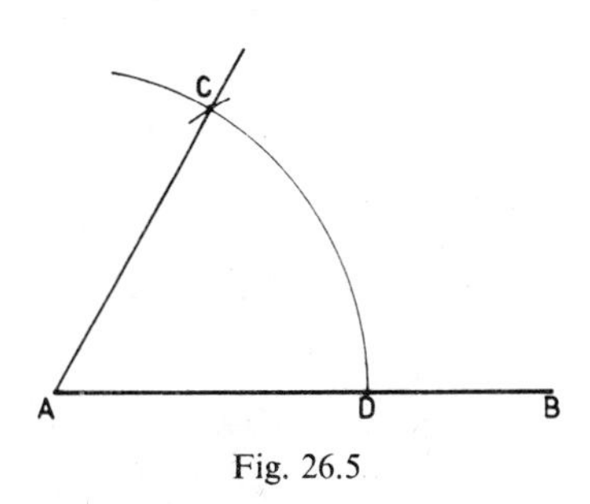

Fig. 26.5

Construction: Draw a line AB. With A as centre and any radius draw a circular arc to cut AB at D. With D as centre and the *same* radius draw a second arc to cut the first arc at C. Join AC. The angle CAD is then 60° (Fig. 26.5).

6) *To bisect a given angle ∠BAC*

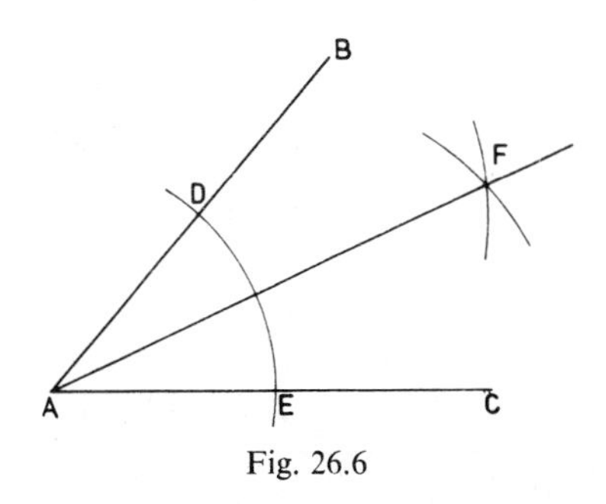

Fig. 26.6

Construction: With centre A and any radius draw an arc to cut AB at D and AC at E. With centres D and E and a radius greater than $\frac{1}{2}$DE draw arcs to intersect at F. Join AF, then AF bisects ∠BAC (Fig. 26.6). Note that by bisecting an angle of 60°, an angle of 30° is obtained. An angle of 45° is obtained by bisecting a right-angle.

7) *To construct an angle equal to a given angle BAC*

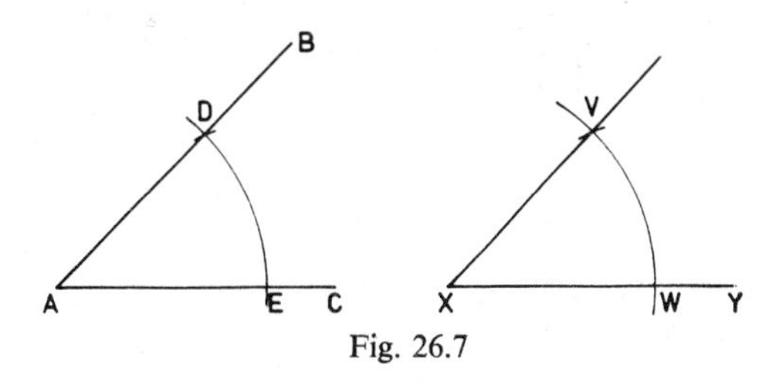

Fig. 26.7

Construction: With centre A and any radius draw an arc to cut AB at D and AC at E. Draw the line XY. With centre X and the same radius draw an arc to cut XY at W. With centre W and radius equal to DE draw an arc to cut the first arc at V. Join VX, then ∠VXW = ∠BAC (Fig. 26.7).

8) *Through a point P to draw a line parallel to a given line AB.*

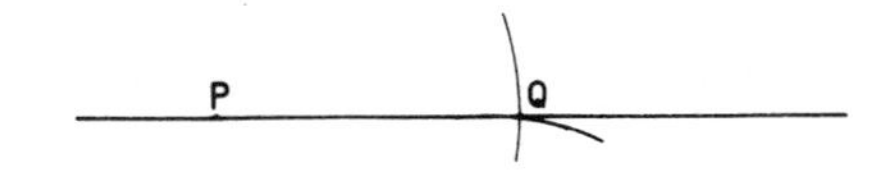

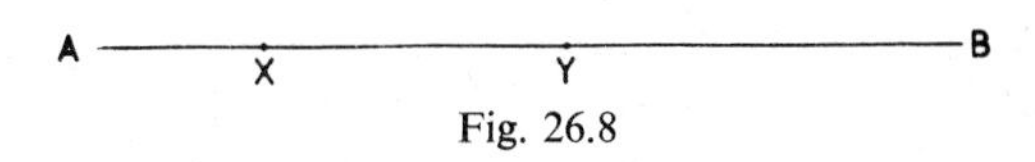

Fig. 26.8

Construction: Mark off any two points X and Y on AB. With centre P and radius XY draw an arc. With centre Y and radius XP draw a second arc to cut the first arc at Q. Join PQ, then PQ is parallel to AB (Fig. 26.8)

9) *To divide a straight line AB into a number of equal parts.*

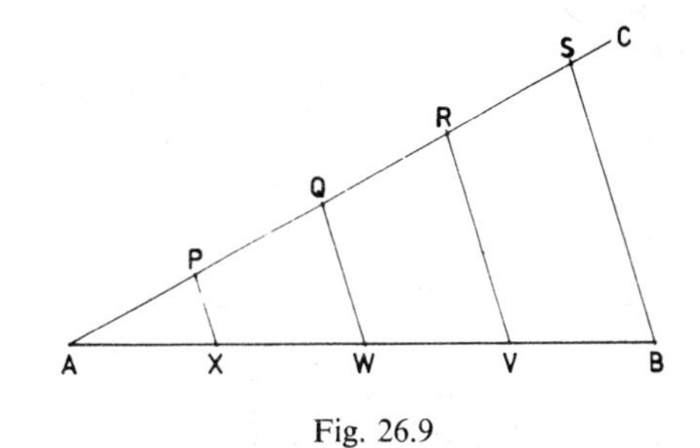

Fig. 26.9

Construction: Suppose that AB has to be divided into four equal parts. Draw AC at any angle to AB. Set off on AC, four equal parts AP, PQ, QR, RS of any convenient length. Join SB. Draw RV, QW and PX each parallel to SB. Then AX=XW=WV=VB (Fig. 26.9).

10) *To draw the circumscribed circle of a given triangle ABC.*

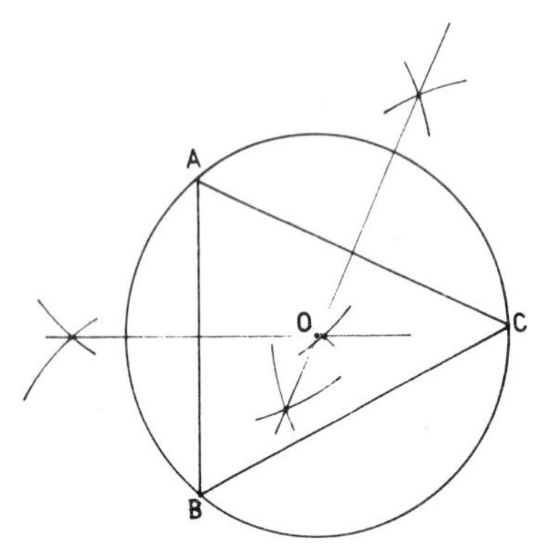

Fig. 26.10

Construction: Construct the perpendicular bisectors of the sides AB and AC (using construction 1) so that they intersect at O. With centre O and radius AO draw a circle which is the required circumscribed circle (Fig. 26.10).

11) *To draw the inscribed circle of a given triangle ABC.*

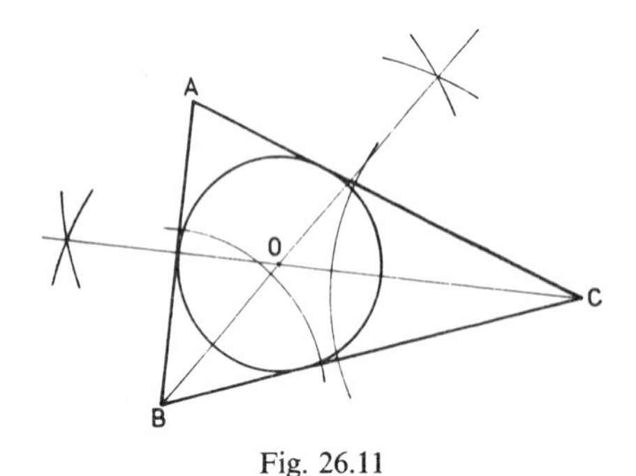

Fig. 26.11

Construction: Construct the internal bisectors of ∠B and ∠C (using construction 6) to intersect at O. With centre O draw the inscribed circle of the triangle ABC (Fig. 26.11).

12) *To draw a triangle whose area is equal to that of a given quadrilateral ABCD.*

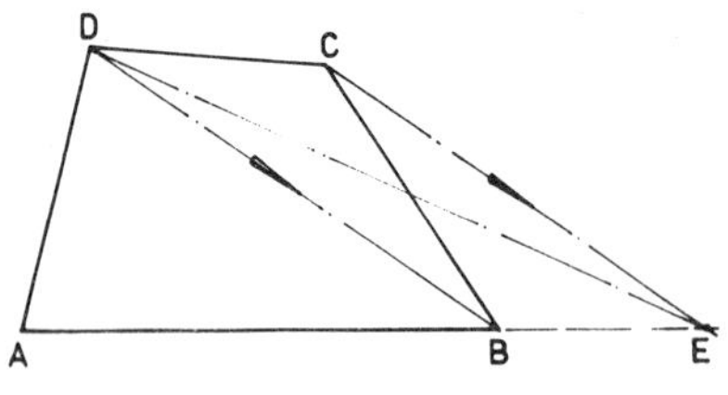

Fig. 26.12

Construction: Join BD and draw CE parallel to BD to meet AB produced at E. Then ADE is a triangle whose area is equal to that of the quadrilateral ABCD (Fig. 26.12).

Proof: As DBE and CDB are equal in area. Add to each of these triangles the area ACB.

13) *To draw a square whose area is equal to that of a given rectangle ABCD.*

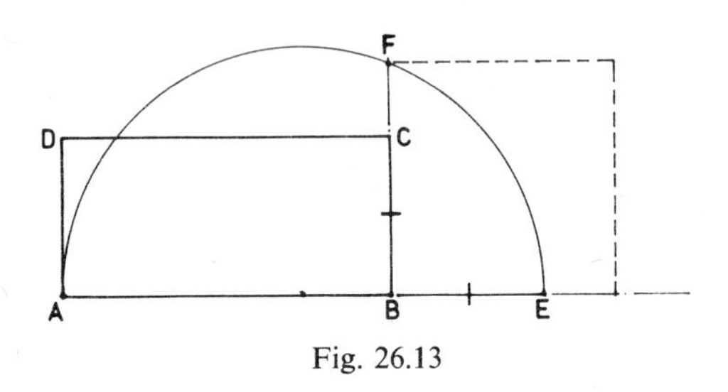

Fig. 26.13

Construction: Produce AB to E so that BC is equal to BE. Draw a circle with AE as diameter to meet BC (or BC produced) at F. Then BF is a side of the required square (Fig. 26.13).

Proof: The diameter of a circle bisects any chord of a circle which is perpendicular to it. Since AE is a diameter and BF is half a chord then $AB.BE=BF^2$.

14) *To draw a tangent to a circle at a given point P on the circumference of the circle.*

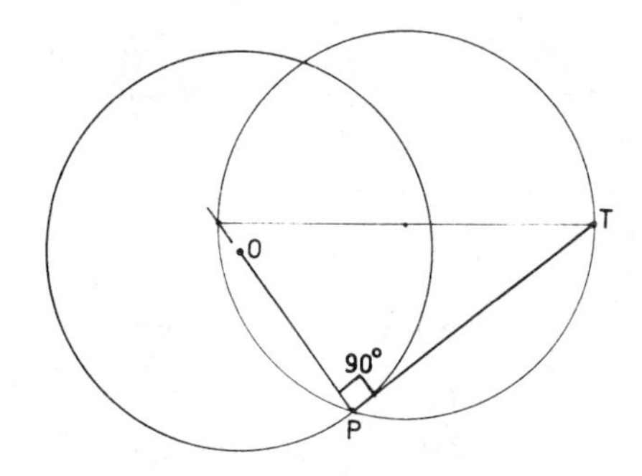

Fig. 26.14

Construction: O is the centre of the given circle. Join OP. Using construction 3 draw the line PT which is perpendicular to OP. PT is the required tangent, since at the point of tangency, a tangent is perpendicular to a radius (Fig. 26.14).

15) *To draw the segment of a circle so that it contains a given angle θ.*

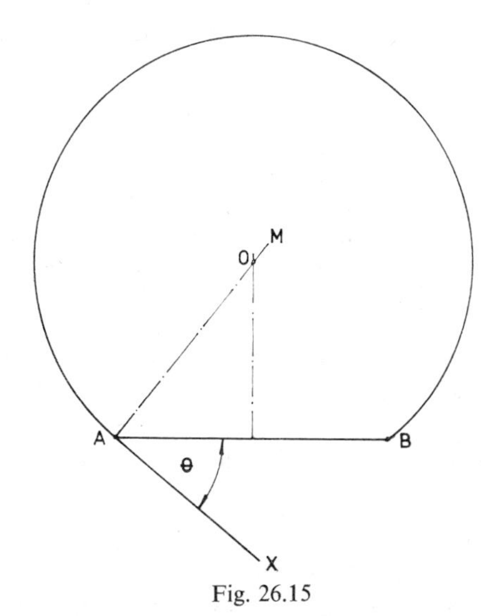

Fig. 26.15

Construction: Draw the lines AB and AX so that $\angle BAX=\theta$. From A draw AM perpendicular to AX. Draw the perpendicular bisector of AB to meet AM at O. With centre O and radius OA draw the circular arc which terminates at A and B. This is the required segment (Fig. 26.15).

16) *To construct a triangle given the lengths of each of the three sides.*

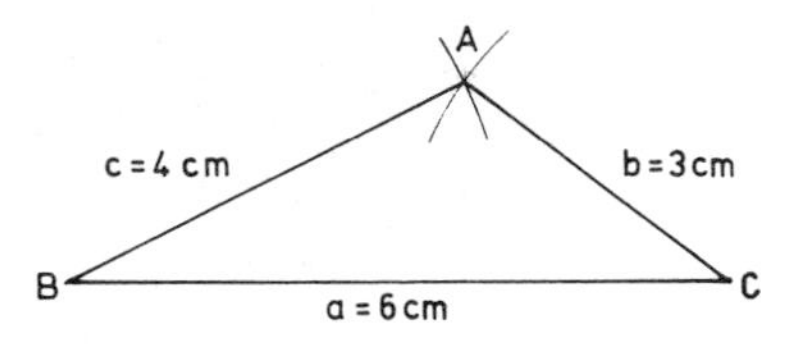

Fig. 26.16

Construction: Suppose $a=6$ cm, $b=3$ cm and $c=4$ cm. Draw BC=6 cm. With centre B and radius 4 cm draw a circular arc. With centre C and radius 3 cm draw a circular arc to cut the first arc at A. Join AB and AC. Then ABC is the required triangle (Fig. 26.16).

17) *To construct a triangle given two sides and the included angle between the two sides.*

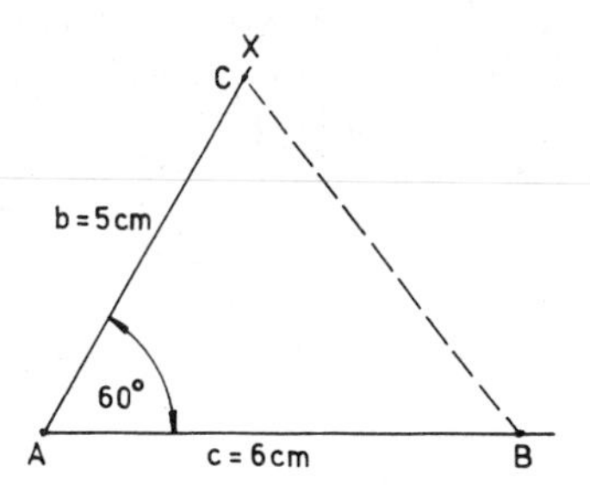

Fig. 26.17

Construction: Suppose $b=5$ cm and $c=6$ m and $\angle A=60°$. Draw AB=6 cm and draw AX such that $\angle BAX=60°$. Along AX mark off AC=5 cm. Then ABC is the required triangle (Fig. 26.17).

18) *To construct a triangle (or triangles) given the lengths of two of the sides and an angle which is not the included angle between the two given sides.*

a) *Construction:* Suppose $a=5$ cm, $b=6$ cm and $\angle B=60°$. Draw BC=5 cm and draw BX such that $\angle CBX=60°$. With centre C and radius of 6 cm describe a circular arc to cut BX at A. Join CA then ABC is the required triangle ABC (Fig. 26.18).
b) Suppose that $a=5$ cm, $b=4{\cdot}5$ cm and $\angle B=60°$. The construction is the same as before but the circular arc drawn

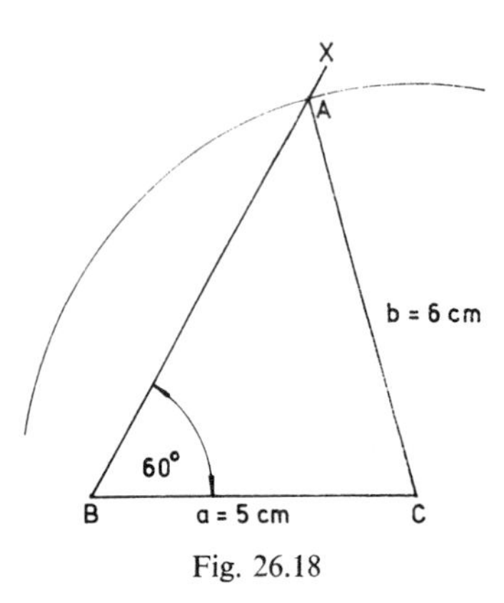

Fig. 26.18

with C as centre now cuts BX at two points A and A^1. This means that there are two traingles which meet the given conditions, i.e. Δs ABC and A^1BC (Fig. 26.19). For this reason this case is often called the *ambiguous case.*

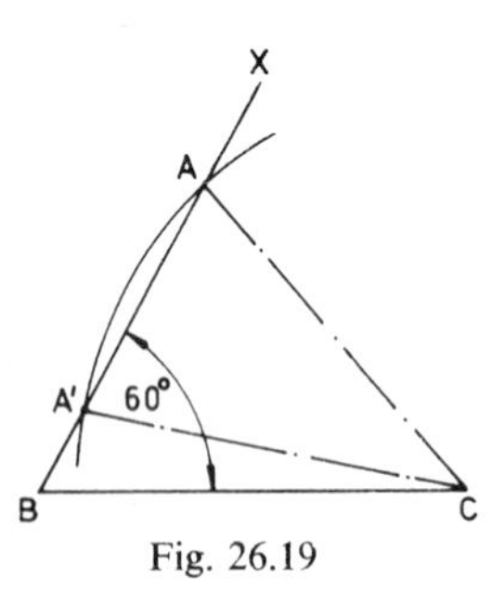

Fig. 26.19

19) *To construct a common tangent to two given circles.*

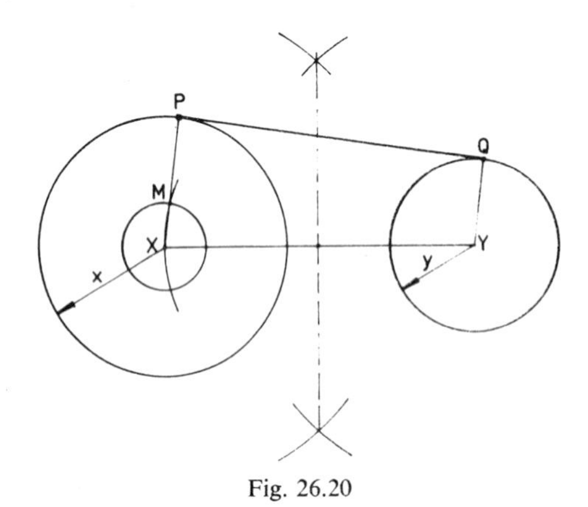

Fig. 26.20

Construction: The two given circles have centres X and Y and radii x and y respectively (Fig. 26.20). With centre X draw a circle whose radius is $(x-y)$. With diameter XY draw an arc to cut the previously drawn circle at M. Join XM and produce to P at the circumference of the circle. Draw YQ parallel to XP, Q being at the circumference of the circle. Join PQ which is the required tangent.

20) *To construct a pair of tangents from an external point to a given circle* (Fig. 26.21)

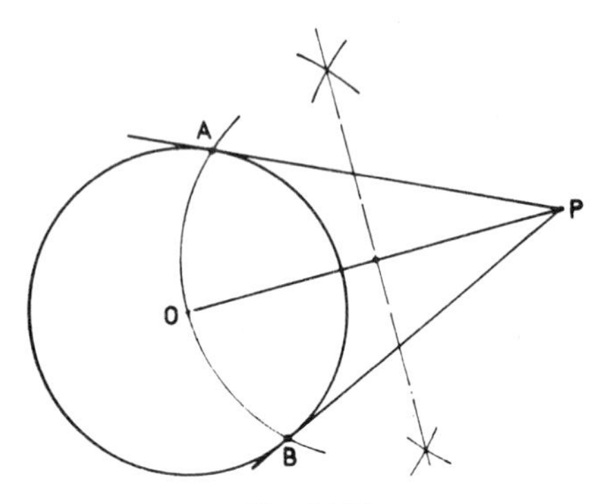

Fig. 26.21

Construction: It is required to draw a pair of tangents from the point P to the circle centre O. Join OP. With OP as diameter draw a circle to cut the given circle at points A and B. Join PA and PB which are the required pair of tangents.

Exercise 96

1) Construct △ABC in which AB = 5 cm, AC = 4 cm and ∠CAB = 45°.

2) Construct the point P inside a triangle which is equi-distant from the three sides of the triangle in question 1.

3) Construct:

a) △ABC in which AB = 9 cm, AC = 7·2 cm and ∠BAC = 60°;

b) the circle through A, B and C.

4) Construct the rhombus ABCD in which AB = 4·6 cm and the diagonal AC = 7·2 cm. Construct also the point P such that CP = 1·3 cm and ∠APC = 90°. Measure AP.

5) Construct:

a) $\triangle ABC$ in which $AB=8$ cm, $\angle A=45°$ and $\angle B=60°$;

b) the bisector of $\angle ACB$;

c) the point E on BC, such that the area of $\triangle BDE$ is one-half of the area of $\triangle BDC$, the point D being the point at which the bisector of $\angle ACB$ cuts AB.

6) Construct:

a) the parallelogram ABCD in which $AB=6$ cm, the diagonal $AC=14$ cm and $\angle ABC=150°$;

b) the rhombus ABXY equal in area to ABCD with $\angle ABX$ obtuse. Measure the length AX.

7) Construct:

a) the rectangle ABCD in which $AB=5{\cdot}8$ cm and the diagonal $AC=7{\cdot}4$ cm;

b) the trapezium AXBC in which BX is parallel to AC and $\angle CAX=60°$. Measure AX.

8) Construct:

a) an equilateral $\triangle ABC$ of side 5 cm;

b) a triangle ABX equal in area to $\triangle ABC$ and such that $\angle ABX=135°$. Measure BX.

9) Construct:

a) the $\triangle ABC$ with $BC=8$ cm, $CA=6$ cm and $AB=4{\cdot}4$ cm;

b) a $\triangle PBC$ equal in area to $\triangle ABC$ and having $PB=PC$;

c) the rhombus BPCQ.

10) Construct:

a) the $\triangle ABC$ with $AB=8$ cm, $BC=6$ cm and $AC=11{\cdot}2$ cm;

b) the incircle of $\triangle ABC$;

c) the points P and Q on the circle which are 1 cm from the side AC.

11) Draw $\triangle ABC$ in which $AB=5$ cm, $BC=6$ cm and $CA=7$ cm. A circle can be drawn to touch BC at C and also to touch BA produced. Construct two lines on each of which the centre of the circle lies and hence draw the circle.

12) Draw an angle AOB of 65°. Find by construction the point P which is 4 cm from OA and 3 cm from OB. Measure OP.

13) Construct in one diagram

a) a $\triangle ABX$ in which $\angle X=90°$, $AB=6$ cm and $BX=2$ cm;

b) a trapezium ABCD in which C is on BX produced and AD is parallel to BC, $AD=4$ cm and $BC=8$ cm;

c) a circle touching AB, BC and CD.

14) A line ST is 9 cm long. Construct the point R so that $\angle SRT=90°$ and $RT=4$ cm. Hence construct on the same diagram a line XY so that the perpendiculars from S and T onto the line XY are 3 cm and 7 cm long respectively.

15) Construct $\triangle ABC$ in which $AB=8$ cm, $AC=7$ cm and $\angle CAB=45°$. Construct the circumcircle of this triangle and hence complete the cyclic quadrilateral ABCD in which $AD=DC$.

16) Construct:

a) the quadrilateral ABCD with $DA=DB=DC$, $AB=5$ cm, $\angle ABC=135°$ and $BC=4$ cm;

b) a circle to touch the line DA at A and to pass through B.

17) Draw a line AB of length 5 cm. By a geometrical construction find a point P on the line, between A and B, such that

$$\frac{AP}{BP}=\frac{2}{3}.$$

18) Construct an equilateral $\triangle ABC$ of side 5 cm. Construct:

a) a square equal in area to $\triangle ABC$;

b) a square of area twice that of the first square; measure the length of the side of each square.

Chapter 27 Loci

A *locus* is a set of positions traced out by a point which moves according to some law. For instance the locus of a point which moves so that it is always 3 cm from a given fixed point is a circle whose radius is 3 cm. It often helps if we mark off a few points according to the given law. By doing this we may gain some idea of what the locus will be. Sometimes three or four points will be sufficient but sometimes ten or more points may be required before the locus can be recognised.

Example

Given a straight line AB of length 6 cm, find the locus of a point P so that $\angle APB$ is always a right angle.

By drawing a number of points P_1, P_2 ... etc so that $\angle AP_1B$, $\angle AP_2B$ ... etc (Fig. 27.1) are all right angles it appears that the locus is a circle with AB as a diameter. We now try to prove that this is so.

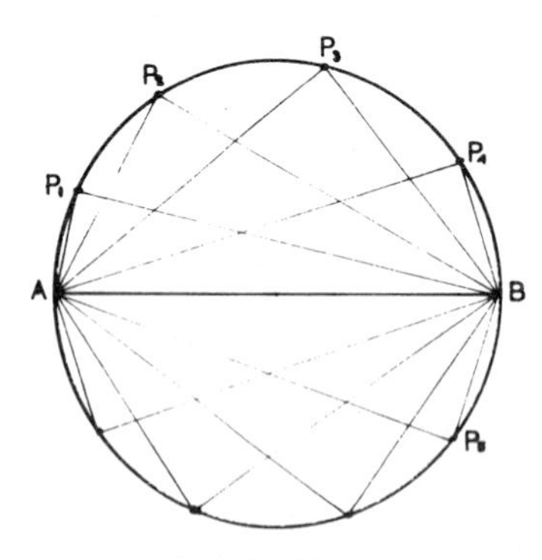

Fig. 27.1

Since the angle in a semi-circle is a right-angle, all angles subtended by the diameter AB at the circumference of the circle will be right-angles. Hence the locus of P is a circle with AB as diameter.

Exercise 97

1) Find the locus of a point P which is always 5 cm from a fixed straight line of infinite length.

2) Find the locus of a point P which moves so that it is always 3 cm from a given straight line AB which is 8 cm long.

3) XYZ is a triangle whose base XY is fixed. If XY = 5 cm and the area of $\triangle XYZ = 10\ \text{cm}^2$ find the locus of Z.

4) Given a square of 10 cm side, find all the points which are 8 cm from two of the vertices. How many points are there?

5) Find all the points which are 5 cm from each of two intersecting straight lines inclined at an angle of 45°. How many points are there?

STANDARD LOCI

The following standard loci should be remembered.

1) The locus of a point equi-distant from two given points A and B is the perpendicular bisector of AB (Fig. 27.2).

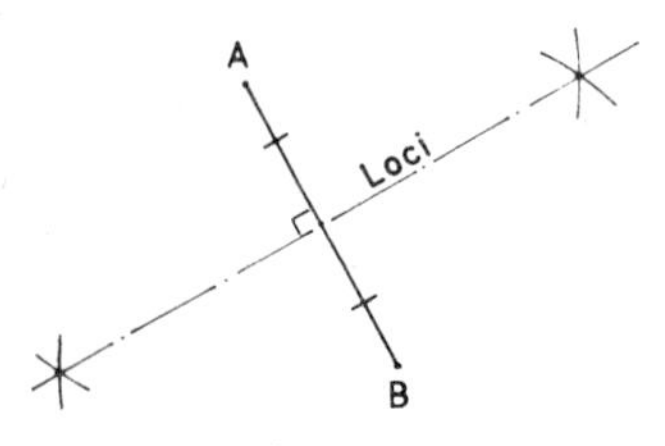

Fig. 27.2

2) The locus of a point equi-distant from the arms of an angle is the bisector of the angle (Fig. 27.3).

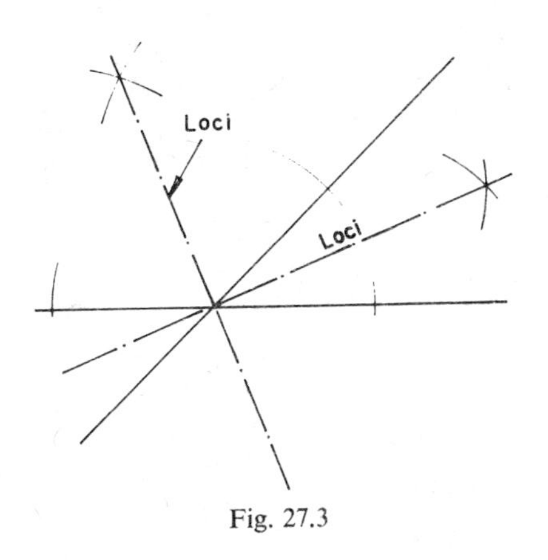

Fig. 27.3

INTERSECTING LOCI

Frequently two pieces of information are given about the position of a point. Each piece of information should then be dealt with separately, since any attempt to comply with the two conditions at the same time will lead to a trial and error method which is not acceptable. Each piece of information will partially locate the point and the intersection of the two loci will determine the required position of the point.

Example

A point P lies 3 cm from a given straight line and it is also equi-distant from two fixed points not on the line and not perpendicular to it. Find the two possible positions of P.

Condition 1. The point P lies 3 cm from the given straight line (AB in Fig. 27.4). To meet this condition draw two straight lines X, Y, and X_1Y_1 parallel to AB and on either side of it.

Condition 2. P is equi-distant from the two fixed points (R and S in Fig. 27.4) To meet this condition we draw the perpendicular

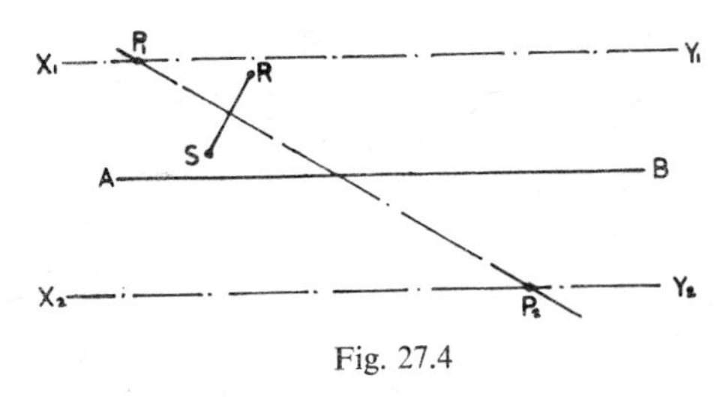

Fig. 27.4

bisector of RS (since $\triangle$RSP must be isosceles).

The intersections of the two loci give the required position of the point P. These are the points P_1 and P_2 in Fig. 27.4.

Exercise 98

1) Find the locus of the centre P of a circle of constant radius 3 cm which passes through a fixed point A.

2) Find the locus of the centre, Q, of a circle of constant radius 2 cm which touches externally a fixed circle, centre B and radius 4 cm.

3) Find the locus of the centre of a variable circle which passes through two fixed points A and B.

4) AB is a fixed line of length 8 cm and P is a variable point. The distance of P from the middle point of AB is 5 cm and the distance of P from AB is 4 cm. Construct a point P so that both of these conditions are satisfied. State the number of possible positions of P.

5) X is a point inside a circle, centre C, and Q is the mid-point of a chord which passes through X. Determine the locus of Q as the chord varies. If CX is 5 cm and the radius of the circle is 8 cm construct the locus of Q accurately and hence construct a chord which passes through X and has its mid-point 3 cm from C.

6) XY is a fixed line of given length. State the locus of a point P which moves so that the size of $\angle$XPY is constant and sketch the locus.

7) AB is a fixed line of 4 cm and R is a point such that the area of $\triangle ABR$ is 5 cm^2. S is the mid-point of AR. State the locus of R and the locus of S.

8) T is a fixed point outside a fixed circle whose centre is O. A variable line through T meets the circle at X and Y. Show that the locus of the mid-point of XY is an arc of the circle on OT as diameter.

9) Chords of a circle, centre C, are drawn through a fixed point A within the circle. Show that the mid-points of all these chords lie on a circle and state the position of the centre of the circle.

10) Draw a circle centre O and radius 4 cm. Construct the locus of the mid-points of all chords of this circle which are 6·5 cm long.

Miscellaneous Exercise

Exercise 99

(All of the type found in O level papers.)

1) In Fig. 1, D is the mid-point of the minor arc BC. Calculate ∠DAC.

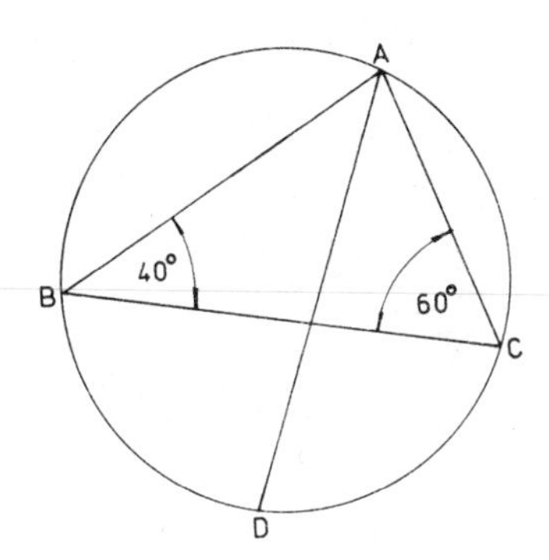

Fig. 1

2) Using rules and compasses only, construct a parallelogram PQRS in which ∠PQR $=45^\circ$, QR $=6$ cm, and PQ $=8$ cm.

3) In an isosceles triangle ABC, ∠A $=140^\circ$ and the bisectors of the exterior angles at B and C meet at O. Calculate ∠BOC.

4) Two lines AD and BC intersect at E and the triangles ABC and BCD are equal in area. Prove that E is the mid-point of AD.

5) Construct a quadrilateral ABCD with AB $=5$ cm, BC $=6$ cm, CD $=7$ cm, AD $=8$ cm and BD $=7{\cdot}5$ cm. Construct a point E on AB produced so that △ADE is equal in area to the quadrilateral. Measure AE.

6) In a circle whose centre is O and whose radius is 15 cm, a chord AB is 10 cm from O. Calculate the length of the chord.

7) A circle whose centre is O has a radius of 9 cm. State the locus of the centres of circles of radius 1 cm which touch this first circle externally.

8) In a regular polygon, three consecutive sides are AB, BC and CD and ∠ABD $=9$ ∠DBC. Calculate the number of sides of the polygon.

9) In the parallelogram ABCD, E is the mid-point of BC and a line through E meets DC at L and AB produced at M. Prove that the trapezium AMLD and the parallelogram ABCD are equal in area.

10) ABCD is a parallelogram and X and Y lie on AD and CD respectively. Prove that triangles ABY and BCX are equal in area.

11) In a triangle ABC, AB $=9$ cm, AC $=7$ cm and BC $=8$ cm. The internal bisector of ∠A meets BC at D and the external bisector meets BC produced at E. Without using tables find DE.

12) In △ABC, AB $=$ AC. The side BA is produced to D so that AB $=$ AD. Prove that ∠BCD is a right-angle.

13) In the cyclic quadrilateral ABCD, 2∠A $=$ 3∠C. Calculate ∠A.

14) In △ABC, ∠B $=2x^\circ$ and ∠C $=4x^\circ$. The inscribed circle of the triangle touches BC, CA and AB at P, Q and R respectively. Find ∠RPQ in terms of x°.

15) Using ruler and compasses only, construct the quadrilateral ABCD in which AB $=4$ cm, ∠B $=90^\circ$, ∠A $=120^\circ$, AD $=5$ cm and CD $=8$ cm. Construct the circle of radius 3 cm which passes through the mid-point of BC and touches CD internally. Measure the length of the tangent from D to this circle.

16) The mid-points of the sides AB, BC, CD and DA of a quadrilateral ABCD are respectively P, Q, R and S. Prove that PQRS is a parallelogram.

17) Construct a triangle PQR with PQ=6 cm, QR=5 cm and ∠Q=40°. Then, using ruler and compasses only, construct a segment of a circle on QR which contains an angle of 40°. Measure the radius of this circle.

18) A cyclic quadrilateral ABCD is such that each angle of △ABD is 60°. The diagonals of the quadrilateral intersect at E. The side DC is produced to F so that CF=BC and BF is joined. Prove:

a) AC is parallel to BF;
b) △s ABC and BDF are congruent;
c) area ABE=area ECD+area BCF.

19) A line PQ is 8 cm long and a circle centre P has a radius of 4 cm. A second circle touches PQ at Q and also touches the second circle. If x cm is the radius of this second circle form an equation in x and hence calculate the radius of this circle.

20) In △ABC, AB=AC and ∠A=50°. A point D is taken on AB produced so that ∠BCD=30°. Calculate ∠BDC.

21) ABCD is a quadrilateral in which ∠A=∠D and ∠B=∠C. Prove that two sides of this quadrilateral are equal.

22) Construct △ABC with AB=5 cm, ∠A=80° and ∠B=70°. Using ruler and compasses only, construct a circle which touches AB at B and which also touches AC.

23) The sides AB and AC or △ABC are equal and ∠BAC=84°. A line AF is drawn parallel to BC. Calculate ∠BAF.

24) Two circles of radii 5 cm and 4 cm touch externally. A common tangent to the circles touches the first circle at P and the second circle at Q. Calculate the length of PQ.

25) A quadrilateral ABCD has ∠A=60° and ∠B:∠C:∠D=1:4:5. Prove that ABCD is a cyclic quadrilateral.

26) The diagonals PR and QS of a cyclic quadrilateral PQRS intersect at K and the tangent at P is parallel to QS. Prove that

a) PQ=PS;

b) $\frac{QK}{KS}=\frac{QR}{RS}$;

c) $\frac{\triangle PKQ}{\triangle PKS}=\frac{QR}{RS}$;

27) ABCD is a square. P is a point on AB and the mid-point of PC is Q. Prove that the area of △DPQ is one-quarter of ABCD.

28) In the quadrilateral ABCD, AB is parallel to DC and ∠DBC=101°, ∠BCD=53° and ∠ADB=25°. The side DA is produced to E. Calculate ∠EAB.

29) Two chords AB and CD of a circle intersect at a point X inside the circle. AX=1·25 cm, XD=8 cm and CX=3 cm. Calculate BX.

30) Using ruler and compasses only, construct the cyclic quadrilateral ABCD which has ∠B=90°, AB=6 cm, BC=4 cm and ∠ACD=60°. Measure AD.

31) △ABC is acute-angled and equilateral triangles ABZ and BCX are described outside it on the sides AB and BC. The lines XA and ZC meet at Y. Prove that

a) AX=CZ;
b) ∠ZYA=60°.

If ZC meets AB at D, prove that ZD.DY=BD.DA.

32) Four exterior angles of a pentagon are 15°, 49°, 82° and 129°. Calculate the fifth exterior angle.

33) From a point O outside a circle, two straight lines OAB and OCD are drawn to cut the circle at A, B, C and D respectively. If OA=6 cm, AB=10 cm, OC=8 cm find the length of CD.

34) Without using a protractor, construct △PQR with PQ=9 cm, PR=7 cm and

$\angle P = 60°$. Construct a circle which touches PQ and PR and whose centre is inside this triangle and 5 cm from Q. Measure the radius of this circle.

35) In the cyclic quadrilateral ABCD, $\angle BAD = 110°$ and $BD = CD$. Calculate the acute angle between BC and the tangent C to the circle.

36) A line XY is parallel to the base BC of a triangle ABC. It meets AB at X and AC at Y. Given that $AX = 3$ cm, $XB = 5$ cm and $AC = 6{\cdot}4$ cm, calculate the length of AY.

37) Construct a triangle PQR in which $PQ = 8$ cm, $QR = 5{\cdot}5$ cm and $\angle Q = 140°$. Construct a point S which is equi-distant from Q and R such that the area of the triangle QRS is equal to 30 cm^2. Measure PS.

38) In a quadrilateral ABCD, $AB = AD$ and $BC = CD$. Prove that AC bisects BD.

39) The sides AB, AC of $\triangle ABC$ are equal. The side BC is produced to a point X and CY bisects $\angle ACX$. If $\angle BAC = 100°$, calculate $\angle YCX$.

40) In a circle of radius $3{\cdot}5$ cm, a diameter AB is produced to a point X, where $BX = 3$ cm. A secant XCD cuts the circle at C and D and $CD = 1$ cm. Calculate the length of XC.

Chapter 28 Trigonometry

THE TRIGONOMETRICAL RATIOS

Consider any angle θ which is bounded by the lines OA and OB as shown in Fig. 28.1. Take any point P on the boundary line OB. From P draw line PM perpendicular to OA to meet it at the point M. Then,

the ratio $\frac{MP}{OP}$ is called the sine of $\angle AOB$

the ratio $\frac{OM}{OP}$ is called the cosine of $\angle AOB$

and

the ratio $\frac{MP}{OM}$ is called the tangent of $\angle AOB$

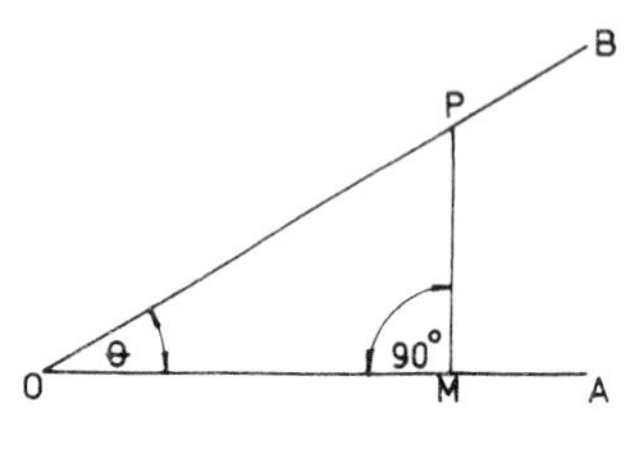

Fig. 28.1

THE SINE OF AN ANGLE

The abbreviation 'sin' is usually used for sine. In any right-angled triangle (Fig. 28.2)

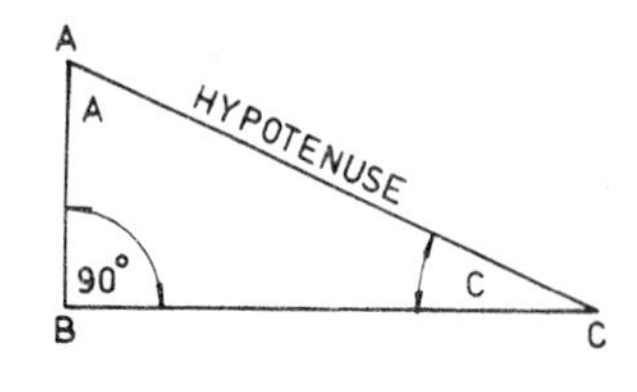

Fig. 28.2

$$\text{the sine of an angle} = \frac{\text{side opposite the angle}}{\text{hypotenuse}}$$

$$\sin A = \frac{BC}{AC}$$

$$\sin C = \frac{AB}{AC}$$

Example

Find by drawing a suitable triangle the value of sin 30°.

Draw the lines AX and AY which intersect at A so that the angle $\angle YAX = 30°$ as shown in Fig. 28.3. Along AY measure off AC equal to 1 unit (say 10 cm) and from C draw CB perpendicular to AX. Measure CB which will be found to be 0·5 units (5 cm in this case).

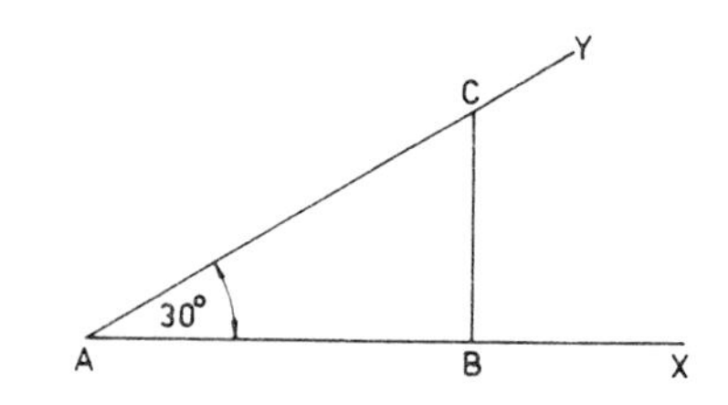

Fig. 28.3

$$\therefore \sin 30° = \frac{5}{10} = 0{\cdot}5$$

Although it is possible to find the sines of angles by drawing this is inconvenient and not very accurate. Tables of sines have been calculated which allow us to find the sine of any angle.

READING THE TABLE OF SINES OF ANGLES

1) *To find sin 12°*. The sine of an angle with an exact number of degrees is shown in the column headed O. Thus sin 12°=0·2079.

2) *To find sin 12°36′*. The value will be found under the column headed 36′. Thus sin 12°36′ =0·2181.

3) *To find sin 12°40′*. If the number of minutes is not an exact multiple of 6 we use the table of mean differences. Now 12°36′=0·2181 and 40′ is 4′ more than 36′. Looking in the mean difference column headed 4 we find the value 11. This is *added* on to the sine of 12°36′ and we have sin 12°40′=0·2181 +0·0011=0·2192

4) *To find the angle whose sine is* 0·1711. Look in the table of sines to find the nearest number *lower* than 0·1711. This is found to be 0·1702 which corresponds to an angle of 9°48′. Now 0·1702 is 0·0009 less than 0·1711 so we look in the mean difference table in the row marked 9° and find 9 in the column headed 3′. The angle whose sine is 0·1711 is then 9°48′+3′=9°51′ or sin 9°51′=0·1711.

Examples

1) Find the length of AB in Fig. 28.4.
AB is the side apposite ∠ACB
BC is the hypotenuse since it is opposite to the right-angle.

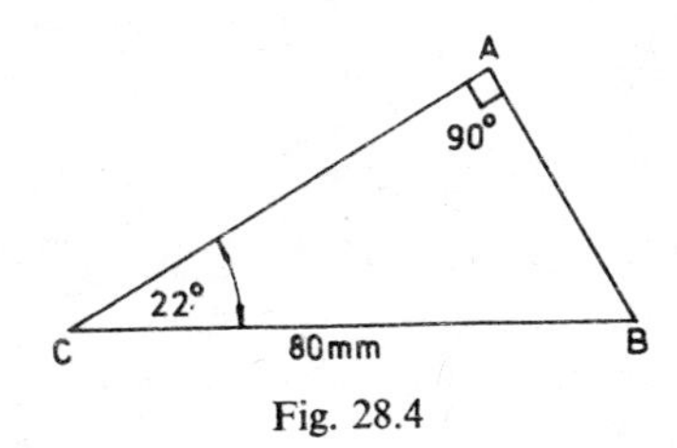

Fig. 28.4

$$\therefore \frac{AB}{BC} = \sin 22°$$

$$AB = BC \times \sin 22° = 80 \times 0{\cdot}3746 = 29{\cdot}97 \text{ mm}$$

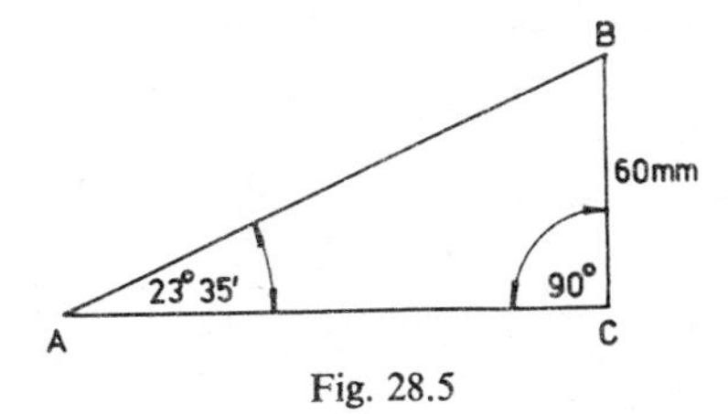

Fig. 28.5

2) Find the length of AB in Fig. 28.5.
BC is the side opposite to ∠BAC and AB is the hypotenuse.

$$\therefore \frac{BC}{AB} = \sin 23°\ 35'$$

$$AB = \frac{BC}{\sin 23°35'} = \frac{60}{0{\cdot}4000} = 150 \text{ mm}$$

3) Find the angles CAB and ABC in △ABC which is shown in Fig. 28.6.

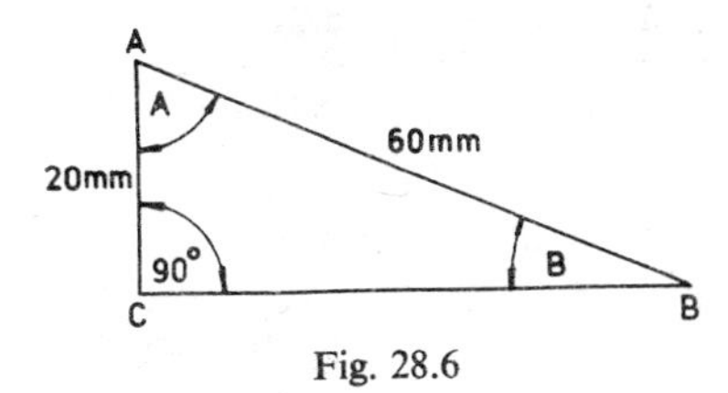

Fig. 28.6

$$\sin B = \frac{AC}{AB} = \frac{20}{60} = 0{\cdot}3333$$

From the sine tables

∠B=19° 28′
∠A=90°−19° 28′=70° 32′

Exercise 100

1) Find, by drawing, the sines of the following angles:

a) 30° b) 45° c) 68°

2) Find by, drawing, the angles whose sines are:

a) $\frac{1}{3}$ b) $\frac{3}{4}$ c) 0·72

3) Use the tables to write down the values of:

a) sin 12° b) sin 18° 12′
c) sin 74° 42′ d) sin 7° 23′
e) sin 87° 35′ f) sin 0° 11′

4) Use the tables to write down the angles whose sines are:

a) 0·1564 b) 0·9135 c) 0·9880
d) 0·0802 e) 0·9814 f) 0·7395
g) 0·0500 h) 0·2700

5) Find the lengths of the sides marked x in Fig. 28.7 the triangles being right-angled.

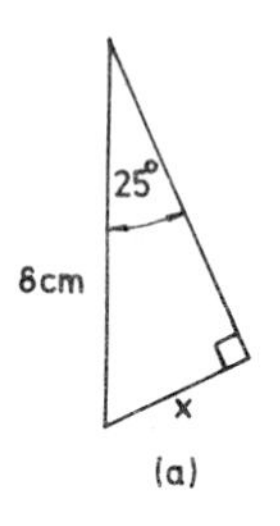

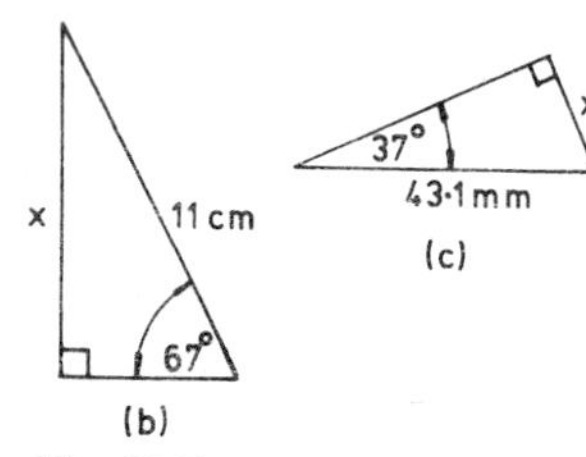

Fig. 28.7

6) Find the angles marked θ in Fig. 28.8, the triangles being right-angled.

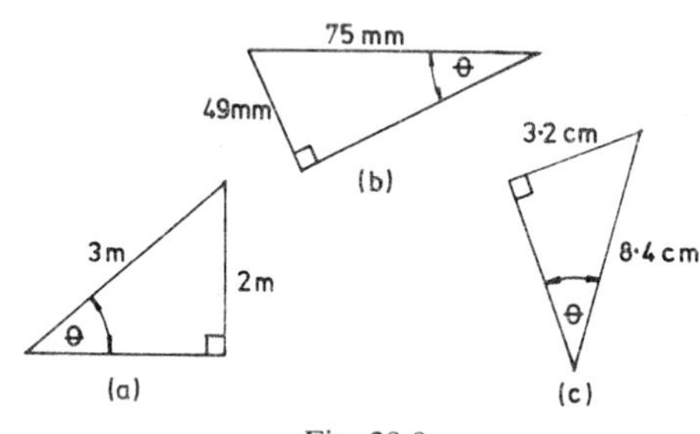

Fig. 28.8

7) In $\triangle ABC$, $\angle C=90°$, $\angle B=23° \; 17'$ and $AC=11{\cdot}2$ cm. Find AB.

8) In $\triangle ABC$, $\angle B=90°$, $\angle A=67° \; 28'$ and $AC=0{\cdot}86$ m. Find BC.

9) An equilateral triangle has an altitude of 18·7 cm. Find the length of the equal sides.

10) Find the altitude of an isosceles triangle whose vertex angle is 38° and whose equal sides are 7.9 m long.

11) The equal sides of an isosceles triangle are each 27 cm long and the altitude is 19 cm. Find the angles of the triangle.

THE COSINE OF AN ANGLE

In any right-angled triangle (Fig. 28.9):
the cosine of an angle

$$= \frac{\text{side adjacent to the angle}}{\text{hypotenuse}}$$

$$\cos A = \frac{AB}{AC}$$

$$\cos C = \frac{BC}{AC}$$

The abbreviation 'cos' is usually used for cosine.

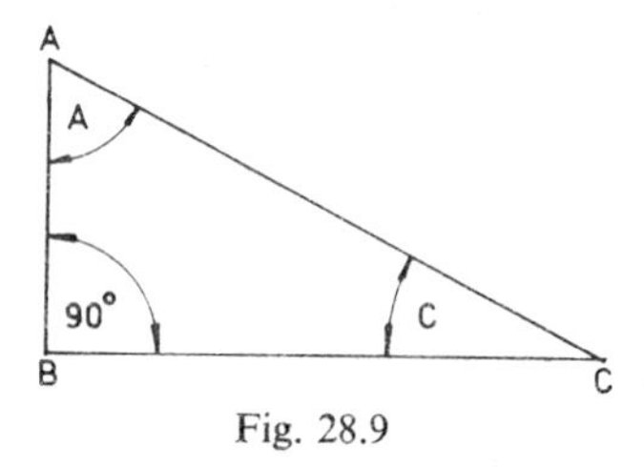

Fig. 28.9

The cosine of an angle may be found by drawing, the construction being similar to that used for the sine of an angle. However, tables of cosines are available and these are used in a similar way to the table of sines except that the mean differences are now *subtracted*.

Examples

1) Find the length of the side BC in Fig. 28.10.
BC is the side adjacent to $\angle BCA$ and AC is the hypotenuse.

$$\therefore \quad \frac{BC}{AC} = \cos 38°$$

$$BC = AC \times \cos 38° = 120 \times 0{\cdot}7880$$
$$= 94{\cdot}56 \text{ mm}$$

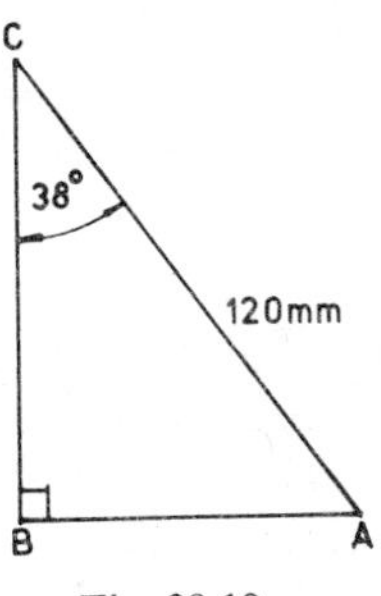

Fig. 28.10

2) Find the length of the side AC in Fig. 28.11.

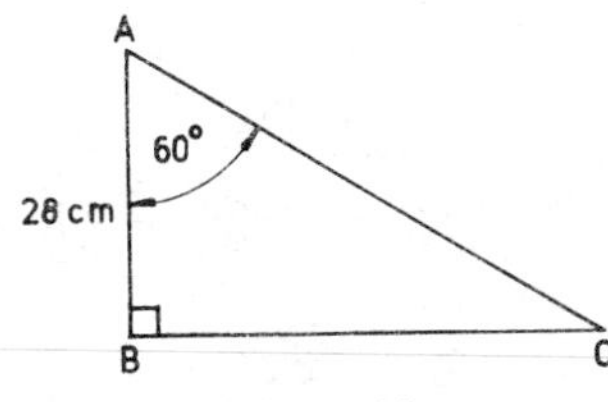

Fig. 28.11

AB is the side adjacent to $\angle$BAC and AC is the hypotenuse.

$$\therefore \frac{AB}{AC} = \cos 60°$$

$$AC = \frac{AB}{\cos 60°} = \frac{28}{0{\cdot}5000} = 56 \text{ cm}$$

3) Find the angle θ shown in Fig. 28.12. Since $\triangle$ABC is isosceles the perpendicular AD bisects the base BC and hence BD=15 mm.

$$\cos \theta = \frac{BD}{AB} = \frac{15}{50} = 0{\cdot}3$$

$$\theta = 72° \ 32'$$

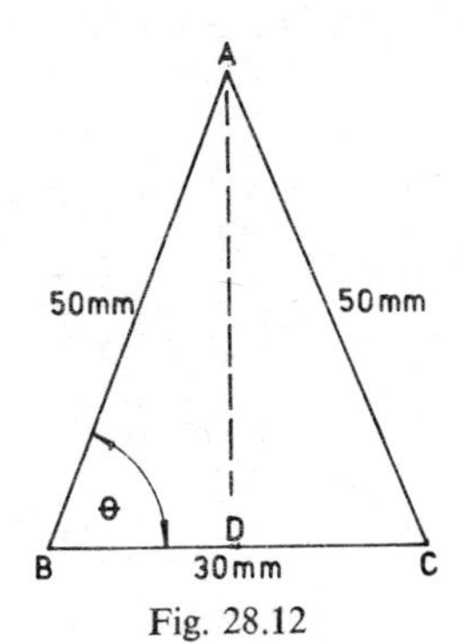

Fig. 28.12

Exercise 101

1) Use the tables to write down the values of:

a) cos 15° b) cos 24° 18′
c) cos 78° 24′ d) cos 0° 11′
e) cos 73° 22′ f) cos 39° 59′

2) Use the tables to write down the angles whose cosines are:

a) 0·9135 b) 0·3420 c) 0·9673
d) 0·4289 e) 0·9586 f) 0·0084
g) 0·2611 h) 0·4700

3) Find the lengths of the sides marked x in Fig. 28.13, the triangles being right-angled.

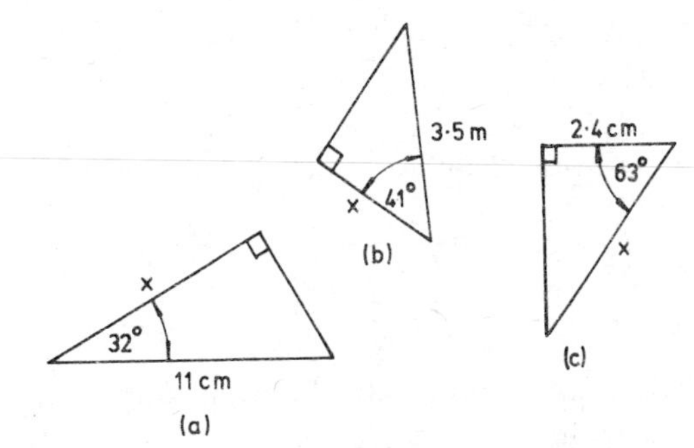

Fig. 28.13

4) Find the angles marked θ in Fig. 28.14, the triangles being right-angled.

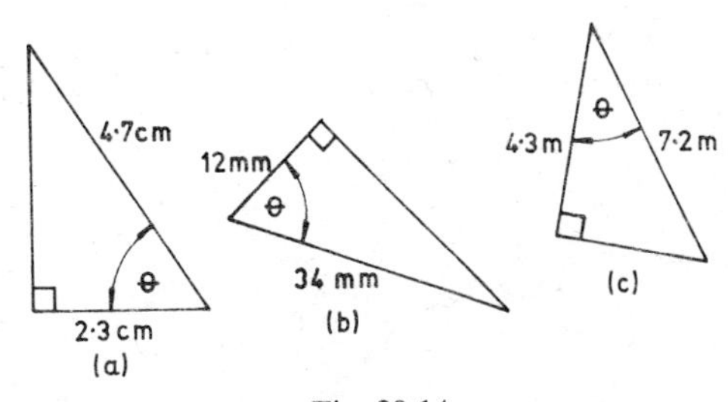

Fig. 28.14

5) An isosceles triangle has a base of 3·4 cm and the equal sides are each 4·2 cm long. Find the angles of the triangle and also its altitude.

6) In $\triangle$ABC, $\angle$C=90°, $\angle$B=33° 27′ and BC=2·4 cm. Find AB.

7) In $\triangle ABC$, $\angle B=90°$, $\angle A=62° 45'$ and AC=4·3 cm. Find AB.

8) In Fig. 28·15, calculate $\angle BAC$ and the length BC.

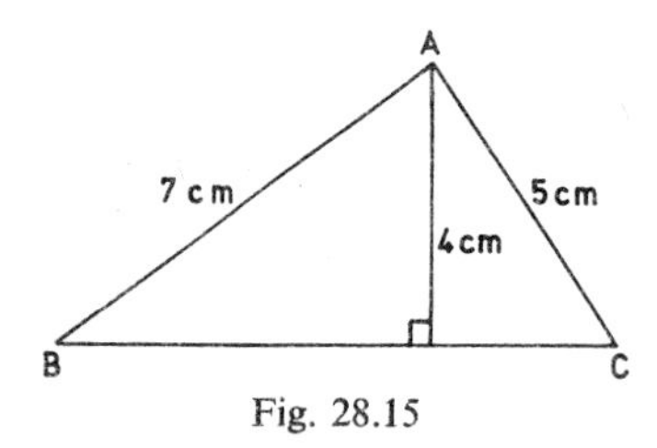

Fig. 28.15

9) In Fig. 28.16 calculate BD, AD, AC and BC.

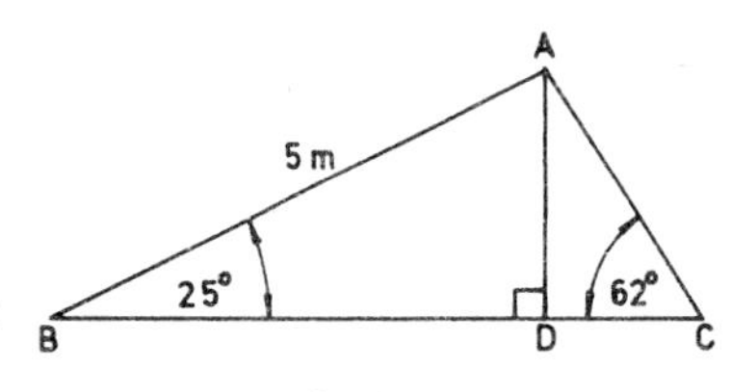

Fig. 28.16

10) Lay out accurately the following angles by using their cosines:
a) 39° b) 70° 6′ c) 18° 11′

THE TANGENT OF AN ANGLE

In any right-angled triangle (Fig. 28.17),

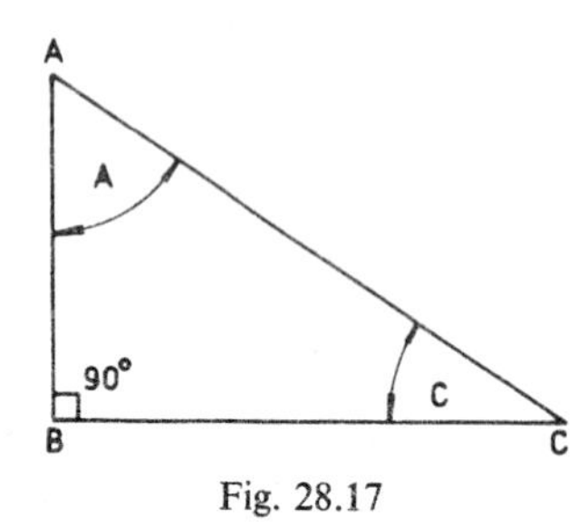

Fig. 28.17

the tangent of an angle

$$=\frac{\text{side opposite to the angle}}{\text{side adjacent to the angle}}$$

$$\tan A=\frac{BC}{AB}$$

$$\tan C=\frac{AB}{BC}$$

The abbreviation 'tan' is usually used for tangent. From the table of tangents, the tangents of angles from 0° to 90° can be read directly. For example,

$$\tan 37°=0{\cdot}7536$$

and

$$\tan 62° 29'=1{\cdot}9196$$

Examples

1) Find the length of the side AB in Fig. 28.18

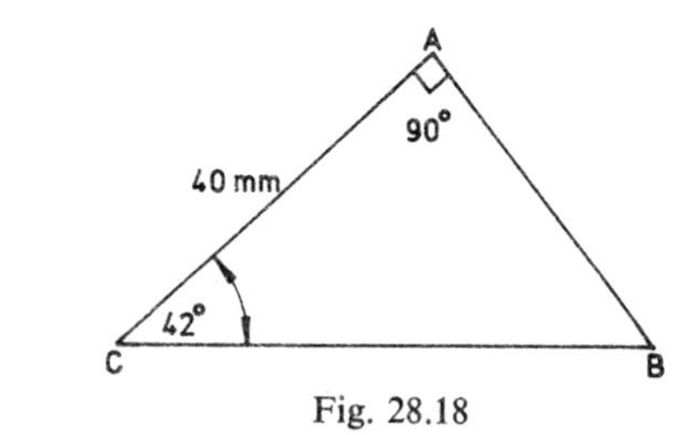

Fig. 28.18

AB is the side opposite $\angle C$ and AC is the side adjacent to $\angle C$. Hence,

$$\frac{AB}{AC}=\tan \angle C$$

$$\frac{AB}{AC}=\tan 42°$$

$$AB=AC\times\tan 42°=40\times 0{\cdot}9004=36{\cdot}02 \text{ mm}$$

2) Find the length of the side BC in Fig. 28.19.

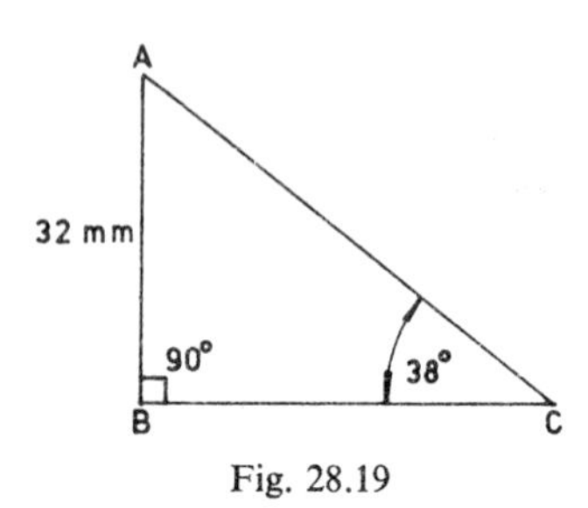

Fig. 28.19

There are two ways of doing this problem.

i) $\dfrac{AB}{BC} = \tan 38°$ or $BC = \dfrac{AB}{\tan 38°}$

$\therefore \; BC = \dfrac{32}{0{\cdot}7813} = 40{\cdot}96$ mm

ii) Since $\angle C = 38°$, $A = 90° - 38° = 52°$

now

$\dfrac{BC}{AB} = \tan A$ or $BC = AB \times \tan A$

$\therefore \; BC = 32 \times 1{\cdot}280 = 40{\cdot}96$ mm

Both methods produce the same answer but method ii) is better because it is quicker and more convenient to multiply than divide. Whenever possible the ratio should be arranged so that quantity to be found is the numerator of the ratio.

Exercise 102

1) Use tables to write down the values of:

a) tan 18° b) tan 32° 24′
c) tan 53° 42′ d) tan 39° 27′
e) tan 11° 20′ f) tan 69° 23′

2) Use tables to write down the angles whose tangents are:

a) 0·4452 b) 3·2709 c) 0·0769
d) 0·3977 e) 0·3568 f) 0·8263
g) 1·9251 h) 0·0163

3) Find the lengths of the sides marked y in Fig. 28.20, the triangles being right-angled.

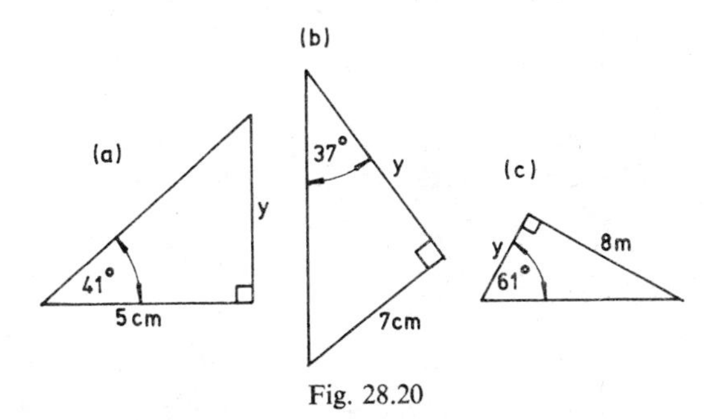

Fig. 28.20

4) Find the angles marked α in Fig. 28.21, the triangles being right-angled.

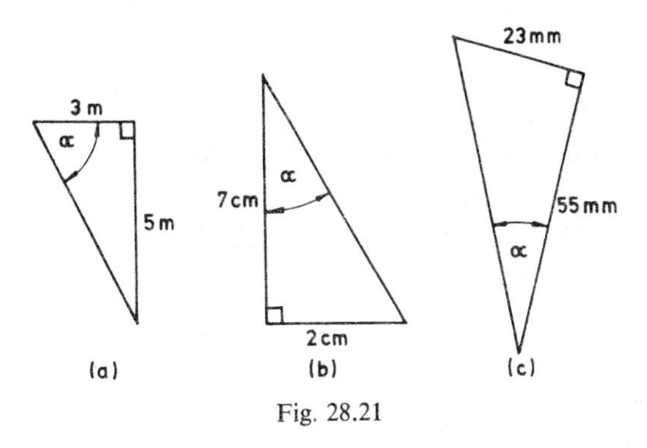

Fig. 28.21

5) An isosceles triangle has a base 10 cm long and the two equal angles are each 57°. Calculate the altitude of the triangle.

6) In $\triangle ABC$, $\angle B = 90°$, $\angle C = 49°$ and $AB = 3{\cdot}2$ cm. Find BC.

7) In $\triangle ABC$, $\angle A = 12°\,23'$, $\angle B = 90°$ and $BC = 7{\cdot}31$ cm. Find AB.

8) Calculate the distance x in Fig. 28.22.

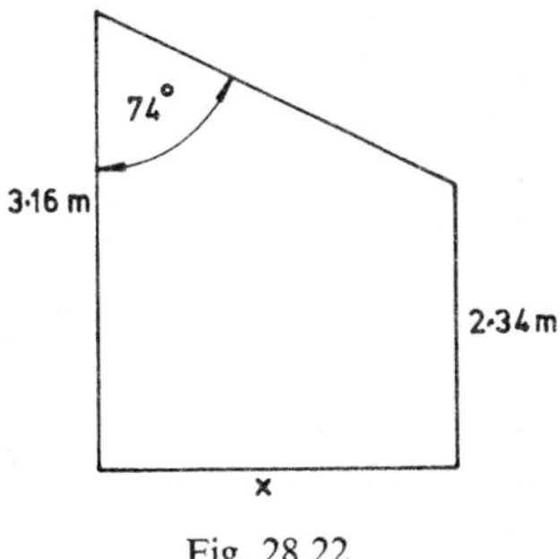

Fig. 28.22

9) Calculate the distance d in Fig. 28.23.

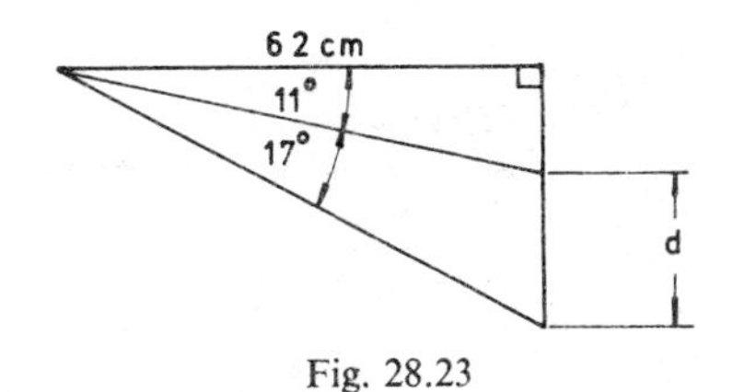

Fig. 28.23

10) Lay out accurately the following angles using their tangents:

a) 18° 16′ b) 37° 11′ c) 68° 19′

LOGARITHMS OF THE TRIGONOMETRICAL RATIOS

Tables are used to find the log of a trig ratio in the same way as to find the ratio itself.

Examples

1) Find the value of $28{\cdot}25 \times \sin 39° 17'$

number	log
28·25	1·4510
sin 39° 17′	$\bar{1}{\cdot}8015$
Answer = 17·88	1·2525

2) Find the angle ∠A given that

$$\cos \angle A = \frac{20{\cdot}23}{29{\cdot}86}$$

number	log
20·23	1·3060
29·86	1·4751
cos A	$\bar{1}{\cdot}8309$

The angle ∠A is found directly from the log cos table:

∠A = 47° 21′

3) If $b = \dfrac{c \sin \angle B}{\sin \angle C}$, find b when $c = 19{\cdot}28$, ∠B = 61° and ∠C = 22° 7′.

number	log
19·28	1·2851
sin 61°	$\bar{1}{\cdot}9418$
	1·2269
sin 22° 7′	$\bar{1}{\cdot}5757$
Answer = 44·79	1·6512

Exercise 103

1) From the tables find the following:

a) log sin 28° 33′ b) log sin 74° 24′
c) log cos 8° 2′ d) log cos 24° 15′
e) log tan 44° 31′ f) log tan 7° 5′

2) From the tables find the following:

a) If log cos ∠A = $\bar{1}{\cdot}7357$ find ∠A.
b) If log sin ∠A = $\bar{1}{\cdot}5813$ find ∠A.
c) If log tan ∠B = 0·5755 find ∠B.
d) If log sin $\phi = \bar{1}{\cdot}3069$ find ϕ.
e) If log cos $\theta = \bar{1}{\cdot}2381$ find θ.
f) If log tan $\alpha = 1{\cdot}5569$ find α.

3) By using logs find the following:

a) If $\cos \angle A = \dfrac{19{\cdot}26}{27{\cdot}58}$ find ∠A.

b) If $\sin \angle B = \dfrac{11{\cdot}23}{35{\cdot}35}$ find ∠B.

c) If $\tan \theta = \dfrac{28{\cdot}13}{17{\cdot}57}$ find θ.

4) If $a = \dfrac{b \sin \angle A}{\sin \angle B}$, find, by using logs, the value of a when $b = 8{\cdot}16$ cm, ∠A = 43° 27′ and ∠B = 37° 11′.

5) If $\cos A = \dfrac{b^2 + c^2 - a^2}{2bc}$, find, by using logs, the value of ∠A when $b = 11{\cdot}23$ cm, $c = 9{\cdot}16$ cm and $a = 8{\cdot}23$ cm.

6) If $\sin \angle C = \dfrac{c \sin \angle B}{b}$ find, by using logs, the value of C when $c = 0{\cdot}323$, ∠B = 29° 8′ and $b = 0{\cdot}517$.

7) If $b^2 = a^2 + c^2 - 2ac \cos \angle B$ find the value of b when $a = 11{\cdot}36$ cm, $c = 8{\cdot}26$ cm and ∠B = 29° 25′.

TRIGONOMETRICAL RATIOS FOR 30°, 60° AND 45°

Ratios for 30° and 60°

Fig 28.24 shows an equilateral triangle ABC with each of the sides equal to 2 units. From

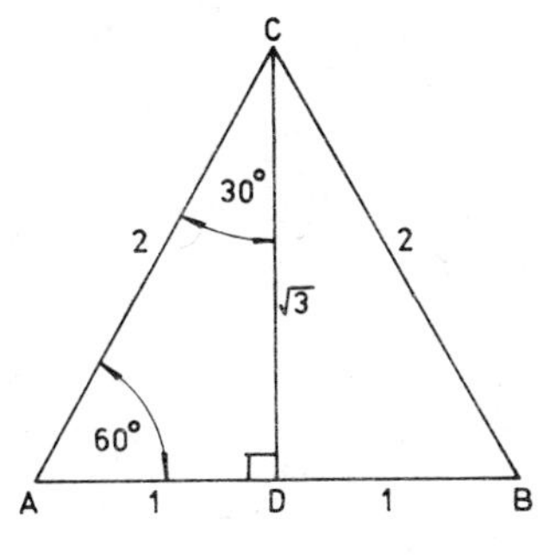

Fig. 28.24

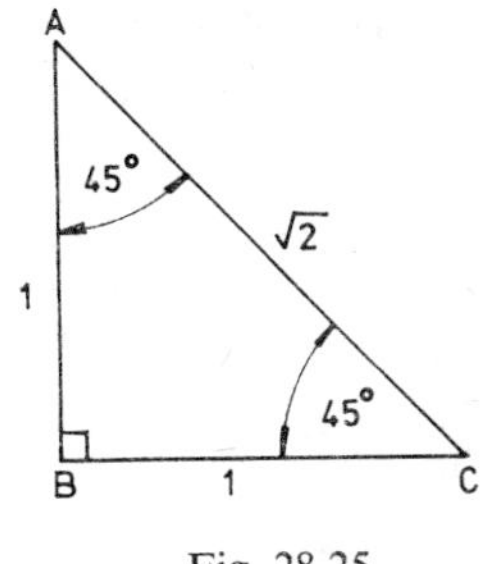

Fig. 28.25

C draw the perpendicular CD which bisects the base AB and also bisects $\angle C$.

In $\triangle ACD$,

$$CD^2 = AC^2 - AD^2 = 2^2 - 1^2 = 3$$
$$\therefore\ CD = \sqrt{3}$$

Since all the angles of $\triangle ABC$ are 60° and $\angle ACD = 30°$

$$\sin 60° = \frac{\sqrt{3}}{2} \qquad \sin 30° = \frac{1}{2}$$

$$\tan 60° = \frac{\sqrt{3}}{1} = \sqrt{3} \qquad \tan 30° = \frac{1}{\sqrt{3}} = \frac{\sqrt{3}}{3}$$

$$\cos 60° = \frac{1}{2} \qquad \cos 30° = \frac{\sqrt{3}}{2}$$

Ratios for 45°

Fig. 28.25 shows a right-angled isosceles triangle ABC with the equal sides each 1 unit in length. The equal angles are each 45°. Now

$$AC^2 = AB^2 + BC^2 = 1^2 + 1^2 = 2$$

$$AC = \sqrt{2}$$

$$\sin 45° = \frac{1}{\sqrt{2}} = \frac{\sqrt{2}}{2}$$

$$\cos 45° = \frac{1}{\sqrt{2}} = \frac{\sqrt{2}}{2}$$

$$\tan 45° = \frac{1}{1} = 1$$

GIVEN ONE RATIO TO FIND THE OTHERS

The method is shown in the following example.

Example
If $\cos \angle A = 0{\cdot}7$, find, without using tables, the values of $\sin \angle A$ and $\tan \angle A$.
In Fig. 28.26 if we make $AB = 0{\cdot}7$ units and $AC = 1$ unit, then

$$\cos \angle A = \frac{0{\cdot}7}{1} = 0{\cdot}7$$

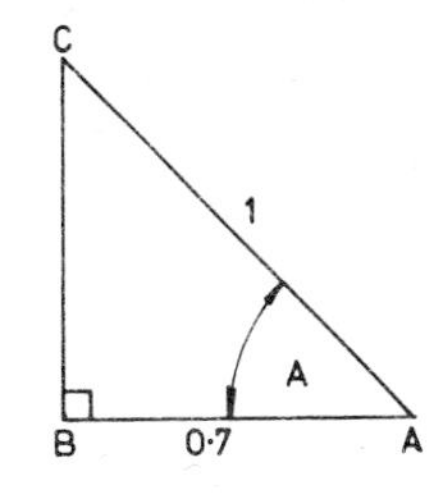

Fig. 28.26

By Pythagoras theorem,

$$BC^2 = AC^2 - AB^2 = 1^2 - 0{\cdot}7^2 = 0{\cdot}51$$

$$BC = \sqrt{0{\cdot}51} = 0{\cdot}7141$$

$$\sin \angle A = \frac{BC}{AC} = \frac{0{\cdot}7141}{1} = 0{\cdot}7141$$

$$\tan \angle A = \frac{BC}{AB} = \frac{0{\cdot}7141}{0{\cdot}7} = 1{\cdot}020$$

THE SQUARES OF THE TRIGONOMETRICAL RATIOS

The square of sin A is usually written as $\sin^2 A$. Thus

$$\sin^2 A = (\sin A)^2$$

and similarly for the remaining trigonometrical ratios. That is,

$$\cos^2 A = (\cos A)^2$$

and

$$\tan^2 A = (\tan A)^2$$

Examples

1) Find the value of $\cos^2 37°$

$\cos 37° = 0{\cdot}7986$
$\cos^2 37° = (0{\cdot}7986)^2 = 0{\cdot}6378$

2) Find the value of $\tan^2 60° + \sin^2 60°$

$\tan 60° = \sqrt{3} \quad \therefore \tan^2 60° = 3$

$\sin 60° = \dfrac{\sqrt{3}}{2} \quad \therefore \sin^2 60° = \dfrac{3}{4}$

$\tan^2 60° + \sin^2 60° = 3 + \dfrac{3}{4} = 3\tfrac{3}{4}$

It is sometimes useful to remember that

$$\sin^2 A + \cos^2 A = 1$$

which may easily be proved by considering a right-angled triangle (Fig. 28.27).

$\sin A = \dfrac{a}{b} \qquad \sin^2 A = \dfrac{a^2}{b^2}$

$\cos A = \dfrac{c}{b} \qquad \cos^2 A = \dfrac{c^2}{b^2}$

$\sin^2 A + \cos^2 A = \dfrac{a^2}{b^2} + \dfrac{c^2}{b^2} = \dfrac{a^2 + c^2}{b^2}$

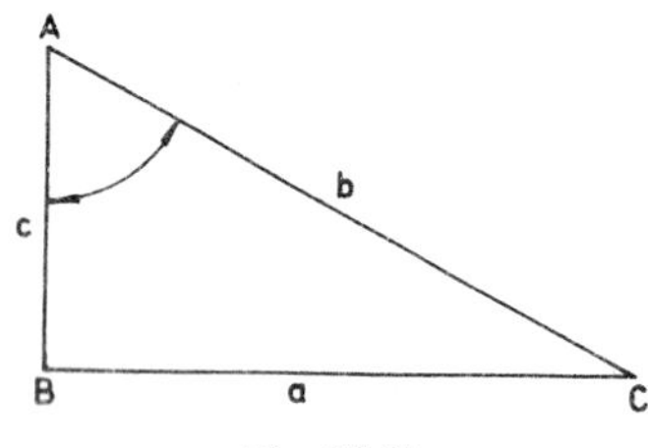

Fig. 28.27

But, by Pythagoras

$a^2 + c^2 = b^2$

$\therefore \sin^2 A + \cos^2 A = \dfrac{b^2}{b^2} = 1$

Example

The angle A is acute and $5 \sin^2 A - 2 = \cos^2 A$. Find the angle A.

Since $\sin^2 A + \cos^2 A = 1$
$\cos^2 A = 1 - \sin^2 A$

$\therefore 5 \sin^2 A - 2 = 1 - \sin^2 A$
$6 \sin^2 A - 3 = 0$
$6 \sin^2 A = 3$
$\sin^2 A = 0{\cdot}5$
$\sin A = \pm\sqrt{0{\cdot}5} = \pm 0{\cdot}7071$

Since A is acute sin A must be positive.

Hence $\sin A = 0{\cdot}7071$
$A = 45°$

Exercise 104

1) If $\sin A = 0{\cdot}3171$ find the values of cos A and tan A without using tables.

2) If $\tan A = \dfrac{3}{4}$ find the values of sin A and cos A without using tables.

3) If $\tan A = 1{\cdot}1504$ find the values of cos A and sin A without using tables.

4) If $\cos A = \dfrac{12}{13}$ find the values of sin A and tan A without using tables.

5) Show that $\cos 60° + \cos 30° = \dfrac{1+\sqrt{3}}{2}$

6) Show that $\sin 60° + \cos 30° = \sqrt{3}$

7) Show that

$$\cos 45° + \sin 60° + \sin 30° = \frac{\sqrt{2}+\sqrt{3}+1}{2}$$

8) a) Find the value of $\cos^2 30°$
b) Find the value of $\tan^2 30°$
c) Find the value of $\sin^2 60°$

9) Evaluate
a) $\cos^2 41°$ b) $\sin^2 27°$ c) $\tan^2 58°$

10) If the angle A is acute and

$2\sin^2 A - \frac{1}{3} = \cos^2 A$ find A.

11) If the angle θ is acute and

$\cos^2\theta + \frac{1}{8} = = \sin^2\theta$ find θ.

ANGLES OF ELEVATION AND DEPRESSION

The *angle of elevation* of an object B from an observer at A (Fig. 28.28) is the angle that AB makes with the horizontal AX.

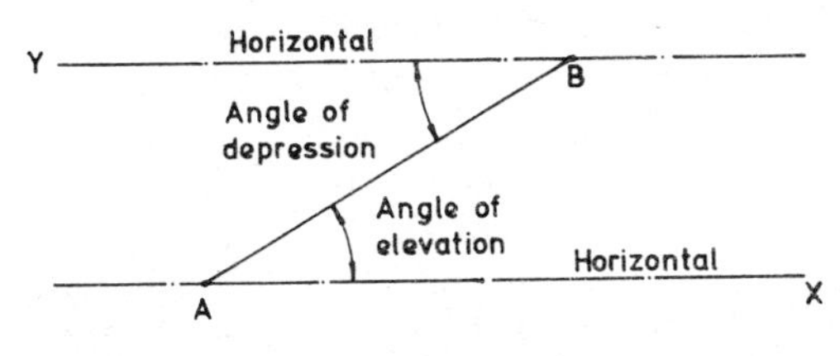

Fig. 28.28

The *angle of depression* of an object B from an observer at A (Fig. 28.28) is the angle that AB makes with the horizontal BY.
Note that both the angle of elevation and the angle of depression are measured from the *horizontal.* As shown in Fig. 28.28 the angle of elevation from A to B equals the angle of depression from B to A.

Examples
1) To find the height of a pylon a surveyor sets up his theodolite some distance from the pylon and finds the angle of elevation to the top of the pylon to be 30°.
He then moves 60 m nearer and finds the angle of elevation to be 42°. Find the height of the pylon assuming the ground to be horizontal.

The conditions of the problem are as shown in Fig. 28.29, where h is the height of the pylon.

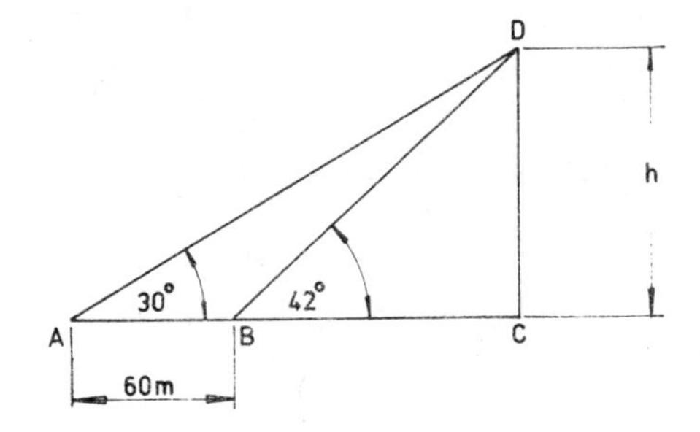

Fig. 28.29

In $\triangle$ADC,

$$\frac{h}{AC} = \tan 30°$$

$$h = AC \times \tan 30° = 0{\cdot}5774 \times AC$$

In $\triangle$DBC

$$\frac{h}{BC} = \tan 42°$$

$$h = BC \times \tan 42° = 0{\cdot}9004 \times BC$$

$$\therefore\ 0{\cdot}5774 \times AC = 0{\cdot}9004 \times BC$$

But

$$AC = AB + BC = 60 + BC$$

$$\therefore\ 0{\cdot}5774 \times (60 + BC) = 0{\cdot}9004 \times BC$$

$$34{\cdot}64 + 0{\cdot}5774 \times BC = 0{\cdot}9004 \times BC$$

$$34{\cdot}64 = 0{\cdot}323 \times BC$$

$$BC = \frac{34{\cdot}64}{0{\cdot}323} = 107{\cdot}3$$

$$h = 0{\cdot}9004 \times BC = 0{\cdot}9004 \times 107{\cdot}3 = 96{\cdot}61 \text{ m}$$

2) A man standing on top of a cliff 50 m high is in line with two buoys whose angles of depression are 18° and 20°. Calculate the distance between the two buoys.

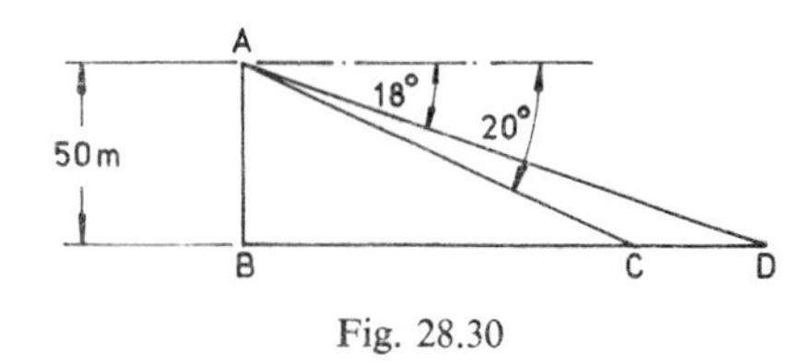

Fig. 28.30

Referring to Fig. 28.30, the positions of the buoys are C and D. In $\triangle ABC$, $\angle BAC = 70°$.

$$\frac{BC}{AB} = \tan \angle BAC \qquad BC = AB \times \tan 70° = 50 \times 2{\cdot}7475 = 137{\cdot}4 \text{ m}$$

In $\triangle ABD$, $\angle BAD = 72°$.

$$\frac{BD}{AB} = \tan \angle BAD \qquad BD = AB \times \tan 72° = 50 \times 3{\cdot}0777 = 153{\cdot}9 \text{ m}$$

$$CD = BD - BC = 153{\cdot}9 - 137{\cdot}4 = 16{\cdot}5 \text{ m}$$

Exercise 105

1) From a point, the angle of elevation of a tower is 30°. If the tower is 20 m distant from the point, what is the height of the tower?

2) A man 1·8 m tall observes the angle of elevation of a tree to be 26°. If he is standing 16 m from the tree, find the height of the tree.

3) A man 1·5 m tall is 15 m away from a tower 20 m high. What is the angle of elevation of the top of tower from his eyes?

4) A man, lying down on top of a cliff 40 m high observes the angle of depression of a buoy to be 20°. If he is in line with the buoy, calculate the distance between the buoy and the foot of the cliff (which may be assumed to be vertical).

5) To find the height of a tower a surveyor stands some distance away from its base and he observes the angle of elevation to the top of the tower to be 45°. He then moves 80 m nearer to the tower and he then finds the angle of elevation to be 60°. Find the height of the tower.

6) A tower is known to be 60 m high. A man using a theodolite stands some distance away from the tower and measures its angle of elevation as 38°. How far away from the tower is he if the theodolite stands 1·5 m above the ground. If the man now moves 80 m further away from the tower what is now the angle of elevation of the tower?

7) A man standing 20 m away from a tower observes the angles of elevation to the top and bottom of a flagstaff standing on the tower as 62° and 60° respectively. Calculate the height of the flagstaff.

8) A surveyor stands 100 km from the base of a tower on which an aerial stands. He measures the angles of elevation to the top and bottom of the aerial as 58° and 56°. Find the height of the aerial.

9) Two points A and B on ground sloping at 10° to the horizontal are in line with Q, the foot of a vertical mast PQ, as shown in Fig. 28.31. If AB = BQ = 20 m and PQ = 25 m calculate

i) the height of P above the level of B;
ii) the distance BP;
iii) the angle $\angle PAB$.

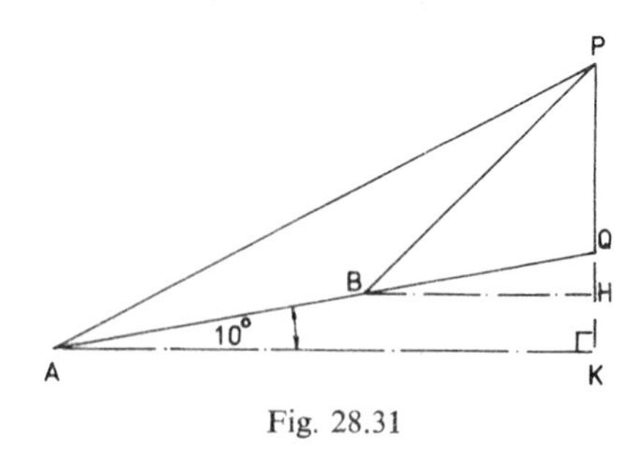

Fig. 28.31

10) A man standing on top of a cliff 80 m high is in line with two buoys whose angles of depression are 17° and 21°. Calculate the distance between the buoys.

BEARINGS

The four cardinal directions are North, South, East and West (Fig. 28.32). The directions NE, NW, SE and SW are frequently used and are as shown in Fig. 28.32.
A bearing of N20°E means an angle of 20° measured from the N towards E as shown in Fig. 28.33. Similarly a bearing of S40°E means an angle 40° measured from the S towards E (Fig. 28.34). A bearing of N50°W means an angle of 50° measured from N towards W (Fig. 28.35). *Bearings quoted in this way are always measured from N and S and never from E and W.*

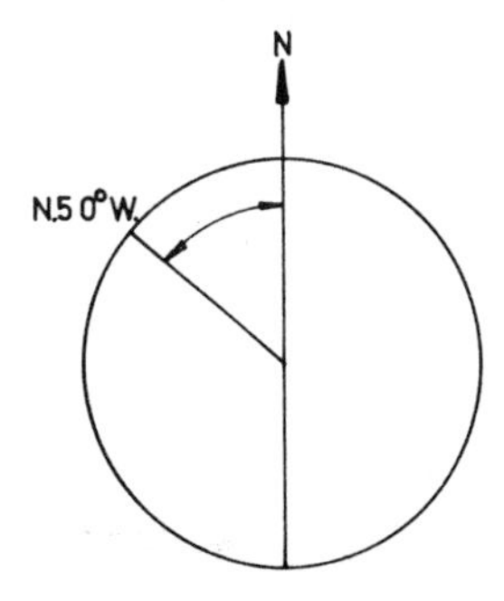

Fig. 28.35

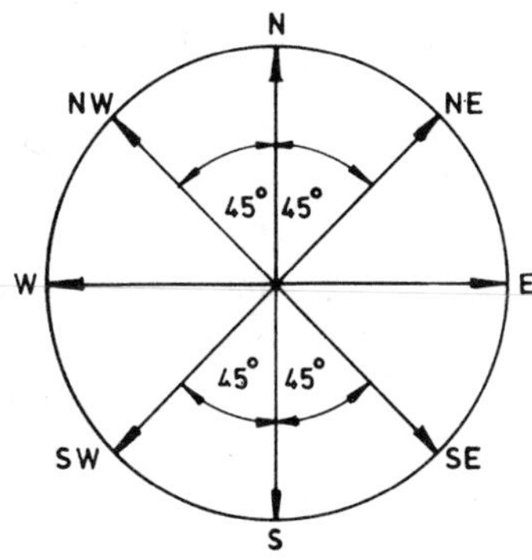

Fig. 28.32

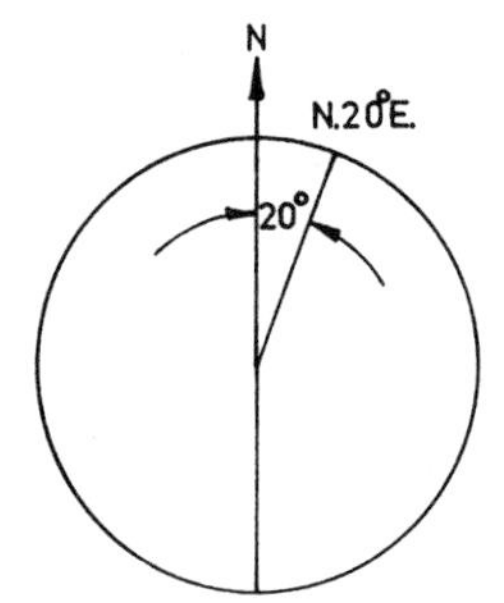

Fig. 28.33

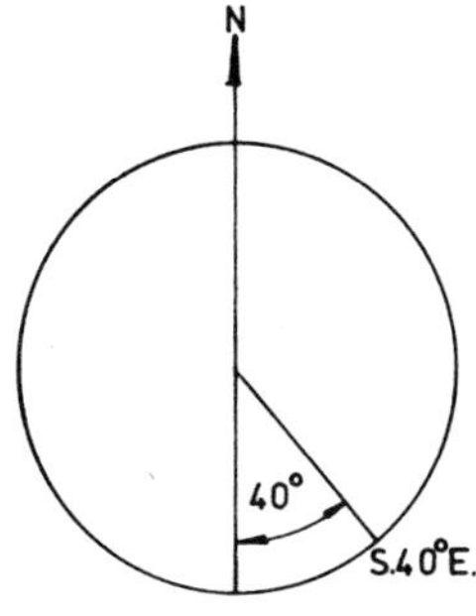

Fig. 28.34

There is a second way of stating a bearing. The angle denoting the bearing is measured from N in a *clockwise* direction, N being reckoned as 0°. Three figures are always stated, for example 005° is written instead of 5°, 035° for 35° etc. East will be 090°, South 180° and West 270°. Some typical bearings are shown in Fig. 28.36.

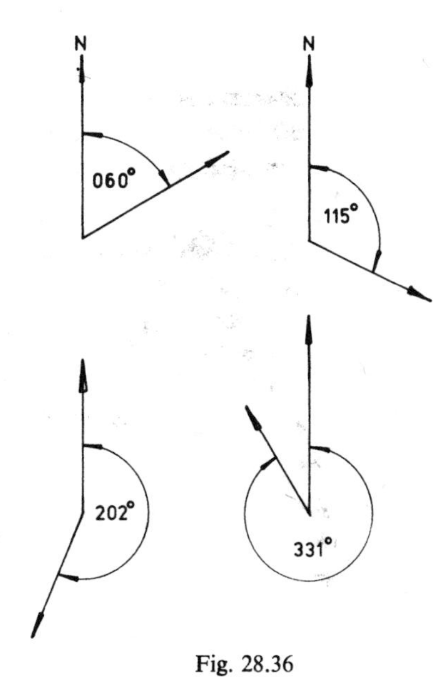

Fig. 28.36

Examples

1) B is a point due east of a point A on the coast. C is another point on the coast and is 6 km due south of A. The distance BC is

1

7 km. Calculate the bearing of C from B.
Referring to Fig. 28.37,

$$\sin \angle B = \frac{AC}{BC} = \frac{6}{7} = 0{\cdot}8572$$

$$\angle B = 59^\circ$$

The bearing of C from B $= 270^\circ - 59^\circ = 211^\circ$

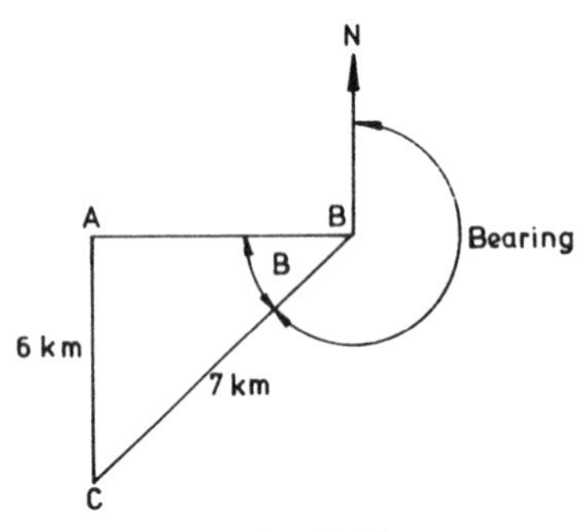

Fig. 28.37

2) B is 5 km due north of P and C is 2 km due east of P. A ship started from C and steamed in a direction N30°E. Calculate the distance the ship had to go before it was due east of B. Find also the distance it is then from B.

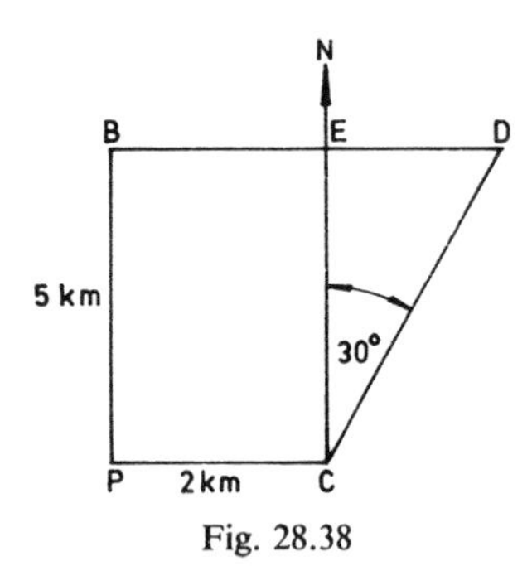

Fig. 28.38

Referring to Fig. 28.38, the ship will be due east of B when it has sailed the distance CD. The bearing N30°E makes the angle $\angle ECD$ equal to 30°. In $\triangle CED$, EC = 5 km and $\angle ECD = 30^\circ$. Hence

$$\frac{EC}{CD} = \cos 30^\circ$$

$$CD = \frac{EC}{\cos 30^\circ} = \frac{5}{\cos 30^\circ} = 5{\cdot}77 \text{ km}$$

Hence the ship had to sail 5·77 km to become due east of B.

$$\frac{ED}{CD} = \sin 30^\circ$$

$$ED = CD \times \sin 30^\circ = 5{\cdot}77 \times 0{\cdot}5000 = 2{\cdot}885 \text{ km}$$

$$BD = BE + ED = 2 + 2{\cdot}885 = 4{\cdot}885 \text{ km}$$

Hence the ship will be 4·885 km due east of B.

3) A boat sails 5 km from a port P on a bearing of 075° and then sails 6 km on a bearing of 050°. Find the distance that the boat is from P and find also its bearing from P.
In Fig. 28.39, PA and AB represent the two courses. Taking north and east as the two reference axes, the problem is to find PB and the angle $\angle BPN$.

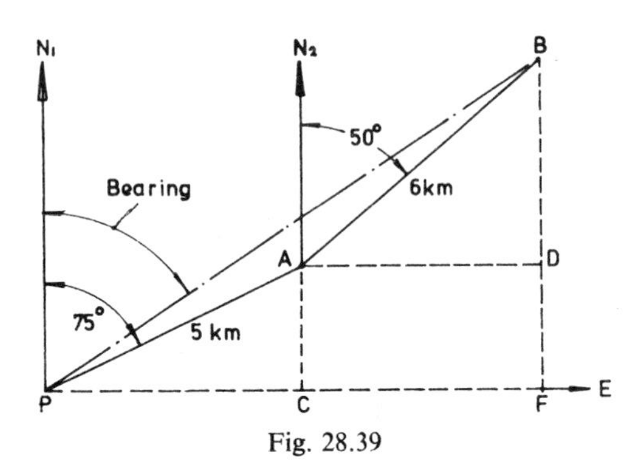

Fig. 28.39

Drawing $\triangle$s PAC and ABD as shown, in $\triangle PAC$, $\angle APC = 15^\circ$ and AP = 5 km. Hence

$$\frac{PC}{AP} = \cos \angle APC$$

$$PC = AP \times \cos \angle APC = 5 \times \cos 15^\circ = 4{\cdot}830 \text{ km}$$

$$\frac{AC}{AP} = \sin \angle APC$$

$$AC = AP \times \sin \angle APC = 5 \times \sin 15^\circ = 1{\cdot}294 \text{ km}$$

in $\triangle ABD$, $\angle BAD = 40°$ and $AB = 6$ km. Hence

$$\frac{AD}{AB} = \cos \angle BAD$$

$$AD = AB \times \cos \angle BAD = 6 \times \cos 40°$$
$$= 4{\cdot}596 \text{ km}$$

$$\frac{BD}{AB} = \sin \angle BAD$$

$$BD = AB \times \sin \angle BAD = 6 \times \sin 40°$$
$$= 3{\cdot}857 \text{ km}$$

In $\triangle PBF$,

$$BF = AC + BD = 1{\cdot}294 + 3{\cdot}857 = 5{\cdot}151 \text{ km}$$

$$PF = PC + CF = PC + AD = 4{\cdot}830 + 4{\cdot}596$$
$$= 9{\cdot}426$$

$$\tan \angle BPF = \frac{5{\cdot}151}{9{\cdot}426}$$

$$\therefore \angle BPF = 28° \, 39'$$

Hence the bearing of B from P is 061° 21′

$$PB^2 = PF^2 + BF^2 \text{ (Pythagoras)}$$
$$= 9{\cdot}426^2 + 5{\cdot}151^2$$
$$\therefore PB = 10{\cdot}74 \text{ km}$$

Exercise 106

1) A ship sets out from a point A and sails due north to a point B, a distance of 120 km. It then sails due east to a point C. If the bearing of C from A is 037° 40′ calculate

a) the distance BC;
b) the distance AC.

2) A boat leaves a harbour A on a course of S60°E and it sails 50 km in this direction until it reaches a point B. How far is B east of A? What distance south of A is B?

3) X is a point due west of a point P. Y is a point due south of P. If the distances PX and PY are 10 km and 15 km respectively, calculate the bearing of X from Y.

4) B is 10 km north of P and C is 5 km due west of P. A ship starts from C and sails in a direction of 330°. Calculate the distance the ship has to sail before it is due west of B and find also the distance it is then from B.

5) A fishing boat places a float on the sea at A, 50 metres due north of a buoy B. A second boat places a float at C, whose bearing from A is S30°E. A taut net connecting the floats at A and C is 80 metres long. Calculate the distance BC and the bearing of of C from B.

6) An aircraft starts to fly from A to B a distance of 140 km, B being due north of A. The aircraft flies on course 18°E of N for a distance of 80 km. Calculate how far the aircraft is then from the line AB and in what direction it should then fly to reach B.

7) X and Y are two lighthouses, Y being 20 km due east of X. From a ship due south of X, the bearing of Y was 055°. Find a) the distance of the ship from Y; b) the distance of the ship from X.

8) An aircraft flies 50 km from an aerodrome A on a bearing of 065° and then flies 80 km on a bearing of 040°. Find the distance of the aircraft from A and also its bearing from A.

9) A boat sails 10 km from a harbour H on a bearing of S30°E. It then sails 15 km on a bearing of N20°E. How far is the boat from H? What is its bearing from H?

10) Three towns A, B and C lie on a straight road running east from A. B is 6 km from A and C is 22 km from A. Another town D lies to the north of this road and it lies 10 km from both B and C. Calculate the distance of D from A and the bearing of D from A.

TRIGONOMETRY AND THE CIRCLE

Many problems involving the circle can be solved by trigonometry. The geometrical theorems needed are as follows:

1) The angle between a tangent and a radius drawn to the point of tangency is 90°.

2) The angle in a semi-circle is 90°.

3) The straight line joining the centres of two circles in contact passes through the point of contact.

Examples

1) Fig. 28.40 shows an end view of two cylinders (of radius 5 cm and 3 cm respectively) in a V shaped slot. The cylinders touch each other and the sides of the slot. Calculate $\angle XYZ$ and the depth YT of the slot.

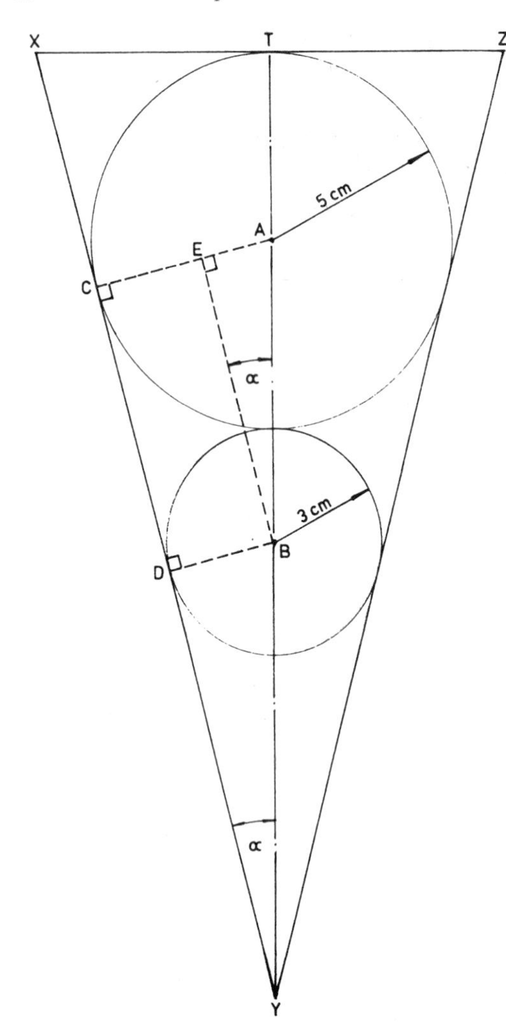

Fig. 28.40

A and B are the centres of the circle.

$$\therefore \quad AB = 5 + 3 = 8 \text{ cm}$$

C and D are the points of tangency.
Join AC and BD and draw BE parallel to XY.

$$\therefore \quad EA = 5 - 3 = 2 \text{ cm}$$

$$\angle AEB = 90^\circ$$

In $\triangle EAB$, $\dfrac{EA}{AB} = \sin\alpha$

$$\therefore \quad \sin\alpha = \frac{2}{8} = 0{\cdot}25$$

$$\alpha = 14^\circ\ 29'$$

$$\angle XYZ = 2\alpha = 2 \times 14^\circ\ 29' = 28^\circ\ 58'$$

In $\triangle DBY$,

$$\frac{DB}{BY} = \sin\alpha$$

$$BY = \frac{DB}{\sin\alpha} = \frac{3}{0{\cdot}25} = 12 \text{ cm}$$

$$TY = TA + AB + BY = 5 + 8 + 12 = 25 \text{ cm}$$

2) Fig. 28.41 shows a dovetail. Two cylinders each 4 cm diameter are held against the dovetail as shown in the diagram. Find the distance M.

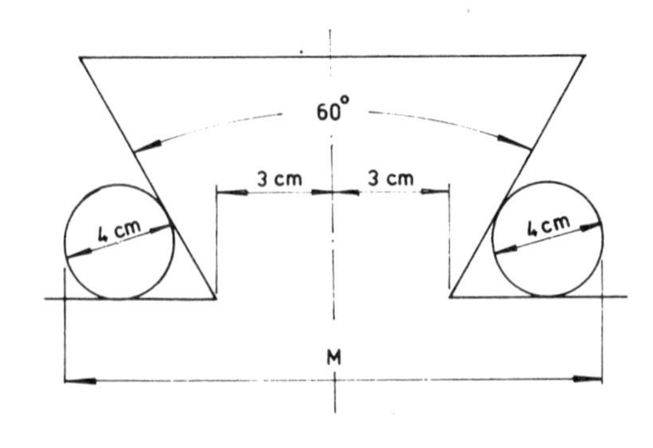

Fig. 28.41

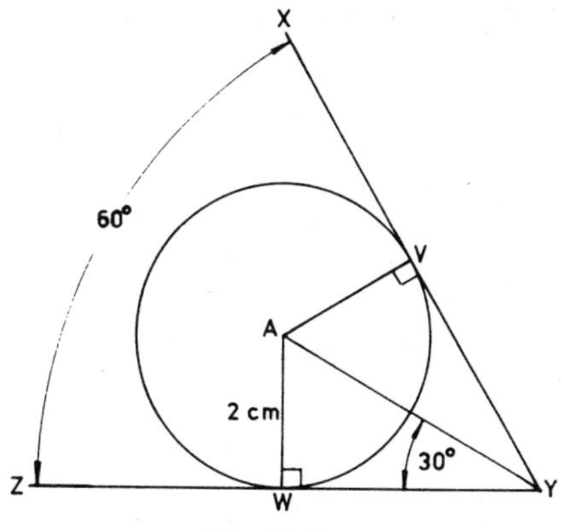

Fig. 28.42

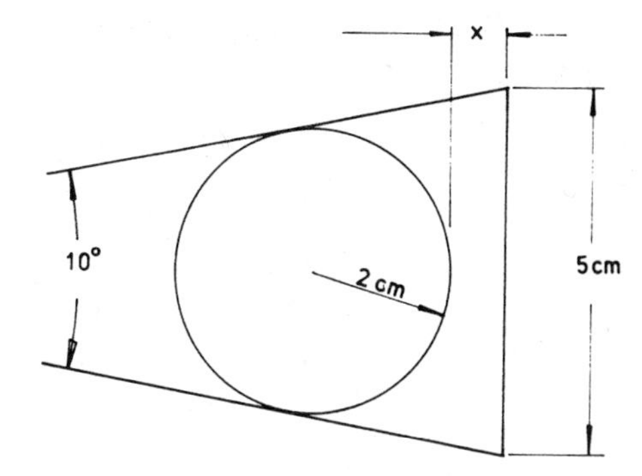

Fig. 28.43

Referring to Fig. 28.42 we need to find WY. The circle touches XY at V and ZY at W.

Since

$$VY = WY \quad \text{(both tangents)}$$

and

$$\angle AWY = \angle AVY = 90^\circ \quad \text{(angle between a radius and a tangent)}$$

$\angle WYV$ is bisected

$$\therefore \angle AYW = 30^\circ$$

$$\frac{AW}{WY} = \tan 30^\circ$$

$$WY = \frac{AW}{\tan 30^\circ} = \frac{2}{0{\cdot}5774} = 3{\cdot}464 \text{ cm}$$

$$\therefore M = 3+3+3{\cdot}464+3{\cdot}464+2+2$$
$$= 16{\cdot}928 \text{ cm}$$

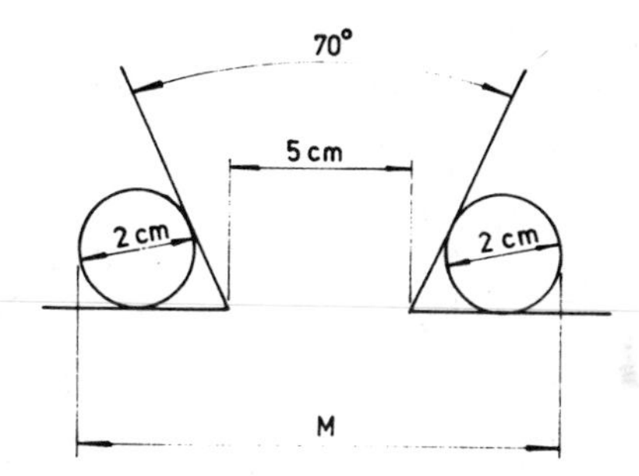

Fig. 28.44

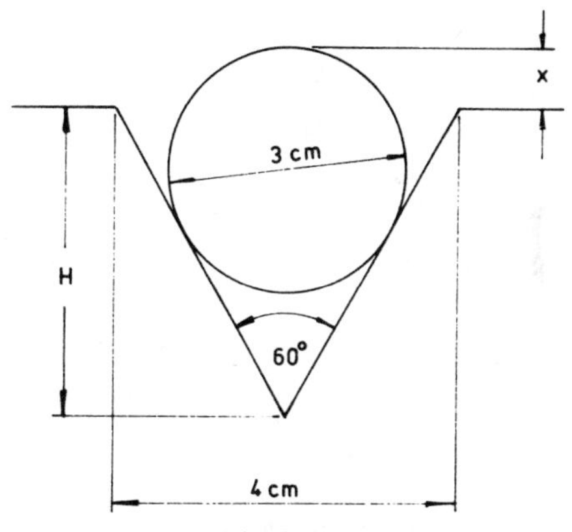

Fig. 28.45

Exercise 107

1) In Fig. 28.43 find the distance x.

2) In Fig. 28.44 find the distance M.

3) In Fig. 28.45 find the distances x and H

4) In Fig. 28.46 find the distance L.

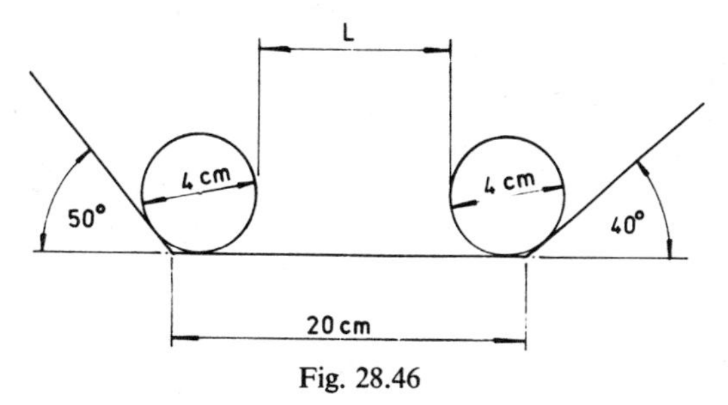

Fig. 28.46

5) In Fig. 28.47 find the angle α and the distance x.

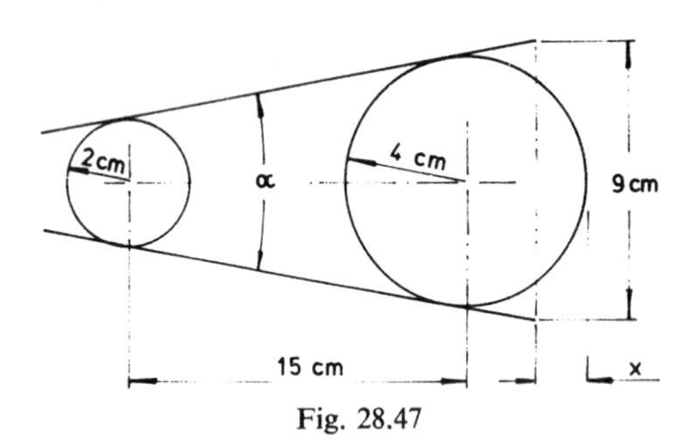

Fig. 28.47

6) Fig. 28.48 consists of two circular arcs joined by straight lines. Find the perimeter of the figure.

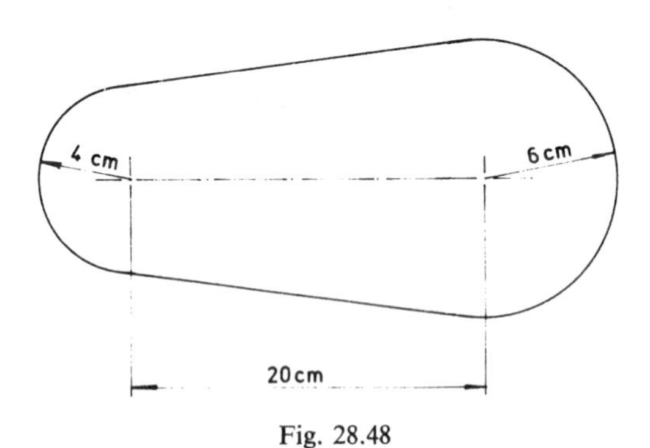

Fig. 28.48

SELF TEST 26

1) In Fig. 28.49, sin x is equal to

a $\frac{h}{p}$ b $\frac{h}{m}$ c $\frac{m}{p}$ d $\frac{p}{h}$

2) In Fig. 28.49, cos x is equal to

a $\frac{h}{p}$ b $\frac{h}{m}$ c $\frac{m}{p}$ d $\frac{p}{h}$

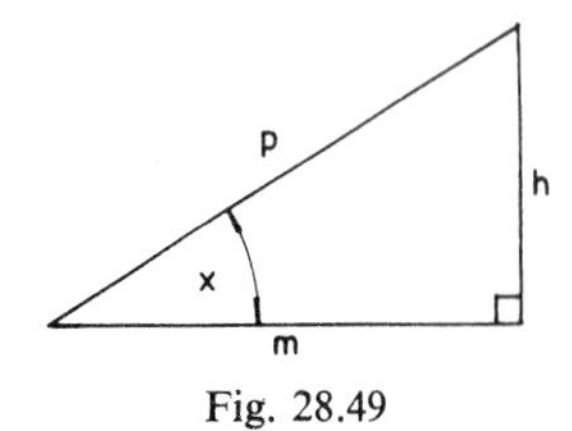

Fig. 28.49

3) In Fig. 28.49, tan x is equal to

a $\frac{h}{p}$ b $\frac{h}{m}$ c $\frac{m}{p}$ d $\frac{p}{h}$

4) In Fig. 28.50, sin A is equal to

a $\frac{a}{b}$ b $\frac{a}{c}$ c $\frac{b}{c}$ d $\frac{c}{a}$

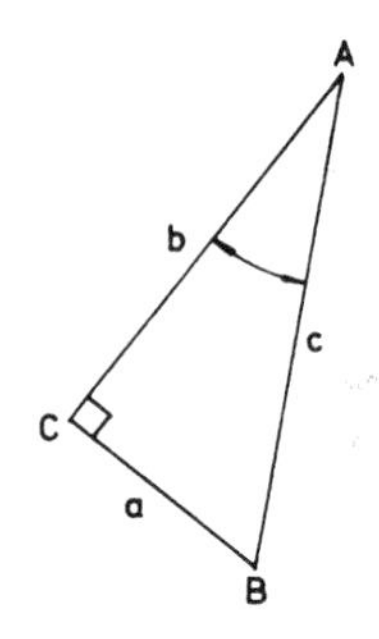

Fig. 28.50

5) In Fig. 28.51, tan x is equal to

a $\frac{q}{p}$ b $\frac{q}{r}$ c $\frac{p}{q}$ d $\frac{r}{q}$

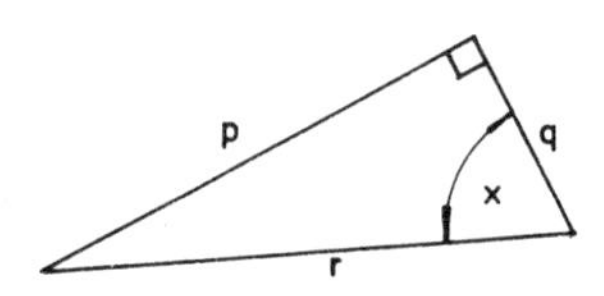

Fig. 28.51

6) In Fig. 28.52, cos y is equal to

a $\frac{s}{t}$ b $\frac{r}{s}$ c $\frac{s}{r}$ d $\frac{t}{s}$

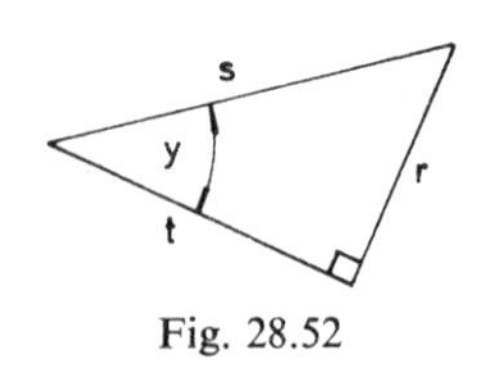

Fig. 28.52

7) The expression for the length AB (Fig. 28.53) is

a $40 \tan 50°$ b $40 \sin 50°$

c $\frac{40}{\tan 50°}$ d $\frac{40}{\sin 50°}$

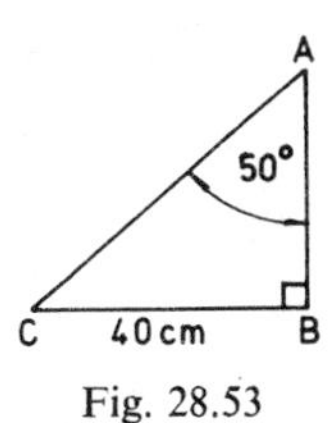

Fig. 28.53

8) The expression for the length AC Fig. 28.53) is

a $40 \sin 50$ b $40 \cos 50°$

c $\dfrac{40}{\sin 50°}$ d $\dfrac{40}{\cos 50°}$

9) In Fig. 28.54, the expression for the side RN is

a $(2 \sin 30° + 3 \sin 60°)$ cm
b $(2 \sin 30° + 3 \cos 60°)$ cm
c $(2 \cos 30° + 3 \cos 60°)$ cm
d $(2 \cos 30° + 3 \sin 60°)$ cm

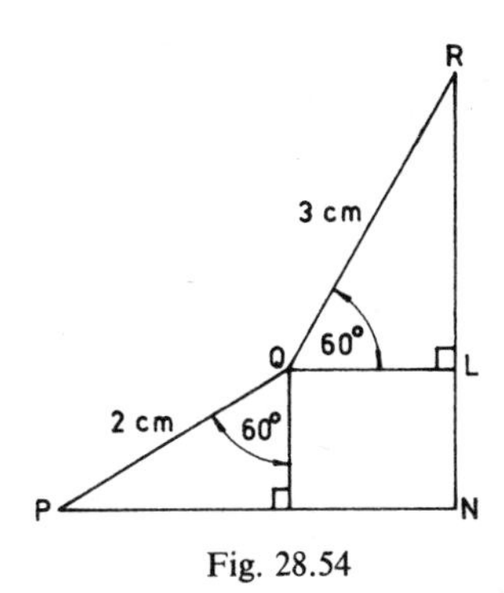

Fig. 28.54

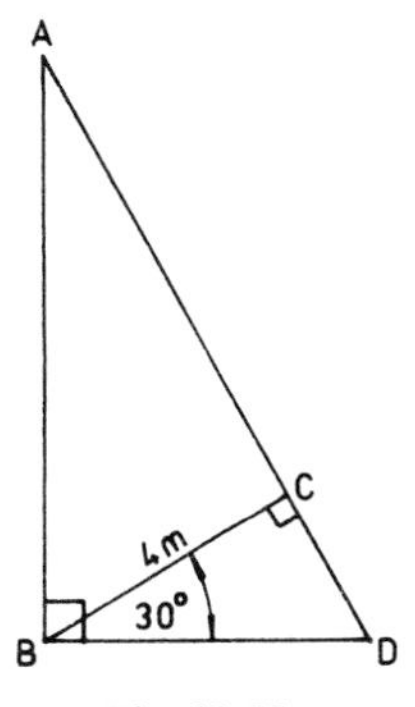

Fig. 28.55

10) In Fig. 28.55, and expression for the length AD is

a $\dfrac{16\sqrt{3}}{3}$ b $\dfrac{16}{\sqrt{3}}$

c $8\sqrt{3}$ d $\dfrac{40\sqrt{3}}{3}$

11) In Fig. 28.56, an expression for the angle θ is

a $\tan \theta = \dfrac{30 - 10\sqrt{3}}{40}$

b $\tan \theta = \dfrac{40}{30 - 10\sqrt{3}}$

c $\tan \theta = \dfrac{20}{50 - 10\sqrt{3}}$

d $\tan \theta = \dfrac{50 - 10\sqrt{3}}{20}$

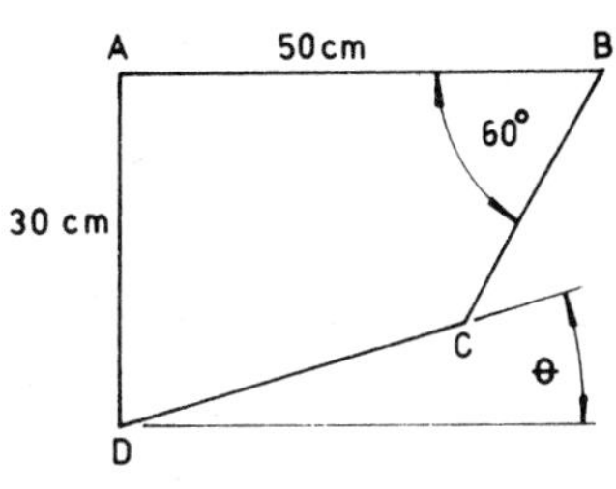

Fig. 28.56

12) The area of the parallelogram PQRS (Fig. 28.57) is

a $30 \sin 70°\ cm^2$ b $30 \cos 70°\ cm^2$
c $60 \sin 70°\ cm^2$ d $60 \cos 70°\ cm^2$

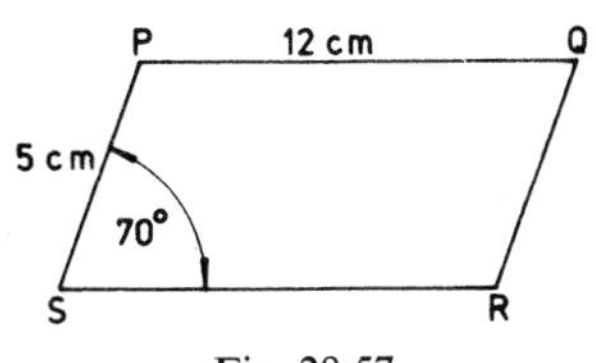

Fig. 28.57

13) In Fig. 28.58, PT and PQ are tangents to the circle whose radius is 2 cm. Hence ∠TPQ is

a 30° b 45° c 60° d 90°

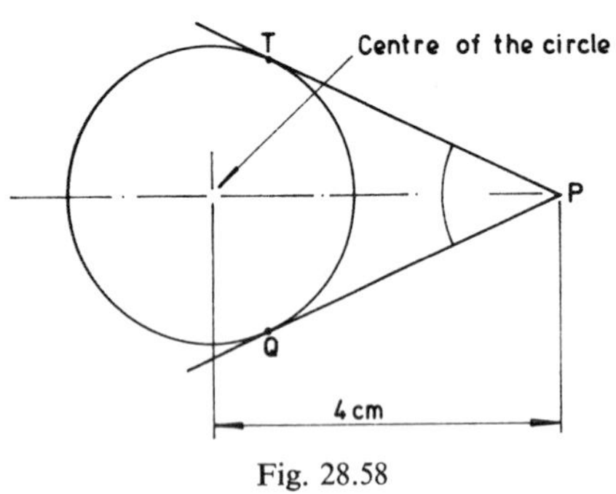

Fig. 28.58

14) log cos 51° 34′ is

a 0·7935 b 0·7947 c $\bar{1}$·7935
d $\bar{1}$·7947

15) A bearing of S40°E can be expressed as a bearing of

a 050° b 040° c 140°
d 130°

16) A man standing on top of a cliff 80 m high is in line with two buoys whose angles of depression are 15° and 20°. The distance between the buoys is given by the expression

a 80(tan 20° − tan 15°)

b 80(tan 75° − tan 70°)

c 80 tan 5°

d $\dfrac{80}{\tan 20° - \tan 15°}$

Chapter 29 The Sine and Cosine Rules

TRIGONOMETRICAL RATIOS BETWEEN 0° AND 360°

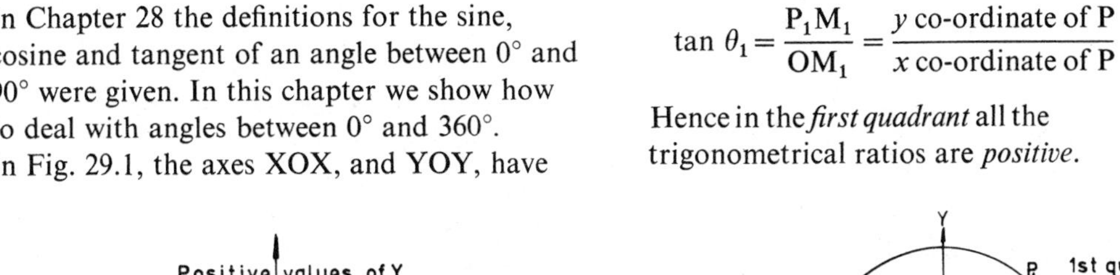

In Chapter 28 the definitions for the sine, cosine and tangent of an angle between 0° and 90° were given. In this chapter we show how to deal with angles between 0° and 360°.
In Fig. 29.1, the axes XOX_1 and YOY_1 have

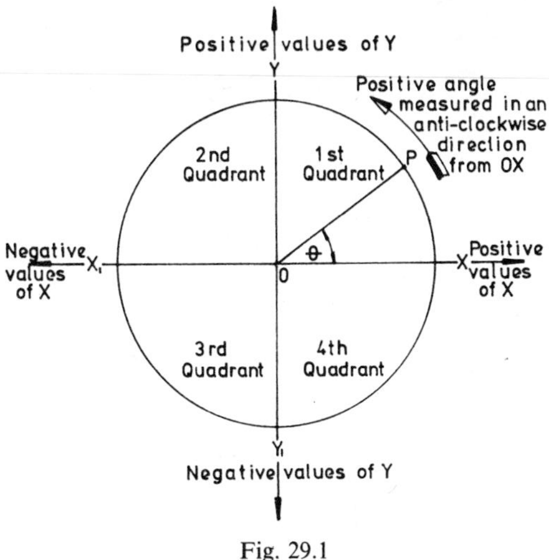

Fig. 29.1

been drawn at right-angles to each other to form the four quadrants. In each of these four quadrants we make use of the sign convention used when drawing graphs.
Now an angle, if positive, is always measured in an anti-clockwise direction from OX and an angle is formed by rotating a line (such as OP) in an anti-clockwise direction. It is convenient to make the length of OP equal to 1 unit. Referring to Fig. 29.2, we see:

In the first quadrant

$$\sin\theta_1 = \frac{P_1M_1}{OP_1} = P_1M_1 = y \text{ co-ordinate of } P_1$$

$$\cos\theta_1 = \frac{OM_1}{OP_1} = OM_1 = x \text{ co-ordinate of } P_1$$

$$\tan\theta_1 = \frac{P_1M_1}{OM_1} = \frac{y \text{ co-ordinate of P}}{x \text{ co-ordinate of P}}$$

Hence in the *first quadrant* all the trigonometrical ratios are *positive.*

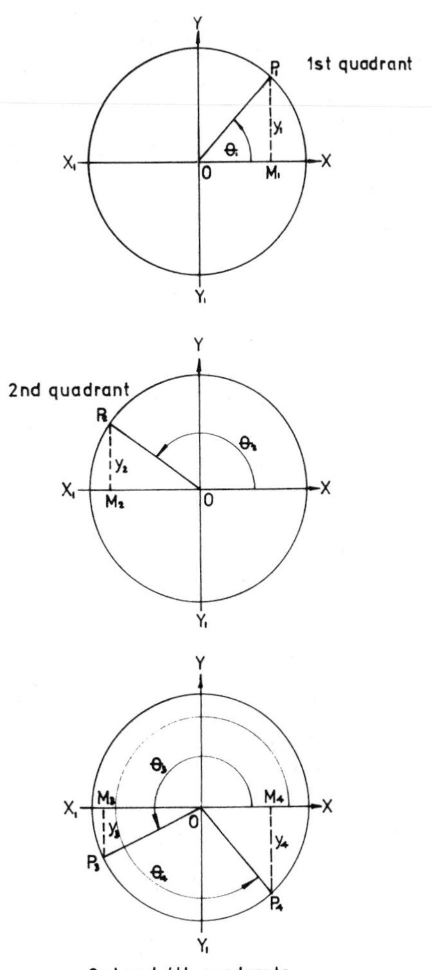

Fig. 29.2

In the second quadrant

$$\sin\theta_2=\frac{P_2M_2}{OP_2}=P_2M_2=y \text{ co-ordinate of } P_2$$

The y co-ordinate of P_2 is positive and hence in the second quadrant the sine of an angle is positive.

$$\cos\theta_2=\frac{OM_2}{OP_2}=OM_2=x \text{ co-ordinate of } P_2$$

The x co-ordinate of P_2 is negative and hence in the second quadrant the cosine of an angle is negative.

$$\tan\theta_2=\frac{P_2M_2}{OM_2}=\frac{y \text{ co-ordinate of } P_2}{x \text{ co-ordinate of } P_2}$$

But the y co-ordinate of P_2 is positive and the x co-ordinate of P_2 is negative, hence the tangent of an angle in the second quadrant is negative.

The trigonometrical tables usually give values of the trigonometrical ratios for angles between 0° and 90°. In order to use these tables for angles greater than 90° we make use of the triangle OP_2M_2, where we see that

$$P_2M_2=OP_2\sin(180-\theta_2)=\sin(180-\theta_2)$$

But $P_2M_2=\sin\theta_2$

$$\therefore \sin\theta_2=\sin(180-\theta_2)$$

Also $OM_2=-OP_2\cos(180-\theta_2)$
$=-\cos(180-\theta_2)$

$$\therefore \cos\theta_2=-\cos(180-\theta_2)$$

Similarly $\tan\theta_2=-\tan(180-\theta_2)$

In the third quadrant by similar considerations

$$\sin\theta_3=-\sin(\theta_3-180°)$$
$$\cos\theta_3=-\cos(\theta_3-180°)$$
$$\tan\theta_3=\tan(\theta_3-180°)$$

In the fourth quadrant,

$$\sin\theta_4=-\sin(360°-\theta_4)$$
$$\cos\theta_4=\cos(360°-\theta_4)$$
$$\tan\theta_4=-\tan(360°-\theta_4)$$

The results are summarised in Fig. 29.3.

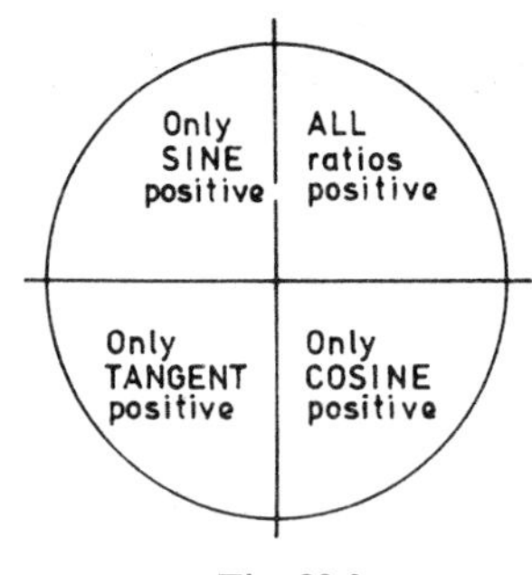

Fig. 29.3

Example

Find the values of sin 158°, cos 158° and tan 158°. Referring to Fig. 29.4,

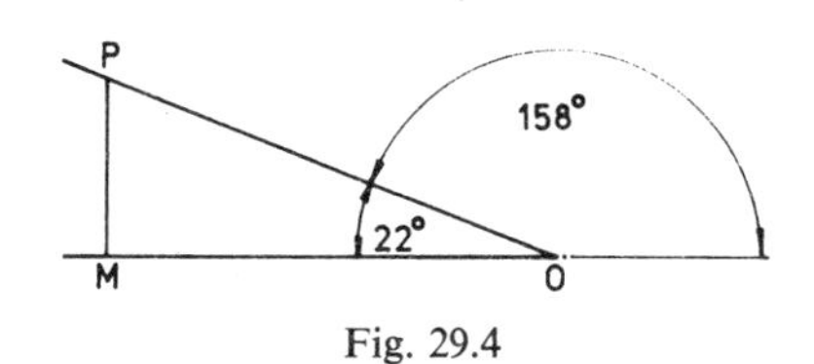

Fig. 29.4

$$\sin 158°=\frac{MP}{OP}=\sin\angle POM$$
$$=\sin(180°-158°)$$
$$=\sin 22°=0{\cdot}3746$$

$$\cos 158°=\frac{OM}{OP}=-\cos\angle POM$$
$$=-\cos(180°-158°)$$
$$=-\cos 22°=-0{\cdot}9272$$

$$\tan 158°=\frac{MP}{OM}=-\tan\angle POM$$
$$=-\tan(180°-158°)$$
$$=-\tan 22°=-0{\cdot}4040$$

The table below may be used for angles in any quadrant.

QUAD-RANT	ANGLE	sin θ=	cos θ=	tan θ=
first	0–90°	sin θ	cos θ	tan θ
second	90°–180°	sin(180°–θ)	—cos(180°–θ)	—tan(180°–θ)
third	180°–270°	—sin(θ–180°)	—cos(θ–180°)	tan(θ–180°)
fourth	270°–360°	—sin(360°–θ)	cos(360°–θ)	—tan(360°–θ)

Examples

1) Find the sine and cosine of the following angles: a) 171° b) 216° c) 289°

a) $\sin 171° = \sin(180° - 171°) = \sin 9°$
$= 0{\cdot}1564$
$\cos 171° = -\cos(180° - 171°) = -\cos 9°$
$= -0{\cdot}9877$

b) $\sin 216° = -\sin(216° - 180°) = -\sin 36°$
$= -0{\cdot}5878$
$\cos 216° = -\cos(216° - 180°)$
$= -\cos 36° = -0{\cdot}8090$

c) $\sin 289° = -\sin(360° - 289°) = -\sin 71°$
$= -0{\cdot}9455$
$\cos 289° = \cos(360° - 289°) = \cos 71°$
$= 0{\cdot}3256.$

2) If $\sin A = \frac{3}{5}$ find the values of cos A.

As shown in Fig. 29.5, the angle A may be in the first or second quadrants. In the first quadrant, by Pythagoras,

$$OM_1^2 = OP_1^2 - P_1M_1^2 = 5^2 - 3^2 = 16$$

$$\therefore OM_1 = 4$$

$$\cos A = \frac{OM_1}{OP_1} = \frac{4}{5}$$

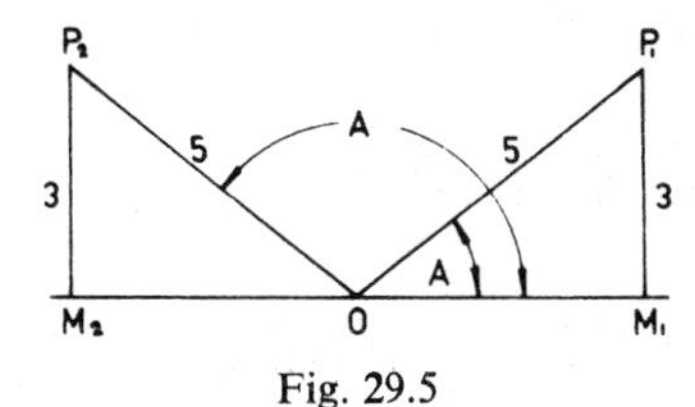

Fig. 29.5

In the second quadrant

$$OM_2 = -4$$

$$\cos A = \frac{OM_2}{M_2P_2} = -\frac{4}{5}$$

3) Find all the angles between 0° and 360° a) whose sine is 0·4676; b) whose cosine is −0·3572

a) The angles whose sines are 0·4676 occur in the first and second quadrants.

In the first quadrant:

$$\sin\theta = 0{\cdot}4676$$
$$\theta = 27°\ 53'$$

In the second quadrant:

$$\sin\theta = \sin(180° - \theta)$$
$$\therefore \sin(180° - \theta) = 0{\cdot}4676$$
$$180° - \theta = 27°\ 53'$$
$$\theta = 180° - 27°\ 53'$$
$$= 152°\ 7'$$

b) The angles whose cosines are −0·3572 occur in the second and third quadrants. In the second quadrant:

$$\cos\theta = -\cos(180° - \theta)$$
$$-\cos(180° - \theta) = -0{\cdot}3572$$
$$\cos(180° - \theta) = 0{\cdot}3572$$
$$180° - \theta = 69°\ 4'$$
$$\theta = 180° - 69°\ 4'$$
$$= 110°\ 56'$$

In the third quadrant

$$\cos\theta = -\cos(\theta - 180°)$$
$$-\cos(\theta - 180°) = -0{\cdot}3572$$
$$\cos(\theta - 180°) = 0{\cdot}3572$$
$$\theta - 180° = 69°\ 4'$$
$$\theta = 180° + 69°\ 4' = 249°\ 4'$$

Exercise 108

1) Find the values of
a) sin 175° 6′ b) cos 108° 12′ c) tan 298° 35′

2) If $\sin A = \frac{a \sin B}{b}$ find the values of A between 0° and 360° when $a = 7{\cdot}26$ mm, $b = 9{\cdot}15$ mm and $B = 18°\ 29'$.

3) If $\cos C = \frac{a^2 + b^2 - c^2}{2ab}$, find the values of C between 0° and 360° when $a = 1{\cdot}26$ cm, $b = 1{\cdot}41$ cm and $c = 2{\cdot}13$ cm.

4) Find the angles in the first and second quadrants
a) whose sine is 0·8158,
b) whose cosine is −0·8817.

5) Evaluate 5 sin 142° − 3 tan 148° + 3 cos 230°.

6) If $\sin A = \frac{12}{13}$ find, without using tables, the values of tan A and cos A, A being acute.

7) Evaluate $\sin A \cos B - \sin B \cos A$ given that $\sin A = \frac{3}{5}$ and $\tan B = \frac{4}{3}$. A and B are both acute angles.

8) The angle A is acute and $\tan A = \frac{15}{8}$. Without using tables find the value of $\sin(180° - A)$. If $B = 90° - A$ find the value of tan B.

9) Use tables to find the angles x and y between 0° and 360° where $\sin x = 0{\cdot}5688$ and $\cos y = 0{\cdot}8774$.

10) The angle A is acute and $\cos A = \frac{60}{61}$. Without using trigonometry tables, find the value of tan A.

THE STANDARD NOTATION FOR A TRIANGLE

In △ABC (Fig. 29.6) the angles are denoted by the capital letters as shown in the diagram. The side a lies opposite the angle A, the side b opposite the angle B and the side c opposite the angle C. This is the standard notation for a triangle.

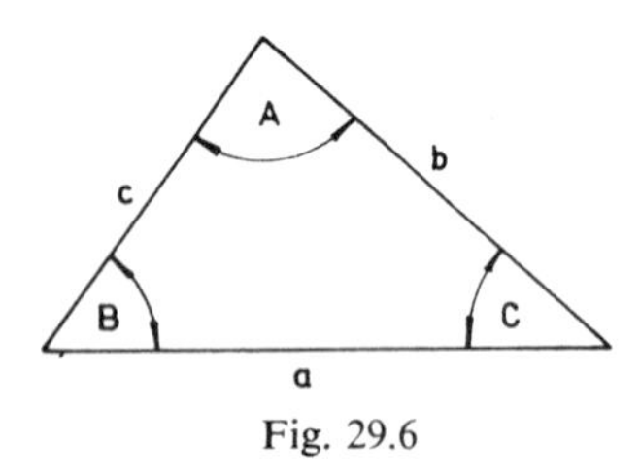

Fig. 29.6

THE SINE RULE

The sine rule is one of the two rules which are used for solving triangles which are *not right-angled.* Using the standard notation for a triangle (Fig. 29.6),

$$\frac{a}{\sin A} = \frac{b}{\sin B} = \frac{c}{\sin C}$$

This formula is proved below for both acute-angled and obtuse-angled triangles.

i) When △ABC is acute angled (Fig. 29.7).

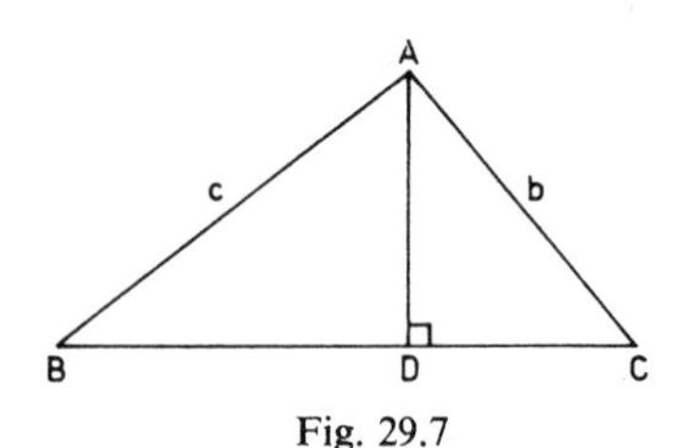

Fig. 29.7

In △ABD,

$$AD = c \sin B$$

In △ADC,

$$AD = b \sin C$$

$$\therefore \; b \sin C = c \sin B$$

or $\dfrac{b}{\sin B} = \dfrac{c}{\sin C}$

Similarly, by drawing the perpendicular from B to AC,

$$\frac{a}{\sin A} = \frac{c}{\sin C}$$

$$\therefore \; \frac{a}{\sin A} = \frac{b}{\sin B} = \frac{c}{\sin C}$$

ii) When △ABC is obtuse-angled (Fig. 29.8).

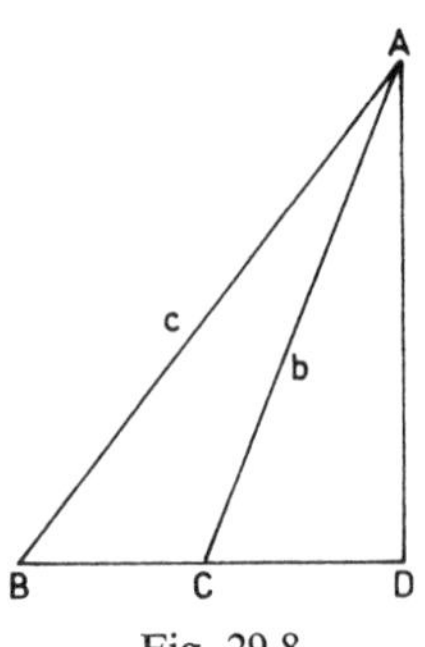

Fig. 29.8

In $\triangle$ACD,

$$AD = b \sin(180° - C) = b \sin C$$

In $\triangle$ABD,

$$AD = c \sin B$$

$$\therefore \quad b \sin C = c \sin B$$

$$\frac{b}{\sin B} = \frac{c}{\sin C}$$

As before,

$$\frac{a}{\sin A} = \frac{b}{\sin B} = \frac{c}{\sin C}$$

Example

Solve $\triangle$ABC given that A=42°, C=72° and b=61·8 mm.
The triangle should be drawn for reference as shown in Fig. 29.9, but there is no need to draw it to scale.

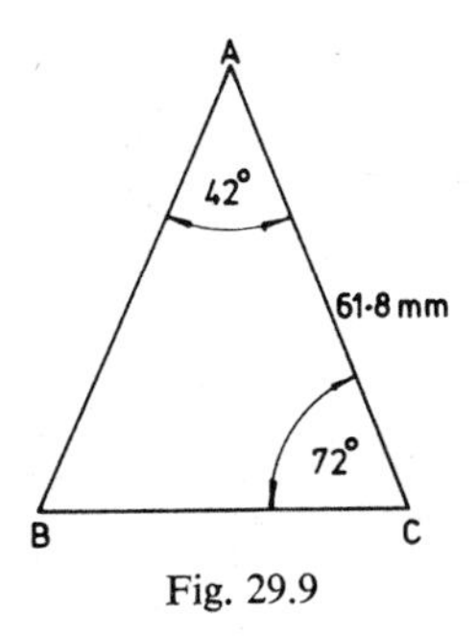

Fig. 29.9

Since A+B+C=180°

$$B = 180° - 42° - 72° = 66°$$

The sine rule states

$$\frac{a}{\sin A} = \frac{b}{\sin B}$$

$$a = \frac{b \sin A}{\sin B}$$

$$= \frac{61{\cdot}8 \times \sin 42°}{\sin 66°}$$

$$= 45{\cdot}27 \text{ mm}$$

number	log
61·8	1·7910
sin 42°	$\bar{1}$·8255
	1·6165
sin 66°	$\bar{1}$·9607
45·27	1·6558

also,

$$\frac{c}{\sin C} = \frac{b}{\sin B}$$

$$c = \frac{b \sin C}{\sin B}$$

$$= \frac{61{\cdot}8 \times \sin 72°}{\sin 66°}$$

$$= 64{\cdot}34 \text{ mm}$$

number	log
61·8	1·7910
sin 72°	$\bar{1}$·9782
	1·7692
sin 66°	$\bar{1}$·9607
64·34	1·8085

The complete solution is:

$\angle$B=66°, a=45·27 mm, c=64·34 mm

A rough check on sine rule calculations may be made by remembering that in any triangle the longest side lies opposite to the largest angle and the shortest side lies opposite to the smallest angle. Thus in the previous example:

smallest angle=42°=A;
shortest side=a=45·27 mm
largest angle=72°=C;
longest side=c=64·34 mm

USE OF THE SINE RULE TO FIND THE DIAMETER OF THE CIRCUMSCRIBING CIRCLE OF A TRIANGLE

Using the notation of Fig. 29.10

$$\frac{a}{\sin A} = \frac{b}{\sin B} = \frac{c}{\sin C} = D$$

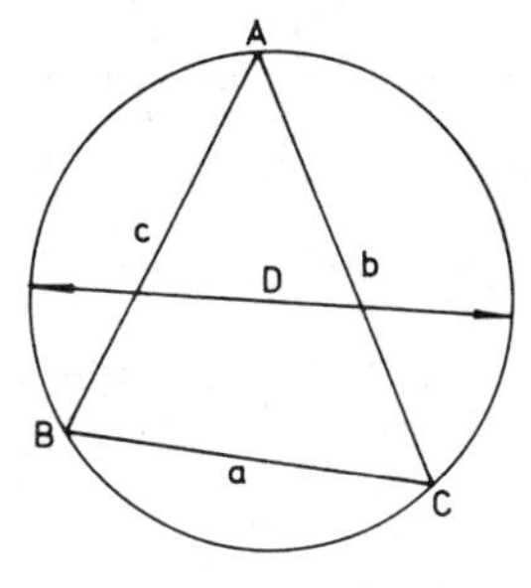

Fig. 29.10

Example

In $\triangle ABC$, $B=41°$, $b=112{\cdot}5$ mm and $a=87{\cdot}63$ mm. Find the diameter of the circumscribing circle. Referring to Fig. 29.11,

$$D=\frac{b}{\sin B}=\frac{112{\cdot}5}{\sin 41°}=171{\cdot}5 \text{ mm}$$

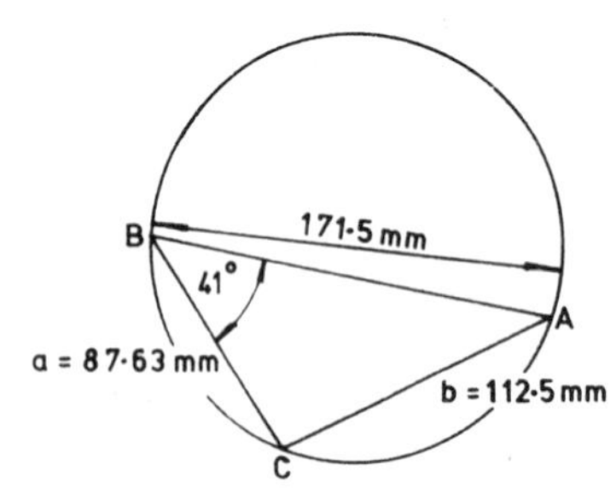

Fig. 29.11

Exercise 109

Solve the following triangles ABC by using the sine rule, given:

1) $A=75°$	$B=34°$	$a=10{\cdot}2$ cm
2) $C=61°$	$B=71°$	$b=91$ mm
3) $A=19°$	$C=105°$	$c=11{\cdot}1$ m
4) $A=116°$	$C=18°$	$a=17$ cm
5) $A=36°$	$B=77°$	$b=2{\cdot}5$ m
6) $A=49°\,11'$	$B=67°\,17'$	$c=11{\cdot}22$ mm
7) $A=17°\,15'$	$C=27°\,7'$	$b=22{\cdot}15$ cm
8) $A=77°\,3'$	$C=21°\,3'$	$a=9{\cdot}793$ m
9) $B=115°\,4'$	$C=11°\,17'$	$c=516{\cdot}2$ mm
10) $a=7$ m	$c=11$ m	$C=22°\,7'$
11) $b=15{\cdot}13$ cm	$c=11{\cdot}62$ cm	$B=85°\,17'$
12) $a=23$ cm	$c=18{\cdot}2$ cm	$A=49°\,19'$
13) $a=9{\cdot}217$ cm	$b=7{\cdot}152$ cm	$A=105°\,4'$

Find the diameter of the circumscribing circle of the following triangles ABC given:

14) $A=75°$	$B=48°$	$a=21$ cm
15) $C=100°$	$B=50°$	$b=90$ mm
16) $A=20°$	$C=102°$	$c=11$ m
17) $A=70°$	$C=35°$	$a=8{\cdot}5$ cm
18) $a=16$ cm	$b=14$ cm	$B=40°$

THE COSINE RULE

This rule is used when we are given:

either 1) two sides of a triangle and the angle between them

or 2) three sides of a triangle.

In all other cases the sine rule is used. The cosine rule states:

either $a^2=b^2+c^2-2bc\cos A$

or $b^2=a^2+c^2-2ac\cos B$

or $c^2=a^2+b^2-2ab\cos C$

When using the cosine rule remember that if an angle is greater than 90° its cosine is negative.

The proof of the cosine rule is given below and is proved for both an acute angle (Fig. 29.12) and an obtuse angle (Fig. 29.13).

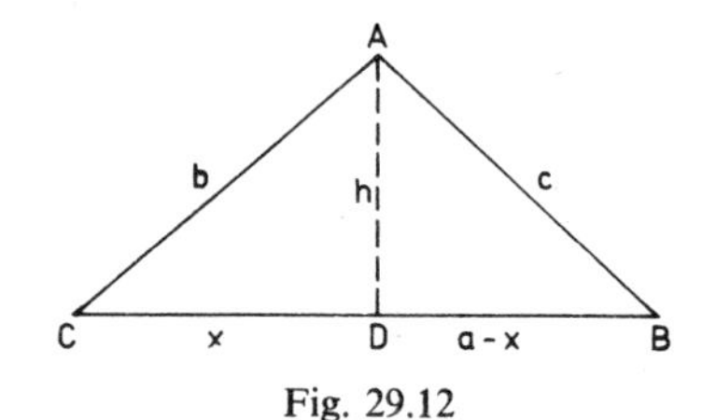

Fig. 29.12

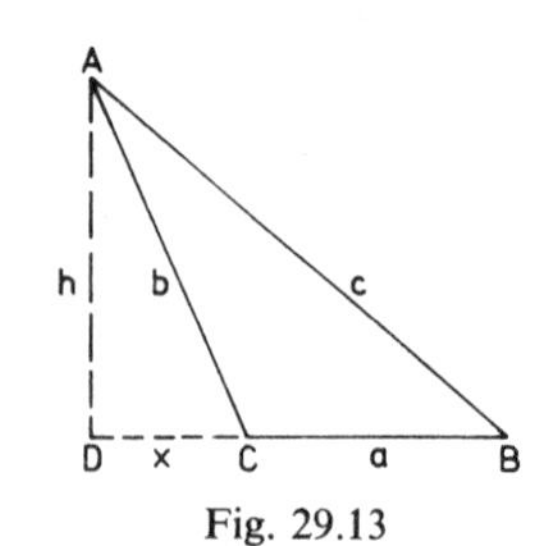

Fig. 29.13

i) Angle C acute

Draw AD perpendicular to BC.

Let $AD=h$ and $CD=x$ then $BD=a-x$

In $\triangle$ADC

$$h^2 = b^2 - x^2$$

In $\triangle$ADB

$$h^2 = c^2 - (a-x)^2$$
$$= c^2 - a^2 + 2ax - x^2$$

$$\therefore\ b^2 - x^2 = c^2 - a^2 + 2ax - x^2$$
$$c^2 = a^2 + b^2 - 2ax$$

But $x = b \cos C$

$$\therefore\ c^2 = a^2 + b^2 - 2ab \cos C$$

ii) Angle C obtuse
Draw AD perpendicular to BC

Let AD$=h$ and CD$=x$ then BD$=a+x$

In $\triangle$ADC

$$h^2 = b^2 - x^2$$

In $\triangle$ABD

$$h^2 = c^2 - (a+x)^2$$
$$= c^2 - a^2 - 2ax - x^2$$
$$b^2 - x^2 = c^2 - a^2 - 2ax - x^2$$
$$c^2 = a^2 + b^2 + 2ax$$

But $x = b \cos(180° - C)$

$$= -b \cos C$$

$$\therefore\ c^2 = a^2 + b^2 - 2ab \cos C$$

If the angle C is acute:

$a^2 + b^2$ is greater than c^2

If the angle C is obtuse:

c^2 is greater than $a^2 + b^2$

we therefore have a simple way of deciding if an angle of a triangle is obtuse or acute.

Examples

1) Solve $\triangle$ABC if $a = 70$ mm, $b = 40$ mm and $C = 64°$.

Referring to Fig. 29.14, to find the side c we use:

$$c^2 = a^2 + b^2 - 2ab \cos C$$
$$= 70^2 + 40^2 - 2 \times 70 \times 40 \times \cos 64°$$
$$= 4044$$
$$c = \sqrt{4044} = 63{\cdot}59 \text{ mm}$$

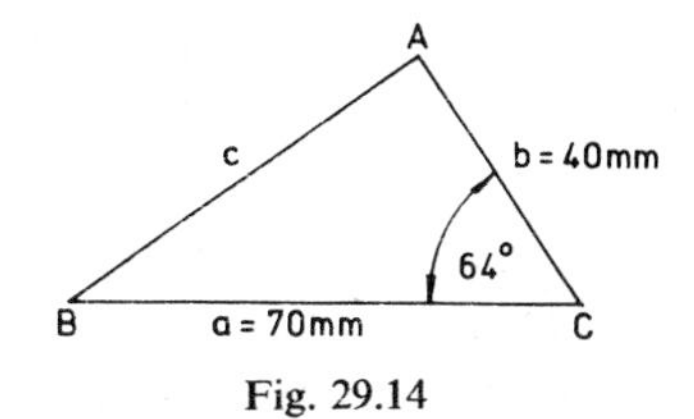

Fig. 29.14

We now use the sine rule to find the angle A:

$$\frac{a}{\sin A} = \frac{c}{\sin C}$$

$$\sin A = \frac{a \sin C}{c} = \frac{70 \times \sin 64°}{63{\cdot}59}$$

$$A = 81°\ 42'$$
$$B = 180° - 81°\ 42' - 64° = 34°\ 18'$$

The complete solution is

$A = 81°\ 42'$, $B = 34°\ 18'$ and $c = 63{\cdot}59$ mm

2) Find the side b in $\triangle$ABC if $a = 160$ mm, $c = 200$ mm and $B = 124°\ 15'$.

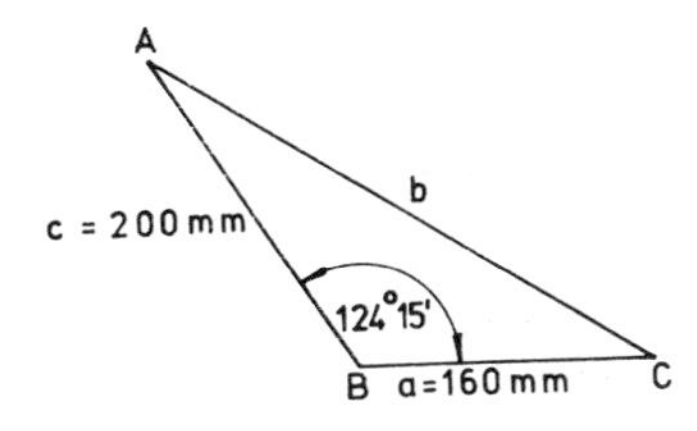

Fig. 29.15

Referring to Fig. 29.15, to find the side b we use:

$$b^2 = a^2 + c^2 - 2ac \cos B$$
$$= 160^2 + 200^2 - 2 \times 160 \times 200 \times \cos 124°\ 15'$$

Now $\cos 124°\ 15' = -\cos(180° - 124°\ 15')$

$$= -\cos 55°\ 45' = -0{\cdot}5628$$
$$\therefore\ b^2 = 160^2 + 200^2 - 2 \times 160 \times 200 \times (-0{\cdot}5628)$$
$$= 160^2 + 200^2 + 2 \times 160 \times 200 \times 0{\cdot}5628$$
$$b = 318{\cdot}7 \text{ mm}$$

3) In $\triangle$ABC, $c = 20$ cm, $b = 12$ cm and $a = 10$ cm. Check if the angle C is acute or obtuse angled and find its magnitude.

The angle C is obtuse angled if

c^2 is greater than a^2+b^2

Now $c^2=20^2=400$
and $a^2+b^2=10^2+12^2=244$
Hence the angle C is obtuse angled.
To find the magnitude ofC we use:

$$c^2=a^2+b^2-2ab \cos C$$

$$20^2=10^2+12^2-2\times10\times12\times\cos C$$

$$400=244-240 \cos C$$

$$240 \cos C=244-400$$

$$\cos C=\frac{244-400}{240}=-\frac{156}{240}=-0{\cdot}6500$$

$$C=180-49° \ 28'=130° \ 32'$$

Exercise 110

Solve the following triangles ABC using the cosine rule, given:

1) $a=9$ cm $\quad b=11$ cm $\quad C=60°$

2) $b=10$ cm $\quad c=14$ cm $\quad A=56°$

3) $a=8{\cdot}16$ m $\quad c=7{\cdot}14$ m $\quad B=37° \ 18'$

4) $a=5$ m $\quad b=8$ m $\quad c=7$ m

5) $a=312$ mm $\quad b=527{\cdot}3$ mm $\quad c=700$ mm

6) $a=7{\cdot}912$ cm $\quad b=4{\cdot}318$ cm $\quad c=11{\cdot}08$ cm

7) If $a=11$ cm, $b=10$ cm and $c=8$ cm find whether the angle A is acute or obtuse and state its magnitude.

8) If $b=90$ mm, $c=85$ mm and $C=40°$ find whether the angle B is acute or obtuse and find its magnitude.

9) If $a=12$ cm, $b=16$ cm and $c=25$ cm, find whether C is acute or obtuse angled and state its magnitude.

10) If $a=28$ cm, $c=24$ cm and $A=45°$ find if the angle B is acute or obtuse and state its magnitude.

SELF TEST 27

1) cos 120° is equal to

a $-\frac{1}{2}$ b $+\frac{1}{2}$ c $+\frac{\sqrt{3}}{2}$

d $-\frac{\sqrt{3}}{2}$

2) sin 150° is equal to

a $-\frac{1}{2}$ b $+\frac{1}{2}$ c $-\frac{\sqrt{3}}{2}$

d $+\frac{\sqrt{3}}{2}$

3) tan 120° is equal to

a $+\sqrt{3}$ b $+\frac{\sqrt{3}}{3}$ c $-\sqrt{3}$

d $-\frac{\sqrt{3}}{3}$

4) sin 240° is equal to

a $-\frac{1}{2}$ b $+\frac{1}{2}$ c $-\frac{\sqrt{3}}{2}$

d $+\frac{\sqrt{3}}{2}$

5) cos 210° is equal to

a $-\frac{1}{2}$ b $+\frac{1}{2}$ c $-\frac{\sqrt{3}}{2}$

d $+\frac{\sqrt{3}}{2}$

6) tan 240° is equal to

a $+\sqrt{3}$ b $-\sqrt{3}$ c $+\frac{\sqrt{3}}{3}$

d $-\frac{\sqrt{3}}{3}$

7) sin 300° is equal to

a $-\frac{1}{2}$ b $+\frac{1}{2}$ c $-\frac{\sqrt{3}}{2}$

d $+\frac{\sqrt{3}}{2}$

8) cos 300° is equal to

a $-\frac{1}{2}$ b $+\frac{1}{2}$ c $-\frac{\sqrt{3}}{2}$

d $+\frac{\sqrt{3}}{2}$

9) tan 330° is equal to

a $+\sqrt{3}$ b $-\sqrt{3}$ c $+\frac{\sqrt{3}}{3}$

d $-\frac{\sqrt{3}}{3}$

10) In Fig. 29.16, given $\angle$A and sides a and b. The angle B is given by the expression

a $b^2=a^2+c^2-2ac\cos B$

b $\cos B=\frac{a^2+c^2-b^2}{-2ac}$

c $\frac{a}{\sin A}=\frac{b}{\sin B}$

d $\frac{a}{b\sin A}$

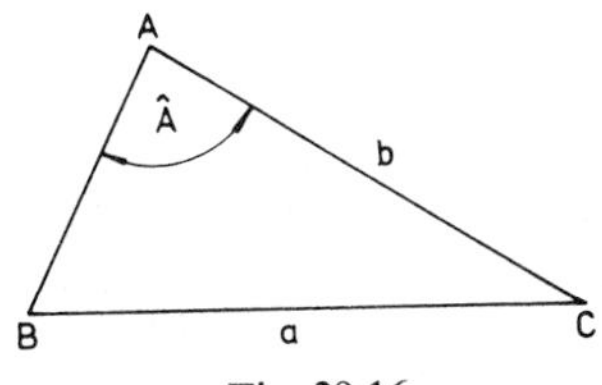

Fig. 29.16

11) In Fig. 29.17, given $\angle$B and sides a and c. The side b is given by the expression

a $b^2=a^2+c^2-2ac\cos B$

b $b^2=a^2+c^2+2ac\cos B$

c $b=\frac{a\sin B}{\sin A}$

d $b=\frac{\sin A}{a\sin B}$

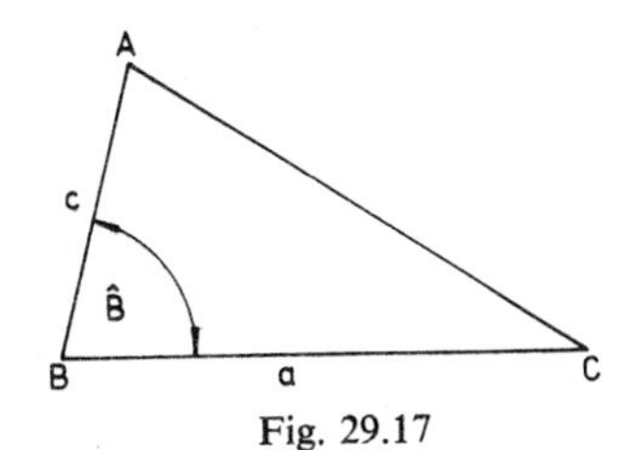

Fig. 29.17

12) If $\sin A=\frac{5}{13}$ when cos A is equal to

a $\frac{12}{13}$ b $\frac{5}{12}$ c $\frac{-12}{13}$ d $\frac{-5}{12}$

13) If $\tan A=\frac{3}{4}$, then cos A is equal to

a $\frac{3}{5}$ b $\frac{4}{5}$ c $\frac{-3}{5}$ d $\frac{-4}{5}$

14) If $\cos A=-\frac{11}{61}$ then sin A is equal to

a $\frac{11}{60}$ b $\frac{-11}{60}$ c $\frac{60}{61}$ d $\frac{-60}{61}$

15) If $\tan A=-\frac{12}{5}$ then cos A is equal to

a $\frac{5}{13}$ b $\frac{12}{13}$ c $\frac{-5}{13}$ d $\frac{-12}{13}$

Chapter 30 Solid Trigonometry

THE PLANE

A plane is a surface such as the top of a table or the cover of a book.

THE ANGLE BETWEEN A LINE AND A PLANE

In Fig. 30.1 the line PA intersects the plane WXYZ at A. To find the angle between PA and the plane draw PL which is perpendicular to the plane and join AL. The angle between PA and the plane is ∠PAL.

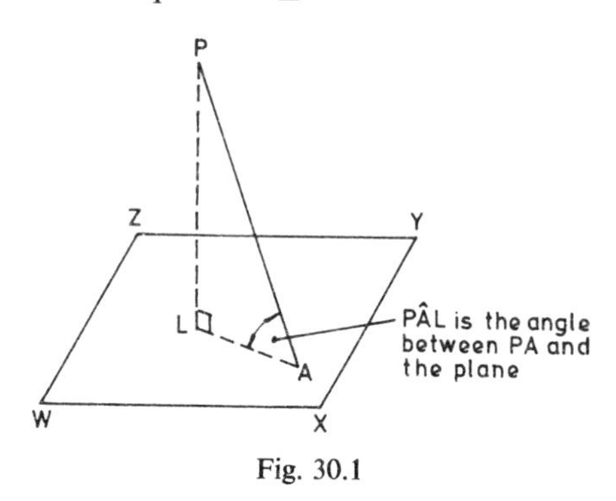

Fig. 30.1

THE ANGLE BETWEEN TWO PLANES

Two planes which are not parallel intersect in a straight line. Examples of this are the floor and a wall of a room and two walls of a room. To find the angle between two planes draw a line in each plane which is perpendicular to the common line of intersection. The angle between the two lines is the same as the angle between the two planes.

Three planes usually intersect at a point as, for instance, two walls and the floor of a room.

SOLVING PROBLEMS IN THREE DIMENSIONS

Problems in solid trigonometry are solved by choosing suitable triangles in different planes. It is essential to make a clear three-dimensional drawing in order to find these suitable triangles. The examples which follow will show the method to be adopted.

Examples

1) Fig. 30.2 shows a small block of wood which has a horizontal rectangular base ABCD in which AD = 7 cm and CD = 4 cm. The end face of the block AXB is vertical and it is an equilateral triangle. The top edge XY is horizontal and it is 4·5 cm long. The sloping end face YDC has YD = YC. Find the angle which the sloping face YDC makes with the base.

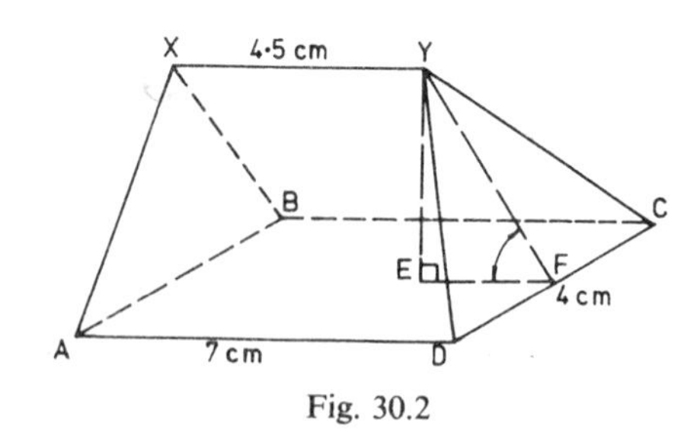

Fig. 30.2

The common line between the planes ABCD and YDC is the line DC. Since △YDC is isosceles (given) then if F is the mid-point of DC then YF is perpendicular to DC. From Y

make the line YE perpendicular to the plane ABCD and join EF. The required angle is then $\angle$YFE.
Since $\triangle$AXB is equilateral

$$EY^2 = 4^2 - 2^2 = 12$$

$$EY = \sqrt{12} = 3{\cdot}464 \text{ cm}$$

$$EF = 7 - 4{\cdot}5 = 2{\cdot}5 \text{ cm.}$$

In $\triangle$YEF

$$\tan \angle YFE = \frac{EY}{EF} = \frac{3{\cdot}464}{2{\cdot}5} = 1{\cdot}386$$

$$\angle YFE = 54° \ 12'$$

2) Figure 30.3 shows a wooden wedge with a horizontal face ABCD. The faces ABCD, ABEF and CDFE are all rectangles.
AB = 12 cm, BE = 9 cm, CE = 5 cm and $\angle$BCE = 50°. Find

1) the angle p that the face ABEF makes with the horizontal;
2) the angle q that the line AE makes with the horizontal.

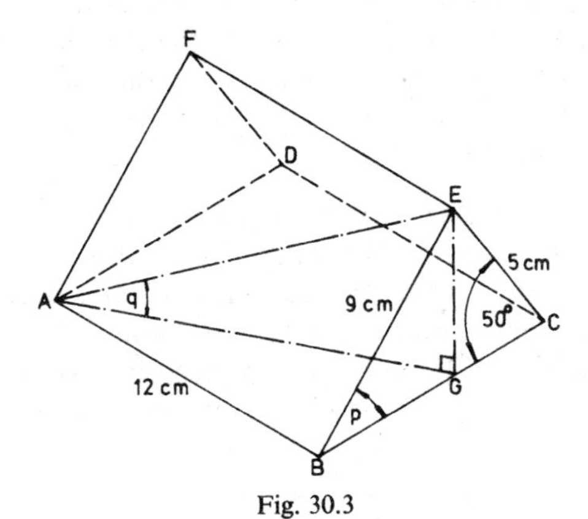

Fig. 30.3

1) In $\triangle$BEC using the sine rule

$$\frac{5}{\sin p} = \frac{9}{\sin 50°}$$

$$\sin p = \frac{5 \times \sin 50°}{9} = 0{\cdot}4255$$

$$p = 25° \ 11'$$

2) The angle q is found by using $\triangle$AEG

$$\tan q = \frac{EG}{AG}$$

In $\triangle$EGB,

$$\frac{EG}{EB} = \sin p$$

$$EG = EB \times \sin p = 9 \times \sin 25° \ 11'$$

$$= 3{\cdot}830 \text{ cm}$$

$$BG = EB \times \cos p = 9 \times \cos 25° \ 11'$$

$$= 8{\cdot}144 \text{ cm}$$

In $\triangle$AGB, using Pythagoras

$$AG^2 = AB^2 + BG^2$$

$$AG^2 = 12^2 + 8{\cdot}144^2 = 210{\cdot}3$$

$$AG = 14{\cdot}61 \text{ cm}$$

$$\therefore \tan q = \frac{EG}{AG} = \frac{3{\cdot}830}{14{\cdot}61}$$

$$q = 14° \ 41'$$

Exercise 111

1) In Fig. 30.4, A and B are points on a straight level road crossing a hillside which slopes at 15° to the horizontal. AH is a path from the road to a house on the hill. Given AB = 55 metres, HBA = 90° and $\angle$HAB = 52°. Calculate
i) HB; ii) the vertical height HM of H above the road; iii) the angle HAM which AH makes with the horizontal.

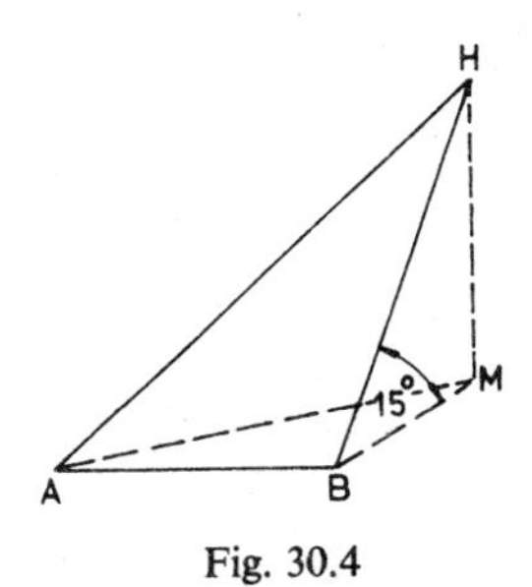

Fig. 30.4

2) The base of the triangular wedge shown in Fig. 30.5 is a rectangle 8 cm long and 6 cm wide. The vertical faces ABC and PQR are equilateral triangles of side 6 cm. Calculate

i) the angle between the diagonals PB and PC;
ii) the angle between the plane PBC and the base;
iii) the angle between the diagonal PC and the base.

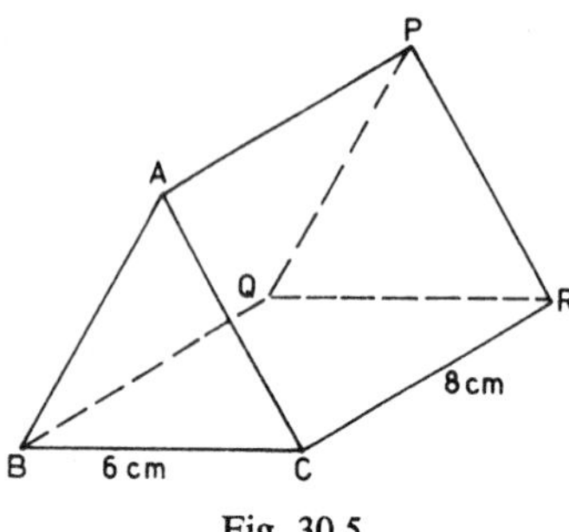

Fig. 30.5

3) Figure 30.6 represents a wooden block in the shape of a triangular prism in which the edges AD, BE and CF are equal and vertical and the base DEF is horizontal. AB = BC = DE = EF = 14 cm and $\angle ABC = \angle DEF = 40°$. The points G and H are on the edges AB and BC respectively. BG = BH = 4 cm and DG = FH = 20 cm.

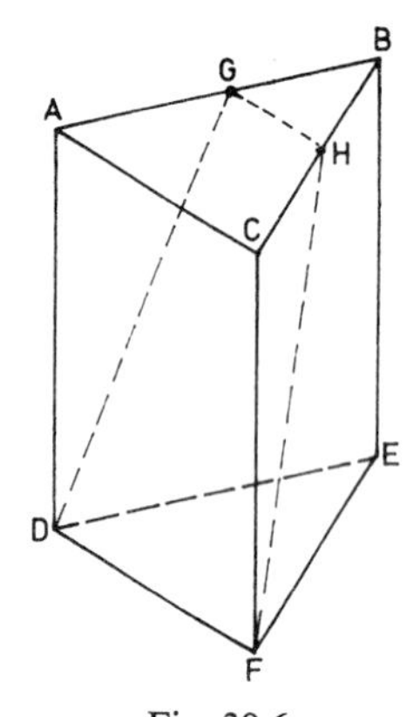

Fig. 30.6

Calculate: a) the length of BE; b) the angle between FH and the base DEF; c) the angle between the plane GHFD and the base DEF; d) the distance between the mid-points of GH and DF.

4) Figure 30.7 shows a shed with a slanting roof ABCD. The rectangular base ABEF rests on level ground and the shed has three vertical sides. Calculate:

a) the angle of inclination of the roof to the ground;
b) the volume of the shed in cubic metres.

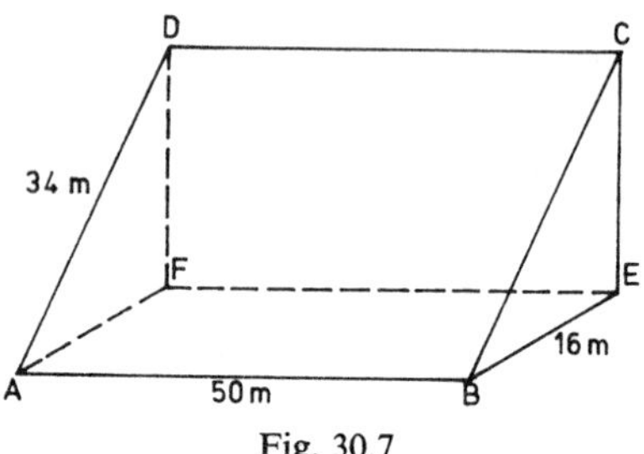

Fig. 30.7

5) The base of a pyramid consists of a regular hexagon ABCDEF of side 4 cm. The vertex of the pyramid is V and VA = VB = VC = VD = VE = VF = 7 cm. Sketch a general view of the solid. Indicate on your diagram the angles p and q described below and calculate the size of these angles:

i) the angle p between VA and the base;
ii) the angle q between the face VCD and the base.

6) In Fig. 30.8, ABCD represents part of a hillside. A line of greatest slope AB is inclined at 36° to the horizontal AE and runs due North from A. The line AC bears 050° (N 50° E) and C is 2500 m East of B.

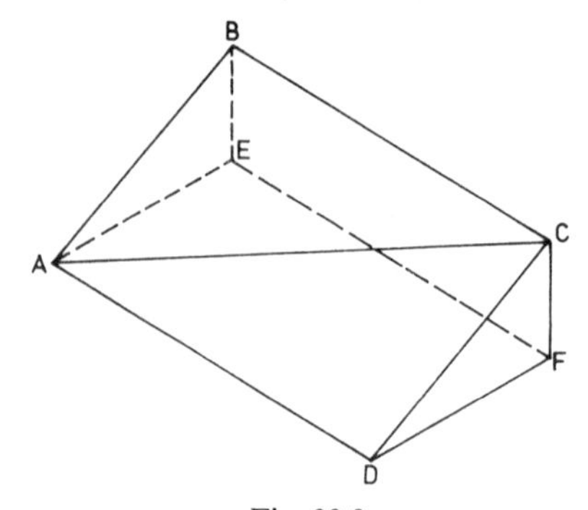

Fig. 30.8

The lines BE and CF are vertical. Calculate:

a) the height of C above A;
b) the angle between AB and AC.

7) Figure 30.9 is a sketch of a roof feature whose horizontal base is a rectangle ABCD 37 m by 23 m and the vertex V is 12 m directly above A. Calculate:

a) the length of VC;
b) the total area in square metres of the inclined surfaces VBC and VCD.

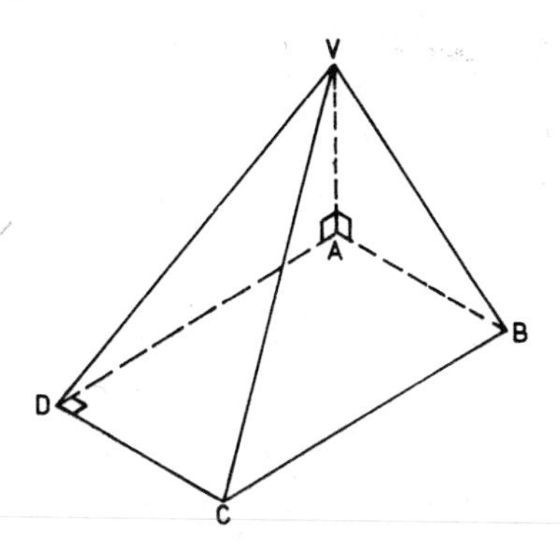

Fig. 30.9

8) Figure 30.10 is a plan view of the roof of a rectangular building, the sloping surfaces of which are all inclined at 34° to the horizontal. Calculate:

a) the height of the ridge XY above the horizontal plane through ABCD;
b) the angle of inclination of AX to the horizontal;
c) the lengths of AX and XY.

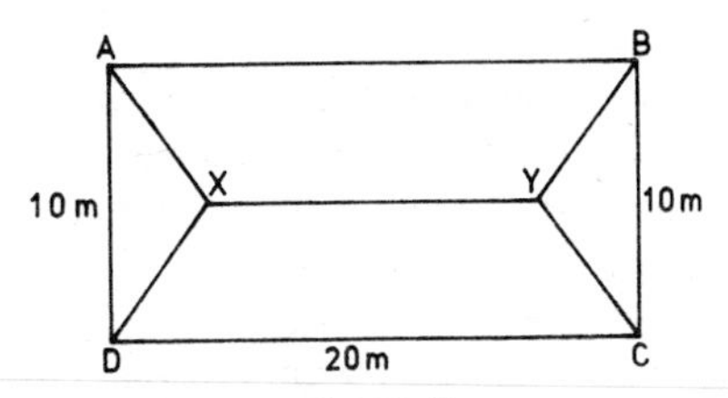

Fig. 30.10

Chapter 31 Maps and Contours

MAP SCALES

Map scales may be expressed as

1) *a ratio*, for example 1 : 25 000 or $\frac{1}{25\,000}$.

No units are involved when a scale is expressed in this way. Any distance measured off the map represents 25 000 times this distance on the ground.

2) *A scale*, for example 1 cm = 1 km. This means that 1 cm on the map represents an actual horizontal distance of 1 kilometre.

Examples

1) The scale of a map is 1 : 50 000. What distance does 3 cm on the map represent?

1 cm on the map represents 50 000 cm on the ground

or $\frac{50\,000}{100} = 500$ m

3 cm on the map represents $3 \times 500 = 1500$ m on the ground.

2) The scale of a map is 1 : 25 000. What distance in kilometres does 1 cm represent?

1 cm on the map represents 25 000 cm on the ground

or $\frac{25\,000}{100 \times 1000} = 0{\cdot}25$ km

3) The scale of a map is 1 cm = 2 km.

a) What distance does 3·5 cm on the map represent?

b) What is the scale expressed as a ratio.

a) 3·5 cm on the map represents $3{\cdot}5 \times 2 = 7$ km on the ground.

b) map scale $= \frac{1\text{ cm}}{2\text{ km}} = \frac{1\text{ cm}}{2 \times 1000 \times 100\text{ cm}}$

$= \frac{1}{200\,000}$

The map scale is 1 : 200 000.

Exercise 112

1) The scale of a map is 1 : 20 000. What distance does 5 cm on the map represent?

2) The scale of a map is 1 : 100 000 what distance does 28 cm on the map represent?

3) The scale of a map is 1 : 50 000. What distance does 1 cm on the map represent?

4) The scale of a map is 1 cm = 5 km. Express this scale as a ratio.

5) The scale of a map is 1 cm = 100 km. Express this scale as a ratio.

6) The scale of a map is 1 cm = 10 km. What distance does 8 cm on the map represent?

7) The scale of a map is 1 : 200 000. What distance on the map represents 150 km?

8) The scale of a map is 1 cm = 50 km. What distance on the map represents 80 km?

9) The distance, measured on a map, between two towns is 8·3 cm. If the map scale is 1 : 25 000 how far are the two towns apart?

10) The scale of a map is 2 cm = 1 km. The distance, measured on the map, between two points is 11·2 cm. How far apart are the two points?

AREAS ON MAPS

When areas are to be compared it is necessary to use the square of the map scale.

Example

1) If a map scale is 1 : 25000 what area does 5 cm^2 represent?

5 cm^2 on the map represents $5 \times (25000)^2$ cm^2 on the ground

or $\frac{5 \times (25000)^2}{100^2}$ m^2 on the ground

$= 312\,500$ m^2 on the ground.

2) A building site with an area of 4000 m^2 is represented on a map by an area of 10 cm^2. What is the map scale expressed as a ratio?

Since 10 cm^2 represents 4000 m^2
1 cm^2 represents 400 m^2
1 cm represents $\sqrt{400} = 20$ m

$$\therefore \text{ map scale} = \frac{1 \text{ cm}}{20 \text{ m}} = \frac{1 \text{ cm}}{20 \times 100 \text{ cm}} = \frac{1}{2000}$$

The map scale is 1 : 2000

3) The scale of a map is 1 cm = 5 km. What area does 8 cm^2 represent? How many cm^2 represents 1 km^2.

8 cm^2 on the map represents 8×5^2 km^2 or 200 km^2

Since 5 km = 1 cm

$$1 \text{ km} = \frac{1}{5} \text{ cm}$$

and $1 \text{ km}^2 = \left(\frac{1}{5}\right)^2 \text{ cm}^2 = \frac{1}{25} \text{cm}^2 = 0{\cdot}04 \text{ cm}^2$

Hence 0·04 cm^2 represents 1 km^2

Exercise 113

1) The scale of a map is 1 : 20000. What area does 3 cm^2 represent?

2) The scale of a map is 1 : 10000. What area does 8 cm^2 represent?

3) A plot of land with an area of 90000 m^2 is represented on a map by an area of 100 cm^2. What is the map scale expressed as a ratio?

4) On a map an area of 187500 m^2 is represented by 3 cm^2. What is the map scale expressed as a ratio?

5) The scale of a map is 1 cm = 2 km. How many cm^2 represents 1 km^2?

6) The scale of a map is 1 cm = 10 km. What area does 7 cm^2 represent?

7) The scale of a map is 10 cm to 1 km. A farmer is offered £400 per hectare for a field. On the map the field is represented by an isosceles triangle LMN in which LM = MN = 12 cm and $\angle$M = 30°. Calculate the price offered for the field (1 hectare = 10000 m^2).

8) A, B and C are three positions on a map whose scale is 10 cm to 1 km. AB = 7 cm, AC = 5 cm and $\angle$BAC = 50°. Calculate

i) the area of the triangle ABC on the map;

ii) the length of BC on the ground in metres;

iii) the actual area represented by the triangle ABC.

9) The scale of a map is 5 cm to 1 km. A field is represented by a trapezium (Fig. 31.1). What is the actual area of the field?

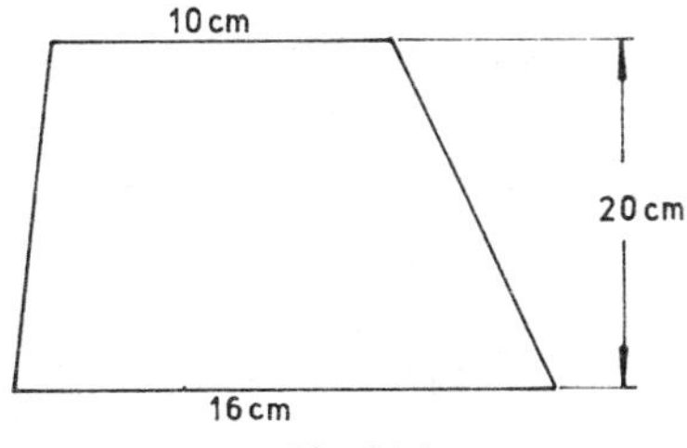

Fig. 31.1

10) A triangular plot of land when measured on a map has sides of 8 cm, 9 cm and 11 cm. If the scale of the map is 1 : 10000 what is the actual area of the plot in hectares. (1 hectare = 10000 m^2).

GRADIENT

When used in connection with roads and railways the gradient is used to denote the ratio of the vertical distance to the corresponding distance measured along the slope. Thus in Fig. 31.2:

$$\text{gradient} = \frac{BC}{AB}$$

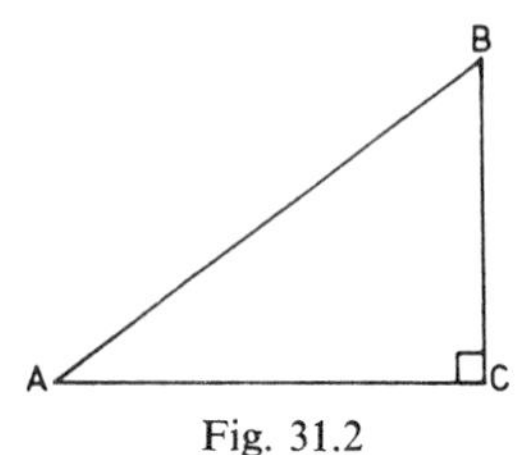

Fig. 31.2

The gradient is usually expressed as a ratio. If a road has a gradient of 1 in 20, we mean that for every 20 m along the slope the road rises 1 m vertically.

Examples

1) Two points P and Q are linked by a straight road with a uniform gradient. If Q is 50 m higher than P and the distance PQ measured along the road is 800 m, what is the gradient of the road.

Referring to Fig. 31.3 we see that:

$$\text{gradient} = \frac{50}{800} = \frac{1}{16}$$

The gradient of the road is 1 in 16.

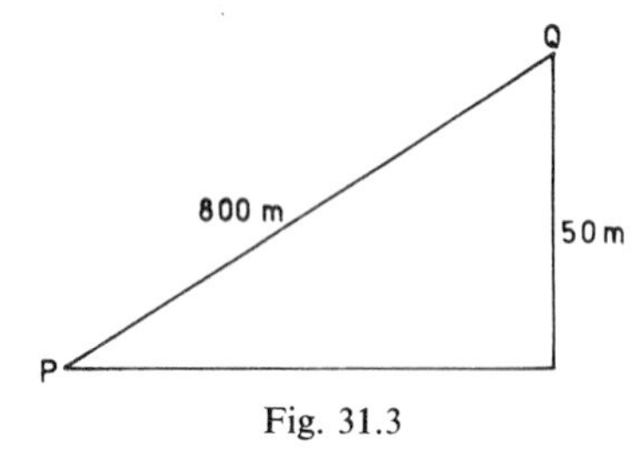

Fig. 31.3

2) A man walks 3 km up a road whose gradient is 1 in 15. How much higher is he than when he started?

From Fig. 31.4,

$$\text{gradient of road} = \frac{x}{3000}$$

$$\therefore \quad \frac{x}{3000} = \frac{1}{15}$$

$$x = \frac{3000}{15} = 200 \text{ m}$$

The man is 200 m than when he started.

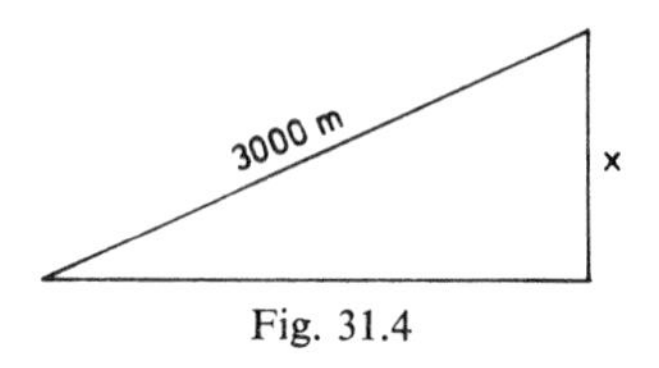

Fig. 31.4

Exercise 114

1) Two points A and B are connected by a straight road with a uniform gradient. If A is 40 m higher than B and the distance AB (measured along the road) is 1 km, what is the gradient of the road?

2) Two points P and Q whose heights differ by 50 m are connected by a straight road 2 km long. What is the gradient of the road?

3) A road has a gradient of 1 in 15. A man walks $1\frac{1}{2}$ km up the road. How much higher is he than when he first started?

4) A man walks up a road whose gradient is 1 in 30. If he walks a distance of 900 m how much higher is he than when he first started?

5) A road has a gradient of 1 in 8. What angle does the road make with the horizontal?

6) A mountain track is inclined at 12° to the horizontal. Express the gradient of the track as $1 : n$.

7) A man walks 100 m up a slope whose gradient is 1 in 20 and then a further 80 m up a slope whose gradient is 1 in 8. How much higher is he than when he first started?

8) Two points P and Q have heights which differ by 100 m. If the gradient of the line joining P and Q is 1 in 12 what is the distance PQ?

CONTOURS

The contour lines on a map join places which have the same height. Figure 31.5 shows a small hill. The bottom view (the plan view) is the one which would be shown on a map.

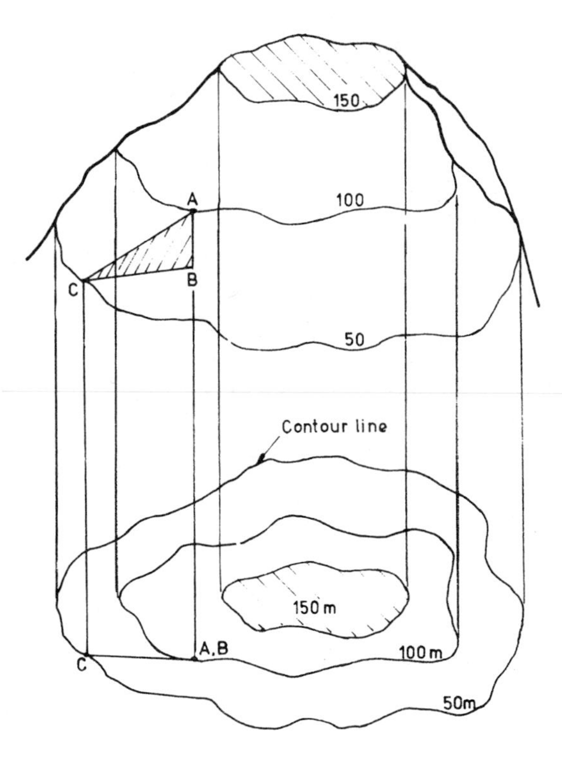

Fig. 31.5

Suppose A and C are two positions on a path up the hill, the path being taken as a straight line. It will be seen that C is on the 50 m contour and A is on the 100 m contour. B is directly below A (within the hill) and on the same contour as C. A and B appear as the same point on the map but the height AB is 50 m. The distance AC measured on the map corresponds to the length BC on the elevation.

The angle of elevation is θ and $\tan\theta = \dfrac{AB}{BC}$.

The gradient is $\dfrac{AB}{AC}$.

Examples

1) Two points P and Q whose heights differ by 50 m are shown 0·60 cm apart on a map whose scale is 1 : 25 000. Calculate the angle of elevation of the line PQ.

0·60 cm on the map represents $0{\cdot}6 \times 25\,000$ cm on the ground

or $\dfrac{0{\cdot}6 \times 25\,000}{100}$ m on the ground

$= 150$ m

From Fig. 31.6,

$$\tan\theta = \frac{50}{150} = 0{\cdot}3333$$

$$\theta = 18^\circ\ 26'$$

$\therefore$ Angle of elevation $= 18^\circ\ 26'$

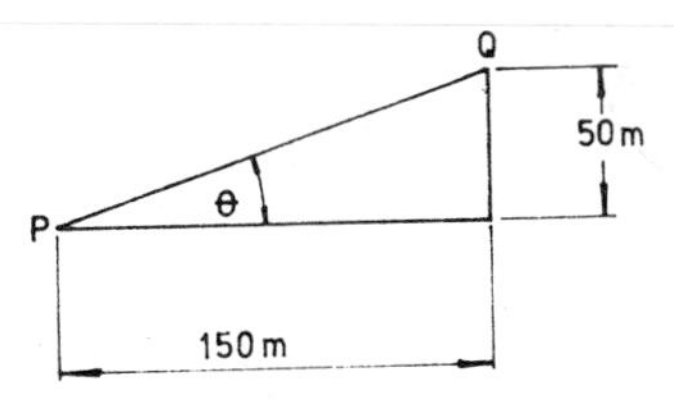

Fig. 31.6

2) Two points A and B are linked by a straight road with a uniform gradient. A is on the 100 m contour and B is on the 125 m contour. If the scale of the map is 10 cm to 1 km and the distance AB on the map is 4 cm find the gradient of the road giving the answer in the form '1 in n'.

The map is shown in Fig. 31.7.

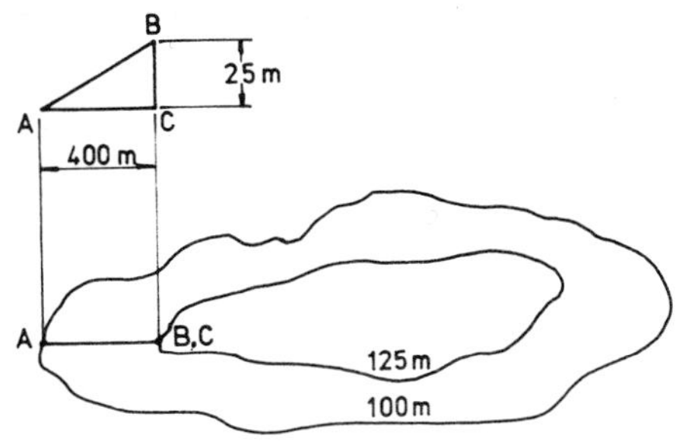

Fig. 31.7

Since the map scale is 10 cm to 1 km, 4 cm represents

$$\frac{4}{10} \times 1 = 0{\cdot}4 \text{ km} = 400 \text{ m on the ground.}$$

Referring to the triangle ABC,

$$\text{Gradient} = \frac{\text{vertical height}}{\text{distance along slope}} = \frac{BC}{AB}$$

$$AB = \sqrt{AC^2 + BC^2} = \sqrt{400^2 + 25^2}$$

$$= 400{\cdot}6 \text{ m}$$

$$\text{Gradient} = \frac{25}{400{\cdot}6}$$

$$= 25 : 400{\cdot}6$$

$$= 1 : 16{\cdot}024$$

Exercise 115

1) Two points A and B whose heights differ by 100 m are shown 1·2 cm apart on a map whose scale is 1 : 10000. Calculate the angle of elevation of the line AB.

2) Two points P and Q whose heights differ by 150 m are shown 0·8 cm apart on a map whose scale is 1 cm = 500 m. Calculate the gradient of the line PQ, giving the answer in the form '1 in n'.

3) A and B are two positions on a map whose scale is 10 cm to 1 km. A is on 500 m contour and B is on the 250 m contour. The distance AB measured on the map is 8 cm. Find the angle of elevation from B to A.

4) The scale of a map is 1 : 10000. A cross-roads X on the 350 m contour is linked with a junction J, on the 600 m contour, by a straight road with a uniform gradient. The distance on the map between X and J is 10·75 cm. Calculate the length (in metres) of the road surface which joins X to J.

5) The points P, Q and R, as seen on a map, form a triangle in which $\angle P = 90°$, PQ = 3 cm and PR = 4 cm. It can also be seen that Q and R lie on the 300 m and 50 m contours respectively. Given that the map scale is 1 : 25000, calculate the angle of elevation from R to Q.

6) X and Y are two positions on a map whose scale is 1 : 10000. X is on the 100 m contour and Y is on the 300 m contour. The distance XY measured on the map is 16 cm. Find the gradient of the line XY.

Chapter 32 Triangle of Velocities

VECTORS

A vector is a quantity which possesses both size (or magnitude) and direction. A velocity requires direction as well as magnitude to describe it and hence it is a vector quantity. Speed is not a vector quantity since when we talk of a speed of 40 km/h no direction is implied, but when we talk of a velocity of 80 km/h in a direction South-East then this is a vector quantity, since a direction has been given.

Any vector can be represented by drawing a straight line, the length of the line (to some chosen scale) representing the magnitude and the direction of the line representing the direction of the vector.

Suppose an aeroplane is flying at 500 km/h in a direction N.E. over a town A. If we decide on a scale of 1 cm = 50 km/h the vector can be represented on a map of the area by a line 10 cm long drawn in a direction N.E. of town A as shown in Fig. 32.1. An arrow is placed on the vector to show that the aeroplane is flying away from A.

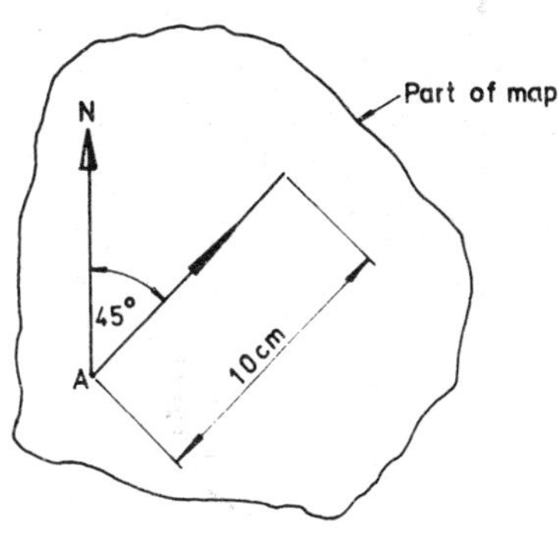

Fig. 32.1

COURSE AND AIRSPEED

The direction in which an aircraft is heading (i.e. pointing) is called its *course*. It is the angle between the longitudinal axis of the aircraft and the direction of North (Fig. 32.2).

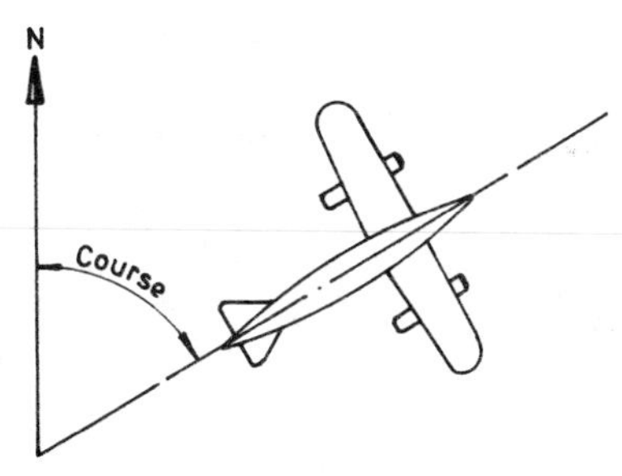

Fig. 32.2

The course is usually given in the three digit notation e.g. 036° or 205°. The airspeed is the speed of the aircraft in still air and the airspeed at which an aircraft is flying may be read directly from the airspeed indicator which is an instrument mounted in the cockpit. The course is measured by means of the aircraft's compass and hence airspeed and course form a vector quantity.

TRACK AND GROUNDSPEED

The track is the direction which the aircraft follows relative to the ground. If the sun is immediately overhead the track is the path of the aircraft's shadow over the ground. When a pilot wishes to fly from aerodrome A to aerodrome B, his track would be the straight line joining A and B. The groundspeed is the speed of the aircraft relative to the ground and it is affected by the direction and speed of the wind. It is the groundspeed which determines the time that a flight will take. Groundspeed and track form a vector quantity just as did airspeed and course.

WIND DIRECTION AND SPEED

The direction of the wind is always given as the direction from which the wind is blowing. The arrow on the wind vector always points in the direction to which the wind blows. Thus a wind of 40 km/h blowing from 060° will be represented as shown in Fig. 32.3.

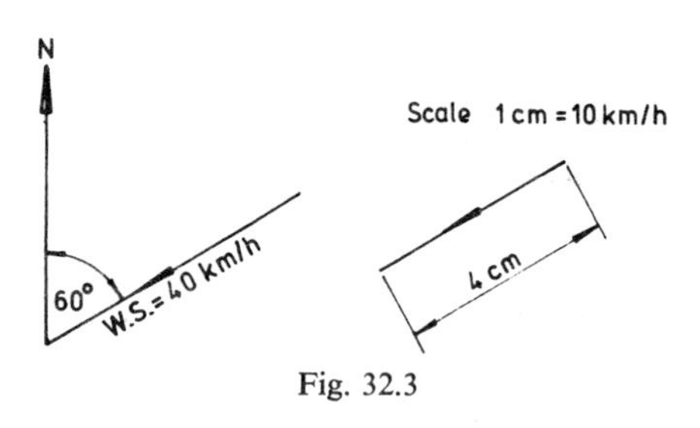

Fig. 32.3

DRIFT

The wind always tries to blow an aircraft off course, that is, it causes *drift*. The drift is the angle between the course and the track (Fig. 32.4). If there is no wind there is no drift and the airspeed and groundspeed are identical as are the course and the track.

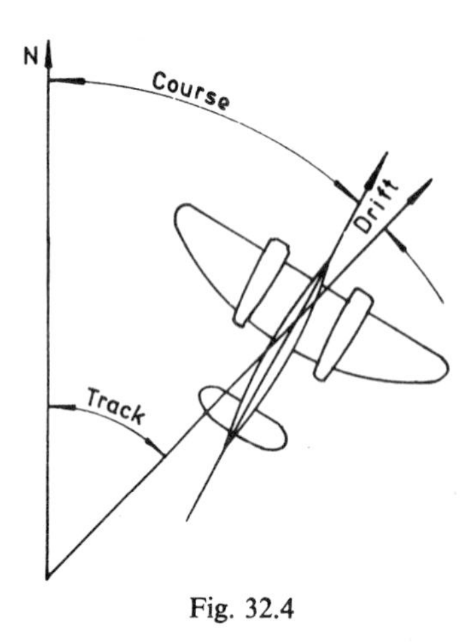

Fig. 32.4

THE TRIANGLE OF VELOCITIES

When an aeroplane is flying it has two velocities as follows:

1) its velocity through the air due to the engine,
2) the velocity due to the wind.

These two velocities combine to give a resultant velocity which is the groundspeed. In order to determine this resultant velocity we draw a triangle of velocities similar to the one shown in Fig. 32.5. In drawing this triangle note carefully that *the arrows on the airspeed and windspeed vectors follow nose to tail.*

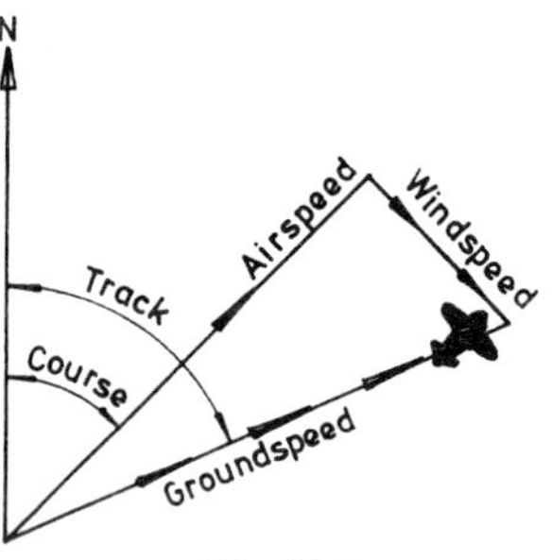

Fig. 32.5

Case 1: Given airspeed and course, wind speed and direction and to find groundspeed and track.

Example. An aircraft whose airspeed is 400 km/h flies on a course of 050°. If the wind blows at 50 km/h from 310° find the track and groundspeed.

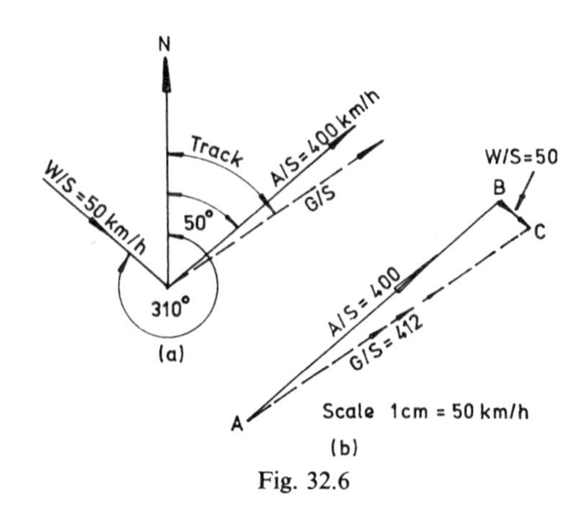

Fig. 32.6

Solution by drawing (Fig. 32.6).

1) Draw diagram (a) which shows the directions of the airspeed (A/S) and the windspeed (W/S). On this diagram only the angles are drawn accurately.

2) To draw the velocity triangle first choose a suitable scale to represent the velocities. In Fig. 32.6(b) a scale of 1 cm = 50 km/h has been chosen. Using set-squares draw AB parallel to the A/S and 8 cm long to represent 400 km/h. From B draw a line parallel to the W/S 1 cm long to give the point C. Join AC. AC represents the groundspeed (G/S) in magnitude and direction. Scale off AC which is found to be 8·24 cm representing 412 km/h. To find the direction of AC (i.e. the track) draw a line parallel to AC onto diagram (a) as shown. Using a protractor the track is found to be 57°.

Solution by calculation

First draw a rough sketch of the velocity triangle. In △ABC (Fig. 32.6),

$$AB = 400, \quad BC = 50 \quad \text{and} \quad \angle ABC = 100^\circ$$

Using the cosine rule,

$$\begin{aligned} b^2 &= a^2 + c^2 - 2ac \cos B \\ &= 50^2 + 400^2 - 2 \times 50 \times 400 \times \cos 100 \\ &= 2500 + 160\,000 - 400 \times (-0{\cdot}1736) \\ &= 162\,500 + 6944 \\ &= 169\,444 \\ b &= 412 \end{aligned}$$

Hence the groundspeed is 412 km/h.
To find the track we first calculate ∠BAC which represents the drift.
Using the sine rule,

$$\frac{a}{\sin A} = \frac{b}{\sin B}$$

$$\frac{50}{\sin A} = \frac{124}{\sin 100^\circ}$$

$$\sin A = \frac{50 \sin 100^\circ}{412}$$

$$A = 6^\circ\ 52' \text{ (or } 7^\circ \text{ to the nearest degree)}$$

$$\therefore \text{ Track} = 57^\circ$$

Case 2: Given airspeed and course, groundspeed and track to find windspeed and direction.

Example. The course and airspeed of an aircraft are 060° and 500 km/h whilst its track and groundspeed are 070° and 520 km/h. Find the speed and direction of the wind.

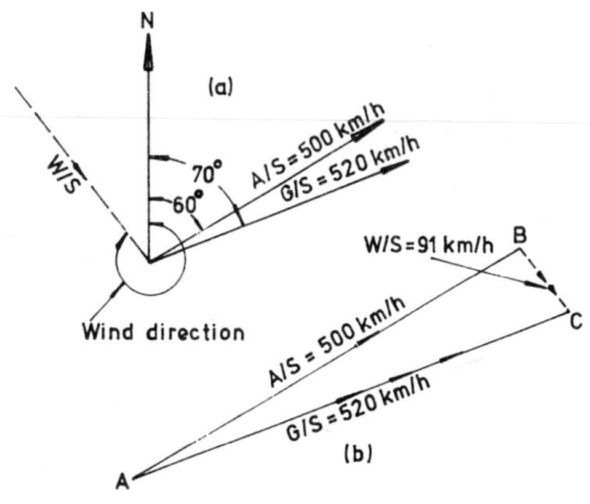

Fig. 32.7

Solution by drawing (Fig. 32.7)

1) Draw diagram (a) which shows the directions of the airspeed (A/S) and the groundspeed (G/S). On this diagram only the angles are drawn accurately.

2) To draw the velocity triangle shown in diagram (b) a scale of 1 cm = 50 km/h has been chosen. Using set squares draw AB, 10 cm long, parallel to A/S and then draw AC, 10·4 cm long, parallel to G/S. Join BC. BC represents the windspeed and scaling BC the windspeed is found to be 91 km/h. To find the wind direction draw a line parallel to BC on diagram (a) as shown. By using a protractor the wind direction is found to be 322°, (that is the wind blows from 322°).

Solution by calculation

In $\triangle ABC$ (Fig. 32.7),

$\angle BAC = 10°$, $AB = 500$ and $AC = 520$.

Using the cosine rule,

$$a^2 = b^2 + c^2 - 2bc \cos A$$
$$= 500^2 + 520° - 2 \times 500 \times 520 \times \cos 10°$$
$$= 250\,000 + 270\,400 - 512\,096$$
$$= 8304$$
$$a = 91{\cdot}11 \text{ km/h.}$$

To find the wind direction we calculate $\angle ABC$. Using the sine rule,

$$\frac{b}{\sin B} = \frac{a}{\sin A}$$

$$\frac{520}{\sin B} = \frac{91{\cdot}11}{\sin 10°}$$

$$\sin B = \frac{520 \times \sin 10°}{91{\cdot}11}$$

$$B = 82°\ 14' \quad \text{or} \quad 97°\ 46'$$

Since B must be greater than C,

$B = 97°\ 46'$ (or 98° to the nearest degree).

The direction from which the wind blows is

$$360° + 60° - 98° = 322°$$

Hence the wind speed is 91 km/h blowing from 322°.

Case 3: Given airspeed, track, windspeed and wind direction to find the course and groundspeed.

Since a navigator knows his airspeed and the track required and also the windspeed and direction this is the case usually met with in practice.

Example. An aircraft has an airspeed of 400 km/h and its track is 55°. The speed of the wind is 50 km/h blowing from 300°. Find the course and groundspeed.

Solution by drawing (Fig. 32.8)

1) On diagram (a) we can draw the wind direction and also the track of 55°.

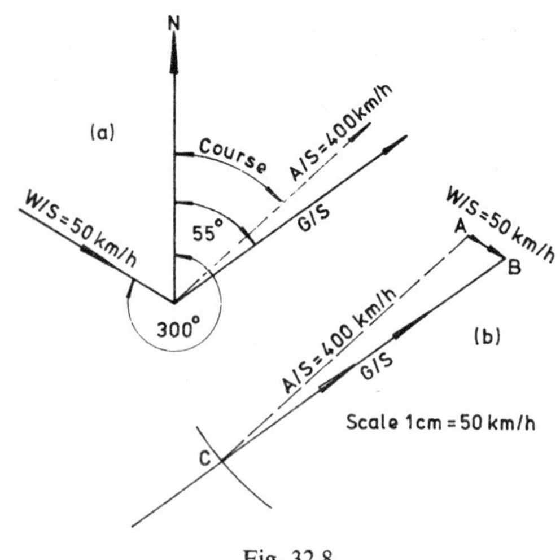

Fig. 32.8

2) Choosing a scale of 1 cm = 50 km/h we draw AB = 1 cm to represent to windspeed (diagram (b)). From B draw a line parallel to G/S. The length of this line is not known as yet. With compasses set at 8 cm (to represent the A/S of 400 km/h) and centre A draw an arc to cut the G/S line at C. BC represents the G/S and by scaling, the G/S = 419. Drawing a line parallel to AC on diagram (a) the course is found to be 48° 30′.

Solution by calculation

In $\triangle ABC$ (Fig. 32.8)

$AC = 400$ and $AB = 50$

The angle ABC is found as shown in Fig. 32.9.

Using the sine rule:

$$\frac{c}{\sin C} = \frac{b}{\sin B}$$

$$\frac{50}{\sin C} = \frac{400}{\sin 65°}$$

$$\sin C = \frac{50 \times \sin 65°}{400}$$

$$C = 6°\ 30'$$

$$A = 180 - 6°\ 30' - 65° = 108°\ 30'$$

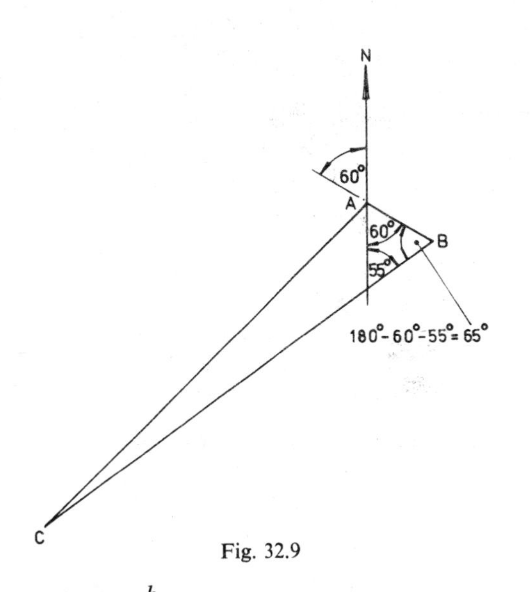

Fig. 32.9

$$\frac{a}{\sin A} = \frac{b}{\sin B}$$

$$a = \frac{b \sin A}{\sin B} = \frac{400 \times \sin 108° \, 30'}{\sin 65°}$$

$$a = 419$$

Hence the course is $55° - 6° \, 30' = 48° \, 30'$ and the groundspeed is 419 km/h.

Exercise 116

Find the track and groundspeed in each of the following:

1) Course 270°, airspeed 300 km/h; wind 135°, 60 km/h.

2) Course 090°, airspeed 500 km/h; wind 010°, 50 km/h.

3) Course 150°, airspeed 400 km/h; wind 070°, 50 km/h.

4) Course 310°, airspeed 350 km/h; wind 110°, 40 km/h.

5) Course 20°, airspeed 300 km/h; wind 320°, 80 km/h.

Find the wind velocity for each of the following:

6) Course 050°, airspeed 280 km/h; track 060°; groundspeed 300 km/h.

7) Course 160°, airspeed 350 km/h; track 180° groundspeed 330 km/h.

8) Course 210°, airspeed 420 km/h; track 190° groundspeed 410 km/h.

9) Course 300°, airspeed 450 km/h; track 320° groundspeed 435 km/h.

10) Course 060°, airspeed 250 km/h; track 040°, groundspeed 260 km/h.

Find the course and groundspeed for each of the following:

11) Track 090°, airspeed 300 km/h; wind 060°, 50 km/h.

12) Track 150°, airspeed 400 km/h; wind 040°, 30 km/h.

13) Track 210°, airspeed 350 km/h; wind 100°, 50 km/h.

14) Track 300°, airspeed 250 km/h; wind 120°, 45 km/h.

15) Track 010°, airspeed 300 km/h; wind 330°, 60 km/h.

16) The pilot of an aircraft sets a north-easterly course at an airspeed of 200 km/h, when the direction of the wind is from the east, so that he actually flies in a direction N 30° E (030°). Find by drawing the speed of the wind.

On another occasion, when the wind is blowing at 100 km/h the pilot sets a northerly course at an airspeed of 150 km/h. If the actual direction in which he flies is again N 30° E (030°), find by drawing the two possible directions from which the wind could be blowing.

17) An aircraft with an airspeed of 200 km/h sets a course of 010°. The wind is blowing at 75 km/h from bearing 325°. Find by accurate drawing, the track and the groundspeed of the aircraft.
With the same wind blowing, another aircraft with an airspeed of 430 km/h has to travel from P to Q. The bearing of Q from P is 253°. Find by accurate drawing, the course to be set.

18) In still air, a carrier pigeon flies at a speed of 50 km/h. When the wind is blowing from the South at 20 km/h find, by drawing, the direction in which the pigeon must head in order that it may travel from point P to point Q. Q is north-west of P.

TIME AND DISTANCE PROBLEMS

Examples

1) At 10.00 hours an aeroplane leaves an aerodrome P to fly to an aerodrome Q which is 250 km away, in a direction 055° from P. The wind is 30 km/h from 300° and the aeroplane is to be flown at 200 km/h. What course should the pilot steer and what is his estimated time of arrival?

We are given an airspeed of 200 km/h, a track of 055°, a windspeed of 30 km/h and a wind direction of 300°. We have to find the course and groundspeed.

The triangle of velocities is shown in Fig. 32.10 from which the course is 47° and the groundspeed is 219 km/h.

$$\text{Time taken to fly 250 km} = \frac{250}{219} = 1{\cdot}14 \text{ hours}$$

$$= 1 \text{ hour 8 min}$$

Estimated time of arrival = 11.08 hours

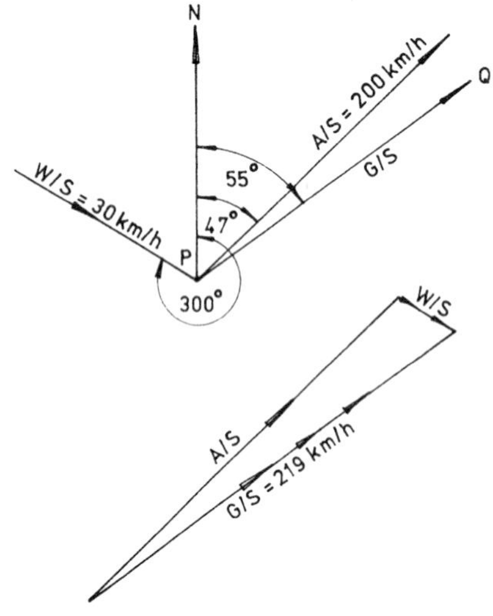

Fig. 32.10

2) Two aircraft P and Q set out from X at the same time. P with an airspeed of 180 km/h sets a course due East and Q with an airspeed of 135 km/h sets a course so as to travel due North. The wind is blowing at 40 km/h from the south-west. Find by accurate drawing,

1) the track of P;
2) the course set by Q.
3) Indicate the positions of P and Q 30 minutes after leaving X and find their distance apart in kilometres.

1) From the triangle of velocities (Fig. 32.11) the track of P is 82°.

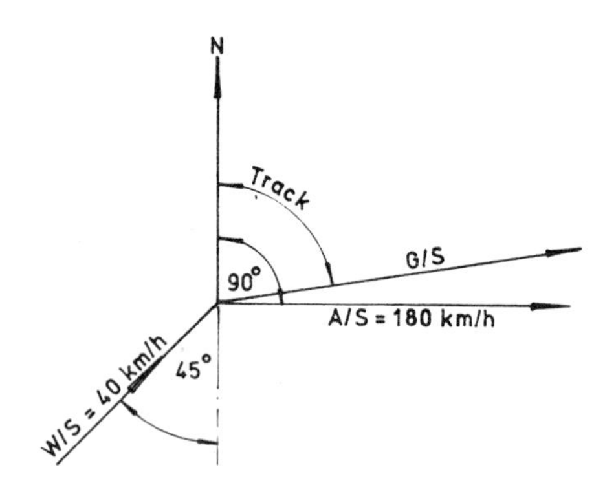

Scale 1 cm = 20 km/h

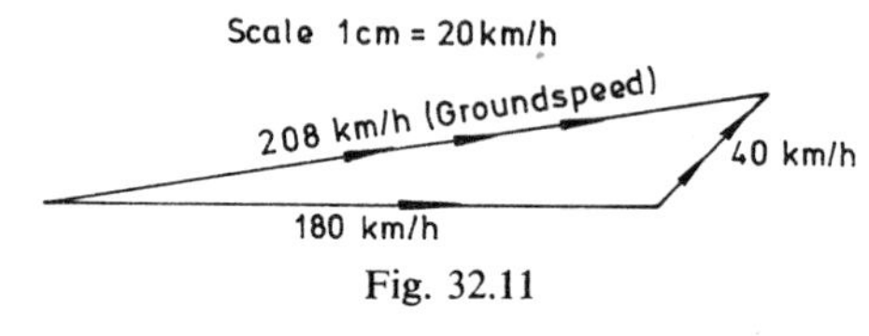

Fig. 32.11

2) From the triangle of velocities (Fig. 32.12) the course set by Q is 348°.

3) From Fig. 32.11, groundspeed of P is 208 km/h with a track of 82°.
From Fig. 32.12, groundspeed of Q is 160 km/h with a track of 0°.
After 30 minutes,
Distance travelled by P = 104 km
distance travelled by Q = 80 km

Figure 32.13 indicates the positions of P and Q at the end of 30 min. By scaling the diagram the distance between them (PQ in the diagram) is 271 km.

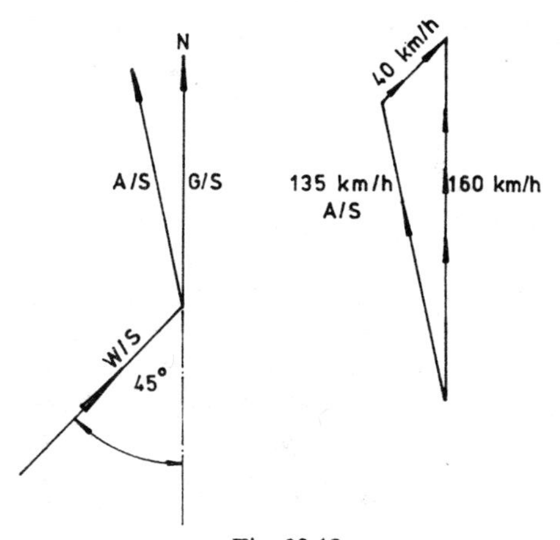

Fig. 32.12

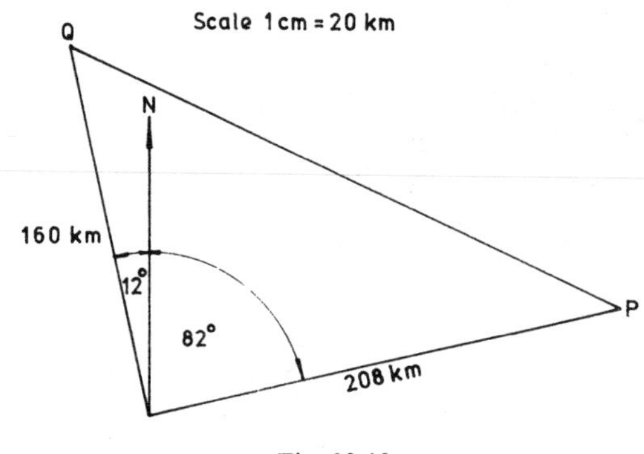

Fig. 32.13

Exercise 117

1) At 10.00 hours an aeroplane is over a point A flying at an airspeed of 250 km/h on a course of 215°. At 10.20 hours it passes over a point B which is 80 km due S.W. of A. What is the wind velocity?

2) An aerodrome Y is 300 km from aerodrome X in a direction 310° from X. A pilot leaves X and is to fly at 200 km/h. If the windspeed is 60 km/h from 240° what course must the pilot steer and how long will the flight take?

3) An aircraft is to fly in a direction 080° from an aerodrome. It is to fly to another aerodrome 600 km away. If the airspeed is to be 270 km/h and the windspeed is 50 km/h from 210° find the course and the time of flight.

4) A light aircraft with an airspeed of 180 km/h leaves X for a destination Y due north of X. The pilot sets his course due north but after flying for 1 hour, a fix shows him that he has been flying on a bearing of 340° and that he is over Z which is 200 km from X and 200 km from Y. By accurate drawing find i) the wind speed and the direction from which it is blowing; ii) the course the pilot must set to fly directly from Z to Y.

5) Two aircraft A and B set out from X at the same time. A with an airspeed of 200 km/h sets a course due west and B, with an airspeed of 150 km/h sets a course so as to travel due south. The wind is blowing at 40 km/h from the south-east. Using accurate drawings find

i) the track of A;
ii) the course set by B;
iii) indicate the positions of A and B, 45 minutes after leaving X and find their distance apart in km.

THE TRIANGLE OF VELOCITIES APPLIED TO BOATS

The motion of a boat through water is very similar to the flight of an aeroplane through air. The corresponding terms are:
for wind velocity read current velocity;
for airspeed read the speed of the boat through the water (i.e. waterspeed);
for groundspeed and track read resultant speed and the direction of motion of the boat.

Example
P is the point on the bank of a river 240 m wide flowing at 2 m/s. R is directly opposite P. A boat travels through the water at 8 m/s.
i) If the boat starts from P and heads directly across the stream how far downstream from R will it reach the bank.
ii) Calculate the speed relative to the banks with which a direct crossing from P to R will be made.

i) The triangle of velocities is shown in Fig. 32.14 from which

$$\tan\theta=\frac{2}{8}=0{\cdot}25$$

$$\theta=14^\circ$$

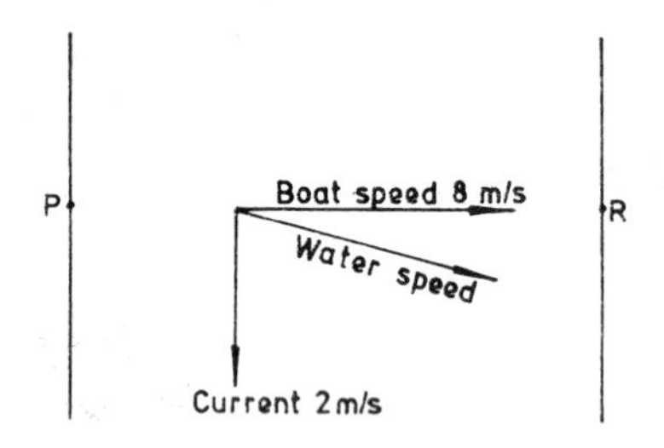

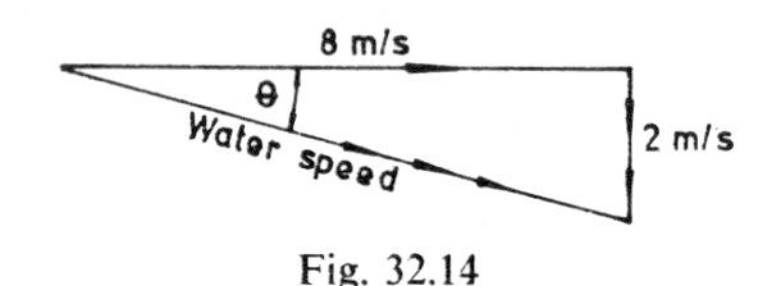

Fig. 32.14

Referring to **Fig.** 32.15, it will be seen that distance downstream from R = 240 tan 14° = 60 m

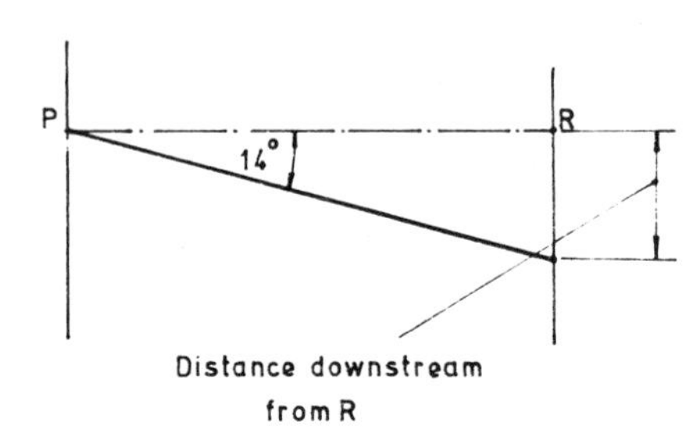

Fig. 32.15

ii)

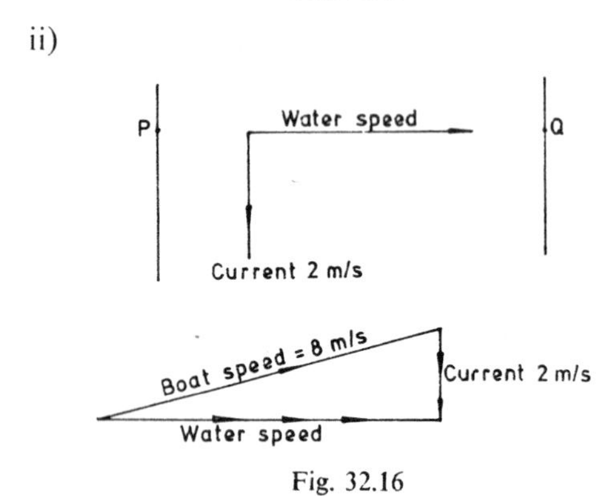

Fig. 32.16

The problem here is to calculate the water-speed. From the triangle of velocities (Fig. 32.16),

$$\text{Waterspeed}=\sqrt{8^2-2^2}=\sqrt{60}=7{\cdot}75 \text{ m/s}.$$

Exercise 118

1) A boat which has a speed of 10 km/h is to be steered across a river in which there is a steady current of 5 km/h flowing parallel to the banks. If the river is 800 m wide and the boat is to cross at right-angles to the bank find the direction in which it should be steered (relative to the bank) and the time taken to cross.

2) A stream flows at 8 km/h. A motor boat can travel at 20 km/h in still water. What angle must the boat make with the direction of the flow of the stream to reach a point directly across the stream.

3) The speed of a boat in still water is 12 km/h. The navigator wishes to travel due east from A to B (Fig. 32.17) in a current which he estimates to be of velocity 5 km/h in a direction 140°. Unknown to the navigator when he sails the actual velocity of the current is 4 km/h in a direction 140°. By drawing show the actual path of the boat and find the speed of the boat and the bearing of its path.

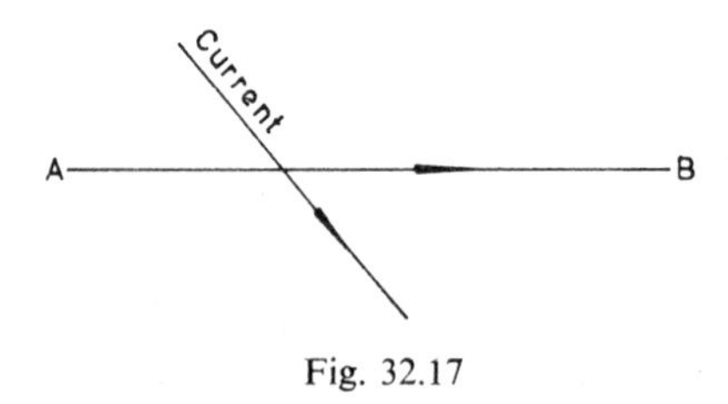

Fig. 32.17

4) A man crosses a river, flowing at 2 m/s by means of a rowing boat which he can propel through still water at 5 m/s. Q is the point on the far bank directly opposite his starting

point P. By accurate drawing or calculation find

i) at what angle to PQ he must head the boat in order to land at Q;

ii) how far downstream from Q he will land if the river is 240 m wide and he heads the boat slightly upstream at 15° to PQ.

5) A river flows at 6 km/h due south between parallel banks.

i) A motor boat whose speed in still water is 9 km/h leaves a point on one bank and steers a course due east. Find by scale drawing the actual path of the boat and state the bearing.

ii) A second motor boat, whose speed in still water is also 9 km/h, leaves a point on one bank and sets a course so as to travel due east. Find by a second scale drawing, the course set by the boat.

6) P is a point on the bank of a river 120 m wide flowing at 1 m/s. R is directly opposite P. A boat travels through the water at 4 m/s.

i) If the boat starts from P and heads directly across the stream how far downstream from R will it reach the bank.

ii) Calculate the speed relative to the banks with which a direct crossing from P to R will be made.

iii) If Q is 120 m upstream from P and on the same bank calculate the average speed of a return journey from P directly upstream to Q and back to P.

7) A man in a boat whose speed in still water is 15 km/h steers due north in a current which flows at 6 km/h from the south east. Find by drawing the actual velocity of the boat as it travels from point A to point B. If the man wishes to return from B to A in what direction should he steer?

Chapter 33 The Sphere, Latitude and Longitude

THE EARTH AS A SPHERE

The earth is usually assumed to be a sphere whose radius is 6370 km.
A circle on the surface of the earth whose centre is at the centre of the earth is called a *great circle*. The shortest distance between any two places on the earth's surface is the minor arc of the great circle passing through them. The semi-circles whose end points are N and S are called *meridians.*

LATITUDE AND LONGITUDE

Any point on the earth's surface can be fixed by using two angles, but we must first fix two reference planes at right-angles to each other. The two reference planes which are used are:
i) The equatorial plane, which is a plane through the earth's centre at right-angles to the polar axis NS. The intersection of this plane with the earth's surface is called the *equator*.
ii) A plane at right-angles to the equatorial plane containing the polar axis NS and also Greenwich. This plane is a meridian plane and its intersection with the earth's surface is called the *Greenwich meridian.*
The *latitude* of the point P (Fig. 33.1) is the angle POQ marked θ in the diagram. It is measured from 0° to 90° N or S of the equator. A circle on the surface of the earth whose centre is not the centre of the earth is called a *small circle*. Small circles whose centres lie on the polar axis NS are *circles of latitude*.
The *longitude* of the point P is the angle POP_1 or QOQ_1 marked ϕ in Fig. 33.1. It is

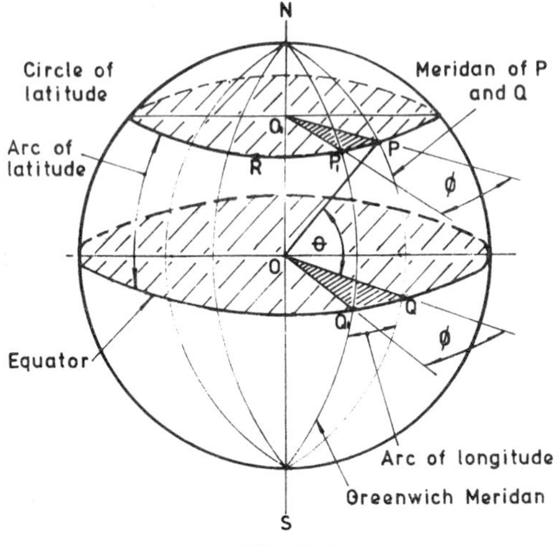

Fig. 33.1

measured east or west of the Greenwich meridian from 0° to 180° each way. The angle ϕ which denotes the longitude is rather like the angle of a slice of an orange.
It should be noted that all meridian circles are great circles but all great circles are not meridian circles. For instance, the equator is a great circle but it is not a meridian circle because it does not contain the poles N and S. All points with the *same* longitude lie on the same meridian circle. All points with the same latitude lie on the same circle of latitude. Thus in Fig. 33.1 the points P and R have the same latitude.

Examples

1) Two places A and B have the same longitude. A has latitude 30° N and B has latitude 15° S. Find the distance between them along their meridian.
This problem boils down to finding the length of the arc AB (Fig. 33.2).

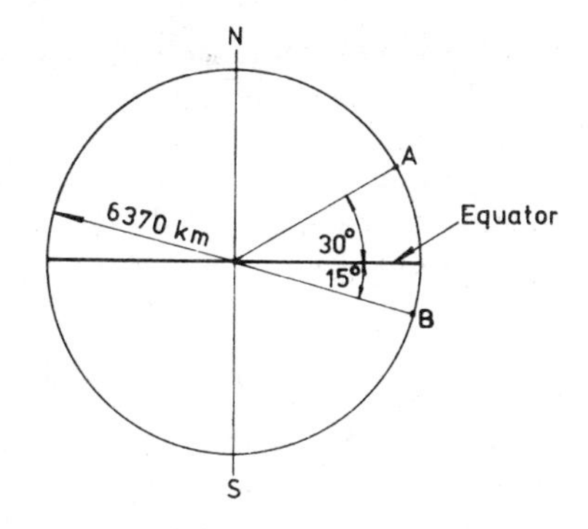

Fig. 33.2

$$\text{Length of arc AB} = 2\pi \times 6370 \times \frac{45°}{360°}$$

$$= 5005 \text{ km}$$

∴ Distance between the places A and B is 5005 km.

2) Two places P and Q are on the equator. P has longitude 20° W and Q has longitude 40° E. What is the distance between them? This problem boils down to finding the length of the arc PQ (Fig. 33.3).

$$\text{Length of arc PQ} = 2\pi \times 6370 \times \frac{60}{360}$$

$$= 6673 \text{ km}$$

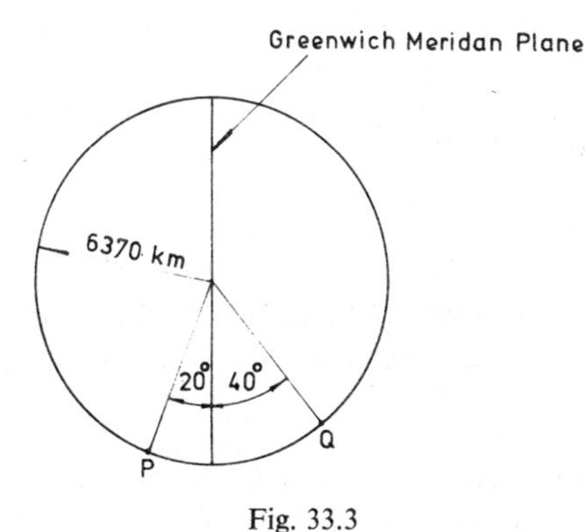

Fig. 33.3

CIRCLES OF LATITUDE

When two places have the same latitude but different longitudes, to find the distance between them we must first find the radius of the circle of latitude.

Referring to Fig. 33.4, O is the centre of the earth and θ is the latitude. The radius of this circle of latitude is AB and

$$AB = R \cos \theta$$

where R is the radius of the earth.

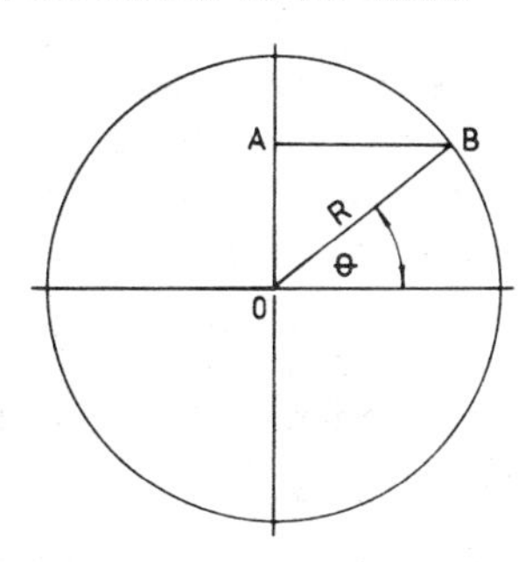

Fig. 33.4

Examples

1) Find the distance along the parallel of latitude between two places which have the latitude 30° N and which differ in longitude by 60°.

If r (Fig. 33.5) be the radius of the circle of latitude then

$$r = R \cos \theta = 6370 \times \cos 30° = 5516 \text{ km}$$

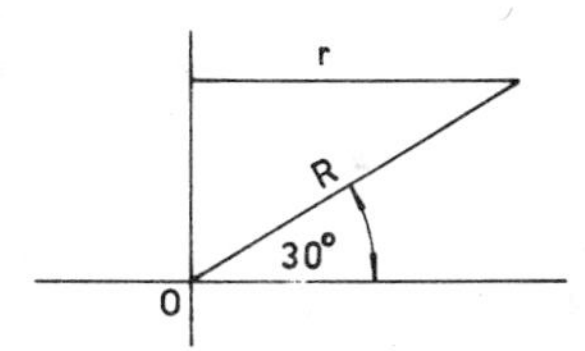

Fig. 33.5

If, in Fig. 33.6, A and B represent the two places then the problem boils down to finding the length of the arc AB.

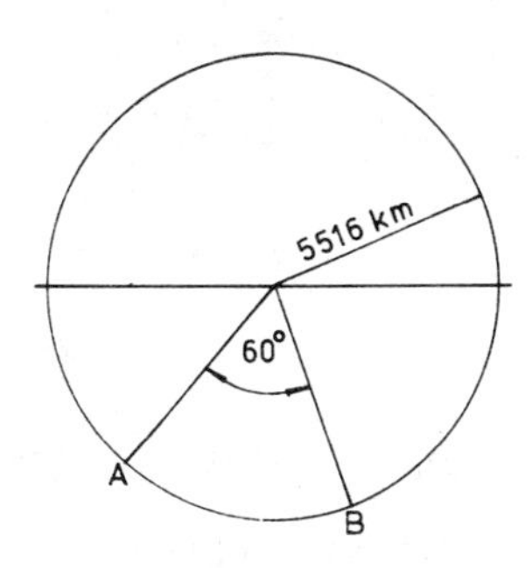

Fig. 33.6

$$\text{Length of arc AB} = 2\pi \times 5516 \times \frac{60}{360}$$

$$= 5777 \text{ km}$$

∴ Distance along the parallel of latitude between the two places is 5777 km.

2) P and Q are both in latitude 50° N and the distance between them, measured along the parallel of latitude, is 2000 km. The longitude of P is 16° 46′ W and Q is situated west of P.
Calculate the longitude of Q.
The first step is to find the radius of the circle of latitude. Thus,

$$r = R \cos \theta = 6370 \times \cos 50° = 4095 \text{ km}$$

The problem now boils down to finding the angle ϕ (Fig. 33.7).

$$2000 = 2\pi \times 4095 \times \frac{\phi}{360}$$

$$\phi = \frac{2000 \times 360}{2\pi \times 4095} = 28°$$

The longitude of Q is
28° + 16° 46′ = 44° 46′ W

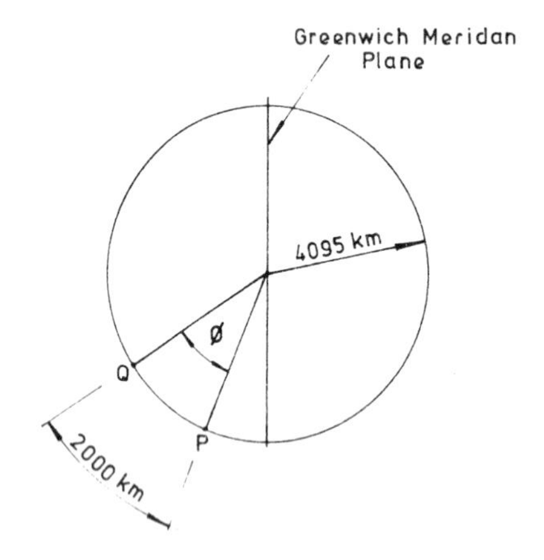

Fig. 33.7

Exercise 119

(Take the radius of the earth as 6370 km)

1) The latitudes of two places A and B are 43° N and 12° S and they both lie on the same meridian. Find the distance between them measured along the meridian.

2) The latitudes of two places P and Q are 17° S and 35° S and they both lie on the same meridian. Find the distance between them measured along the meridian.

3) The latitudes of two places are 30° 52′ N and 9° 8′ S and they both lie on the same meridian. Find the distance between them measured along the meridian.

4) Find the radius of the circle of latitude 30° S.

5) Find the distance along the parallel of latitude between two places both having a latitude of 60° N which differ in longitude by 40°.

6) The position of Leningrad is Lat. 60° N., Long. 34° E.

i) Calculate the distance from Leningrad to the North Pole.

ii) Calculate the longitude of the point P that you will reach if you travel this distance due East from Leningrad.

7) A classroom globe has a radius of 22·5 cm. Calculate, in centimetres, the distance, measured along their line of latitude, between two cities both in latitude 50° 30′ N and in longitudes 5° E and 13° 30′ E.

8) A ship steamed 500 km due east from A (Latitude 20° N., Longitude 40° W.) to B and then 300 km due N from B to C. Calculate the latitude and longitude of C to the nearest minute.

9) A ship sails due south from a port in latitude 24° S, longitude 46° 20′ W to a point in latitude 38° S. Find the distance that the ship has travelled. The ship now turns and travels the same distance due east along the line of latitude 38° S. Calculate the new longitude.

10) A ship leaves a port P which lies in latitude 20° N. It sails due east through 30°

of longitude and then due south to Q, which lies on the equator. Calculate the distance it has travelled. On the return journey it sails due west through 30° of longitude and then due N back to P. Find the difference in length between the outward and inward journeys.

11) Two places P and Q are both in latitude 61° 30′ N. Calculate the distance, in km, of either P or Q from the North Pole. Calculate

i) the distance along the latitude line from P to Q given that their longitudes are 44° W and 31° E respectively;

ii) the latitude, correct to the nearest minute, of a place 900 km due south of Q.

12) A ship is in the Atlantic Ocean in a position latitude 40° N, longitude 65° W. It sails along the parallel of latitude to a point whose longitude is 22° W. Calculate how far it has sailed.
It then sails due south for 6000 km. Calculate its new latitude.

13) Calculate the distance along the parallel of latitude between the points (64° 39′ N, 31° 28′ E) and (64° 39′ N, 28° 55′ W). Calculate also the distance, measured along the meridian, of the North Pole from either point.

14) Two points T and R on the earth's surface are each on the parallel of latitude 49° N. Their longitudes are 10° E and 135° E respectively. Calculate the distance of T from R by the shorter route along the parallel of latitude, and the great circle distance of either point from the North Pole.

15) i) The distance measured along a parallel of latitude, between two points of longitude 24° W and 28° W respectively is 300 km. Calculate the latitude of either point.

ii) Calculate the great circle distance of Vladivostock (43° 10′ N, 132° E) from the equator.

SELF TEST 28

1) In Fig. 1, $\cos\theta$ is equal to

a $\frac{p}{r}$ b $\frac{r}{p}$ c $\frac{q}{r}$ d $\frac{r}{q}$

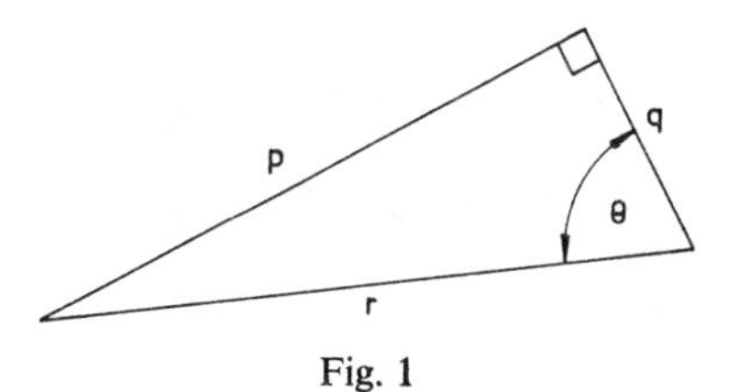

Fig. 1

2) In Fig. 2, the expression for the length BC is

a 30 tan 50° b $\frac{30}{\tan 50°}$

c 30 sin 50° d $\frac{30}{\sin 50°}$

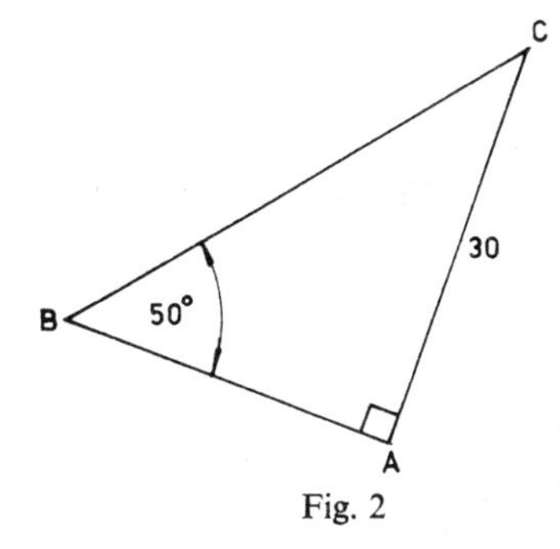

Fig. 2

3) In Fig. 3, the expression for the length DC is

a 30 sin 70° + 42 cos 44°

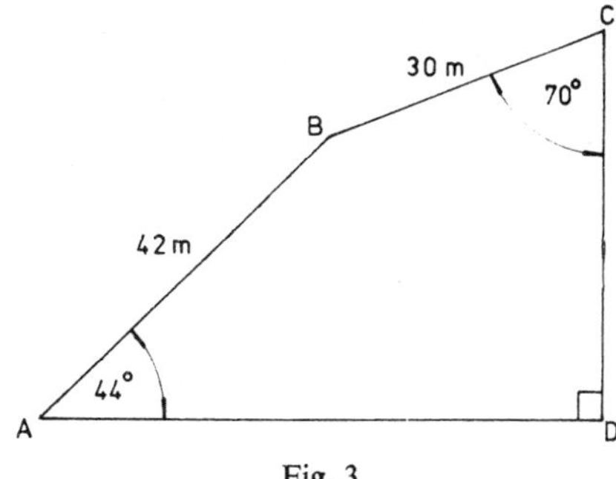

Fig. 3

b $30 \cos 70° + 42 \sin 44°$

c $\dfrac{30}{\sin 70°} + \dfrac{42}{\cos 44°}$

d $\dfrac{30}{\cos 70°} + \dfrac{42}{\sin 44°}$

4) In Fig. 4, AC is equal to

a $10(\sqrt{3}+1)$ b $\dfrac{40}{\sqrt{3}} + 10$

c $\dfrac{80\sqrt{3}}{3}$ d $\dfrac{80}{\sqrt{3}}$

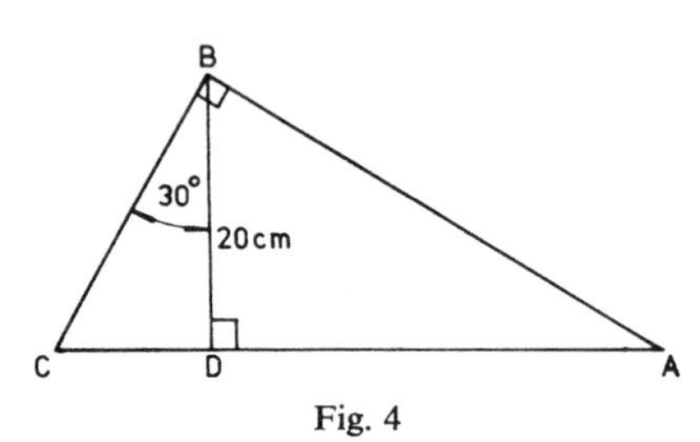

Fig. 4

5) In Fig. 5, the expression that may be used to find the angle x is

a $\tan x = \dfrac{8-3\sqrt{3}}{7}$ b $\sin x = \dfrac{8-3\sqrt{3}}{7}$

c $\tan x = \dfrac{10-3\sqrt{3}}{5}$

d $\sin x = \dfrac{10-3\sqrt{3}}{5}$

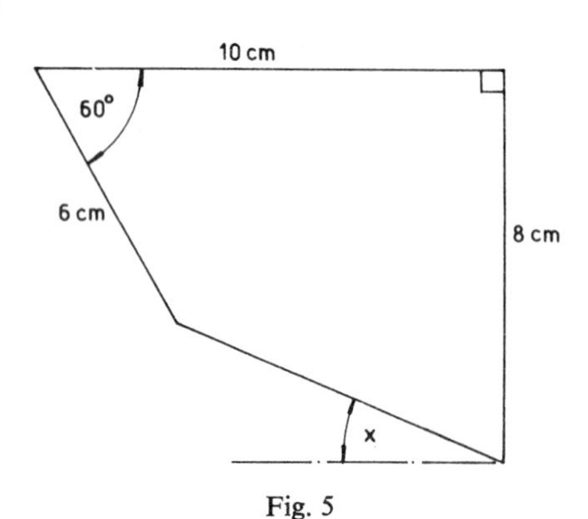

Fig. 5

6) The area of the parallelogram in Fig. 6 is

a 30 cm^2 b 15 cm^2

c $15\sqrt{3} \text{ cm}^2$ d $30\sqrt{3} \text{ cm}^2$

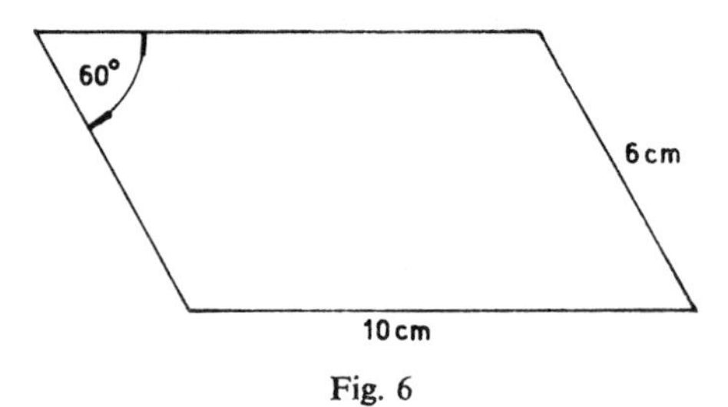

Fig. 6

7) In Fig. 7, PT and PQ are tangents to the circle. The radius of the circle is

a $8 \sin 25°$ b $8 \sin 50°$

c $8 \tan 25°$ d $8 \tan 50°$

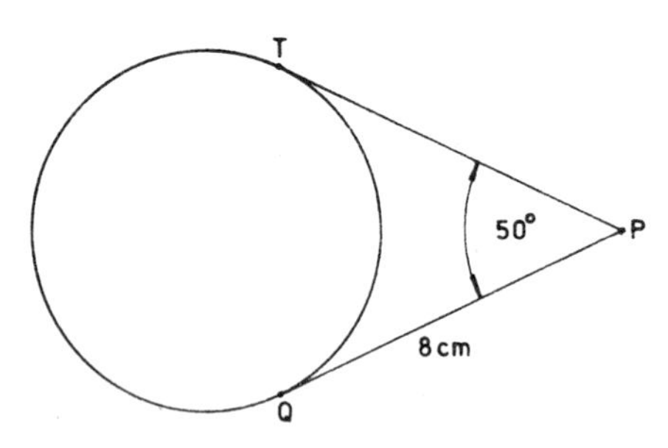

Fig. 7

8) In Fig. 8, PQ is a common tangent to two circles that touch externally. Hence

a $\sin x = \dfrac{1}{5}$ b $\tan x = \dfrac{1}{5}$

c $\cos x = \dfrac{1}{5}$

d none of these is correct

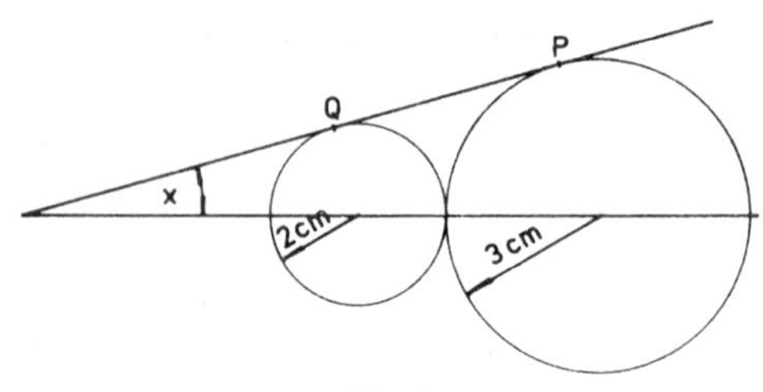

Fig. 8

9) The angle A is acute and $\tan A = \frac{12}{5}$.

Hence, sin A is equal to

a $\frac{13}{12}$ b $\frac{13}{5}$ c $\frac{12}{13}$ d $\frac{5}{13}$

10) The angle A is obtuse and $\sin A = \frac{4}{5}$. The value of cos A is

a $\frac{3}{5}$ b $\frac{3}{4}$ c $-\frac{3}{4}$ d $-\frac{3}{5}$

11) If $\tan A = \sqrt{3}$ then $\cos(180 - A)$ is equal to

a $\frac{1}{2}$ b $-\frac{1}{2}$ c $\frac{\sqrt{3}}{2}$ d $-\frac{\sqrt{3}}{2}$

12) A pyramid has a square base of side 6 cm and it is 4 cm high, the vertex being above the centre of the base. The angle of inclination of one of the faces to the base is equal to

a $36° 52'$ b $53° 7'$
c $33° 42'$ d $56° 19'$

13) cos 150° is equal to

a $-\frac{1}{2}$ b $\frac{1}{2}$

c $\frac{\sqrt{3}}{2}$ d $-\frac{\sqrt{3}}{2}$

14) In Fig. 9 given $\angle A = 140°$ and the length of the sides b and c. The side a is given by the expression

a $\sqrt{b^2 + c^2 + 2bc \cos 140°}$

b $\sqrt{b^2 + c^2 + 2bc \cos 40°}$

c $\frac{b \sin 40°}{\sin B}$ d $\frac{c \sin 140°}{\sin C}$

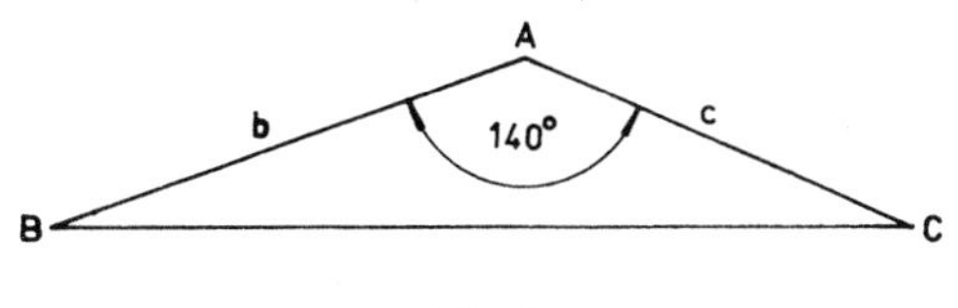

Fig. 9

15) If the angle A is obtuse and $\sin A = \frac{60}{61}$ then cos A is equal to

a $\frac{11}{61}$ b $\frac{11}{60}$ c $-\frac{11}{61}$ d $-\frac{11}{60}$

16) The pyramid in Fig. 10 has a square base of side 8 cm and it is 3 cm high. If the vertex is directly above the centre of the base then the length of the edge AB is

a 5 b $\sqrt{89}$ c $\sqrt{41}$ d $\sqrt{39}$

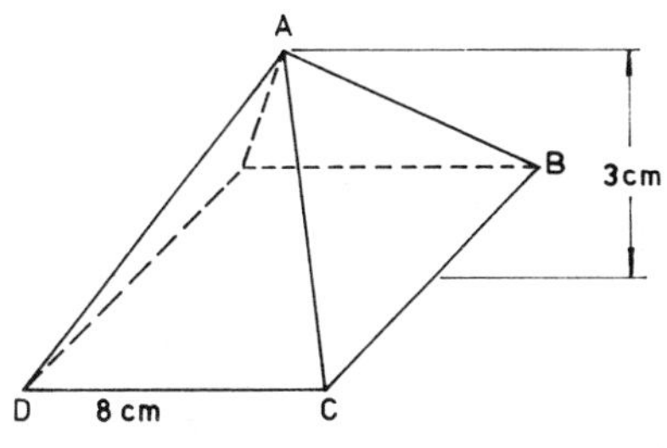

Fig. 10

17) If $\log \cos C = \bar{1}.8129$ the C is equal to

a $49° 30'$ b $49° 33'$ c $49° 24'$
d $49° 27'$

18) A bearing of N 50° W can be expressed as

a 050° b 040° c 310° d 220°

19) To find the height of a tower a surveyor stands some distance from its base and finds the angle of elevation to the top of the tower is 30°. He moves 150 m nearer to the base and finds the angle of elevation is now 60°. If the ground is horizontal, the height of the tower is

a $75\sqrt{3}$ metres b $50\sqrt{3}$ metres
c 75 metres d 50 metres

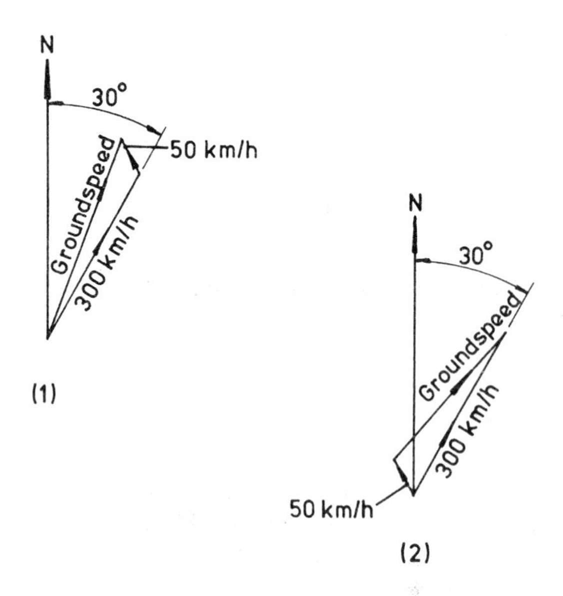

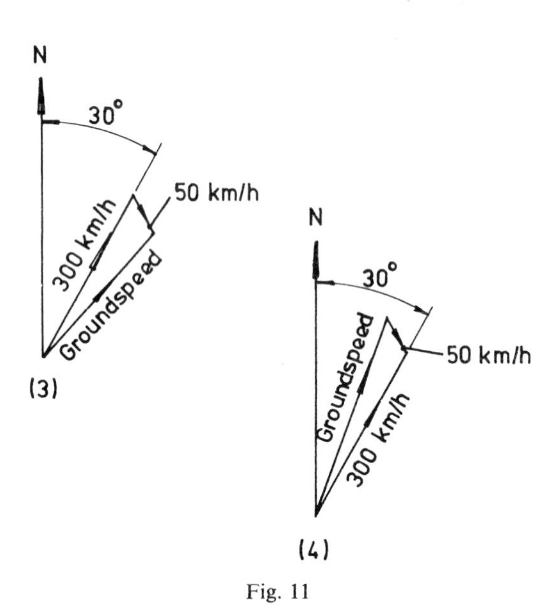

Fig. 11

20) A man on top of a cliff 20 m high observes the angle of depression of a buoy at sea as 15°. The distance of the buoy from the foot of the cliff is

a $20 \tan 15°$ b $\dfrac{20}{\tan 15°}$

c $20 \sin 15°$ d $\dfrac{20}{\sin 15°}$

21) An aeroplane has an airspeed of 300 km/h and a course of 030°. The windspeed is 50 km/h from 330°. The vector diagram to find the ground speed and track is one of the diagrams in Fig. 11. It is

a diagram (1) b diagram (2)
c diagram (3) d diagram (4)

22) A motor boat has a speed of 20 km/h through still water. The current of the stream is 6 km/h in a direction parallel to the banks. In order to cross the river at right-angles to the banks the boat will steer in a direction shown in one of the diagrams in Fig. 12. It is

a diagram (1) b diagram (2)
c diagram (3) d diagram (4)

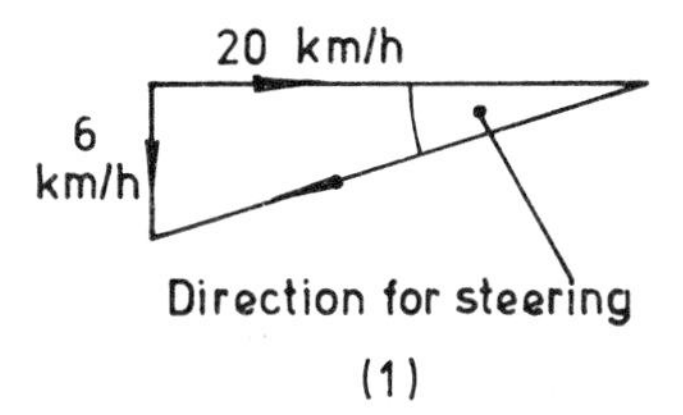

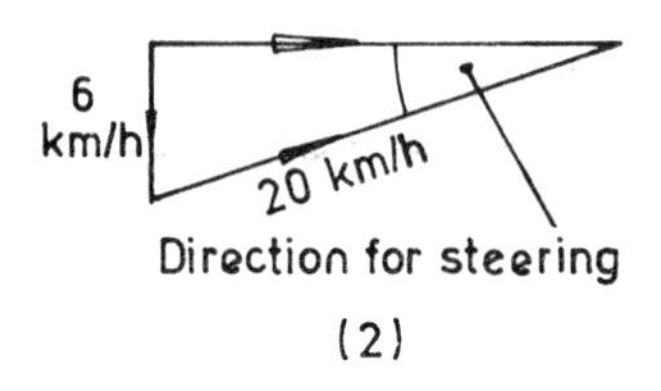

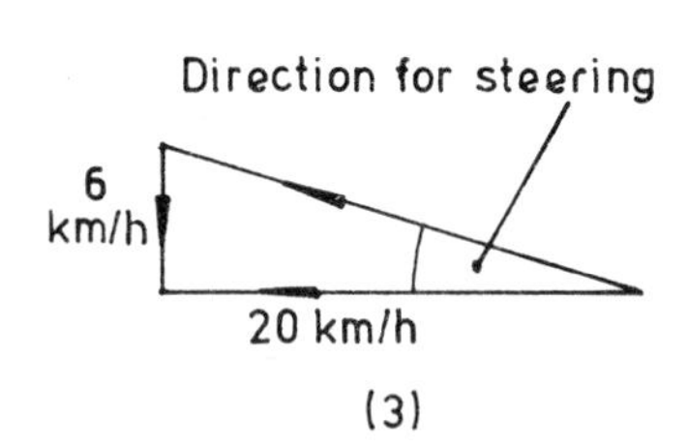

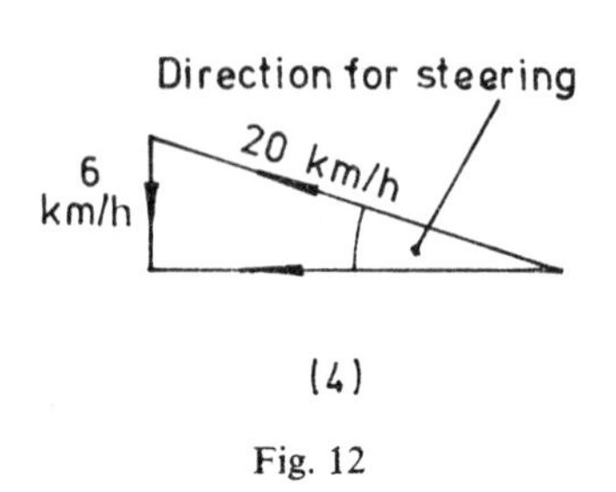

Fig. 12

23) A point A is in latitude 65° N and longitude 35° E. The distance between A and the North Pole along a meridian is given by the expression

a $2\pi \times 6370 \times \frac{35}{360}$ b $2\pi \times 6370 \times \frac{55}{360}$

c $2\pi \times 6370 \times \frac{65}{360}$ d $2\pi \times 6370 \times \frac{25}{360}$

(The radius of the earth is 6370 km)

24) Two points A and B are in lat. 50° N long. 30° E and lat. 50° N and long. 50° W. The distance between the points A and B measured along a circle of latitude is given by the expression

a $2\pi \times 6370 \cos 50° \times \frac{80}{360}$

b $2\pi \times 6370 \cos 40° \times \frac{70}{360}$

c $2\pi \times 6370 \cos 50° \times \frac{20}{360}$

d $2\pi \times 6370 \cos 40° \times \frac{20}{360}$

Miscellaneous Exercise

Exercise 120

(All taken from past 'O' Level papers.)

1) The hypotenuse of a right-angled triangle is 15 cm and one angle of the triangle is 63°. Calculate the shortest side.

2) In a triangle ABC, $\angle A=57°$, $\angle C=44°$ and $BC=8{\cdot}3$ cm. Calculate AC.

3) Use tables to evaluate the cube root of

$$\frac{\tan 38° \, 34'}{35{\cdot}71}.$$

4) In $\triangle ABC$, $AB=x$ cm, $BC=5$ cm, $AC=(x-1)$ cm and $\angle B=60°$. Find x.

5) The angle A is obtuse and $\tan A=-\frac{5}{12}$.

Without using tables evaluate $\frac{\sin A}{1-\cos A}$.

6) Two angles of a triangle are 71° and 79° and the shortest side is 25 cm. Find the longest side.

7) The angle A is acute and

$2\sin^2 A+\frac{2}{3}=\cos^2 A$. Find sin A.

8) The angle A is an obtuse angle and

$\sin A=\frac{7}{25}$. Without using tables find cos A.

9) The angle A is acute and $\sin A=0{\cdot}9474$. Use tables to find the value of $\tan(180°-A)$.

10) Find the area of a right-angled triangle which has one angle equal to 30° and whose hypotenuse is 20 cm.

11) A pyramid is 150 m high and rests on a square base of side 250 m, the vertex being above the centre of the base. Calculate the inclination of one of the faces to the ground.

12) In $\triangle ABC$, $AB=15$ cm, $AC=12$ cm and $BC=17$ cm. Calculate $\angle A$.

13) A line AB is drawn on level ground and is 56 m long. Two vertical poles AD and BC are erected with $AD=55$ m and $BC=22$ m. Calculate DC, the angle of inclination of DC to the horizontal and the area of the trapezium ABCD. A point E is taken in AB with $AE=x$ metres, so that the angles of elevation of the tops of the poles from E are equal. Calculate x.

14) Without using tables, calculate the value of $2\sin^2 A+\cos^2 A$ when $\sin A=\frac{3}{5}$.

15) In the quadrilateral ABCD, $DB=11$ cm, $\angle ADB=31°$, $\angle ABD=90°$, $\angle DBC=46°$ and $\angle BDC=62°$. Calculate a) BC; b) AB; c) the area of the quadrilateral.

16) The vertices of an equilateral triangle lie on a circle of radius 2·5 cm. Calculate the length of a side of the triangle.

17) The angle A is acute and $\tan A=\frac{15}{8}$.

Without using tables find the value of $\sin(180°-A)$. If $B=90°-A$ find the value of tan B.

18) Two circles of radii 5 cm and 4 cm touch externally. Two common tangents touch the first circle at P and R and the second circle at Q and S. PQ and RS are produced to meet at T. Find $\angle QTS$.

19) Calculate the length of the shorter diagonal of a rhombus of side 9 cm, if one angle of the rhombus is 45°.

20) In $\triangle ABC$, $AB=7$ cm, $AC=9$ cm and $BC=8$ cm. Calculate i) $\cos \angle ABC$, ii) the length of the perpendicular from A to BC. The triangle ABC is laid on a horizontal plane and rotated about BC, as an axis, until A is in the position A′ at a height of 5 cm above the horizontal plane. Calculate iii) the angle between the plane A′BC and the horizontal plane, iv) the angle between A′B and the horizontal plane.

21) Calculate the radius of a circle in which a chord of length 36 cm subtends an angle of 100° at the centre.

22) An angle x is acute and $\log \cos x = \bar{1}{\cdot}75$
Evaluate i) $6{\cdot}85 \cos x$ iii) $\sqrt{10} \cos x$
ii) $(\cos x)^2$ iv) $\log \sin x$

23) A fishing boat places a float on the sea at A, 50 m due North of a buoy B. A second boat places a float at C whose bearing from A is South 30° East. A taut net connecting A and C is 80 m long. Calculate the distance from C to B.

24) The hypotenuse of a right-angled triangle is four times the smallest side of the triangle. Calculate the smallest angle of the triangle.

25) The lines OA and OB are at right angles and OC is a third line perpendicular to the plane of the other two. $OA=OB=OC$ and P is a point on the line AB such that $AP=\frac{1}{3}AB$. If $AB=6$ cm i) show that $OC=3\sqrt{2}$ and calculate the length of CP; ii) calculate the angle between the planes ABC and AOB and iii) calculate the angle between the line CP and the plane AOB.

26) A horizontal rectangular platform ABCD is hinged to a vertical wall along the line AB. The mid-point of AB is E and $AB=60$ m, $BC=20$ m. The platform is maintained in a horizontal position by two straight chains joining C and D to a point F on the wall, vertically above E. Each chain is 50 m long. Calculate EF, $\angle ECF$ and the inclination of the plane of the two chains to the platform.

27) ABCD is a horizontal rectangular parade ground whose length AB is 120 m. A vertical flagpole TX of height 40 m stands with its foot X on CD. If $\angle ABX=50°$ and $\angle BAX=40°$, calculate
a) the lengths of BX and BC;
b) the angle of elevation from B of T, the top of the flagpole.

28) Figure 1 shows a triangular prism. The triangular faces of the prism are parallel and the other three faces are rectangles. The edges AC and CF are 3 cm and 4 cm long respectively whilst $\angle BAC$ and $\angle BCA$ are 30° and 40° respectively. Calculate:
a) the length of the edge AB;
b the angle $\angle AFB$.

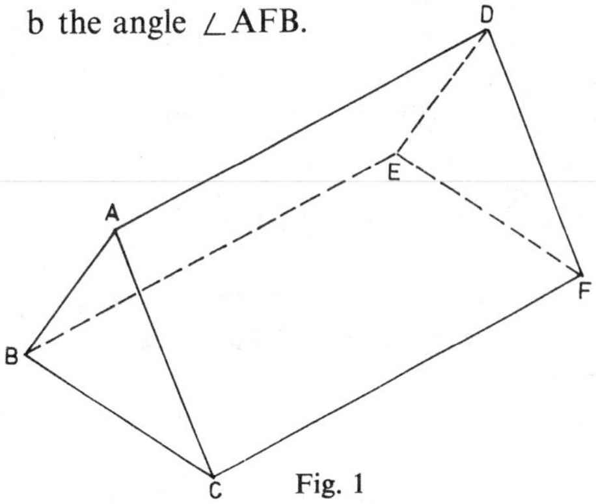

Fig. 1

29) From a point due South of a hill the angle of elevation of the summit of the hill is 23° and from a point due west of the hill the angle of elevation of the summit is 17°. The distance between the points of observation is 1000 m. Find
a) the height of the hill;
b) the bearing of the first mentioned point of observation from the second.

30) a) The point C is 6 m above level ground and 20 m measured horizontally from a vertical pole AB where B is at ground level. If $\angle ACB=64°$, calculate the length of AB.
b) A vertical wall of length 50 m and height 8 m runs due N and S. Find the area of the shadow of the wall cast on level ground by the sun shining from W at an elevation of 27°.

31) Two points A and B which are 1·64 cm apart as measured on the map are on the 350 m and 200 m contours respectively. If the elevation of A from B is 10° 26′ calculate a) the actual horizontal distance between A and B, b) the scale of the map giving your answer in centimetres to the kilometre correct to the nearest whole number.

32) The Alaska-Canada border 1126 km long runs directly along a line of longitude. Taking the earth as a sphere of radius 6373 km find the angle which the border subtends at the centre of the earth. Find also the angle which the border subtends at the South Pole.

33) Calculate the distance along the line of latitude between two points A (Latitude 57° N, Longitude 35° W) and B (Latitude 57° N, Longitude 75° W). Assume the earth is a sphere of radius 6373 km and take $\pi = 3{\cdot}142$.

34) The total length of a circle of latitude in the northern hemisphere is 8497 km. Calculate the latitude of the places on this circle. (Take $\pi = 3{\cdot}142$ and the radius of the earth as 6373 km.)

35) Leningrad is in latitude 60° N and longitude 30° E. Helsinki is in latitude 60° N and longitude 25° E. Calculate the distance between the cities measured along the circle of latitude. (Take $\pi = \frac{22}{7}$ and the radius of the earth as 6373 km.)

36) An aircraft flies from an airfield in lat. 36° S, long. 138° E to an airfield in lat. 36° N, long. 138° E, at an average speed of 480 km/h. Calculate the time of flight in hours. If the aircraft now flies due West at the same average speed, find its position $2\frac{1}{2}$ hours later.

37) A man crosses a river flowing at 1 m/s by means of a rowing boat which he propels through still water at 2·5 m/s. Q is the point on the far bank directly opposite his starting point P. Find (by calculation or drawing)

a) at what angle to PQ he must head the boat in order to land at Q;

b) how far downstream from Q he will land if the river is 40 m wide and he heads the boat upstream at 15° to PQ.

Chapter 34 Sets

SETS AND SUBSETS

A set is a collection of objects which must be clearly defined by describing it or by listing its elements or members. To make it clear that we are speaking of a mathematical set chain brackets or braces are used.
If the number of elements of a set is small enough to be listed they may be written inside chain brackets thus: $\{1, 3, 5, 7, 9\}$. This is the set of positive odd numbers less than 10. It is usual to use a capital letter to denote the set thus: $A=\{1, 3, 5, 7, 9\}$. It is possible to have a set with no elements like this $\{\ \}$. This is the **null set** which is indicated by the symbol ϕ.
$\{1, 3, 5\}$ is a **subset** of $\{1, 3, 5, 7, 9\}$ because each of the three elements in $\{1, 3, 5\}$ is an element of $\{1, 3, 5, 7, 9\}$. To indicate that one set is a subset of another set the symbol $\subset$ is used. Thus $\{5\} \subset \{3, 7, 5, 9\}$ and $\{5, 7\} \subset \{3, 7, 5, 9\}$. These two statements could be written $\{5\} \subset \{5, 7\} \subset \{3, 7, 5, 9\}$.
Note that the set $\{3, 5, 7, 9\}$ is the same as the set $\{3, 7, 5, 9\}$ since the order in which the elements are written is immaterial.

PROPER SUBSETS

Out of the set $\{7, 8, 9\}$ we can select either 0 elements, 1 element, 2 elements or three elements as follows:
1) $\{\ \}$ 2) $\{7\}$ 3) $\{8\}$ 4) $\{9\}$
5) $\{7, 8\}$ 6) $\{7, 9\}$ 7) $\{8, 9\}$
8) $\{7, 8, 9\}$.

The two extremes $\{\ \}$ and $\{7, 8, 9\}$ are, for purposes of uniformity, regarded as subsets of the original set. Every set is a subset of itself and the null set is a subset of every set. Sets numbered 1 to 7 above are called **proper subsets** because each of them has at least one element less than the original set.
Two sets are equal if every member of one set is a member of the other set. Thus

$$A=\{6, 7, 8, 9\} \qquad B=\{8, 6, 9, 7\}.$$

Every element of A is an element of B. That is

$$A\subset B \quad \text{or} \quad B\subset A \quad \text{or} \quad A=B.$$

In the set $A=\{3, 4, 5, 6, 7\}$ the fact that 5 is an element of A is written $5 \in A$. In the set $B=\{3, 7, 11\}$ 5 is not an element of B. This fact is written $5 \notin B$.
The universal set $\mathscr{E}$ is the overall set from which subsets are formed. Thus the set of even numbers is a subset of the universal set of whole numbers.

VENN DIAGRAMS

Venn diagrams are useful for representing sets. The rectangle (Fig. 34.1) $\mathscr{E}$ represents the universal set of whole numbers and the circle E represents the set of even numbers. The Venn diagram shows that $E\subset\mathscr{E}$.

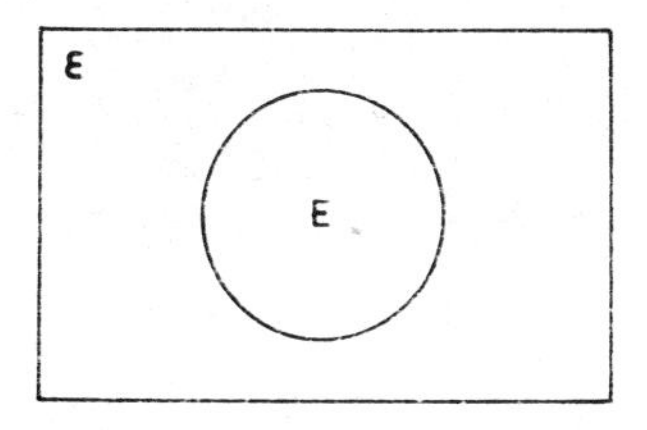

Fig. 34.1

INTERSECTION

In Fig. 34.2 the set of all the elements which are in both the sets A and B, called the intersection of A and B is represented by the shaded portion. The intersection of set A and set B is written $A \cap B$. Thus if $A=\{1, 3, 5, 7, 9\}$ and $B=(7, 9, 11)$ then $A \cap B=\{7, 9\}$.

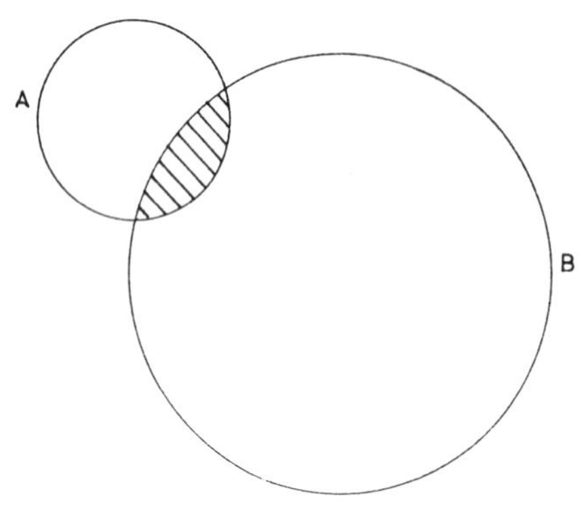

Fig. 34.2

Example. Of 150 people questioned 66 drank tea, 57 drank coffee and 58 drank milk. If 13 drank tea and coffee, 10 drank tea and milk and 11 drank coffee and milk how many drank all three?

In Fig. 34.3 let x be the number of people who drank all three beverages. This is represented by the intersection of all three circles.

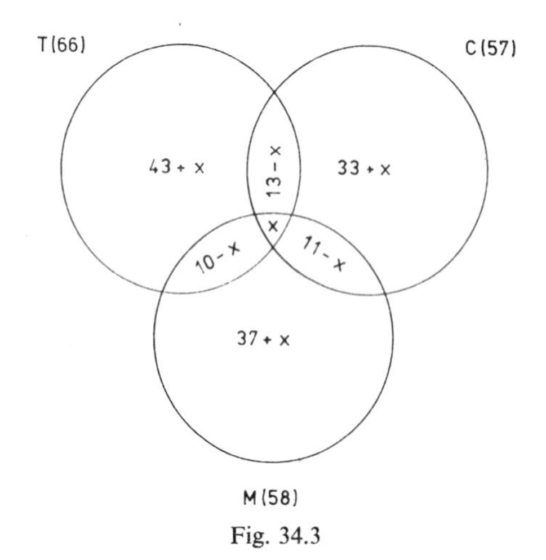

Fig. 34.3

Since 13 drank tea and coffee

$13-x$ drank tea and coffee only

10 drank tea and milk

$10-x$ drank tea and milk only

11 drank coffee and milk

$11-x$ drank coffee and milk only

66 drank tea $66-(10-x)-x-(13-x)$
$=43+x$ drank tea only

57 drank coffee $57-(13-x)-(11-x)-x$
$=33+x$ drank coffee only

and 58 drank milk $58-(10-x)-x-(11-x)$
$=37+x$ drank milk only.

$$\therefore (43+x)+(33+x)+(37+x)+(10-x) +(13-x)+(11-x)+x=150.$$

$$147+x=150$$

$$x=3$$

Hence 3 people drank all three beverages.

UNION

If $A=\{3, 4, 5\}$ and $B=\{6, 7, 8\}$ then by putting the two sets together we obtain $C=\{3, 4, 5, 6, 7, 8\}$. The set C is obtained by the union of sets A and B and we write $C=A \cup B$.

If $S=\{1, 3, 5, 7\}$ and $R=\{7, 9, 11\}$ then $S \cup R=\{1, 3, 5, 7, 9, 11\}$ which is illustrated by the shaded portion of the Venn diagram shown in Fig. 34.4. Elements which occur in more than one of the sets, in this case 7, are listed only once in the union of the sets.

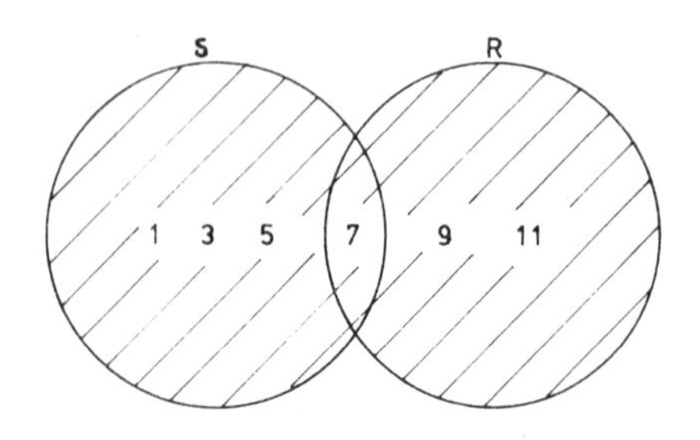

Fig. 34.4

COMPLEMENT

In the Venn diagram (Fig. 34.5) $\mathscr{E}=\{1, 2, 3, 4, 5, 6$. and A is the subset $\{1, 3, 5\}$. The remaining elements of $\mathscr{E}$ form the complement of A which is written A′.

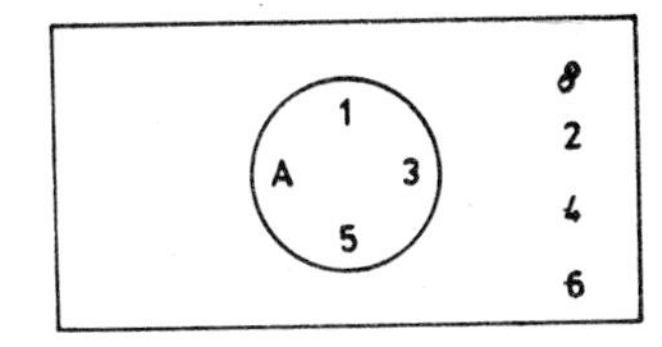

Fig. 34.5

Thus $A'=\{2, 4, 6\}$. Similarly the shaded part of Fig. 34.6. represents the complements of the set A.

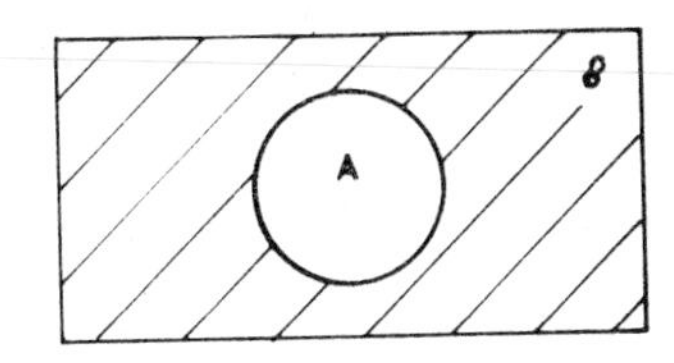

Fig. 34.6

DIFFERENCE OF TWO SETS

If A and B are two sets, $A-B$ is the set difference of A and B. If $A=\{1, 3, 5, 7, 9, 11\}$ and $B=\{5, 7, 9\}$ then $A-B=\{1, 3, 11\}$. The set $A-B$ is the shaded portion of Fig. 34.7.

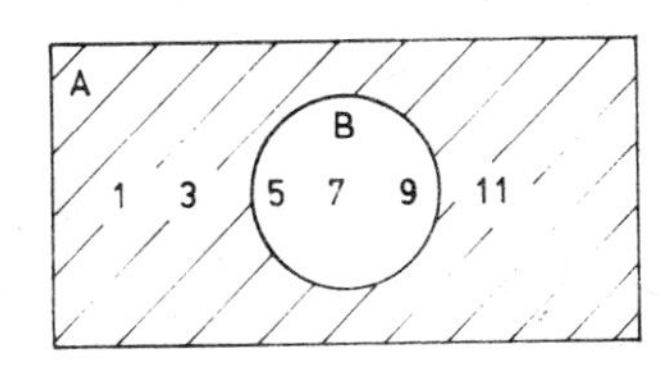

Fig. 34.7

In Fig. 34.8 A′ is the shaded part but also $\mathscr{E}-A$ is the shaded part. Hence

$$A'=\mathscr{E}-A.$$

It can also be shown that

$$A-B=A\cap B'.$$

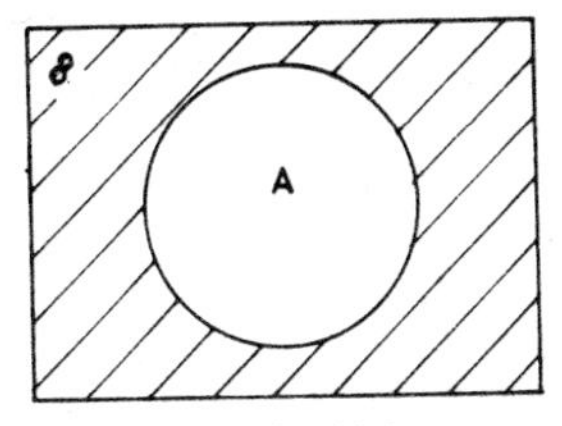

Fig. 34.8

In Fig. 34.9 the shaded part is $A\cup B$. The unshaded part is the complement of $A\cup B$ which is written $(A\cup B)'$
Also $(A\cap B)'=A'\cup B'$.

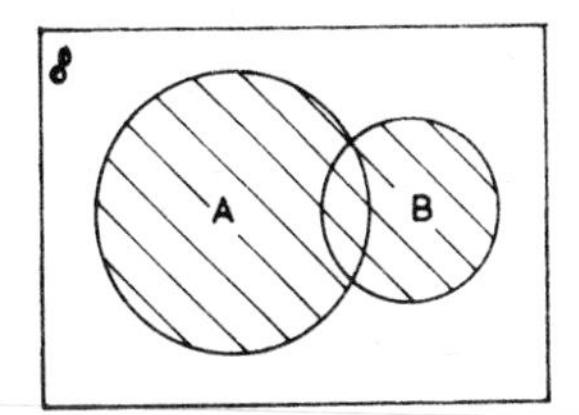

Fig. 34.9

THE NUMBER OF ELEMENTS IN A SET

If A is any set then $n(A)$ indicates the number of elements in A. It can easily be shown that

$$n(A\cup B)=n(A)+n(B)-n(A\cap B).$$

Example. In a school of 150 students 85 take Physics and 115 take Mathematics. Each student takes at least one of these subjects. How many take both?
Let P = set of physics students
and M = set of mathematics students

$$n(P\cup M)=n(P)+n(M)-n(P\cap M)$$
$$150=85+115-n(P\cap M)$$
$$n(P\cap M)=200-150=50.$$

Hence 50 take both subjects as illustrated in the Venn diagram of Fig. 34.10.
It follows that

$$n(A\cup B\cup C)=n(A)+n(B)+n(C)-n(A\cap B)-n(A\cap C)-n(B\cap C)+n(A\cap B\cap C).$$

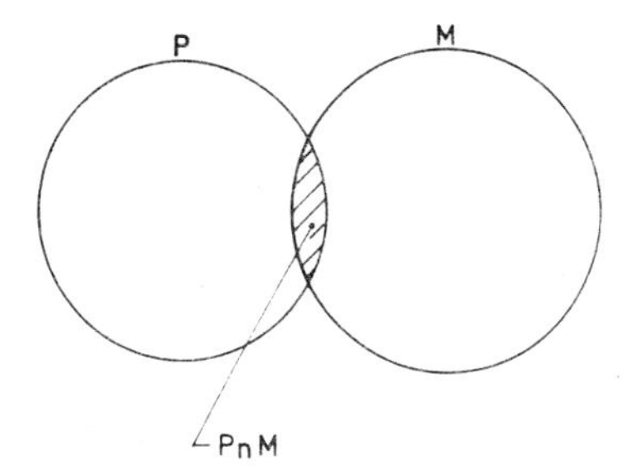

Fig. 34.10

Example. Out of three recreations, gardening, reading and playing sport 120 people were invited to state in which they were interested. 60 were interested in gardening, 86 were interested in reading and 64 were interested in playing sport. 32 gardened and read, 38 gardened and played sport whilst 34 read and played sport. How many were interested in all three activities.

$$n(G \cup R \cup S) = n(G) + n(R) + n(S) - n(G \cap R) - n(G \cap S) - n(R \cap S) + n(G \cap R \cap S)$$

$$120 = 60 + 86 + 64 - 32 - 38 - 34 + n(G \cap R \cap S)$$

$$120 = 106 + n(G \cap R \cap S)$$

$$n(G \cap R \cap S) = 14.$$

Hence 14 people were interested in all three activities.

LAWS FOR INTERSECTIONS AND UNIONS

1) $A \cup B = B \cup A$
(Compare with $a + b = b + a$ in ordinary algebra).

2) $A \cup (B \cup C) = (A \cup B) \cup C$
(Compare with $a + (b + c) = (a + b) + c$)

3) $A \cap (B \cup C) = (A \cap B) \cup (A \cap C)$
(Compare with $a \times (b + c) = a \times b + a \times c$)

Exercise 121

1) List the elements of the following: a) all odd whole numbers from 1 to 15 inclusive b) all the whole numbers between 1 and 80 inclusive which are perfect squares c) all the whole numbers between 1 and 50 inclusive which are multiples of 5.

2) If $A = \{3, 6, 9, 12, 15\}$, $B = \{3, 6, 7, 12\}$ and $C = \{3, 12\}$ which of the following are correct: a) $A \subset B$ (b) $B \subset A$ c) $C \subset A$ d) $C \subset B$ e) $C \subset B \subset A$?

3) If $A = \{1, 3\}$, $B = \{1, 3, 5, 7\}$ and $C = \{1, 3, 5\}$ is the statement $A \subset B \subset C$ correct? If not what is the correct statement?

4) How many subsets of $\{a, b, c, d\}$ are there? How many of these are proper subsets?

5) If $A \subset B$ does it imply that if $x \in A$ then $x \in B$?

6) If $A = \{2, 4, 6, 8, 10\}$ which of the following are correct: a) $2 \in A$ b) $8 \notin A$ c) $7 \in A$ d) $5 \notin A$ e) $B \subset A$ if $B = \{2, 4, 6\}$ f) $2 \in B$ g) $5 \in B$ h) $3 \notin B$?

7) In Fig. 34.11 state the meanings of the shaded areas.

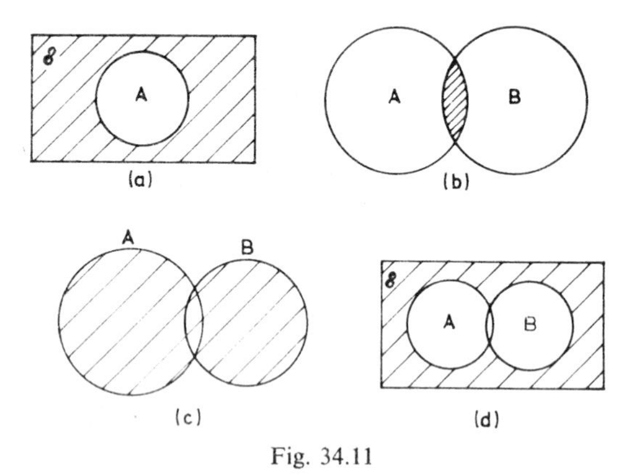

Fig. 34.11

8) Out of 136 students in a school, 60 take French, 100 take Chemistry and 48 take Physics. If 28 take French and Chemistry, 44 take Chemistry and Physics and 20 take French and Physics how many take all three subjects?

9) In a school 42% of the students take Physics, 35% take Chemistry, 30% take Botany and 20% take none of these subjects. If 9% take Botany and Physics, 10% take Botany and Chemistry and 11% take Physics and Chemistry find

a) the percentage of students that take all three subjects
b) the percentage that take Physics only
c) the percentage that take Botany and Chemistry only.

10) In an examination 60 candidates offered Maths, 80 offered English and 50 offered Physics. If 20 offered Maths and English, 15 English and Physics, 25 Maths and Physics and 10 offered all three how many candidates entered the examinations?

11) If $\mathscr{E} = \{1, 2, 3, 4 \ldots 10\}$ $A = \{1, 3, 5, 6\}$ and $B = \{2, 4, 6, 8\}$ write down

a) B' b) $A \cap B'$ c) $A - B$.

12) If $\mathscr{E} = \{3, 4, 5, 6, 7, 8, 9, 10, 11, 12\}$
$A = \{3, 4, 5, 6\}$
$B = \{5, 6, 7, 8\}$

find the following:

a) $A \cap B$ b) $B \cap A$ c) $A \cup B$
d) $A' \cup B'$ e) $(A \cap B)'$ f) $A \cap A'$

13) If $n(A) = 20$, $n(B) = 30$ and $n(A \cap B) = 10$ find $n(A \cup B)$.

14) If $n(A) = 50$, $n(B) = 60$, $n(C) = 40$, $n(A \cap B) = 20$, $n(A \cap C) = 25$, $n(B \cap C) = 15$ and $n(A \cup B \cup C) = 100$ find $n(A \cap B \cap C)$.

15) If $A \subset C$ and $B \subset C$ is $A \cap B = \phi$? Does $A \cup B \subset C$?

16) For the sets $A = \{1, 2, 3, 4\}$, $B = \{1, 2, 4\}$ and $C = \{1, 2, 3, 4, 5\}$ show that

$$A \cap (B \cup C) = (A \cap B) \cup (A \cap C)$$

Chapter 35 Matrices

Matrices are used frequently in physics and statistics and they are particularly useful when used in connection with linear simultaneous equations.

Consider the equations:

$$5x_1+8x_2=21 \qquad \ldots(1)$$

$$3x_1+2x_2=7 \qquad \ldots(2)$$

By arranging the coefficients of x_1 and x_2 in the way in which they occur in the equations we obtain the array $\begin{pmatrix}5 & 8\\3 & 2\end{pmatrix}$

This array is an example of a matrix.

If the number of rows in the matrix is m and the number of columns is n then the matrix is said to be of order $m\times n$. For the present we shall concern ourselves with matrices not larger than 2×2.

TYPES OF MATRICES

1) *Row matrix.* This is a matrix having only one row. Thus $(3 \quad 5)$ is a row matrix.

2) *Column matrix.* This is a matrix having only one column. Thus $\begin{pmatrix}1\\6\end{pmatrix}$ is a column matrix.

3) *Null matrix.* This is a matrix with all its elements zero. Thus $\begin{pmatrix}0 & 0\\0 & 0\end{pmatrix}$ is a null matrix.

4) *Square matrix.* This is a matrix having the same number of rows and columns. Thus $\begin{pmatrix}2 & 1\\6 & 3\end{pmatrix}$ is a square matrix.

5) *Diagonal matrix.* This is a square matrix in which all the elements are zero except the diagonal elements. Thus $\begin{pmatrix}2 & 0\\0 & 3\end{pmatrix}$ is a diagonal matrix. Note that the diagonal in a matrix always runs from upper left to lower right.

6) *Unit matrix.* This is a diagonal matrix in which the diagonal elements equal 1. A unit matrix is usually denoted by the symbol I. Thus

$$I=\begin{pmatrix}1 & 0\\0 & 1\end{pmatrix}$$

ADDITION AND SUBTRACTION OF MATRICES

Two matrices may be added or subtracted provided they are of the same order. Addition is done by adding together the corresponding elements of each of the two matrices. Thus

$$\begin{pmatrix}3 & 5\\6 & 2\end{pmatrix}+\begin{pmatrix}4 & 7\\8 & 1\end{pmatrix}=\begin{pmatrix}3+4 & 5+7\\6+8 & 2+1\end{pmatrix}=\begin{pmatrix}7 & 12\\14 & 3\end{pmatrix}$$

Subtraction is done in a similar fashion except the corresponding elements are subtracted. Thus

$$\begin{pmatrix}6 & 2\\1 & 8\end{pmatrix}-\begin{pmatrix}4 & 3\\7 & 5\end{pmatrix}=\begin{pmatrix}6-4 & 2-3\\1-7 & 8-5\end{pmatrix}=\begin{pmatrix}2 & -1\\-6 & 3\end{pmatrix}$$

MULTIPLICATION OF MATRICES

1) *Scalar multiplication.* A matrix may be multiplied by a number as follows:

$$3\begin{pmatrix}2 & 1\\6 & 4\end{pmatrix}=\begin{pmatrix}3\times2 & 3\times1\\3\times6 & 3\times4\end{pmatrix}=\begin{pmatrix}6 & 3\\18 & 12\end{pmatrix}$$

2) *General matrix multiplication.* Two matrices can only be multiplied together if the number of columns in the one is equal to the number of rows in the other. The multiplication is done by multiplying a row by a column as shown below.

$$\begin{pmatrix}2 & 3\\4 & 5\end{pmatrix}\times\begin{pmatrix}5 & 2\\3 & 6\end{pmatrix}=\begin{pmatrix}2\times5+3\times3 & 2\times2+3\times6\\4\times5+5\times3 & 4\times2+5\times6\end{pmatrix}$$

$$=\begin{pmatrix}19 & 22\\35 & 38\end{pmatrix}$$

$$\begin{pmatrix}3 & 4\\2 & 5\end{pmatrix}\times\begin{pmatrix}6\\7\end{pmatrix}=\begin{pmatrix}3\times6+4\times7\\2\times6+5\times7\end{pmatrix}=\begin{pmatrix}46\\47\end{pmatrix}$$

MATRIX NOTATION

It is usual to denote matrices by capital letters. Thus

$$A=\begin{pmatrix}3 & 1\\7 & 4\end{pmatrix}\quad\text{and}\quad B=\begin{pmatrix}2\\3\end{pmatrix}.$$

Generally speaking matrix products are *non-commutative*, that is

$A\times B$ does not equal $B\times A$

Examples

1) Form $C=A+B$ if

$$A=\begin{pmatrix}3 & 4\\2 & 1\end{pmatrix}\quad\text{and}\quad B=\begin{pmatrix}2 & 3\\4 & 2\end{pmatrix}$$

$$C=\begin{pmatrix}3 & 4\\2 & 1\end{pmatrix}+\begin{pmatrix}2 & 3\\4 & 2\end{pmatrix}=\begin{pmatrix}5 & 7\\6 & 3\end{pmatrix}$$

2) Form $Q=RS$ if

$$R=\begin{pmatrix}1 & 2\\3 & 4\end{pmatrix}\quad\text{and}\quad S=\begin{pmatrix}3 & 1\\5 & 6\end{pmatrix}$$

$$Q=\begin{pmatrix}1 & 2\\3 & 4\end{pmatrix}\begin{pmatrix}3 & 1\\5 & 6\end{pmatrix}=\begin{pmatrix}13 & 13\\29 & 27\end{pmatrix}$$

Note that just as in ordinary algebra the multiplication sign is omitted so we omit it in matrix algebra.

TRANSPOSITION OF MATRICES

When the rows of a matrix are interchanged with its column the matrix is said to be *transposed.* If the original matrix is A, the transpose is denoted by A'. Thus

$$A=\begin{pmatrix}3 & 4\\5 & 6\end{pmatrix}\qquad A'=\begin{pmatrix}3 & 5\\4 & 6\end{pmatrix}$$

INVERTING A MATRIX

If $AB=I$ (I is the unit matrix) then B is called the *inverse or reciprocal* of A. The inverse of A is usually written A^{-1} and hence

$$AA^{-1}=I$$

If

$$A=\begin{pmatrix}a & b\\c & d\end{pmatrix}\qquad A^{-1}=\frac{1}{ad-bc}\begin{pmatrix}d & -b\\-c & a\end{pmatrix}$$

Example.

If $A=\begin{pmatrix}4 & 1\\2 & 3\end{pmatrix}$ form A^{-1}

$$A^{-1}=\frac{1}{4\times3-1\times2}\begin{pmatrix}3 & -1\\-2 & 4\end{pmatrix}$$

$$=\frac{1}{10}\begin{pmatrix}3 & -1\\-2 & 4\end{pmatrix}=\begin{pmatrix}0{\cdot}3 & -0{\cdot}1\\-0{\cdot}2 & 0{\cdot}4\end{pmatrix}$$

To check

$$AA^{-1}=\begin{pmatrix}4 & 1\\2 & 3\end{pmatrix}\begin{pmatrix}0{\cdot}3 & -0{\cdot}1\\-0{\cdot}2 & 0{\cdot}4\end{pmatrix}=\begin{pmatrix}1 & 0\\0 & 1\end{pmatrix}$$

EQUALITY OF MATRICES

If two matrices are equal then their corresponding elements are equal. Thus if

$$\begin{pmatrix}a & b\\c & d\end{pmatrix}=\begin{pmatrix}e & f\\g & h\end{pmatrix}$$

then $a=e$, $b=f$, $c=g$ and $d=h$

SOLUTION OF SIMULTANEOUS EQUATIONS

Consider the two simultaneous equations

$3x+2y=12 \quad \ldots(1)$

$4x+5y=23 \quad \ldots(2)$

We may write these equations in matrix form as follows:

$$\begin{pmatrix}3 & 2\\4 & 5\end{pmatrix}\begin{pmatrix}x\\y\end{pmatrix}=\begin{pmatrix}12\\23\end{pmatrix}$$

If we let

$$A=\begin{pmatrix}3 & 2\\4 & 5\end{pmatrix},\quad X=\begin{pmatrix}x\\y\end{pmatrix},\quad K=\begin{pmatrix}12\\23\end{pmatrix}$$

Then $AX=K$

and $X=A^{-1}K$

$$A^{-1}=\frac{1}{3\times5-2\times4}\begin{pmatrix}5 & -2\\-4 & 3\end{pmatrix}=\begin{pmatrix}\frac{5}{7} & -\frac{2}{7}\\-\frac{4}{7} & \frac{3}{7}\end{pmatrix}$$

$$\therefore \begin{pmatrix}x\\y\end{pmatrix}=\begin{pmatrix}\frac{5}{7} & -\frac{2}{7}\\-\frac{4}{7} & \frac{3}{7}\end{pmatrix}\begin{pmatrix}12\\23\end{pmatrix}=\begin{pmatrix}2\\3\end{pmatrix}$$

$\therefore$ The solutions are $x=2$ and $y=3$

Exercise 122

1) If $A=\begin{pmatrix}3 & 2\\4 & 5\end{pmatrix}$ and $B=\begin{pmatrix}2 & 1\\3 & 3\end{pmatrix}$ form

a) $A+B$ b) $A-B$ c) AB d) BA e) A' f) B' g) A^{-1} h) B^{-1}

2) If $\begin{pmatrix}1 & 2\\3 & 4\end{pmatrix}\begin{pmatrix}2\\3\end{pmatrix}=k\begin{pmatrix}16\\36\end{pmatrix}$ find k.

3) Find the values of a and b if

$$\begin{pmatrix}a & 2\\3 & b\end{pmatrix}\begin{pmatrix}3\\4\end{pmatrix}=\begin{pmatrix}14\\21\end{pmatrix}$$

4) If $P=\begin{pmatrix}2 & 1\\3 & 5\end{pmatrix}$ find PP'

5) Solve the following simultaneous equations:

a) $x+3y=7 \quad \ldots(1)$
$2x+5y=12 \quad \ldots(2)$

b) $4x+3y=24 \quad \ldots(1)$
$2x+5y=26 \quad \ldots(2)$

c) $2x+7y=11 \quad \ldots(1)$
$5x+3y=13 \quad \ldots(2)$

VECTORS

A vector is a quantity which has both magnitude and direction, for instance a velocity of 20 km/h due east is a vector quantity and so is a force of 50 newtons acting downwards.

The point P (Fig. 35.1) has the rectangular co-ordinates $x=4$ and $y=3$ usually written (4, 3) for brevity. If OP is a vector the length OP represents the magnitude and the arrow the direction and it may be denoted by $\overrightarrow{OP}$.

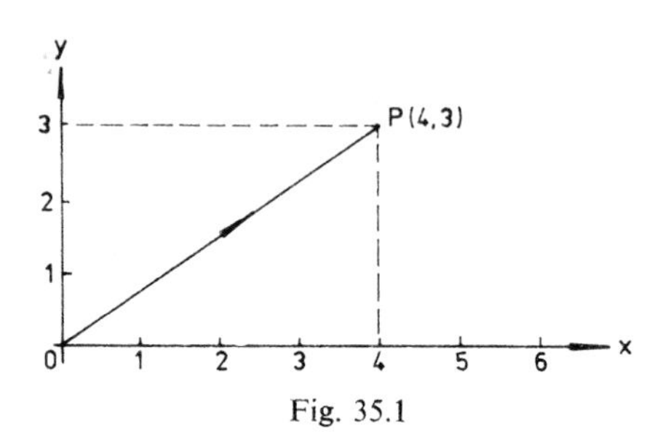

Fig. 35.1

The column matrix $\begin{pmatrix}4\\3\end{pmatrix}$ can be thought of as a column vector representing the carriage from the origin to the point P where the co-ordinates are (4, 3).

In general if the co-ordinates of P are (x, y) the column vector corresponding to $\overrightarrow{OP}$ is $\begin{pmatrix}x\\y\end{pmatrix}$. The magnitude (or length) of $\overrightarrow{OP}$ is the distance from O to P and is found by using Pythagoras. In the example of Fig. 35.1 the magnitude of $\overrightarrow{OP}$ is 5.

ADDITION OF VECTORS

Two vectors may be added by drawing a parallelogram of vectors (Fig. 35.2). If OA represents the vector P and OB the vector Q then P+Q is represented by the vector $\overrightarrow{OC}$.

We may write $\overrightarrow{OA}+\overrightarrow{OB}=\overrightarrow{OC}$ or P+Q=R.

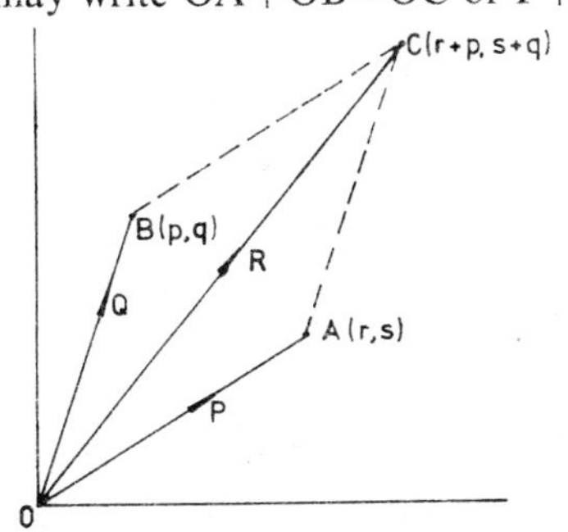

Fig. 35.2

In Fig. 35.2 $P=\begin{pmatrix} r \\ s \end{pmatrix}$, $Q=\begin{pmatrix} p \\ q \end{pmatrix}$ and hence

$$R=P+Q=\begin{pmatrix} r \\ s \end{pmatrix}+\begin{pmatrix} p \\ q \end{pmatrix}=\begin{pmatrix} r+p \\ s+q \end{pmatrix}$$

Example. If $U=\begin{pmatrix} 2 \\ 1 \end{pmatrix}$ and $V=\begin{pmatrix} 1 \\ 3 \end{pmatrix}$ find R=U+V and determine the magnitude of R.

$$R=\begin{pmatrix} 2 \\ 1 \end{pmatrix}+\begin{pmatrix} 1 \\ 3 \end{pmatrix}=\begin{pmatrix} 3 \\ 4 \end{pmatrix}$$

Since the co-ordinates of point P (Fig. 35.3) are (3, 4) the length from O to P is $\sqrt{3^2+4^2}=5$. Hence the magnitude of $\overrightarrow{OP}$ is 5.

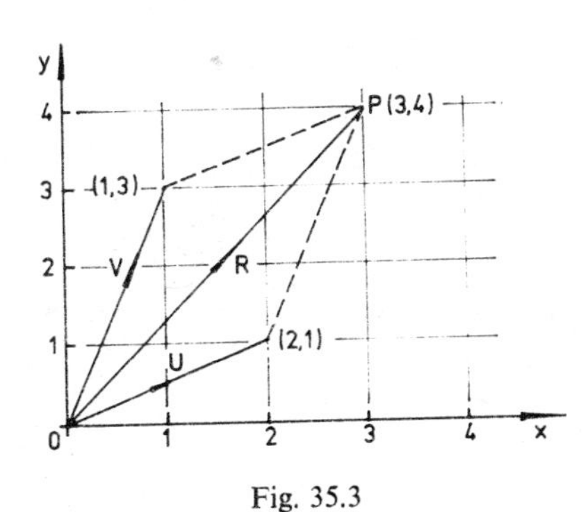

Fig. 35.3

TRANSFORMATIONS

If a point P is given a new position P′ the point has undergone a transformation. If a column vector is pre-multiplied by a 2×2 matrix it is transformed into another column vector.

Example. A point P has the co-ordinates (3, 2). The matrix which transforms P into P′ is $\begin{pmatrix} 2 & 0 \\ 1 & 1 \end{pmatrix}$. Find the co-ordinates of P′.

To find the co-ordinates of P′ the column vector $\begin{pmatrix} 3 \\ 2 \end{pmatrix}$ is pre-multiplied by the matrix $\begin{pmatrix} 2 & 0 \\ 1 & 1 \end{pmatrix}$ thus

$$\begin{pmatrix} 2 & 0 \\ 1 & 1 \end{pmatrix}\begin{pmatrix} 3 \\ 2 \end{pmatrix}=\begin{pmatrix} 6 \\ 5 \end{pmatrix}$$

Hence the co-ordinates of P′ are (6, 5).

The same transformation could have been accomplished by using a diagonal 2×2 matrix. Thus

$$\begin{pmatrix} 2 & 0 \\ 0 & 2{\cdot}5 \end{pmatrix}\begin{pmatrix} 3 \\ 2 \end{pmatrix}=\begin{pmatrix} 6 \\ 5 \end{pmatrix}$$

Most transformations are made by utilising a diagonal matrix.

Example. M is the matrix which transforms the point (2, 5) into the point (4, 8). Find M.

Let $M=\begin{pmatrix} a & 0 \\ 0 & b \end{pmatrix}$

Then $\begin{pmatrix} a & 0 \\ 0 & b \end{pmatrix}\begin{pmatrix} 2 \\ 5 \end{pmatrix}=\begin{pmatrix} 4 \\ 8 \end{pmatrix}$

$$\begin{pmatrix} 2a \\ 5b \end{pmatrix}=\begin{pmatrix} 4 \\ 8 \end{pmatrix}$$

Hence $2a=4 \qquad a=2$

$5b=8 \qquad b=1{\cdot}6$

$$M=\begin{pmatrix} 2 & 0 \\ 0 & 1{\cdot}6 \end{pmatrix}$$

The main transformations are:

1) The matrix $\begin{pmatrix}1 & 0\\0 & -1\end{pmatrix}$ which transforms $\begin{pmatrix}x\\y\end{pmatrix}$ into its reflection in the x-axis (or in the line $y=0$).

Example. $\triangle$ABC has vertices which have the co-ordinates (1, 1), (3, 1) and (1, 4) respectively. Show in a diagram the reflection in the x-axis of $\triangle$ABC.

Point A is transformed into A′ thus

$$\begin{pmatrix}1 & 0\\0 & -1\end{pmatrix}\begin{pmatrix}1\\1\end{pmatrix}=\begin{pmatrix}1\\-1\end{pmatrix}$$

Hence the co-ordinates of A′ are (1, −1). Similarly the co-ordinates of B′ are found to be (3, −1) and for C′ (1, −4). The reflection is shown in Fig. 35.4).

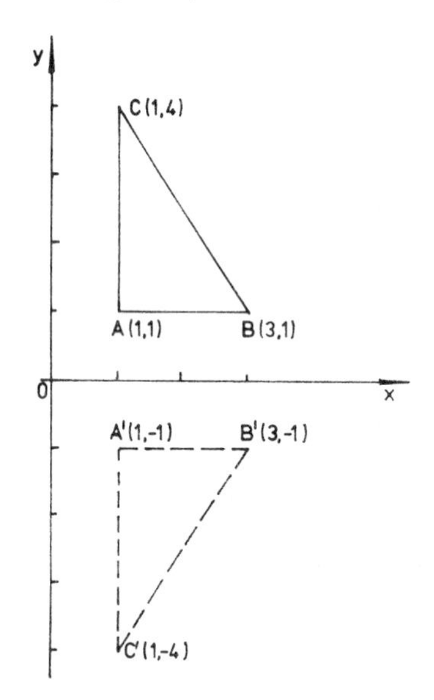

Fig. 35.4

2) The matrix $\begin{pmatrix}-1 & 0\\0 & 1\end{pmatrix}$ which transforms $\begin{pmatrix}x\\y\end{pmatrix}$ into its reflection in the y axis (or in the line $x=0$).

Example. Show the reflection of $\triangle$ABC in the previous example in the y-axis.

Point A is transformed into A′ thus

$$\begin{pmatrix}-1 & 0\\0 & 1\end{pmatrix}\begin{pmatrix}1\\1\end{pmatrix}=\begin{pmatrix}-1\\1\end{pmatrix}$$

Hence the co-ordinates of A′ are (−1, 1). Similarly the co-ordinates of B′ and C′ are (−3, 1) and (−1, 4) respectively. The reflection is shown in Fig. 35.5

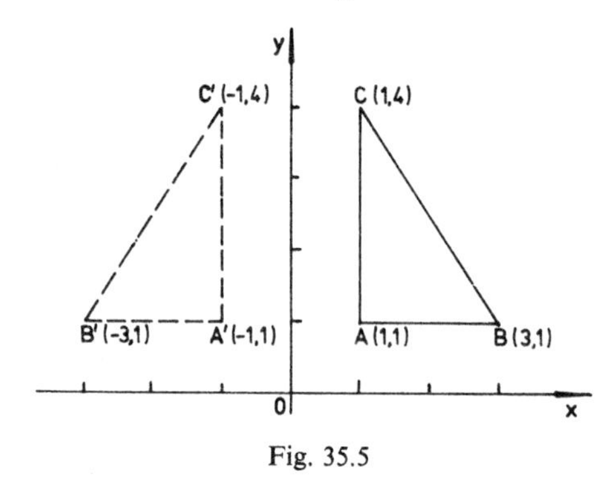

Fig. 35.5

3) The matrix $\begin{pmatrix}-1 & 0\\0 & -1\end{pmatrix}$ which transforms $\begin{pmatrix}x\\y\end{pmatrix}$ into its reflection in the origin.

Example. Transform $\triangle$ABC (see previous example) into its reflection in the origin.

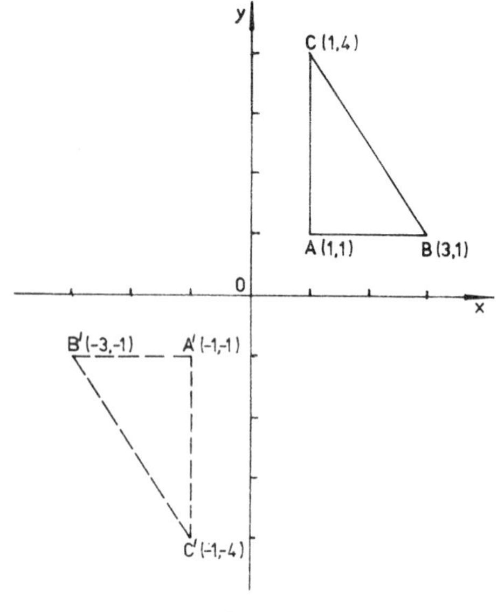

Fig. 35.6

Point A is transformed into A′ thus

$$\begin{pmatrix}-1 & 0\\ 0 & -1\end{pmatrix}\begin{pmatrix}1\\1\end{pmatrix}=\begin{pmatrix}-1\\-1\end{pmatrix}$$

The co-ordinates of A′ are therefore $(-1, 1)$ and similarly the co-ordinates of B′ and C′ are found to be $(-3, -1)$ and $(-1, -4)$ respectively. The reflection is shown in Fig. 35.6.

4) The matrix $\begin{pmatrix}0 & 1\\ 1 & 0\end{pmatrix}$ which transforms $\begin{pmatrix}x\\y\end{pmatrix}$ into its reflection in the line $y=x$. This transformation reverses the co-ordinates of a point.

Example. If a point P has the co-ordinates (3, 2) find its reflection in the line $y=x$.
Let P′ be the new point. Its co-ordinates are found thus

$$\begin{pmatrix}0 & 1\\ 1 & 0\end{pmatrix}\begin{pmatrix}3\\2\end{pmatrix}=\begin{pmatrix}2\\3\end{pmatrix}$$

Hence P′ has the co-ordinates (2, 3).

5) The matrix $\begin{pmatrix}\cos\alpha & -\sin\alpha\\ \sin\alpha & \cos\alpha\end{pmatrix}$ which rotates the point $\begin{pmatrix}x\\y\end{pmatrix}$ through the angle α counterclockwise about the origin.

Example. A point P has the co-ordinates (3, 4). Find its new co-ordinates if it is rotated through 30° counterclockwise about the origin.
The new co-ordinates are found by premultiplying the column vector $\begin{pmatrix}3\\4\end{pmatrix}$ by the matrix $\begin{pmatrix}\cos 30^\circ & -\sin 30^\circ\\ \sin 30^\circ & \cos 30^\circ\end{pmatrix}$

$=\begin{pmatrix}0{\cdot}866 & -0{\cdot}5\\ 0{\cdot}5 & 0{\cdot}866\end{pmatrix}$. Thus

$$\begin{pmatrix}0{\cdot}866 & -0{\cdot}5\\ 0{\cdot}5 & 0{\cdot}866\end{pmatrix}\begin{pmatrix}3\\4\end{pmatrix}=\begin{pmatrix}0{\cdot}598\\4{\cdot}964\end{pmatrix}$$

As can be seen from Fig. 35.7 the line OP has been rotated 30° anticlockwise to the new position OP′.

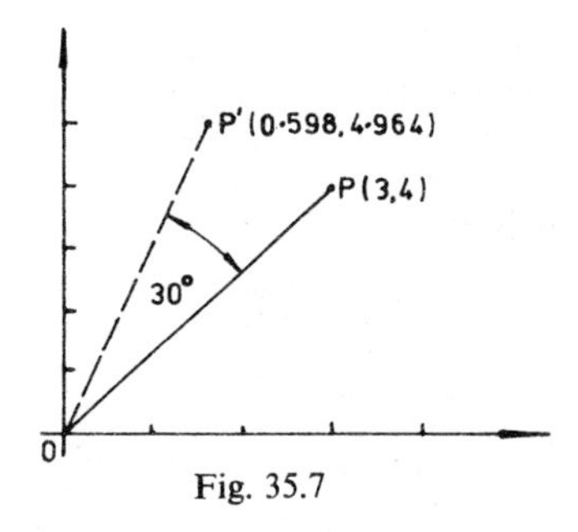

Fig. 35.7

A double transformation may be made into a single operation by multiplying the matrices, defining the two transformations, together. To find the matrix giving the single operation *pre-multiply* the matrix for the first transformation by the matrix for the second transformation.

Example. Express as a single matrix the operation $\begin{pmatrix}1 & 3\\ 0 & 1\end{pmatrix}$ followed by $\begin{pmatrix}1 & 0\\ 4 & 1\end{pmatrix}$
If M is the single matrix then

$$M=\begin{pmatrix}1 & 0\\ 4 & 1\end{pmatrix}\begin{pmatrix}1 & 3\\ 0 & 1\end{pmatrix}=\begin{pmatrix}1 & 3\\ 4 & 13\end{pmatrix}$$

Notice that the reflection shown in Fig. 35.6 is really a double transformation i.e. a reflection in the x-axis followed by a reflection in the y-axis (or vice versa). Hence the single matrix for the operation is

$$\begin{pmatrix}-1 & 0\\ 0 & 1\end{pmatrix}\begin{pmatrix}1 & 0\\ 0 & -1\end{pmatrix}=\begin{pmatrix}-1 & 0\\ 0 & -1\end{pmatrix}$$

which is the matrix which transforms $\begin{pmatrix}x\\y\end{pmatrix}$ into its reflection in the origin.

Exercise 123

1) $U=\begin{pmatrix}1\\4\end{pmatrix}$ and $V=\begin{pmatrix}3\\2\end{pmatrix}$ find $R=U+V$ and and find the magnitude of R.

2) If a vector $P=\begin{pmatrix}2\\6\end{pmatrix}$ and a vector $Q=\begin{pmatrix}4\\2\end{pmatrix}$ find, by drawing the appropriate parallelogram, $R=P+Q$.

3) P is the point (2, 4). If O is the origin calculate the distance OP.

4) The vertices of the square ABCD are respectively (0, 1), (2, 1), (2, 3) and (0, 3). Write down the transformation of these points whose matrix is $\begin{pmatrix}1 & 0\\2 & 1\end{pmatrix}$.

5) A point P has the co-ordinates (4, 2). The matrix which transforms P into P′ is $\begin{pmatrix}1 & 4\\0 & 1\end{pmatrix}$. Find the co-ordinates of P′.

6) The vertices of a triangle ABC have the co-ordinates (2, 1), (5, 1) and (5, 4) respectively. Find the co-ordinates of the reflection of $\triangle$ABC in the x-axis.

7) The vertices of a rectangle ABCD have the co-ordinates (1, 1), (3, 1), (3, 4) and (1, 4) respectively. What are the co-ordinates of the reflection of ABCD in the y-axis.

8) The point P is rotated through an angle of 60° counterclockwise about the origin to a point P_1. If P has the co-ordinates (2, 3) find the co-ordinates of P_1.

9) Express as a single matrix the operation $\begin{pmatrix}1 & 0\\2 & 1\end{pmatrix}$ followed by $\begin{pmatrix}1 & 4\\0 & 1\end{pmatrix}$.

10) Express as a single matrix the operation $\begin{pmatrix}1 & 4\\0 & 1\end{pmatrix}$ followed by a reflection in the x-axis.

Chapter 36 Number Systems

THE DECIMAL SYSTEM

In the ordinary decimal system (sometimes known as the dernary system) the digits 0 to 9 are used. Consider the number 5623. It means

$$5 \times 1000 + 6 \times 100 + 2 \times 10 + 3 \times 1$$

Remembering that $10^0 = 1$ we may write

$$5623 = 5 \times 10^3 + 6 \times 10^2 + 2 \times 10^1 + 3 \times 10^0$$

Thus $80\,321 = 8 \times 10^4 + 0 \times 10^3 + 3 \times 10^2 + 2 \times 10^1 + 1 \times 10^0$

Now consider the decimal fraction 0·3813. It means

$$\frac{3}{10} + \frac{8}{100} + \frac{1}{1000} + \frac{3}{10\,000}$$

Hence $0{\cdot}3813 = 3 \times 10^{-1} + 8 \times 10^{-2} + 1 \times 10^{-3} + 3 \times 10^{-4}$

Finally consider the number 736·58. It means

$$7 \times 10^2 + 3 \times 10^1 + 6 \times 10^0 + 5 \times 10^{-1} + 8 \times 10^{-2}$$

Note that the decimal point indicates the change from positive powers of 10 to negative powers of 10.

OTHER NUMBER SYSTEMS

It is possible to have a number system which works on the powers of any number. One of the most popular of these systems is the binary (bi meaning two) which operates to the base 2. It will be noticed that in the decimal system the greatest digit used is 9 which is one less than 10. Thus in the binary system the greatest digit that can be used is 1 which is 1 less than the base 2. Hence: A number system in the base 8 would have 7 as its greatest digit.
A number system in the base 3 would have 2 as its greatest digit.
The binary number 1011 (often written 1011_2, the small number 2 indicating the base) means

$$1 \times 2^3 + 0 \times 2^2 + 1 \times 2' + 1 \times 2^\circ$$

The number $0{\cdot}1101_2$ means

$$1 \times 2^{-1} + 1 \times 2^{-2} + 0 \times 2^{-3} + 1 \times 2^{-4}$$

The number $101{\cdot}11_2$ means

$$1 \times 2^2 + 0 \times 2^1 + 1 \times 2^0 + 1 \times 2^{-1} + 1 \times 2^{-2}$$

Likewise the number $734{\cdot}26_8$ means

$$7 \times 8^2 + 3 \times 8^1 + 4 \times 8^\circ + 2 \times 8^{-1} + 6 \times 8^{-2}$$

Example. Express 211_3 as a number in base 10.

$$211_3 = 2 \times 3^2 + 1 \times 3^1 + 1 \times 3^\circ$$
$$= 2 \times 9 + 1 \times 3 + 1 \times 1 = 18 + 3 + 1 = 22$$

ADDITION AND SUBTRACTION

In the binary system (numbers in base 2)

$1 + 0 = 1 \quad 0 + 1 = 1 \quad 1 + 1 = 0$ carry 1

Examples

1) Find $110\,101_2 + 11\,111_2$

$$\begin{array}{r} 110\,101 \\ 11\,111 \\ \hline 1010\,100 \\ \hline \end{array}$$

$110\,101_2 + 11\,111_2 = 1010\,100_2$

2) Find $10111_2 + 1011_2 + 1101_2$

$$\begin{array}{r} 10111 \\ 1011 \\ 1101 \\ \hline 101111 \\ \hline \end{array}$$

In a number system in base 3 we have

$1+0=1 \qquad 1+1=2 \qquad 1+2=0$ carry 1

Examples

1) Find $212_3 + 121_3$

$$\begin{array}{r} 212 \\ 121 \\ \hline 1110 \\ \hline \end{array} \qquad 212_3 + 121_3 = 1110_3$$

2) Find $1220_3 + 2212_3 + 111_3$

$$\begin{array}{r} 1220 \\ 2212 \\ 111 \\ \hline 12020 \\ \hline \end{array} \qquad 1220_3 + 2212_3 + 111_3 = 12020_3$$

Similar considerations apply to any other number system.

Example. Find $637_8 + 56_8$

$$\begin{array}{r} 637 \\ 56 \\ \hline 715 \\ \hline \end{array} \qquad 637_8 + 56_8 = 715_8$$

Subtraction is done similarly.

Examples

1) Find $11011_2 - 1101_2$

$$\begin{array}{r} 11011 \\ 1101 \\ \hline 1110 \\ \hline \end{array} \qquad 11011_2 - 1101_2 = 1110_2$$

2) Find $62_8 - 37_8$

$$\begin{array}{r} 62 \\ 37 \\ \hline 23 \\ \hline \end{array} \qquad 62_8 - 37_8 = 23_8$$

CONVERSION FROM ONE BASE TO ANOTHER

It is difficult to change directly from one base to another without going through 10 first because the calculation would have to be done in a strange system. A simple method of converting a number in base 10 to a number in any other base is shown in the following examples.

Examples

1) Convert 234_{10} to binary.

	REMAIN-DER	POWER OF 2	EXPLANATION
234			
117	0	2^0	$234 \div 2 = 117$ remainder 0
58	1	2^1	$117 \div 2 = 58$ remainder 1
29	0	2^2	$58 \div 2 = 29$ remainder 0
14	1	2^3	$29 \div 2 = 14$ remainder 1
7	0	2^4	$14 \div 2 = 7$ remainder 0
3	1	2^5	$7 \div 2 = 3$ remainder 1
1	1	2^6	$3 \div 2 = 1$ remainder 1
0	1	2^7	$1 \div 2 = 0$ remainder 1

Hence $234_{10} = 11101010_2$

2) Convert 413_{10} into its equivalent in base 8.

	REMAIN-DER	POWER OF 8	EXPLANATION
413			
51	5	8^0	$413 \div 8 = 51$ remainder 5
6	3	8^1	$51 \div 8 = 6$ remainder 3
0	6	8^2	$6 \div 8 = 0$ remainder 6

Hence $413_{10} = 635_8$

3) Express $(101_8 - 101_2)$ as a number in base eight.

$$101_2 = (1 \times 2^2 + 0 \times 2^1 + 1 \times 2^0)_{10} = 5_{10}$$
$$5_{10} = 5_8$$

$$\therefore\ (101_8 - 101_2) = 101_8 - 5_8 = 74_8$$

Exercise 124

1) Express each of the following in base 10
a) 326_8 b) 2120_3 c) 11011_2

2) Find 11011_2+1101_2 as a number in base 2.

3) Find $110111_2+10011_2+111_2$ as a number in base 2.

4) Find 212_3+22_3 as a number in base 3. What is the number in base 10?

5) Find 673_8+256_8 as a number in base 8.

6) Find 5321_8-677_8 as a number in base 8.

7) Find 10110_2-111_2 as a number in base 2.

8) Convert 169_{10} into a binary number.

9) Convert 3672_{10} into an octal number (i.e. in base 8).

10) Convert 358_{10} into numbers in base a) 2, b) 3 and c) 8.

11) Express $325_{10}+325_8$ as a number in base 8.

12) Express 212_3-110_2 as a number in base 2.

13) Express 222_8-222_3 as a number in base 3.

14) Express 111_8+111_2 as a number in base 8.

MODULAR ARITHMETIC

Suppose that a watch, when fully wound, goes for 32 hours and the time is now 9 a.m. When will the watch stop? To find the answer to this problem we perform the operations $(9+32)\div 24=1$ and remainder 17. That is the watch will stop at 17.00 hours (or 5 p.m.) on the second day.
In this example we were interested in both the day and time when the watch stopped and we note that the remainder in the calculation gives the time. However in many problems we are interested only in the remainder.

Example. The hand on the clock face of an instrument rotates through 2910°. Find the angle that it is to its original position.

Every 360° brings the hand back to its original position and in order to find the final angle we perform the operation

$2910\div 360=8$ and remainder 30.

The remainder gives the required angle which is 30°.

MODULO 5 SYSTEM

Figure 36.1 shows a clock face with 5 spaces on it. If the hand is on 3 where will it be if we add 4 more spaces? The answer is, of course, 2. We write

$3+4\equiv 2 \pmod 5$

which is read as "$3+4$ is congruent to 2 modulo 5."

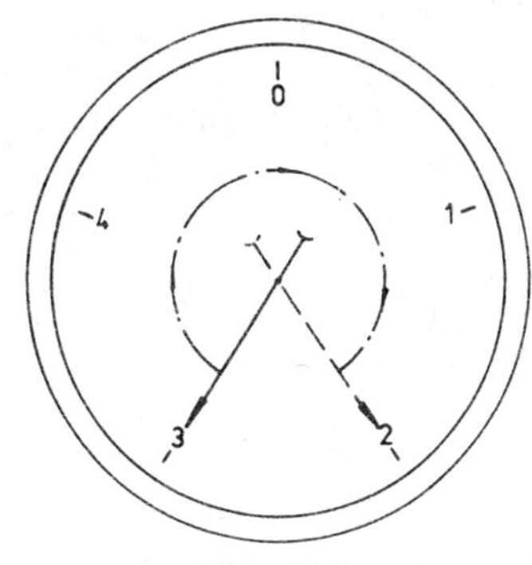

Fig. 36.1

The modulo 5 system uses only the numbers 0, 1, 2, 3 and 4 and an addition table in the modulo 5 system is shown below.

+	0	1	2	3	4
0	0	1	2	3	4
1	1	2	3	4	0
2	2	3	4	0	1
3	3	4	0	1	2
4	4	0	1	2	3

To find $2+4 \pmod 5$ look along the row marked 2 and in the column marked 4 we find the number 1.
Hence $2+4\equiv 1 \pmod 5$

Example. What is $1-3 \pmod 5$?
To find the solution consider the equation $3+x=1$. This suggests that to find $(1-3)$ mod 5 the congruence $3+x\equiv 1$ has to be solved. Looking at the addition table above we see that $x=3$. Thus $1-3\equiv 3 \pmod 5$.
A multiplication table may also be made:

×	0	1	2	3	4
0	0	0	0	0	0
1	0	1	2	3	4
2	0	2	4	1	3
3	0	3	1	4	2
4	0	4	3	2	1

Thus by finding where row 3 and column 2 meet we have $3\times 2\equiv 1 \pmod 5$. Note that $4\times 4=16$ and $16\div 5=3$ remainder 1 so that $4\times 4\equiv 1 \pmod 5$.

Example. What is $32\times 18 \pmod 5$

$$32\equiv 2 \pmod 5$$

and $$18\equiv 3 \pmod 5$$

$$32\times 18\equiv 2\times 3\equiv 1 \pmod 5$$

OTHER MODULAR SYSTEMS

There is nothing special about the modulo 5 system—any number may be used as the modulus. If the modulus is 6 then only the numbers 0, 1, 2, 3, 4 and 5 are used etc.

Examples
1) $5+4\equiv 3 \pmod 6$

2) $9+8\equiv 6 \pmod{11}$

3) $5\times 7\equiv 3 \pmod 8$

4) $10\times 8\equiv 2 \pmod{13}$

When solving congruences such as $x+3\equiv 1 \pmod 5$ it is not possible to transpose, that is we *cannot* say $x\equiv 1-3$.

Examples
1) If $x+3\equiv 2 \pmod 5$ find x.
This problem boils down to "in the modulo 5 system what number added to 3 makes 2.
The answer is clearly 4 and hence $x=4$.
Check: $4+3\equiv 2 \pmod 5$.
An alternative way of solving the congruence is to find a number such that $3+?\equiv 0 \pmod 5$. The number is 2 and hence

$$x+3+2\equiv 2+2 \pmod 5$$
$$x+0\equiv 4 \pmod 5$$
or $$x=4$$

2) If $2x+3\equiv 6 \pmod 7$ find x.
Adding 4 to each side we get

$$2x+3+4\equiv 6+4 \pmod 7$$
$$2x+0\equiv 3 \pmod 7$$
$$2x\equiv 3 \pmod 7$$

The problem now boils down to "what do we multiply 2 by in order to make $1x$." The answer is 4 because

$$2\times 4\equiv 1 \pmod 7$$
$$(2\times 4)x\equiv 3\times 4$$
$$x\equiv 5$$

Check: $(2\times 5)+3\equiv 3+3\equiv 6 \pmod 7$

3) If $2x\equiv 4 \pmod 6$ find x
The solutions are $x=2$ and $x=5$ because

$$2\times 2\equiv 4 \pmod 6 \quad \text{and} \quad 2\times 10\equiv 4 \pmod 6.$$

(Whenever the modulus is not a prime number it is possible to have more than 1 solution although sometimes no solution is possible see next example.)

4) Solve $4x\equiv 3 \pmod 6$
It is impossible to find a solution because 4 times any number will not give 3 in the modulo 6 system. Try them

$$4\times 0\equiv 0 \pmod 6 \qquad 4\times 1\equiv 4 \pmod 6$$
$$4\times 2\equiv 2 \pmod 6 \qquad 4\times 3\equiv 0 \pmod 6$$
$$4\times 4\equiv 4 \pmod 6 \qquad 4\times 5\equiv 2 \pmod 6$$

Similarly $4x\equiv 1$ and $4x\equiv 5$ have no solutions.

Exercise 125

1) In the modulo 5 system find
a) $2+4$ b) $3+4$ c) $2-4$ d) $1-2$
e) 3×3 f) 4×2 g) 27×16 h) 11×9.

2) In the modulo 11 system find
a) $7+9$ b) $2+10$ c) $8-10$ d) $3-7$
e) 8×7 f) 2×8 g) 15×18 h) 16×19.

3) Perform the following computations in the modulo 7 system
a) $3+4+6$ b) $2+3+4+5+6$
c) $(3\times 5)+(2\times 6)$ d) $(6\times 5)+(3\times 4)$

4) If $65\equiv 5 \pmod{n}$ find all the possible values of n.

5) Solve the following congruences for x:
a) $x+4\equiv 3 \pmod{5}$ b) $x+6\equiv 3 \pmod{7}$
c) $x+9\equiv 2 \pmod{11}$

6) Solve the following congruences for x:
a) $2x\equiv 4 \pmod{5}$ b) $2x+3\equiv 6 \pmod{7}$
c) $5x+3\equiv 1 \pmod{11}$ d) $2x\equiv 4 \pmod{8}$
e) $2x\equiv 4 \pmod{12}$.

Chapter 37 Statistics

RECORDING INFORMATION

Suppose that in a certain factory the number of persons employed on various jobs is as given in the following table:

TYPE OF PERSONNEL	NUMBER EMPLOYED	PERCENTAGE
Machinists	140	35
Fitters	120	30
Clerical staff	80	20
Labourers	40	10
Draughtsmen	20	5
Total	400	100

The information in the table can be represented pictorially in several ways:

1) *The pie chart* (Fig. 37.1) displays the proportions as angles (or sector areas), the complete circle representing the total number employed. Thus for machinists the angle is $\frac{140}{400} \times 360 = 126^\circ$ and for fitters $\frac{120}{400} \times 360 = 108^\circ$ etc.

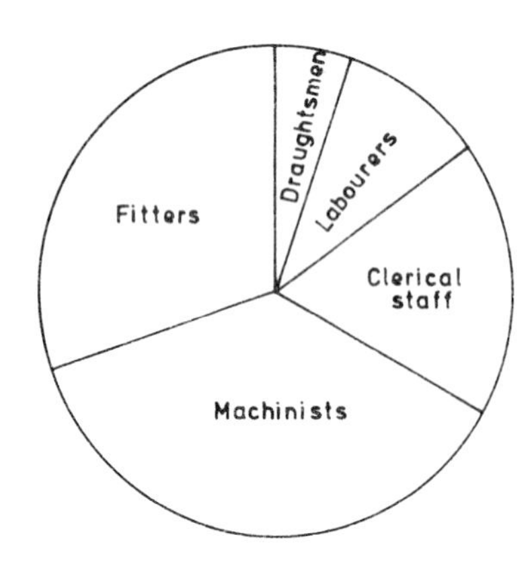

Fig. 37.1 The Pie Chart

2) *The bar chart* (Fig. 37.2) relies on heights (or areas) to convey the proportions the total height of the diagram representing 100%.

Draughtsmen 5%
Labourers 10%
Clerical staff 20%
Fitters 30%
Machinists 35%

Fig. 37.2 100% Bar Chart

3) *The horizontal bar chart* (Fig. 37.3) gives a better comparison of the various types of personnel employed but it does not readily display the total number employed in the factory.

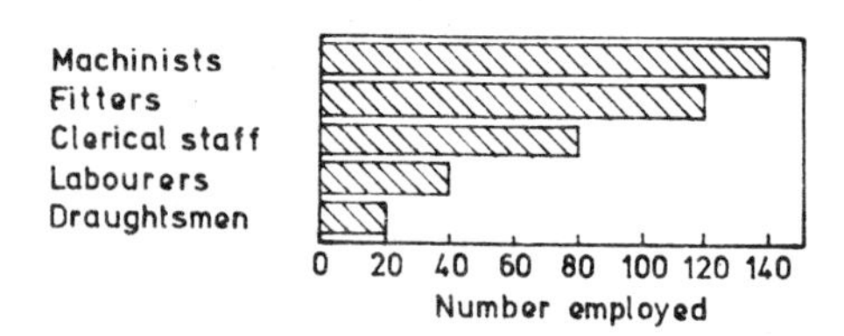

Fig. 37.3 Horizontal Bar Chart

FREQUENCY DISTRIBUTIONS

Suppose we measure the diameters of 100 ball bearings. We might get the following readings in millimetres:

15·02	15·03	14·98	14·98	15·00	15·01	15·00
15·01	14·99	15·02	14·99	15·03	15·02	15·01
15·01	15·02	15·04	14·98	15·01	15·02	14·98
15·01	15·01	15·01	14·99	15·00	15·00	15·00
15·01	15·01	15·01	15·03	14·98	14·99	14·99
14·98	15·00	14·97	15·00	15·02	15·00	15·01
15·00	14·99	15·00	15·00	15·02	14·96	15·01
14·98	15·01	15·00	15·01	15·00	15·01	14·99
15·01	15·00	14·99	15·02	14·99	15·01	15·00
15·01	15·00	14·99	14·98	14·97	14·99	15·00
14·98	14·97	15·00	14·99	14·98	15·03	14·99
15·03	15·00	14·99	14·97	15·02	15·03	15·03
14·99	15·00	14·99	14·99	14·96	15·04	14·99
15·01	15·00	15·00	15·00	15·03	14·98	14·99
15·01	14·99					

These figures do not mean very much as they stand and so we rearrange it into what is called a frequency distribution. To do this we collect all the 14·96 mm readings together, all the 14·97 mm readings and so on. A tally chart (Table 1) is the best way of doing this. Each time a measurement arises a tally mark is placed opposite the appropriate measurement. The fifth tally mark is usually made in an oblique direction thus tying the tally marks into bundles of five to make counting easier. When the tally marks are complete the marks are counted and the numerical value recorded in the column headed 'frequency'. The frequency is the number of times each measurement occurs. From Table 1 it will be seen that the measurement 14·96 occurs twice (that is, it has a frequency of 2), the measurement 14·97 occurs four times (a frequency of 4) and so on.

TABLE 1

MEASURE-MENT (mm)	NUMBER OF BALL BEARINGS WITH THIS MEASUREMENT	FREQUENCY
14·96	11	2
14·97	1111	4
14·98	1111 1111 1	11
14·99	1111 1111 1111 1111	20
15·00	1111 1111 1111 1111 111	23
15·01	1111 1111 1111 1111 1	21
15·02	1111 1111	9
15·03	1111 111	8
15·04	11	2

THE HISTOGRAM

The frequency distribution becomes even more understandable if we draw a diagram to represent it. The best type of diagram is the histogram (Fig. 37.4) which consists of a set of rectangles, each of the same width, whose heights represent the frequencies. On studying the histogram the pattern of the variation is easily understood, most of the values being grouped near the centre of the diagram with a few values more widely dispersed.

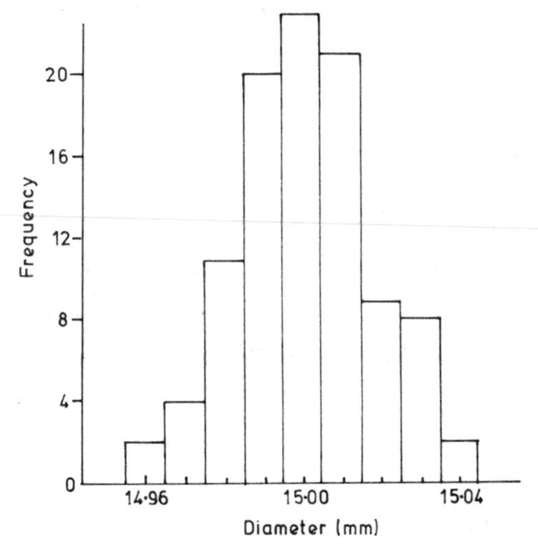

Fig. 37.4

GROUPED DATA

When dealing with a large number of observations it is often useful to group the data into classes or categories. We can then determine the number of items which belong to each class thus obtaining a class frequency. Table 2 shows the results of a test given to 200 students (the maximum mark being 20.)

TABLE 2

MARK	FREQUENCY
1–5	22
6–10	55
11–15	93
16–20	30

From Table 2 we see that the first class consists of marks from 1 to 5. 22 students obtained marks in this range and the class frequency is therefore 22. A histogram of this data may be drawn (Fig. 37.5) by using the mid-points of the class intervals as the centres of the rectangles.

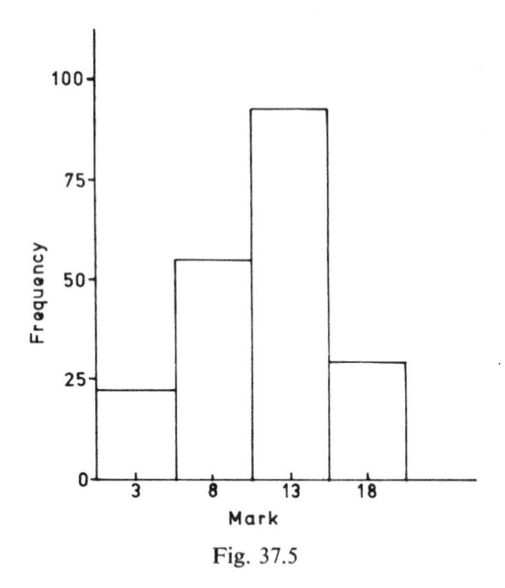

Fig. 37.5

STATISTICAL AVERAGES

In statistics the following kinds of average are used.

1) *The arithmetic mean.* This is the commonest type of average and it is determined by adding up all the items in the set and dividing the result by the number of items. Thus

$$\text{Arithmetic mean} = \frac{\text{the total of the items}}{\text{the number of items}}$$

Example. The marks of a student in five examinations were 84, 90, 72, 60 and 74. Find the arithmetic mean of his marks.

$$\text{Arithmetic mean} = \frac{84+90+72+60+74}{5}$$

$$= \frac{380}{5} = 76$$

The arithmetic mean of a frequency distribution must take into account the frequencies as well as the measured observations.

If $x_1, x_2, x_3 \ldots$ are the measured observations and $f_1, f_2, f_3 \ldots$ are the corresponding frequencies then the arithmetic mean is

$$\bar{x} = \frac{x_1 f_1 + x_2 f_2 + x_3 f_3 \ldots}{f_1 + f_2 + f_3 \ldots}$$

Example. Find the arithmetic mean of the data in Table 1.

x	f	xf
14·96	2	29·92
14·97	4	59·88
14·98	11	164·78
14·99	20	299·80
15·00	23	345·00
15·01	21	315·21
15·02	9	135·18
15·03	8	120·24
15·04	2	30·08
Totals	100	1500·09

The arithmetic mean is

$$\bar{x} = \frac{1500{\cdot}09}{100} = 15{\cdot}0009 \text{ mm}$$

The arithmetic mean of a grouped distribution is found by taking the values of x as the class mid-points.

Example. Find the arithmetic mean for the information given in Table 2.

CLASS	x	f	xf
1–5	3	22	66
6–10	8	55	440
11–15	13	93	1209
16–20	18	30	540
Totals		200	2255

Hence the arithmetic mean is

$$\bar{x} = \frac{2255}{200} = 11{\cdot}275 \text{ marks}$$

2) *The median.* If a distribution is arranged so that all the items are arranged in ascending (or descending) order of size the median is the value which is half-way along the series. Generally there will be an equal number of items below and above the median. If there is an even number of items the median is found by taking the average of the two middle items.

Examples

1) The median of 3, 4, 4, 5, 6, 8, 8, 9, 10 is 6

2) The median of 3, 3, 5, 7, 9, 10, 13, 15 is $\frac{1}{2}(7+9)=8$.

Since, in a histogram, area corresponds to frequency the median divides the histogram into two equal areas (Fig. 37.6). Hence the median occurs at half of the total frequency.

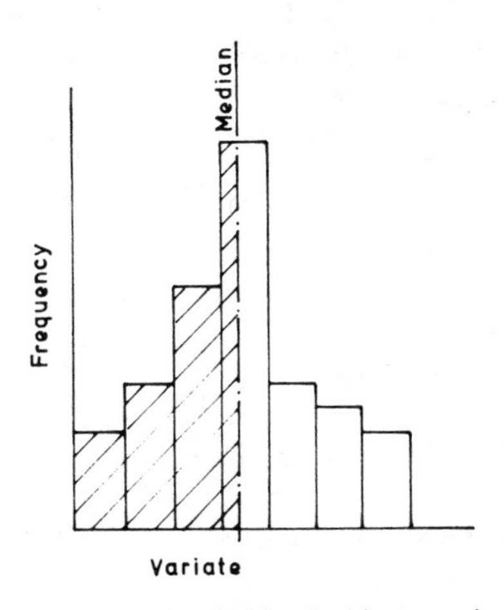

Fig. 37.6 The median divides the histogram into two equal areas and hence the shaded area corresponds to half of the total frequency.

However the easiest way of finding the median of a frequency distribution is to draw a cumulative frequency curve as shown in the next example.

Example. The data in the table below relates to the mass in grammes of packets of chemical. By drawing a cumulative frequency diagram find the median of this distribution.

MASS (GRAMMES)	FREQUENCY
119	5
120	9
121	19
122	25
123	18
124	4

The cumulative frequencies are given below:

MASS (GRAMMES)	CUMULATIVE FREQUENCY
not more than 119·5	5
,, ,, ,, 120·5	5+9=14
,, ,, ,, 121·5	14+19=33
,, ,, ,, 122·5	33+25=58
,, ,, ,, 123·5	58+18=76
,, ,, ,, 124·5	76+4=80

Notice that the mass of 119 grammes includes all the masses between 118·5 and 119·5 grammes. Likewise the mass 120 grammes includes all the masses between 119·5 grammes and 120·5 grammes.

The cumulative frequency curve is shown in Fig. 37.7. Since the total frequency is 80 the median corresponds to a frequency of 40, that is, 121·8 grammes.

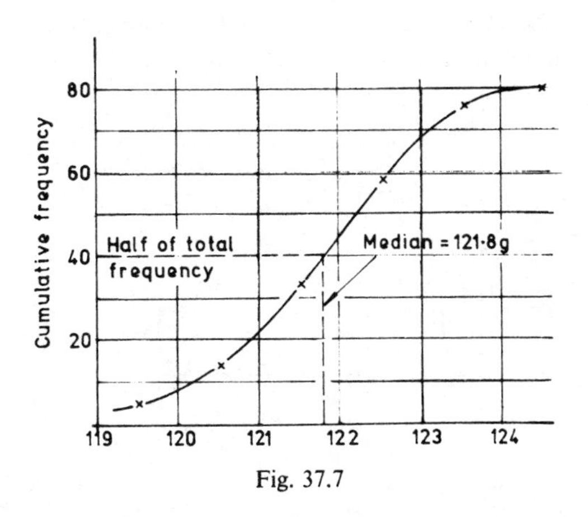

Fig. 37.7

3) *The mode.* The observation which occurs most frequently in a distribution is called the mode. A frequency curve may be drawn to represent a frequency distribution by joining up the tops of the rectangles in the histogram (Fig. 37.8). The mode is then the value of the variate corresponding to the maximum point on the curve.

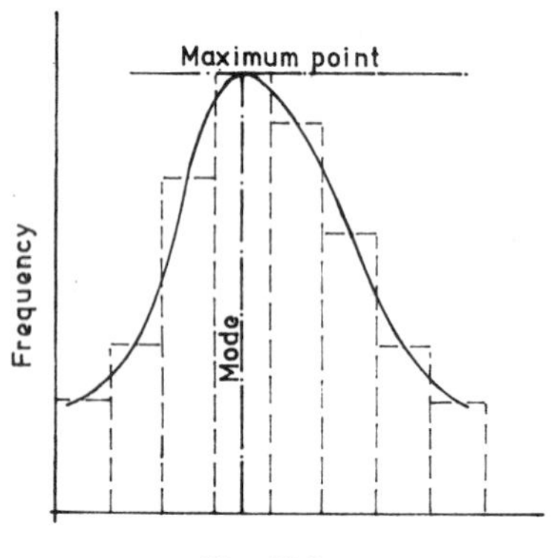

Fig. 37.8

Example
The heights of a group of boys are measured to the nearest centimetre with the results as follows:

Height	157	158	159	160	161
Frequency	20	36	44	46	39

Height	162	163	164	165	166	167
Frequency	30	22	17	10	4	2

Find the mode.
By constructing the histogram and hence the frequency curve (Fig. 37.9) the mode is found to be 159·7 cm. It is worth while noting that the model class is 159·5 to 160·5 cm.

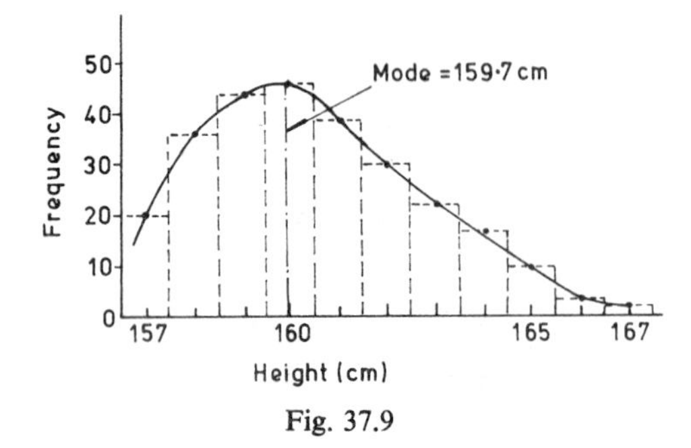

Fig. 37.9

When the frequency distribution curve is symmetrical the mean, median and mode have the same value but for unsymmetrical curves they have different values (Fig. 37.10).

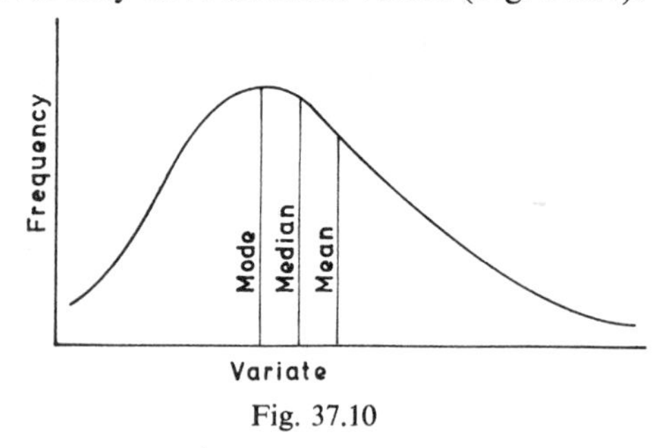

Fig. 37.10

Exercise 126

1) A building contractor surveying his labour force finds 30% are engaged on factories,' 40% are engaged on house building and 30% are working on public works (schools, hospitals etc).

a) Draw a pie chart of this information.
b) Present the information on a single vertical bar chart.

2) An industrial organisation gives an aptitude test to all applicants for employment. The results for 150 people taking the test were:

Score (out of 10)	1	2	3	4	5
Frequency	6	12	15	21	35

Score (out of 10)	6	7	8	9	10
Frequency	24	20	10	6	1

Draw a histogram.

3) The marks obtained in a school examination in mathematics by 80 students were as follows:

77	68	60	73	79	96	74	74
85	84	68	79	62	78	53	87
75	75	74	88	67	89	76	75
76	82	69	73	97	61	62	65
63	68	77	60	78	75	78	61
72	90	94	93	85	95	88	72
81	62	75	71	76	60	57	63
73	88	82	59	65	79	73	78
67	76	78	85	71	83	80	95
86	93	66	75	75	71	65	62

Arrange this information in a frequency table taking the classes as 50–54, 55–59, 60–64 etc. Hence draw a histogram to represent the frequency distribution.

4) The diameters of 40 steel bars were measured with the following results (all in millimetres):

24·98	24·96	24·97	24·98	24·99
25·02	24·99	25·01	25·03	25·01
25·00	25·02	25·00	25·02	25·01
25·02	25·01	24·97	24·98	25·01
25·03	25·05	24·95	24·98	24·99
24·99	25·02	24·97	25·04	25·00
24·97	25·04	25·00	25·00	24·99
25·01	25·03	25·03	25·02	25·01

Draw up a frequency table and hence draw a histogram to represent the frequency distribution.

5) A cricketer in 8 innings made the following scores: 19, 8, 23, 0, 17, 32, 18 and 3. What is his mean score?

6) The lengths of 100 pieces of wood, all nominally the same length, were measured with the following results.

Length (cm)	29·5	29·6	29·7	29·8	29·9
Frequency	2	4	11	18	31

Length (cm)	30·0	30·1	30·2	30·3	30·4
Frequency	22	8	2	1	1

Find the mean length.

7) The table below gives a frequency distribution for the lifetime of cathode ray tubes. Calculate the mean lifetime of these tubes.

Lifetime (hours)	300–399	400–499	500–599
Frequency	28	90	117

Lifetime (hours)	600–699	700–799	800–899
Frequency	152	134	127

Lifetime (hours)	900–999	1000–1099	1100–1199
Frequency	100	62	10

8) Find the median for the sets of numbers:

a) 2, 3, 3, 5, 6, 7, 8
b) 4, 4, 6, 8, 9, 11, 14, 16
c) 8, 9, 8, 7, 9, 6, 10, 7, 9

9) Find the median for the following frequency distribution:

Variate	18–26	27–35	36–44	45–53
Frequency	6	10	18	23

Variate	54–62	63–71	72–80
Frequency	11	9	3

10) The wages of 60 employees of a company are as follows. Find the median.

Wage £	20	21	22	23	24	25	26
Frequency	7	9	16	13	9	5	1

11) The table below gives the number of marks scored by 30 children in a class in a particular test

Number of marks	0	1	2	3	4	5
Frequency	1	6	5	6	4	8

State the mode of this distribution, What is the median?

12) Find the mode of the frequency distribution shown in question 10.

Miscellaneous Exercises

Exercise 127

1) Write down the inverse of the matrix

$$\begin{pmatrix} 2 & -1 \\ 2 & 4 \end{pmatrix}$$

2) 101_8 is a number in base 8 and 101_2 is a number in base 2. Express $(101_8 - 101_2)$ as a number in base 8.

3) Solve the equations
$2x - y = 10$
$3x + 2y = 8$

4) $\mathscr{E}$ = natural numbers
$A = \{1, 3, 5, 7, 9, 11\}$
$B = \{1, 2, 3, 5, 8, 13\}$

Write down 1) $A \cup B$; 2) $A \cap B'$.

5) The table below shows the marks scored by 50 students in a test.

Number of marks	0	1	2	3	4	5
Frequency	1	12	14	15	7	1

a) Find the mode of this distribution.
b) Find the median.

6) In a college 100 students take Maths, 140 take English and 70 take Physics. If 30 take Maths and English, 20 English and Physics and 40 Maths and Physics and 15 take all three how many students are there in all?

7) $26_8 + 51_8 = x_{10}$. Find x.

8) If $M = \begin{pmatrix} 2 & 4 \\ 1 & 0 \end{pmatrix}$ and $N = \begin{pmatrix} 2 & 1 \\ 0 & 3 \end{pmatrix}$ find $MN - NM$.

9) The mean of n numbers is 20. If the same numbers together with 30 have a mean of 22 find n.

10) If $\begin{pmatrix} 2 & x \\ 4 & y \end{pmatrix}\begin{pmatrix} 5 \\ 2 \end{pmatrix} = \begin{pmatrix} 14 \\ 30 \end{pmatrix}$ find x and y.

11) Matrices A and B are given as follows:

$$A = \begin{pmatrix} 1 & -1 \\ 2 & 3 \end{pmatrix} \quad B = \begin{pmatrix} 3 & 1 \\ -2 & 1 \end{pmatrix}$$

a) Find the products BA and AB.
b) Find the matrix X such that $AX = B$.

12) A point P is rotated through an angle of 45° anticlockwise about the origin to a point P_1. The point P_1 is then reflected in the line $y = 0$ and becomes P_2. If P has the co-ordinates (2, 1) find
a) the co-ordinates of P_1
b) the co-ordinates of P_2.

13) If a set $A = \{2, 7, 9\}$, set $B = \{1, 5, 9\}$ and set $C = \{7, 2, 8\}$ find
a) $A \cup B$ b) $B \cap C$
c) $A \cap B$ d) $A \cup C$.

14) Perform the following computations in the modulo 12 system:
a) $8 + 9$ b) $7 - 4$
c) 8×7 d) 62×123

15) Packets of a chemical were weighed with the following results:

Weight (kg)	0·996	0·997	0·998	0·999
Frequency	1	6	25	72

Weight (kg)	1·000	1·001	1·002
Frequency	93	69	27

Weight (kg)	1·003	1·004
Frequency	6	1

Find the mean weights of the packets.

16) Solve the following congruence for x:

$5x + 6 \equiv 1 \pmod 7$

17) If $\mathscr{E}=(1, 3, 5, 7, 9, 11, 13, 15)$ and $A=\{1, 2, 3, 4, 5, 6\}$ and $B=\{3, 5, 7, 9\}$ find $A \cap B'$.

18) The table below gives the distribution of marks in a test for which the maximum mark was 20.

Marks	1–4	5–7	8	9	10	11	12	13–15	16–20
Frequency	11	20	23	24	30	25	21	33	13

a) Find the median
b) Calculate the mean mark.

19) If $8x+6 \equiv 1 \pmod{11}$ find x.

20) Find 583_8+111_2 as a number in base 10.

SELF TEST 29

1) If $56_8+17_8=x_{10}$ then x is

a 59 b 61 c 73 d 75

2) If $65=5 \bmod n$ then n is

a 0 b 5 c 10 d 13

3) If the circles in Fig. 1 represent the sets P, Q and R then the shaded area is represented by

a $P \cap Q \cap R$ b $P' \cap Q \cap R$
c $P' \cap (Q \cup R)$ d $Q \cap R$

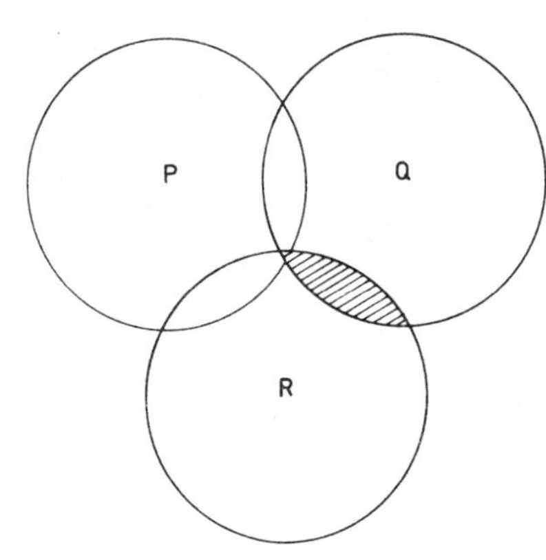

Fig. 1

4) A matrix $\begin{pmatrix} 1 & 0 \\ 0 & -1 \end{pmatrix}$ represents the transformation of

a a rotation through 90° clockwise
b a rotation through 90° anticlockwise
c a reflection in the line $x=0$
d a reflection in the line $y=0$

5) The mean of n numbers is 9. If the same numbers together with 14 have a mean of 10, then n is

a 4 b 5 c 9 d 10

6) $n(A \cup B)$ is equal to

a $nA+nB$ b $nA-nB$
c $nA+nB+n(A \cap B)$
d $nA+nB-n(A \cap B)$

7) If $A=\{1, 3, 6, 9, 12, 15\}$, $B=\{3, 6, 7, 12\}$ and $C=\{3, 9\}$ then

a $A \subset B$ b $B \subset A$ c $C \subset A$
d $C \subset B$

8) If the circles in Fig. 2 represent the sets A and B then the shaded area represents

a $A \cup B$ b $(A \cup B)'$ c $A \cap B$
d $(A \cap B)'$

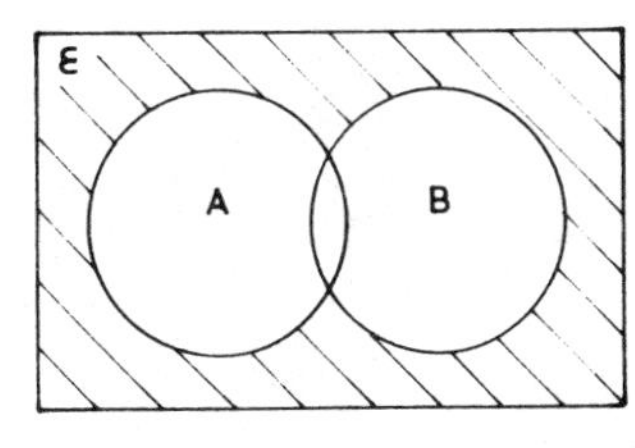

Fig. 2

9) If $\mathscr{E}=\{1, 2, 3, 4, 5, 6, 7, 8, 9, 10\}$, $A=\{1, 3, 5\}$ and $B=\{1, 2, 4, 6, 8\}$ then $(A \cup B)'$ is equal to

a $\{7, 9, 10\}$ b ϕ
c $\{1, 2, 3, 4, 5, 6, 8\}$ d $\{1\}$

10) If $M=\begin{pmatrix}2 & 3\\4 & 5\end{pmatrix}$ and $N=\begin{pmatrix}3 & 1\\2 & 2\end{pmatrix}$ then MN is

a $\begin{pmatrix}12 & 8\\22 & 14\end{pmatrix}$ b $\begin{pmatrix}10 & 14\\12 & 16\end{pmatrix}$

c $\begin{pmatrix}6 & 3\\8 & 10\end{pmatrix}$ d $\begin{pmatrix}5 & 4\\6 & 7\end{pmatrix}$

11) If $A=\begin{pmatrix}a & b\\c & d\end{pmatrix}$ then A^{-1} is equal to

a $\begin{pmatrix}\frac{1}{a} & \frac{1}{b}\\ \frac{1}{c} & \frac{1}{d}\end{pmatrix}$ b $\begin{pmatrix}c & d\\a & b\end{pmatrix}$

c $\frac{1}{ad-bc}\begin{pmatrix}d & -b\\-c & a\end{pmatrix}$

d $\frac{1}{ad-bc}\begin{pmatrix}b & -d\\-a & c\end{pmatrix}$

12) $U=\begin{pmatrix}x\\y\end{pmatrix}$ and $V=\begin{pmatrix}a\\b\end{pmatrix}$. If $R=U+V$ then the magnitude of R is

a $\sqrt{x^2+y^2}$ b $\sqrt{a^2+b^2}$

c $\sqrt{(x+a)^2+(y+b)^2}$

d $\sqrt{(x+a)+(y+b)}$

13) Expressed as a single matrix the operation $\begin{pmatrix}1 & 0\\3 & 2\end{pmatrix}$ followed by $\begin{pmatrix}4 & 1\\1 & 0\end{pmatrix}$ is

a $\begin{pmatrix}5 & 1\\4 & 2\end{pmatrix}$ b $\begin{pmatrix}4 & 1\\14 & 3\end{pmatrix}$ c $\begin{pmatrix}7 & 2\\1 & 0\end{pmatrix}$

d $\begin{pmatrix}4 & 2\\5 & 1\end{pmatrix}$

14) 101_3-101_2 is equal to

a 0 b 5_{10} c 12_{10} d 12_8

15) If $6x+5\equiv 3 \pmod{11}$ then x is equal to

a 7 b 0 c 3 d 5

16) The mode of the set of numbers 1, 3, 3, 5, 7, 8, 9, 9, 9, is

a 7 b 6 c 9, d 5

17) The median of 3, 4, 4, 4, 5, 6, 8, 8, 8, 8, is

a 5·5 b 8 c 4 d 5·8

Chapter 38 Arithmetical and Geometrical Series

SERIES

A set of numbers which are connected by some definite law is called a *series* or *progression*. Each of the numbers forming the set is called a term of the series. The following sets of numbers are examples of series:
1, 3, 5, 7 . . . (each term is obtained by adding 2 to the previous term)
1, 3, 9, 27 . . . (each term is three times the preceding term)

SERIES IN ARITHMETICAL PROGRESSION

A series in which each term is obtained by adding or subtracting a constant amount is called an arithmetical progression (often abbreviated to A.P.) Thus in series 1, 4, 7, 10 . . . the difference between each term and the preceding one is 3. The difference between one term and the preceding term is called the *common difference*. Thus for the above series the common difference is 3.
Some further examples of series in A.P. are
0, 5, 10, 15 . . . (common difference is 5)
6, 8, 10, 12 . . . (common difference is 2)

GENERAL EXPRESSION FOR A SERIES IN A.P.

Let the first number in the series be a and the common difference be d. The series can then be written as

$$a, a + d, a + 2d, a + 3d \dots$$

The first term is a
The second term is $a + d$
The third term is $a + 2d$
The fourth term is $a + 3d$
Notice that the coefficient of d is always one less than the number of the term. Thus
the 8th term is $a + 7d$
the 19th term is $a + 18d$

Examples

1) Find the 9th and 18th terms of the series 1, 5, 9 . . . The 1st term is 1 and hence $a = 1$
The common difference is 4 and hence $d = 4$
The 9th term $= a + 8d = 1 + 8 \times 4 = 33$
The 18th term $= a + 17d = 1 + 17 \times 4 = 69$

2) The 5th term of a series in A.P. is 19 and the 11th term is 43. Find the 20th term.
The 5th term is $a + 4d = 19$ . . . . . (1)
The 11th term is $a + 10d = 43$ . . . (2)
Subtracting equation (1) from (2)

$$6d = 24$$
$$d = 4$$

Substituting for d in equation (1)

$$a + 4 \times 4 = 19$$
$$a = 3$$

The 20th term is $a + 19d = 3 + 19 \times 4 = 79$

3) Find the number of the term which is 65 in the series 2, 5, 8 . . .
Here $a = 2$ and $d = 3$
Let the nth term be 65 then

$$a + (n - 1)d = 65$$
$$2 + 3(n - 1) = 65$$
$$3(n - 1) = 63$$
$$n - 1 = 21$$
$$n = 22$$

Hence the 22nd term is 65

4) Insert five arithmetic means between 2 and 26. This question poses the problem of finding

five numbers between 2 and 26 such that
2 ? ? ? ? ? 26 forms a series in A.P.
The 1st term is 2 hence $a = 2$
the 7th term is 26 hence $a + 6d = 26$

$$2 + 6d = 26$$
$$6d = 24$$
$$d = 4$$

2nd term is $a + d = 2 + 4 = 6$
3rd term is $a + 2d = 2 + 8 = 10$
4th term is $a + 3d = 2 + 12 = 14$
similarly the 5th term is 18 and the 6th term is 22.
The series is 2, 6, 10, 14, 18, 22, 26.

SUM OF A SERIES IN A.P.

The sum of a series in A.P. is

$$S = \frac{n}{2}(a + l)$$

Where n = the number of terms in the series
a = the 1st term in the series
l = the last term in the series

Example

Find the sum of the series 5, 9, 13 . . . if the series consists of 12 terms.
We have $a = 5$ and $d = 4$ with $n = 12$
$$l = a + 11d = 5 + 44 = 49$$

Hence the sum is

$$S = \frac{12}{2}(5 + 49) = 6 \times 54 = 324$$

Exercise 128

1) Find the 10th term of the series 1, 4, 7, 10 . . .

2) Find the 8th term of the series 5, 11, 17 . . .

3) The 3rd term of a series in A.P. is 32 and the 7th term is 16. Find the 12th term.

4) The 7th term of a series in A.P. is 32 and the 15th term is 72. Find the 24th term.

5) Find the number of the term which is 47 in the series 2, 5, 8 . . .

6) Insert six arithmetic means between 7 and 42.

7) Find the sum of the series 8, 11, 14 . . . If there are 10 terms in the series.

8) Find the sum of the series in arithmetical progression which has a 1st term of 60 and a 9th and last term of 96.

9) A series in A.P. has its 5th term equal to 25 and its 9th term equal to 41. If the series has 14 terms calculate its sum.

10) A series in A.P. has a sum of 120. If its 3rd term is 7 and its 7th term is 15 find the number of terms in the series.

SERIES IN GEOMETRICAL PROGRESSION

A series in which each term is obtained from the preceding term by multiplying or dividing by a constant quantity is called a *geometric progression* or simply a G.P. The constant quantity is called the common ratio for the series. In the series 1, 2, 4, 8, 16 . . ., each successive term is formed by multiplying the preceding term by 2. The series is therefore in G.P. with a common ratio of 2.
Some further examples of series in G.P. are:
2, 10, 50, 250 . . . (common ratio of 5)
8, 24, 72, 216 . . . (common ratio of 3)
24, 12, 6, 3 . . . (common ratio of $\frac{1}{2}$)

GENERAL EXPRESSION FOR A SERIES IN G.P.

Let the first term be a and the common ratio be r. The series can be represented by a, ar, ar^2, ar^3 . . .
The first term is a
The second term is ar
The third term is ar^2
The fourth term is ar^3
Notice that the index of r is always one less than the number of the term in the series.
Thus the 9th term is ar^8 (the index of r is 8)
the 20th term is ar^{19} (the index of r is 19)

Examples

1) Find the 7th term of the series 2, 6, 18 . . .
The 1st term is 2 and the common ratio is 3, that is $a = 2$ and $r = 3$.
The 7th term is $ar^6 = 2 \times 3^6 = 1458$

2) The 1st term of a series in G.P. is 19 and the 6th term is 27. Find the 10th term.
The first term is $a = 19$
The sixth term is $ar^5 = 27$ . . . . . (1)
Substituting for a in (1),

$$19r^5 = 27$$

$$r = \sqrt[5]{\frac{27}{19}} = 1{\cdot}073$$

The tenth term is $ar^9 = 19 \times 1{\cdot}073^9 = 35{\cdot}75$

3) Insert four geometric means between 1·8 and 11·2. This is equivalent to putting four terms between 1·8 and 11·2 so that the six numbers form a series in G.P.
Thus 1·8, ?, ?, ?, ?, 11·2 must be a G.P.
The first term is $a = 1{\cdot}8$
The sixth term is $ar^5 = 11{\cdot}2$ . . . . . (1)
Substituting for a in (1),

$$1{\cdot}8^5 = 11{\cdot}2$$

$$r = \sqrt[5]{\frac{11{\cdot}2}{1{\cdot}8}} = 1{\cdot}442$$

Hence the second term is $ar = 1{\cdot}8 \times 1{\cdot}442 = 2{\cdot}596$
the third term is $ar^2 = 1{\cdot}8 \times 1{\cdot}442^2 = 2{\cdot}596 \times 1{\cdot}442$
$= 3{\cdot}774$
the fourth term is $ar^3 = 3{\cdot}774 \times 1{\cdot}442^3 = 5{\cdot}399$
the fifth term is $ar^4 = 5{\cdot}399 \times 1{\cdot}442^4 = 7{\cdot}785$
The required series is 1·8, 2·596, 3·774, 5·399, 11·2

SUM OF A SERIES IN G.P.

The sum of a series in G.P. is

$$S = \frac{a(r^n - 1)}{r - 1}$$

which is a convenient form if r is greater than 1.

If r is less than 1 it is better to use the following expression:

$$S = \frac{a(1 - r^n)}{1 - r}$$

In both expressions:
a = the first term in the series
r = the common ratio
n = the number of terms in the series

Examples

1) Find the sum of the series 2, 6, 18 . . . which has six terms.
Since the common ratio, r, is 3 (obtained from $\frac{6}{2}$ or $\frac{18}{6}$)
we shall use the expression:

$$S = \frac{a(r^n - 1)}{r - 1}$$

In this problem $a = 2$ and $n = 6$. Substituting these two values in the expression for S we get,

$$S = \frac{2 \times (3^6 - 1)}{3 - 1} = \frac{2 \times (729 - 1)}{2} = 728$$

2) Find the sum of the series 71·2, 65·50, 60·26 . . . which has eight terms.

$$\text{Here } r = \frac{65{\cdot}50}{71{\cdot}20} = 0{\cdot}92$$

Hence we shall use the expression:

$$S = \frac{a(1 - r^n)}{1 - r}$$

Since $a = 71{\cdot}2$ and $r = 0{\cdot}92$ we have

$$S = \frac{71{\cdot}2 \times (1 - 0{\cdot}92^8)}{1 - 0{\cdot}92} = \frac{71{\cdot}2 \times (1 - 0{\cdot}5132}{0{\cdot}08}$$

$$= \frac{71{\cdot}2 \times 0{\cdot}4868}{0{\cdot}08} = 433{\cdot}25$$

Exercise 129

1) In a series in G.P. the first term is 5 and the common ratio is 2. Find the eighth and fourteenth terms.

2) Find the seventh term of the series 3·12, 48 . . .

3) Find the eighth term of the series 1·1, 1·21, 1·331 . . .

4) Find the sixth term of the series 14·6, 11·826, 9·579 . . .

5) The first term in a series in G.P. is 24 and the fifth term is 39. Find the eleventh term.

6) The third term in a series in G.P. is 17 and the seventh term is 42. Find the ninth term.

7) Insert six geometric means between 9·2 and 27·6.

8) Find the sum of the series 9·8, 6·86, 4·802 . . . which has eight terms.

9) Find the sum of the series 5, 8, 12·8 . . . which has seven terms.

10) The first term of a series in G.P. is 3.4 and the fifth term is 19.8. If the series has ten terms find its sum.

Chapter 39 Compound Interest. The Remainder Theorem

Compound interest is different from simple interest in that the interest which is added also attracts interest. Suppose that a sum of £P is invested at $r\%$ per annum for n years. The initial value is £P and after one year the value is

$$£\left(P + \frac{Pr}{100}\right) \text{ or } £P\left(1 + \frac{r}{100}\right)$$

In the second year the interest will be paid on

$$£P\left(1 + \frac{r}{100}\right)$$

Hence after the second year the value is

$$£\left[P\left(1 + \frac{r}{100}\right) + P\left(1 + \frac{r}{100}\right) \times \frac{r}{100}\right]$$

$$= £P\left(1 + \frac{r}{100}\right)^2$$

It can be seen that the value or amount after n years is

$$£P\left(1 + \frac{r}{100}\right)^n$$

Thus the amounts after one, two, three . . . years form a G.P. whose first term is P and whose common ratio is $\left(1 + \frac{r}{100}\right)$

Examples

1) Calculate the value of £2500 invested at 5% compound interest after eight years. Here P = £2500, $r = 5$ and n = 8. Substituting these values in the formula gives

$$A = £2500\left(1 + \frac{5}{100}\right)^8 = £2500 \times 1{\cdot}05^8$$

$$= £3693.$$

The calculation is best done by logarithms as follows:

Number	Operation	Log
2500		3·3979
$1{\cdot}05^8$	$8 \times 0{\cdot}0212$	0·1696
		3·5675

2) A sum of $£x$ is invested for four years at 8% per annum and the value of the investment amounts to £2720 at the end of this period. Calculate x.

The amount after four years is $£x\left(1 + \frac{8}{100}\right)^4 = 1{\cdot}08^4 x$

Therefore $1{\cdot}08^4\, x = 2720$

$x = 2000$

3) £5000 is invested for n years at 9% per annum compound. At the end of this period the value of the investment amounts to £6475. Find n.

We have

$$5000 \times 1{\cdot}09^n = 6475$$

The easiest way of finding n is to use logarithms:

$$\text{Log } 5000 + n \log 1{\cdot}09 = \log 6475$$

$$3{\cdot}6990 + 0{\cdot}0374n = 3{\cdot}8112$$

$$0{\cdot}3741n = 0{\cdot}1122$$

$$n = \frac{0{\cdot}1122}{0{\cdot}3704} = 3 \text{ years}$$

Exercise 130

1) Calculate the value of £3000 invested at 12% compound interest after five years.

2) £8000 is invested at 7% per annum for eight years. What is the value of the investment at the end of this period?

3) Find the Compound Interest on £328 for 3

years at 4 per cent per annum to the nearest penny.

4) Calculate the total interest on £840 invested at 5% per annum at compound interest for 2 years.

5) Find the sum which must be invested at 5% per annum compound interest to yield an amount of £441 at the end of 2 years.

6) Find how long it would take for £700 to amount to £1025 at 10% per annum compound interest.

7) How much must be invested at 9% per annum compound interest to yield an amount of £1036 at the end of 3 years?

8) How long will it take for £1200 to amount to £1684 at 7% per annum compound interest?

THE REMAINDER THEOREM

An expression like $5x^4 + 7x^3 - 3x^2 + 2x - 9$ is called a *polynomial* expression. Suppose we wish to find the remainder when this expression is divided by $(x - 3)$. This could be done by long division but the work is speeded up by using the remainder theorem.
The *remainder theorem* states that when a polynomial expression is divided by $(x - a)$ the remainder is obtained by writing a for x in the given expression.
Thus to find the remainder when $5x^4 + 7x^3 - 3x^2 + 2x - 9$ is divided by $(x - 3)$ all we have to do is to substitute $x = 3$ in the given polynomial expression. Hence

$$\text{Remainder} = 5 \times 3^4 + 7 \times 3^3 - 3 \times 3^2 + 2 \times 3 - 9 = 564$$

Example

Find the value of k if the remainder when the polynomial $3x^4 + kx^3 - 6x^2 + 5x + 8$ is divided by $(x - 2)$ is 74.
Substituting $x = 2$ in the expression we have

$$\text{Remainder} = 3 \times 2^4 + k \times 2^3 - 6 \times 2^2 + 5 \times 2 + 8 = 42 + 8k$$

Therefore $42 + 8k = 74$
$8k = 32$
$k = 4$

The remainder theorem may also be used to help factorise polynomial expressions. If the remainder is zero when a polynomial expression is divided by $(x - a)$ then $(x - a)$ is a factor of the expression.

Examples

1) Show that $(x - 5)$ is a factor of $x^3 + 3x^2 - 2x - 190$ Substituting $x = 5$ in the given expression we have,

$$\text{Remainder} = 5^3 + 3 \times 5^2 - 2 \times 5 - 190 = 0$$

Since the remainder is zero when $x = 5$ is substituted in the given expression then $(x - 5)$ is a factor of it

2) Show by using the remainder theorem that $(2x - 3)$ is a factor of $2x^2 - 11x + 12$
We may write $(2x - 3)$ as $2(x - 1{\cdot}5)$ and hence $(2x - 3)$ is a factor if the remainder is zero when $x = 1{\cdot}5$ is substituted in $2x^2 - 11x + 12$

$$\text{Remainder} = 2 \times 1{\cdot}5^2 - 11 \times 1{\cdot}5 + 12 = 0$$

Hence $(2x - 3)$ is a factor of $2x^2 - 11x + 12$

Exercise 131

1) Find the remainder when $x^4 + 5x^2 - 7x + 8$ is divided by $x - 5$.

2) Find the remainder when $16x^4 - 8x^3 + 4x^2 - 2x + 3$ is divided by $2x - 5$.

3) Find the value of k if the remainder is 60 when $2x^4 + kx^3 - 11x^2 + 4x + 12$ is divided by $x - 3$.

4) Show that $(x + 3)$ and $(x - 2)$ are both factors of $x^3 - 7x + 6$. What is the third factor?

5) Find the values of p and q if $(x - 1)$ and $(x + 2)$ are both factors of $px^3 + 4x^2 + qx - 6$.

Miscellaneous Exercises

Exercise 132 (*miscellaneous*)

1) Write down the 17th term of the series 3, 6, 12 . . .

2) The sum of the first four terms of an arithmetic progression is 10 and the sum of the four terms from the 7th to the 10th inclusive is 42. Find the sum of the ten terms.

3) The 3rd term of a G.P. exceeds the 2nd term by $\frac{35}{22}$ and the 2nd term exceeds the first by $\frac{7}{11}$. Calculate the 1st term and the common ratio.

4) If £100 is invested at 5% per annum compound interest, find the sums to which it will amount (a) at the end of one year, (b) at the end of two years.

5) Find the sum which must be invested at 8% per annum compound interest to yield an amount of £629.86 at the end of 3 years.

6) Find the remainder when $2x^3 - 3x^2 - 8x + 13$ is divided by $(x - 2)$.

7) Find the sum of all the terms from the 11th to the 30th inclusive of the series 5, 16, 27 . . .

8) The 2nd term of a G.P. is $\frac{8}{9}$ and the 6th terms is $4\frac{1}{2}$. Find the value of the common ratio and the 8th term.

SELF TEST 30

1) In the series 2, 4, 6 . . . the common difference is

a 4 b 2 c 12 d 3

2) In the series a, $a + d$, $a + 2d$ . . . the 10th term is

a $a + 11d$ b $a + 10d$ c $a + 9d$

3) The 11th term of the series 3, 5, 7 . . . is

a 23 b 21 c 25

4) The series 2, 5, 8 . . . consists of 8 terms. The sum of these 8 terms is

a 100 b 112 c 124

5) In the series 4, 12, 36 . . . the common ratio is

a 4 b 12 c 36 d 3

6) The 9th term of the series a, ar, ar . . . is

a ar^{10} b ar^9 c ar^8

7) The 8th term of the series 3, 6, 12 . . . is

a 384 b 192 c 96

8) The sum of the series 3, 9, 27 . . . which consists of 9 terms is

a 29523 b 8746 c 88572

9) £500 invested at 10% compound interest for 2 years becomes

a £600 b £700 c £605
d £665.50

10) A sum of £x is invested for 3 years at 10% per annum compound and the value of the investment amounts to £1331 at the end of this period. Hence x is equal to

a £1024 b £1100 c £1000

11) When $(x - 2)$ is divided into $3x^2 + 2x + 7$ the remainder is

a 0 b 23 c −23 d 15

12) The factors of $x^3 + 6x^2 + 11x + 6$ are

a $(x - 1)(x - 2)(x - 3)$
b $(x + 1)(x + 2)(x + 3)$
c $(x + 1)(x - 2)(x + 3)$
d $(x - 1)(x + 2)(x + 3)$

Chapter 40 Stock, Shares, Bankruptcy

SHARES

Many people at some time or another will have money to invest. It could be deposited in a bank, a building society or in the National Savings Bank where it will earn a fixed rate of interest. It could also be invested in a company by buying some of its shares.

Companies obtain their capital by the issue of shares to the public. Shares have a nominal value of 25p, £1, £5, etc. which cannot be divided into fractional amounts. If the company does well the price of the shares will appreciate to a value above the nominal price. On the other hand if the company does badly the price will fall below the nominal value. Shares can be bought and sold through the Stock Exchange. Companies issue different kinds of shares, two of the most important kinds being as follows:

Preference shares which carry a fixed rate of interest. The holders of these shares have first call on any profits the company may make in order that payments due to them can be made.

Ordinary shares, the dividends on which vary according to the amount of profit that the company makes. The directors of the company decide what dividend shall be paid. The dividends payable on the shares are a percentage of the nominal value of the shares.

Examples

1) A man buys 200 Emperor 25p shares at 44p. How much does he pay for the shares?

25p is the nominal value of the shares
44p is the price he actually pays for the shares

Amount paid $= 200 \times 44\text{p} = 8800\text{p} = £88$

2) Tom Jones sells 500 Tuxo £5 shares at £4. How much does he get?

£5 is the nominal value of the shares
£4 is the price at which he sells the shares
Amount received $= 500 \times £4 = £2000$

3) A man owns 300 Ludo shares. The company declares a dividend of 32p per share. How much does he get in dividends?

Amount payable $= 300 \times 32\text{p} = 9600\text{p} = £96$

4) A man holds 600 Snake 50p shares which he bought for 35p. The declared dividend is 8% of the nominal value of the shares. How much does he get in dividends and what is the yield per cent on his investment?

Dividend per share $= 8\%$ of $50\text{p} = 4\text{p}$
Amount obtained in dividends $= 600 \times 4\text{p}$
$= 2400\text{p} = £24$

$$\text{Yield per cent} = \frac{\text{dividend per share}}{\text{price paid per share}} \times 100$$
$$= \frac{4}{35} \times 100 = 11{\cdot}4\%$$

5) The paid up capital of a company consists of 50 000 6% preference shares of £1 each and 200 000 ordinary shares of 50p each. The profits available for distribution are £11 973. What dividend can be paid to the ordinary share holders? What is the dividend per cent?

Preference dividend $= 6\%$ of £50 000 $= £3000$
Amount of profit remaining
$= £11\,973 - £3000 = £8973$

$$\text{Dividend per share} = £\,\frac{8973}{200\,000}$$
$$= £0{\cdot}448 \text{ or } 4{\cdot}48\text{p}$$

$$\text{Dividend per cent} = \frac{4{\cdot}48}{50} \times 100 = 8{\cdot}96\%$$

Exercise 133

1) Find the cost of the following shares
   a) 400 Oak £2 shares at 314p
   b) 900 Staite 50p shares at 39p
   c) 280 Mako 25p shares at 59p

2) Find the amount raised by selling the following shares
   a) 180 Kneck 80p shares at 97p
   b) 350 Truck £5 shares at 87p
   c) 490 Mill 50p shares at 117p

3) Calculate the dividend received from the following investments
   a) 300 Proof shares; dividend is 8p per share
   b) 450 Shock shares; dividend is 18p per share
   c) 750 Stairways shares; dividend is 85p per share

4) Calculate the amount of dividend received from the following investments
   a) 300 Well 75p shares; declared dividend is 10%
   b) 500 Toomet £5 shares; declared dividend is 5%
   c) 1200 Boom 50p shares; declared dividend is 16%

5) Calculate the yield per cent on the following investments
   a) 200 Salto £3 shares bought at 250p; declared dividend is 9%
   b) 500 Melting 40p shares bought at 85p; declared dividend is 5%
   c) 85 Penn £2 shares bought at 98p; declared dividend is 3%

6) Calculate the yearly income from 500 Imperial £2 shares at 150p when a dividend of 8% is declared. What is the yield per cent on the money invested?

7) Find the cost, income and yield per cent from 900 £1 shares at 82p when a dividend of 7% is declared.

8) 500 Flag 25p shares are sold for £160. What is the cash value of each share?

9) How much profit does a man make when he sells 900 Emblem £3 shares, bought at 265p, sold for 390p?

10) The paid up capital of a company consists of 100 000 8% preference shares of £2 each and 250 000 ordinary shares of £1 each. The profits available for distribution are £27 000. What is the dividend per cent that can be paid to the ordinary shareholders?

11) The issued capital of a company is 200 000 8% preference £1 shares, 150 000 6% £2 preference shares and 500 000 ordinary 25p shares. If the profits available for distribution to the shareholders is £92 000, what dividend per share is payable to the ordinary shareholders?

12) The paid up capital of a company consists of 90 000 7% preference shares and 110 000 ordinary 50p shares. If a dividend of 10% is declared, what is the profit of the company?

STOCK

Stock is issued by the Government and Local Authorities when they need cash. This stock is issued at a fixed rate of interest and it can be redeemed after a certain number of years. Stock is always issued in £100 units but an investor does not necessarily need to buy whole units of stock. The price of stock varies in the same way as does the price of shares.

If you look in the financial section of your newspaper you will see statements like this:

Treasury 5% 1986–89 52

This means that £100 worth of Treasury stock paying 5% interest and redeemable between 1986 and 1989 can be bought for £52.

Example

How much 9% Treasury stock at 70 can be bought for £280? How much in dividends are payable and what is the yield per cent?

£70 buys £100 stock

£280 buys £ $\frac{100}{70} \times 280 = £400$ stock

Interest payable $= 9\%$ of £400 $= £36$

$$\text{Yield per cent} = \frac{\text{Interest payable}}{\text{Amount paid for stock}} \times 100$$

$$= \frac{36}{280} \times 100 = 12{\cdot}86\%$$

BROKERAGE

Brokerage is the commission charged by a broker for the purchase or sale of stocks and shares. It is calculated as a percentage of the sum for which they are bought or sold.

When buying add the brokerage to the price paid
When selling deduct the brokerage from the sum received

Example

Find the change in income resulting from selling £2500 5% stock at 105 and investing the proceeds in 3% stock at 55. Brokerage is $1\frac{1}{2}\%$.

Original income = 5% of £2500 = £125 per annum
Amount received for stock sold =

$$£2500 \times \frac{105}{100} = £2625$$

Brokerage = $1\frac{1}{2}\%$ of £2625 = £39·37
Amount received for stock sold less brokerage = £2625 − £39·37
= £2585·63
Amount available for reinvestment =

$$£2585{\cdot}63 \times \frac{100}{101\frac{1}{2}} = £2547{\cdot}41$$

$$\text{Stock bought at 55} = £2547{\cdot}41 \times \frac{100}{55}$$

$$= £4631{\cdot}67$$

New Income = 3% of £4631.67 = £138·95
Change in income = £138·95 − £125 = £13·95

Exercise 134

1) Calculate the amount of stock that can be bought in each case
a) £300 invested in 5% Crewe at 75
b) £250 invested in Treasury stock at 69
c) £700 invested in Devon at 120

2) Calculate the amount of interest received each year from the following
a) £800 stock at $3\frac{1}{2}\%$
b) £1200 stock at 5%
c) £500 stock at 4%

3) Calculate the amount of 6% Cambridge stock that can be bought for £600 if the price is 88. What is the interest earned per annum and what is the yield per cent?

4) How much does it cost to buy £300 of Railway stock at 80?

5) What are the proceeds from selling £400 of Dorset 3% stock at 75?

6) A man sells £800 worth of Argentine stock at 30 and buys German 8% stock at 110. How much German stock does he buy?

7) A man buys £500 of a certain stock at 75 and sells when they reach 57. How much does he lose on the transaction?

8) An investor sold £5000 of 4% American stock at 80 and invested the proceeds in French 6% stock at 90. Calculate the amount of French stock bought and the change in income.

9) Find the change in income which results from selling £4000 of 5% stock at 102 and investing the proceeds in 4% stock at 92. Brokerage is 2% on each transaction.

10) A 6% stock is quoted at 130. Allowing for brokerage at $1\frac{1}{2}\%$, calculate
a) How much stock can be bought for £2000
b) How much is realised by the sale of £2000 of this stock
c) The net income from £2000 stock less income tax at 35%

BANKRUPTCY

A person or company is *insolvent* when the liabilities of the person or company exceed the assets. Under certain conditions the

person or company may be declared bankrupt. In this case the creditors (the people who are owed money) may appoint a trustee to sell the debtor's (the person or company owing the money) assets. After expenses and certain prior claims have been met the remainder of the proceeds will be distributed amongst the creditors.
To do this a dividend is declared which is worked out as follows:

$$\text{Dividend} = \frac{\text{Net assets}}{\text{total liabilities}}$$

Examples

1) A bankrupt has liabilities of £10 000 and assets of £2000. Find the dividend. How much will a creditor owed £3000 be paid?

$$\text{Dividend} = £\frac{2000}{10\,000} = £0{\cdot}20 \text{ in the £1 or}$$

20p in the £1

This means that each creditor will receive 20p for each £1 that he is owed.

Amount received by creditor owed £3000 = £3000 × £0·20 = £600

Secured creditors do not rank for dividend unless the security does not realise the amount of the claim. In this case the balance of the claim ranks for dividend in the usual way. If however the security raises more than the amount of the claim the residue is added to the assets thus increasing the amount of the dividend.

2) A company, declared bankrupt, owes £85 873 to fully secured creditors and £98 748 to unsecured creditors. If the assets of the company realise £150 786 net, find the dividend payable to the unsecured creditors.

Net assets	£150 786
Secured creditors	£ 85 873
Available for dividend	£64 913

$$\text{Dividend} = £\frac{64\,913}{98\,748} = £0{\cdot}65 \text{ in the £1 or}$$

65p in the £1

Exercise 135

1) Find the dividends payable in the following cases

a) Net assets £20 000. Creditors £50 000
b) Net assets £7532. Creditors £82 516
c) Net assets £190 632. Creditors £567 826

2) Find the amounts paid to the following creditors

a) Dividend 30p in the £1. Creditor owed £7000
b) Dividend 8p in the £1. Creditor owed £6378
c) Dividend 19p in the £1. Creditor owed £15 678

3) The total assets of a company are sold for £358 907 but the expenses incurred are £28 712. If the company owes £739 054 to unsecured creditors find the amount of the dividend which can be paid.

4) The liabilities of a man in bankruptcy are £25 832 to fully secured creditors and £48 798 to unsecured creditors. If his assets realise £35 746 net what dividend can be paid?

5) A bankrupt owed £8572 to his ordinary (unsecured) creditors and £1750 to his secured creditors. If the expenses of the winding up were £125 and a dividend of 20p in the £1 was declared, how much were his assets?

SELF TEST 31

In each of the following state the letter (or letters) corresponding to the correct answer (or answers).

1) A man buys 500 Empress 50p shares at 40p. He therefore pays for the shares
a £50 b £250 c £40 d £200

2) A person sells 400 Tuxedo £2 shares for 150p. He gets
a £60 b £600 c £80 d £800

3) A man holds 500 Ladder 80p shares which he bought for 70p. The declared dividend is 8%. He therefore receives
a £32 b £28 c £40 d £56

4) A person holds 800 Persona 50p shares which he bought for 75p. The declared dividend is 10%. The yield per cent on the shares is

a 10% b $6\frac{2}{3}$% c 15% d $7\frac{1}{2}$%

5) How much 8% stock at 75 can be bought for £150?

a £200 b £75 c £150 d £1200

6) Calculate the interest received per annum from £800 stock at 5%.

a £50 b £400 c £40 d £500

7) A man sells £600 French stock at 50 and buys Belgian stock at 40. How much Belgian stock does he buy?

a £600 b £1200 c £750 d £400

8) A bankrupt has liabilities of £7000 and assets of £3500. How much will a creditor owed £2000 be paid?

a £2000 b £1000 c £500 d £700

Revision Summary

In this summary the important facts and formulae which are needed for examination purposes are given. These should be memorised.

Percentages

$$\text{Profit } \% = \frac{\text{selling price} - \text{cost price}}{\text{cost price}} \times 100$$

$$\text{Loss } \% = \frac{\text{cost price} - \text{selling price}}{\text{cost price}} \times 100$$

Simple interest

$$I = \frac{PRT}{100}$$

Factors

$$a^2 + 2ab + b^2 = (a+b)^2$$
$$a^2 - 2ab + b^2 = (a-b)^2$$
$$a^2 - b^2 = (a+b)(a-b)$$

Indices

$$a^m \times a^n = a^{m+n}$$

$$a^m \div a^n = a^{m-n}$$

$$(a^m)^n = a^{mn}$$

$$a^{\frac{m}{n}} = \sqrt[n]{a^m}$$

$$a^0 = 1$$

$$a^{-m} = \frac{1}{a^m}$$

Logarithms

$$\log (ab) = \log a + \log b$$

$$\log \frac{a}{b} = \log a - \log b$$

$$\log a^m = m \log a$$

Quadratic equation

If $ax^2 + bx + c = 0$

then $x = \dfrac{-b \pm \sqrt{b^2 - 4ac}}{2a}$

Areas

Area of rectangle = length × breadth

Area of parallelogram
= base × vertical height

Area of triangle
$= \frac{1}{2} \times$ base × vertical height

Area of trapezium
= average of parallel sides × distance between them

Area of circle $= \pi \times (\text{radius})^2$

Circumference of circle
$= 2 \times \pi \times \text{radius} = \pi \times \text{diameter}$

Volumes

Volume of any solid having a uniform cross-section = cross-sectional area × length of solid

Volume of cylinder
$= \pi \times (\text{radius})^2 \times$ length of cylinder

Volume of cone
$= \frac{1}{3} \times \pi \times (\text{radius})^2 \times$ vertical height of cone

Volume of sphere $= \frac{4}{3} \times \pi \times (\text{radius})^3$

Volume of pyramid

$=\frac{1}{3}\times$ area of base $\times$ vertical height

Volume of prism

$=$cross-sectional area $\times$ length of prism

Surface areas of solids

Surface area of cone

$=\pi\times$ radius $\times$ slant height of cone

Surface area of sphere $=4\times\pi\times(\text{radius})^2$

Surface area of cylinder

$=2\times\pi\times$ radius $\times$ height of cylinder

Density and specific gravity

$$\text{Density}=\frac{\text{Mass}}{\text{Volume}}$$

$$\text{Specific gravity}=\frac{\text{density of substance}}{\text{density of water}}$$

The specific gravity of a substance is the same as its density if this is given in the units of grammes per cubic centimetre (g/cm^3).

Flow of water

Quantity of water flowing

$=$cross-sectional area of pipe $\times$ speed of flow

Similar solids

The *surface areas* of similar solids are proportional to the squares of their linear dimensions. If A_1 and A_2 are the areas of two similar solids and h_1 and h_2 are their respective heights then

$$\frac{A_1}{A_2}=\left(\frac{h_1}{h_2}\right)^2$$

The *volumes* of similar solids are proportional to the cubes of their linear dimensions. If V_1 and V_2 are the volumes of two similar solids and h_1 and h_2 are their respective heights then

$$\frac{V_1}{V_2}=\left(\frac{h_1}{h_2}\right)^3$$

Graphs

The equation of a straight line is $y=ax+b$

Variation

1) Direct variation. If $y\propto x$ then $y=kx$

2) Inverse variation. If $y\propto\frac{1}{x}$ then $y=\frac{k}{x}$

3) Joint variation. If $p\propto t$ and $p\propto q$ then $p=ktq$

4) Variation as the sum of two parts. If $y\propto x$ and $y\propto x^2$ then $y=ax+bx^2$

Differentiation

If $y=ax^n$ $\quad\frac{dy}{dx}=nax^{n-1}$

$$\text{Velocity}=\frac{ds}{dt}$$

$$\text{Acceleration}=\frac{dv}{dt}$$

For max. or min. $\frac{dy}{dx}=0$

Integration

$$\int ax^n\,dx=\frac{ax^{n+1}}{n+1}+c$$

$$\text{Area under a curve}=\int_a^b y\,dx$$

Distance $=\int v\,dt$ where $v=$velocity

Velocity $=\int a\,dt$ where $a=$acceleration

$$\text{Volume of solid of revolution}=\int_a^b \pi y^2\,dx$$

Geometric axioms and theorems

1) The total angle on a straight line is 180°.

2) When two straight lines intersect the opposite angles are equal.

3) When two parallel lines are cut by a transversal the corresponding angles are equal, the alternate angles are equal and the interior angles are supplementary.

4) The sum of the angles of a triangle are equal to 180°.

5) In every triangle the greatest angle lies opposite to the longest side and the smallest angle is opposite to the shortest side.

6) The exterior angle of a triangle is equal to the sum of the opposite interior angles.

7) In an isosceles triangle the perpendicular to the base bisects the angle between the two equal sides and also bisects the base.

8) Two triangles are congruent if:
  1) One side and two angles in one triangle are equal to one side and two similarly located angles in the other triangle.
  2) Two sides and the angle between them in one triangle are equal to two sides and the angle between them in the second triangle.
  3) Three sides of one triangle equal the three sides of the second triangle.
  4) In right-angled triangles the hypotenuses are equal and one other side in each triangle are also equal.

9) Two triangles are similar if:
  1) Two angles in the one triangle equal two angles in the other.
  2) Three sides in one triangle are proportional to three sides in the other.
  3) Two sides in one triangle are proportional to the corresponding sides in the other triangle and the angles between these sides are equal.

10) Triangles having equal bases and equal heights are equal in area.

11) The ratio of the areas of similar triangles is equal to the ratio of the squares on corresponding sides.

12) The internal bisector of an angle of a triangle divides the opposite side in the ratio of the sides containing the angle.

13) The external bisector of an angle of a triangle divides the opposite side externally in the ratio of the sides containing the angle.

14) The theorem of Pythagoras states that in a right-angled triangle the square on the hypotenuse is equal to the sum of the squares on the other two sides.

15) In any quadrilateral the sum of its angles is 360°.

16) A parallelogram has both pairs of opposite sides equal and parallel. The angles which are opposite to each other are equal. The diagonals bisect each other and the diagonals also bisect the parallelogram.

17) A rectangle has all the properties of a parallelogram but in addition the diagonals are equal in length.

18) A rhombus has all its sides equal in length. It has all the properties of a parallelogram but in addition the diagonals bisect at right angles and the diagonal bisects the angle through which it passes.

19) A square has all the properties of a parallelogram, rectangle and rhombus.

20) In a convex polygon having n sides the sum of the interior angles is $(2n-4)$ right-angles. The sum of the exterior angles is 360°.

21) Parallelograms having equal bases and equal heights are equal in area.

22) The area of a triangle is half the area of a parallelogram drawn on the same base and between the same parallels.

23) If the diameter of a circle is at right-angles to a chord then it divides the chord into two equal parts.

24) Chords which are equal in length are equi-distant from the centre of the circle.

25) If two chords intersect inside or outside a circle the product of the segments of one chord is equal to the product of the segments of the other chord.

26) If a triangle is inscribed in a semi-circle the angle opposite the diameter is a right-angle.

27) Angles in the same segment of a circle are equal.

28) The angle which an arc of a circle subtends at the centre is twice the angle which the arc subtends at the circumference.

29) The opposite angles of a cyclic quadrilateral are equal to 180°.

30) A tangent to a circle is at right-angles to a radius drawn to the point of tangency.

31) If from a point outside a circle, tangents are drawn to the circle then the tangents are equal in length.

32) The angle between a chord and a tangent equals one-half of the angle at the centre subtended by the chord.

33) The angle between a tangent and a chord equals the angle at the circumference subtended by the chord.

34) A line which cuts a circle at two points is called a secant. If from a point outside a circle a secant and a tangent are drawn then the square on the tangent is equal to the rectangle contained by the whole secant and that part of it which lies outside the circle.

Trigonometry

$$\text{sine} = \frac{\text{opposite}}{\text{hypotenuse}} \quad \text{cosine} = \frac{\text{adjacent}}{\text{hypotenuse}}$$

$$\text{tangent} = \frac{\text{opposite}}{\text{adjacent}}$$

$$\sin^2 A + \cos^2 A = 1$$

$$\frac{a}{\sin A} = \frac{b}{\sin B} = \frac{c}{\sin C}.$$

This is the sine rule.

$$a^2 = b^2 + c^2 - 2bc \cos A.$$

This is the cosine rule.

Sets

$A \subset B$ means that A is a subset of B.
$a \in B$ means that a is an element of set B.
$\mathscr{E}$ is the universal set from which all subsets are formed.
$A \cap B$ is the intersection of A and B.
$A \cup B$ is the union of A and B.
A' is the complement of A. $A' = \mathscr{E} - A$.
$A - B = A \cap B'$
$(A \cap B)' = A' \cup B'$
$n(A \cup B) = n(A) + n(B) - n(A \cap B)$
$n(A \cup B \cup C) = n(A) + n(B) + n(C) - n(A \cap B) - n(A \cap C) - n(B \cap C) + n(A \cap B \cap C)$
$A \cup B = B \cup A$
$A \cup (B \cup C) = (A \cup B) \cup C$
$A \cap (B \cup C) = (A \cap B) \cup (A \cap C)$

Matrices

$$\begin{pmatrix} a & b \\ c & d \end{pmatrix} + \begin{pmatrix} u & v \\ w & x \end{pmatrix} = \begin{pmatrix} a+u & b+v \\ c+w & d+x \end{pmatrix}$$

$$\begin{pmatrix} a & b \\ c & d \end{pmatrix} \begin{pmatrix} u & v \\ w & x \end{pmatrix} = \begin{pmatrix} au+bw & av+bx \\ cu+dw & by+dx \end{pmatrix}$$

If $A = \begin{pmatrix} a & b \\ c & d \end{pmatrix}$ $\quad A^{-1} = \dfrac{1}{ad-bc} \begin{pmatrix} d & -b \\ -c & a \end{pmatrix}$

If $AX = K$ then $X = A^{-1}K$

If $U = \begin{pmatrix} a \\ b \end{pmatrix}$ and $V = \begin{pmatrix} x \\ y \end{pmatrix}$ then $U + V = \begin{pmatrix} a+x \\ b+y \end{pmatrix}$

Number systems

The number 4325 in base x means $5 \times x^0 + 2 \times x^1 + 3 \times x^2 + 4 \times x^3$.

Statistics

$$\text{Arithmetic mean} = \frac{x_1 f_1 + x_2 f_2 + x_3 f_3 + \cdots}{f_1 + f_2 + f_3 \ + \cdots}$$

Answers

ANSWERS TO CHAPTER 1

Exercise 1

1) 13 2) 10 3) 57 4) 7 5) 35 6) 15
7) 45 8) 74 9) 13 10) 20

Exercise 2

1) $3\frac{13}{14}$ 2) 5 3) $2\frac{1}{2}$ 4) $\frac{5}{6}$ 5) $\frac{2}{3}$ 6) $2\frac{1}{2}$

7) $1\frac{2}{5}$ 8) $\frac{2}{3}$ 9) $\frac{1}{6}$ 10) $\frac{3}{25}$

Exercise 3

1a) (i) 24·8658 (ii) 24·87 (iii) 25
b) (i) 0·008357 (ii) 0·00836 (iii) 0·0084
c) (i) 4·9785 (ii) 4·98 (iii) 5
d) 22 e) 35·60
f) (i) 28 388 000 (ii) 28 000 000
g) (i) 4·1498 (ii) 4·150 (iii) 4·15 h) 9·20
2a) (i) 5·1499 (ii) 5·150 (iii) 5·15
b) (i) 35·29 (ii) 35·3
c) (i) 0·00498 (ii) 0·0050 (iii) 0·005
d) (i) 8·408 (ii) 8·41 (iii) 8·4
e) (i) 0·853 (ii) 0·85
3a) 1·31 b) 0·016 c) 190 d) 37·65
e) 20·6

Self Test 1

1) b 2) a 3) b 4) c 5) b, c 6) b
7) d 8) a 9) c 10) b 11) c 12) d
13) c 14) d 15) c 16) true 17) false
18) false 19) true 20) false 21) true
22) true 23) true 24) true 25) true
26) false 27) true 28) false 29) true
30) true

ANSWERS TO CHAPTER 2

Exercise 5

1a) 71 p b) 63 p c) $58\frac{1}{2}$ p
2a) £0·06 b) £0·72$\frac{1}{2}$ c) £0·97$\frac{1}{2}$
3a) £10·06 b) £215·58 c) £5·42$\frac{1}{2}$ d) £2·35
e) £2·00$\frac{1}{2}$
4) £1·10 5) £9·50 6) £9·00
7a) £4·05$\frac{1}{2}$ b) £2·63 c) £12·00$\frac{1}{2}$
8a) £181·10 b) £70·35 c) £91·50
d) £217·50
9a) £66·36 b) £12·45$\frac{1}{2}$ c) £148·12$\frac{1}{2}$
10a) £75·79 b) £48·57 c) £113·12
11) 13 p 12) $21\frac{1}{2}$ p 13) £1·30$\frac{1}{2}$ 14) £2·16$\frac{1}{2}$

Exercise 6

1) 80·7 2) 3000 3) 64·96 4) £109·852
5) 6·013 6) 61·538 7) \$63.08 8) £5·22
9a) 2·594 b) 8·76 c) 11·11
10) 615·70 11) \$16·00 12) £2·76

Self Test 2

1) True 2) False 3) True 4) True
5) False 6) True 7) True 8) False
9) True 10) True 11) True 12) True
13) False 14) False 15) False

ANSWERS TO CHAPTER 3

Exercise 7

1) $\frac{8}{3}$ 2) $\frac{2}{3}$ 3) $\frac{3}{1}$ 4) $\frac{10}{3}$ 5) $\frac{2}{3}$ 6) $\frac{3}{5}$
7) $\frac{15}{2}$ 8) $\frac{1}{6}$ 9) $\frac{2}{75}$ 10) $\frac{3}{20}$ 11) $\frac{25}{2}$
12) $\frac{25}{4}$ 13) 150 m 14) 192 km/h
15a) $\frac{2}{5}$ b) $\frac{2}{3}$ c) $\frac{50}{1}$ d) $\frac{5}{12}$

Exercise 8

1) £500 and £300 2) £64 and £16 3) £280
4) £50, £40 and £30
5) 26·47 kg; 35·30 kg; 13·24 kg
6) 84 mm; 294 mm; 462 mm 7) £75
8) 36 kg; 44 kg; 80 kg 9) 13:5:1
10) £6258 and £4470

Self Test 3

1) True 2) False 3) True 4) True
5) True 6) False 7) True 8) True
9) False 10) True 11) True 12) False
13) b 14) d 15) a 16) b 17) c 18) c

ANSWERS TO CHAPTER 4

Exercise 9

1) 30% 2) 55% 3) 36% 4) 80%
5) 62% 6) 63% 7) 81·3% 8) 66·7%
9) 72·3% 10) 2·7% 11) $\frac{8}{25}$ 12) $\frac{39}{50}$
13) $\frac{3}{50}$ 14) $\frac{6}{25}$ 15) $\frac{63}{200}$ 16) $\frac{241}{500}$ 17) $\frac{1}{40}$
18) $\frac{1}{80}$ 19) $\frac{79}{2000}$ 20) $\frac{201}{1000}$

Exercise 10

1a) 10 b) 24 c) 6 d) 2·4 e) 21·315
f) 2·516
2a) 12·5% b) 20% c) 16 d) 16·3%
e) 45·5%
3) 60%; 27 4) 115 cm 4) $88\frac{2}{3}$ cm
6a) 2·08% b) 3·08% c) 87·76%
7) 39 643 8) 9 825

Exercise 11

1) 25%
2a) 20% b) $16\frac{2}{3}$%
3a) $13\frac{1}{3}$% b) 9·95%
4) 20% 5) $33\frac{1}{3}$% 6) $12\frac{1}{2}$% 7) 17·65%
8) 10%

Exercise 12

1a) £6·00 b) £4·35 2a) £16·00 b) 56p
3) £12·00 4a) £3·40 b) £59·20 5) £64·00
6) £2·85 7) £3·00 8) £20·00 9) £500
10) £12150

Exercise 13

1) £6·30 2) £200 3) £19·00 4) £6·00
5) £96·00 6) 30% 7) 4% 8) 6·72%
9) £17·00 10) 62·5%

Self Test 4

1) True 2) False 3) True 4) False
5) False 6) True 7) True 8) True
9) False 10) True 11) True 12) False
13) False 14) True 15) a, b, d 16) b
17) c, d 18) a 19) c 20) b 21) b
22) c 23) b 24) b 25) d

ANSWERS TO CHAPTER 5

Exercise 14

1) £126 2) £20 3) £30·50 4) 4 years
5) 5 months 6) $2\frac{1}{2}$ years 7) 7% 8) 9%
9) £320 10) £0·30 11) £66 12) £378·56
13) £1 054·25 14) 6 years 15) £11 600

Self Test 5

1) d 2) b 3) a 4) c 5) d 6) b
7) d 8) a 9) a 10) c

ANSWERS TO CHAPTER 6

Exercise 15

1) 80 p; 94 p 2) 6 hours 3) £32·50
4) 7 hours 5) £20 6) £1 225 7) £21·60
8) 6 hours

Exercise 16

1) £13 2) 730 3) 2·2 p 4) £11·81
5) 440 6) Tariff 2 7) £15·00 8) 300
9) Tariff 2; £2·90

Exercise 17

1) £63·00 2) £120 3) 90 p in the £
4) £8 500; 93 in the £ 5) £202 308 6) 2·9 p
7) £15 600; 3 p in the £; £3·78
8) £6·45 inc. $25\frac{1}{2}$ p dec.

Exercise 18

1) £360 2) £200 3) £721·50 4) £397·50
5) £637·50

Self Test 6

1) b 2) c 3) a 4) d 5) b 6) c
7) a 8) d 9) b 10) d

ANSWERS TO CHAPTER 7

Exercise 19

1) 22·24 2) $12\frac{1}{8}$ kg 3) 16 p 4) 8·58 p
5) 97·6 6) 2 000 7) $14\frac{1}{2}$ years 8) 9·13 p
9) 40 10) 76 11) 63 12) £32·90; 65·8 p

Exercise 20

1) 75 km/h 2) 4 hours 3) 350 km
4) 26 km/h 5) 75 km/h 6) 80 km/h
7) 5 hours 8) 38·4 km/h

Self Test 7

1) d 2) b 3) a 4) b 5) d 6) a
7) a 8) c 9) c 10) b

ANSWERS TO MISCELLANEOUS EXERCISE

Exercise 21

1) 57·15 francs 2) $\frac{3}{4}$ 3) 125 grammes
4) 2·16 5) 63 p 6) 22 litres 7) 3 years
8) 550 9) £15·75 10) £3·00; 21p 11) $1\frac{1}{3}$
12) £13·$12\frac{1}{2}$ 13) 0·123 14) $2\frac{2}{3}$ minutes
15) $56\frac{1}{4}$ litres 16) £2·10 17) £129·80
18) $1\frac{1}{2}$ 19) 23·5 francs 20) £10·50
21) £1 669·50 22) £34 23) 315 marks
24) $4\frac{1}{2}$ 25) 162·5 kg 26) 0·56 27) 36 litres
28) £450 29) $16\frac{1}{2}$ p 30) £120 31) £2·00
32) 830 33) $\frac{1}{4}$ 34) 79·12 35) £85·00; £59·50
35) £85·00; £59·50 36) £64·84 37) 15
38) £7·84 39) £5·52 40) 7 months
41) 11·1% 42) £9·30 43) 80 44) 153 cm
45a) £614·25 b) £771·75 46) £3135, £1097·25
47) 19·6%; 11 804; 520
48) 1 989 765; 29p 49) £18 000; 35p; 3·9p
50) £4·20; £17·70; 5·25 p.m. 51) £280; £12; 7·2%
52) £300; $62\frac{2}{3}$% 53) £42·45
54) £1·20; £7·92; 18 p 55) 28%

ANSWERS TO CHAPTER 8

Exercise 22

1a) 5 b) 0 c) 4 d) 16 e) 18 f) 4
g) 1 h) 8 i) 2 j) 2 k) 4 l) 30
2a) 79 b) 5 c) $\frac{7}{8}$ d) $\frac{20}{17}$ e) 1 f) $\frac{11}{6}$
3a) 9 b) 64 c) 64 d) 1 152 e) 7
f) 3 456 g) 46 656 h) 36 864 i) 36
j) 108 k) 18 l) $\frac{1}{3}$

Exercise 23

1) +15 2) −12 3) −32 4) 14 5) −5
6) −9 7) −14 8) 5 9) 5 10) −20
11) −16 12) −1 13) 0 14) −8
15) −20 16) 2 17) 3 18) 14 19) 4
20) 1 21) −5 22) 7 23) −12 24) −42
25) −42 26) 42 27) 42 28) −48 29) 4
30) 120 31) −27 32) −3 33) −3 34) 3
35) 3 36) −2 37) −1 38) $\frac{1}{2}$ 39) −1
40) −2 41) −12 42) −1 43) 2 44) −8
45) −3 46) −72

Exercise 24

1) $18x$ 2) $2x$ 3) $-3x$ 4) $-6x$
5) $-5x$ 6) $5x$ 7) $9a$ 8) $12m$ 9) $5b^2$
10) ab 11) $14xy$ 12) $-3x$ 13) $-6x^2$
14) $7x-3y+6z$
15) $9a^2b-3ab^3+4a^2b^2+11b^4$
16) $1{\cdot}2x^3+0{\cdot}3x^2+6{\cdot}2x-2{\cdot}8$
17) $9pq-0{\cdot}1qr$
18) $-0{\cdot}4a^2b^2-1{\cdot}2a^3-5{\cdot}5b^3$
19) $10xy$ 20) $12ab$ 21) $12m$ 22) $4pq$
23) $-xy$ 24) $6ab$ 25) $-24mn$
26) $-12ab$ 27) $24pqr$ 28) $60abcd$ 29) $2x$
30) $\frac{-4a}{7b}$ 31) $\frac{-5a}{8b}$ 32) $\frac{a}{b}$ 33) $\frac{2a}{b}$ 34) $2b$
35) $3xy$ 36) $-2ab$ 37) $2ab$ 38) $\frac{7ab}{3}$
39) a^2 40) $-b^2$ 41) $-m^2$ 42) p^2
43) $6a^2$ 44) $5X^2$ 45) $-15q^2$ 46) $-9m^2$
47) $9pq^2$ 48) $-24m^3n^4$ 49) $-21a^3b$
50) $10q^4r^6$ 51) $30mnp$ 52) $-75a^3b^2$
53) $-5m^5n^4$

Exercise 25

1) $3x+12$ 2) $2a+2b$ 3) $9x+6y$ 4) $\frac{x}{2}-\frac{1}{2}$
5) $10p-15q$ 6) $7a-21m$ 7) $-a-b$
8) $-a+2b$ 9) $-3p+3q$ 10) $-7m+6$
11) $-4x-12$ 12) $-4x+10$ 13) $-20+15x$
14) $2k^2-10k$ 15) $-9xy-12y$ 16) $ap-aq-ar$
17) $4abxy-4acxy+4dxy$ 18) $3x^4-6x^3y+3x^2y^2$
19) $-14P^3+7P^2-7P$ 20) $+2m-6m^2+4mn$
21) $5x+11$ 22) $14-2a$ 23) $x+7$
24) $16-17x$ 25) $7x-11y$ 26) $\frac{7y}{6}-\frac{3}{2}$
27) $-8a-11b+11c$ 28) $7x-2x^2$ 29) $3a-9b$
30) $-x^3+18x^2-9x-15$

Self Test 8

1) True 2) True 3) False 4) True
5) True 6) False 7) True 8) True
9) True 10) False 11) True 12) True
13) True 14) False 15) False 16) True
17) False 18) True 19) False 20) True
21) True 22) False 23) False 24) True
25) True 26) False 27) False 28) True
29) False 30) True 31) True 32) False
33) True 34) True 35) True 36) True
37) False 38) False 39) True 40) False
41) False 42) True 43) True 44) False
45) True 46) True 47) False 48) True
49) False 50) False

ANSWERS TO CHAPTER 9

Exercise 26

1) p^2q 2) ab^2 3) $3mn$ 4) b 5) $3xyz$
6) $2(x+3)$ 7) $4(x-y)$ 8) $5(x-1)$
9) $4x(1-2y)$ 10) $m(x-y)$ 11) $x(a+b+c)$
12) $\frac{1}{2}\left(x-\frac{y}{4}\right)$ 13) $5(a-2b+3c)$ 14) $ax(x+1)$
15) $\pi r(2r+h)$ 16) $3y(1-3y)$ 17) $ab(b^2-a)$
18) $xy(xy-a+by)$ 19) $5x(x^2-2xy+3y^2)$
20) $3xy(3x^2-2xy+y^4)$ 21) $I_0(1+\alpha t)$
22) $\frac{1}{3}\left(x-\frac{y}{2}+\frac{z}{3}\right)$ 23) $a(2a-3b)+b^2$
24) $x(x^2-x+7)$ 25) $\frac{m^2}{pn}\left(1-\frac{m}{n}+\frac{m^2}{pn}\right)$

26) $\frac{xy}{a}\left(\frac{x}{2}-\frac{2y}{5a}+\frac{y^2}{a^2}\right)$ 27) $\frac{l^2m}{5}\left(\frac{m}{3}-\frac{1}{4}+\frac{lm}{2}\right)$
28) $\frac{a^2}{2x^3}\left(a-\frac{b}{2x}-\frac{c}{3}\right)$

Exercise 27

1) x^2+3x+2 2) x^2+4x+3 3) $x^2+9x+20$
4) $2x^2+11x+15$ 5) $3x^2+25x+42$
6) $5x^2+21x+4$ 7) $6x^2+16x+8$
8) $10x^2+17x+3$ 9) $21x^2+41x+10$
10) x^2-4x+3 11) x^2-6x+8
12) $x^2-9x+18$ 13) $2x^2-9x+4$
14) $3x^2-11x+10$ 15) $4x^2-33x+8$
16) $6x^2-16x+8$ 17) $6x^2-17x+5$
18) $21x^2-29x+10$ 19) x^2+2x-3
20) $x^2+5x-14$ 21) $x^2-2x-15$
22) $2x^2+x-10$ 23) $3x^2+13x-30$
24) $3x^2+23x+30$ 25) $6x^2+x-15$
26) $12x^2+4x-21$ 27) $6x^2-x-15$
28) $3x^2+5xy+2y^2$ 29) $2p^2-7pq+3q^2$
30) $6v^2-5uv-6u^2$ 31) $6a^2+ab-b^2$
32) $5a^2-37a+42$ 33) $6x^2-xy-12y^2$
34) x^2+2x+1 35) $4x^2+12x+9$
36) $9x^2+42x+49$ 37) x^2-2x+1
38) $9x^2-30x+25$ 39) $4x^2-12x+9$
40) $4a^2+12ab+9b^2$ 41) $x^2+2xy+y^2$
42) $P^2+6PQ+9Q^2$ 43) $a^2-2ab+b^2$
44) $9x^2-24xy+16y^2$ 45) $M^2-4MN+4N^2$
46) x^2-1 47) x^2-9 48) x^2-49
49) $4x^2-25$ 50) $9x^2-49$ 51) $4x^2-1$
52) a^2-b^2 53) $9x^2-4y^2$ 54) $25a^2-4b^2$

Exercise 28

1) $(x+y)(a+b)$ 2) $(p-q)(m+n)$ 3) $(ac+d)^2$
4) $(2p+q)(r-2s)$ 5) $2(a-b)(2x+3y)$
6) $(x^2+y^2)(ab-cd)$ 7) $(mn-pq)(3x-1)$
8) $(k^2l-mn)(l-1)$ 9) $(x+3)(x+1)$
10) $(x+4)(x+2)$ 11) $(x+5)(x+4)$
12) $(x-2)(x-1)$ 13) $(x-4)(x-2)$
14) $(x-4)(x-3)$ 15) $(x+5)(x-3)$
16) $(x+7)(x-4)$ 17) $(x+7)(x-1)$
18) $(x-4)(x+3)$ 19) $(x-7)(x+2)$ 20) $(x-y)^2$
21) $(a+3b)(a+b)$ 22) $(p-q)(p-8q)$
23) $(m+3n)(m-8n)$ 24) $(3p-2)(p+1)$
25) $(2x+3)(x+5)$ 26) $(m+2)(3m-14)$
27) $(2x+1)(2x-6)=2(2x+1)(x-3)$
28) $(5a-3)(2a+5)$ 29) $(3x+1)(7x+10)$
30) $(2p+3)(13p-3)$ 31) $(3x-7)(2x+5)$
32) $(3p-q)(2p+3q)$ 33) $(2a+3b)(a+2b)$
34) $(6a-5b)(5a-3b)$ 35) $(4x+y)(3x-2y)$
36) $(2x+3)^2$ 37) $(x+y)^2$ 38) $(3x+1)^2$
39) $(p+2q)^2$ 40) $\left(\frac{1}{x}+\frac{1}{y}\right)^2$ 41) $\left(\frac{m}{2}+\frac{1}{3}\right)^2$
42) $(5x-2)^2$ 43) $(m-n)^2$ 44) $(a-\frac{1}{2})^2$
45) $(x-2)^2$ 46) $\left(5-\frac{2}{R}\right)^2$ 47) $\left(\frac{1}{x}-1\right)^2$
48) $(2x+y)(2x-y)$ 49) $(m+n)(m-n)$
50) $(x+\frac{1}{3})(x-\frac{1}{3})$ 51) $(3p+2q)(3p-2q)$
52) $\left(\frac{1}{x}+\frac{1}{y}\right)\left(\frac{1}{x}-\frac{1}{y}\right)$ 53) $(11p+8q)(11p-8q)$
54) $(1+b)(1-b)$ 55) $(y+\frac{3}{4}x)(y-\frac{3}{4}x)$
56) $(x+y+q)(x+y-q)$ 57) $(a+p+q)(a-p-q)$
58) $3(x+1)(x-3)$

Exercise 29

1) $(a-b)(a-b-2x)$ 2) $(x+y)(3+x+y)$
3) $5(3m-n)(3m-n-a)$ 4) $(x-y)(x+y+1)$
5) $(a+b)(a-b-3)$ 6) $(x+y)[3(x-y)-2]$
7) $2[(m+n)(m-n+2)]$ 8) $\pi R(R+2h)$
9) $y(2y+3a)(2y-3a)$ 10) $\pi l(R+r)(R-r)$
11) $5x(2x+3y)(2x-3y)$ 12) $\frac{1}{3}\pi r^2(2r+h)$
13) $(a+c)(a-2b)$ 14) $(x-1+2y)(x-1-2y)$
15) $y(2p-3)(x-1)$ 16) $(a+b)(a-b-1)$
17) $2p(3p+1)(3p-1)$ 18) $(x-y)[3(x+y)+2]$
19) $4(x-y)[2(x-y)+1]$ 20) $(a-b)(a+b+3)$

Self Test 9

1) True 2) False 3) True 4) True
5) False 6) True 7) True 8) False
9) False 10) True 11) False 12) True
13) False 14) True 15) True 16) False
17) True 18) False 19) True 20) True
21) True 22) False 23) True 24) True
25) True 26) True 27) False 28) True
29) True 30) True 31) $(x-3)$ 32) $(x-7)$
33) $(2x+5)$ 34) $(5x-3)$ 35) $(3a+2b)$
36) $(2x-5y)$ 37) $x+1)$ 38) $(x-1)$
39) $(3p-5)$ 40) $(x+2y)$ 41) $(p-q-r)$
42) $(x-y+3)$ 43) $(a-b)$ 44) $(x-y-1)$
45) $(3x+1)(3x-1)$

ANSWERS TO CHAPTER 10

Exercise 30

1) $\frac{b}{c}$ 2) $\frac{9s^2}{2t}$ 3) $\frac{8acz}{3y^3}$ 4) $\frac{3}{8}$ 5) $\frac{5}{6}$
6) $\frac{16(a+2b)}{9(a+b)}$ 7) $\frac{9qs}{pr}$ 8) $\frac{21b^2}{10ac}$ 9) ac^2b^2
10) $\frac{3}{2}$ 11) $\frac{1}{x+2y}$ 12) $\frac{5x}{4(x+4)}$ 13) $a-b$
14) $\frac{1}{3a-2b}$ 15) $\frac{1}{4x+5}$ 16) $\frac{1}{2x-3}$ 17) $\frac{1}{x+3}$
18) $a+b$ 19) $a(a-3x)$ 20) $2x-y$

Exercise 31

1) $12x$ 2) $6xy$ 3) $12ab$ 4) abc
5) $36m^2n^2p^2q$ 6) $10a^2b^4$ 7) $(m-n)^2$
8) $(x+1)(x+3)^2$ 9) x^2-1 10) $9a^2-b^2$
11) $\frac{47x}{60}$ 12) $\frac{a}{36}$ 13) $\frac{1}{2q}$ 14) $\frac{32}{15y}$
15) $\frac{9q-10p}{15pq}$ 16) $\frac{9x^2-5y^2}{6xy}$ 17) $\frac{15xz-4y}{5z}$
18) $\frac{40-11x}{40}$ 19) $\frac{19m-n}{7}$ 20) $\frac{a+11b}{4}$

21) $\frac{8n-3m}{3}$ 22) $\frac{5x-2}{20}$ 23) $\frac{x-14}{12}$
24) $\frac{13x-21}{30}$ 25) $\frac{1}{x-5}$ 26) $\frac{4x+3}{(2x+1)(2x-1)}$
27) $\frac{3}{x-3}$ 28) $\frac{3}{x(x+5)}$ 29) $\frac{-2x}{x^2-9}$
30) $\frac{2x^2+4xy-3x+3y}{(x+y)^2(x-y)}$

Self Test 10

1) a 2) a and c 3) b 4) a and d 5) c
6) b 7) d 8) d 9) b 10) c 11) c
12) a

ANSWERS TO CHAPTER 11

Exercise 32

1) $x=5$ 2) $t=7$ 3) $q=2$ 4) $x=20$
5) $q=-3$ 6) $x=3$ 7) $y=6$ 8) $m=12$
9) $x=2$ 10) $x=3$ 11) $p=4$ 12) $x=-2$
13) $x=-1$ 14) $x=4$ 15) $x=2$ 16) $x=6$
17) $m=2$ 18) $x=-8$ 19) $d=6$ 20) $x=5$
21) $x=3$ 22) $m=5$ 23) $x=-\frac{29}{5}$ 24) $x=2$
25) $x=\frac{45}{8}$ 26) $x=-2$ 27) $x=-15$
28) $x=\frac{50}{47}$ 29) $m=-1{\cdot}5$ 30) $x=\frac{15}{28}$
31) $m=1$ 32) $x=2{\cdot}5$ 33) $t=6$ 34) $x=4{\cdot}2$
35) $y=-70$ 36) $x=\frac{5}{3}$ 37) $x=13$
38) $x=-10$ 39) $m=\frac{25}{26}$ 40) $y=\frac{9}{7}$
41) $x=\frac{25}{3}$ 42) $x=3{\cdot}5$ 43) $x=20$ 44) $x=13$
45) $x=-53$ 46) $x=4$ 47) $p=\frac{13}{4}$
48) $m=3$ 49) $x=\frac{15}{4}$ 50) $x=\frac{7}{2}$

Exercise 33

1) $(x-5)$ years 2) $(3a+8b)$ pence
3) $(5x+y+z)$ hours 4) $2(l+b)$ mm
5) £$(a-x)$; £$(b+x)$ 6) $(120+x)$ minutes
7) £$\left(Y+\frac{nx}{100}\right)$ 8) £$\frac{nx}{100\text{m}}$ 9) £$\frac{13(12a+3b)}{25}$
10) $(P+Qn)$ kg 11) $6a+4b+c$
12) £$(Mx+Ny+Pz)$ 13) £$(49u+3v)$
14) $\frac{\mathbf{100\ mN}}{100M+mn}$ 15) £$\left(u+\frac{xy}{100}\right)$
16) £$(X+Y+Z)\left(1-\frac{g}{100}\right)$ 17) $a(m-a)$
18) £$\frac{x}{100}(M-Q)$ 19) £$\frac{49X}{30}$
20) £$\left(a+b+\frac{cn}{100}\right)$

Exercise 34

1) £3, £9; £12 2) 2 and 10 3) £50
4) 5 and 4 5) 5 at 8 p and 10 at 5p
6) 10 at £1·00 and 8 at £1·25 7) 15, 16 and 17
8) 15m × 16·5m
9) A carries 12 people; B carries 8 people.
10) 3 750 litres 11) £172 12) 80 13) 3
14) 12·5 and 25 litres/min
15) AB=120mm; BC=80mm and AC=60mm

Self Test 11

1) True 2) False 3) True 4) True
5) False 6) False 7) False 8) False
9) True 10) True 11) False 12) True
13) False 14) False 15) False 16) True
17) False 18) False 19) True 20) False
21) True 22) False 23) False 24) False
25) False 26) a 27) c 28) b 29) b
30) a, b 31) c, d 32) d 33) b 34) d
35) c 36) b 37) c 38) c 39) b 40) c

ANSWERS TO CHAPTER 12

Exercise 35

1) 21·98 2) 28·26 3) 162 000 4) 0·04
5) 8 6) 20 7) 977 8) 540 9) 5·25
10) $2\frac{5}{6}$

Exercise 36

1) 88 2) $\frac{3}{8}$ 3) 6·5 4) 6 5) 2 6) 7·5
7) −1·25 8) 216 9) 13·5 10) 32 11) 4
12) −2·25 13) 36 14) 0·8 15) $2\frac{1}{6}$

Exercise 37

1) $\frac{c}{\pi}$ 2) $\frac{\text{S}}{\pi n}$ 3) $\frac{c}{P}$ 4) $\frac{A}{\pi r}$ 5) $\frac{v^2}{2g}$ 6) $\frac{\text{I}}{\text{PT}}$
7) $\frac{a}{x}$ 8) $\frac{\text{E}}{\text{I}}$ 9) ax 10) $\frac{\text{PV}}{\text{R}}$ 11) $\frac{0{\cdot}866}{d}$
12) $\frac{\text{ST}}{s}$ 13) $\frac{33\,000\ \text{H}}{\text{PAN}}$ 14) $\frac{4\text{V}}{\pi d^2}$ 15) $p+14{\cdot}7$
16) $\frac{v-u}{a}$ 17) $\frac{n-p}{c}$ 18) $\frac{y-b}{a}$ 19) $5(y-17)$
20) $\frac{\text{H}-\text{S}}{\text{L}}$ 21) $\frac{b-a}{c}$ 22) $\frac{\text{B}-\text{D}}{1{\cdot}28}$ 23) $\frac{\text{R}(\text{V}-2)}{\text{V}}$
24) $\text{C}(\text{R}+r)$ 25) $\frac{\text{S}}{\pi r}-r$ 26) $\frac{\text{H}}{wS}+t$
27) $2pc+n$ 28) $\text{D}-\frac{\text{TL}}{12}$ 29) $\frac{Vr}{V-2}$
30) $\frac{\text{SF}}{S-P}$ 31) $\frac{V^2}{2g}$ 32) $\frac{w^2}{k^2}$ 33) $\frac{t^2g}{4\pi^2}$
34) $\frac{4\pi^2\text{W}}{gt^2}$ 35) $\frac{Pr}{V^2+gr}$ 36) $\frac{z^2y}{1-z^2}$ 37) $\frac{2-k}{k-3}$
38) $\frac{3-5a}{4a}$ 39) $\frac{2ka}{aV^2+2k}$ 40) $\frac{2s-dn(n-l)}{2n}$
41) $\frac{c^2+4h^2}{8h}$ 42) $\frac{xD}{x+h}$ 43) $\frac{p(D^2+d^2)}{D^2-d^2}$

Self Test 12

1) b 2) d 3) a 4) b 5) d 6) b
7) d 8) a, c 9) b 10) c 11) c
12) b, d 13) c, d 14) a, b 15) d 16) c
17) d 18) c 19) b 20) c, d

ANSWERS TO CHAPTER 13

Exercise 38

1) 1, 2 2) 4, 5 3) 4, 1 4) 7, 3 5) $\frac{1}{2}$, $\frac{3}{4}$
6) 3, 2 7) 4, 3 8) 3, 4 9) $-1{\cdot}5$, 1
10) -1, -2 11) 6, 3 12) 2, $-1{\cdot}5$

Exercise 39

1) 15 and 12 2) 16 3) £120 and £200
4) 9 and 7 5) 40 and 75 6) 300
7) 48, 64; 40 8) 30p 9) 15, 21 10) 12

Self Test 13

1) c 2) a 3) b 4) d 5) c 6) a
7) a 8) c 9) a, b 10) b 11) b, c
12) b, c

ANSWERS TO MISCELLANEOUS EXERCISE

Exercise 40

1) $-\dfrac{9}{4}$ 2) $\dfrac{at^2}{2}-ut$
3a) $(3z+5)(3z-5)$ b) $(3a+2b)(a-4c)$
4) $\dfrac{3}{x+2}$ 5) $f=\dfrac{4(s-u)^2}{t^2}$ 6) $\dfrac{2x}{x^2-1}$
7) $-x+2y+6z$ 8) $7(1+3a)(1-3a)$
9) $7t(t-2)$ 10) $\dfrac{c}{b-a}$ 11) $(3x+10)(2x-7)$
12) $2y(y-x)$ 13) $\dfrac{5y-1}{4y-3}$ 14) $\dfrac{x+1}{6}$
15) $25x^2-70x+49=(5x-7)^2$
16i) 16 ii) 0 iii) 6
17) $(a+3)(a+1)$
18a) $(a+b+2c)(a+b-2c)$ b) $(2+a)(x+2y)$
19) $\dfrac{2}{3}$ 20) $\dfrac{4}{2-x}$ 21) $\dfrac{1}{2(x-2)}$
22i) $(2x+y)(2x-y)$ ii) $(a+b+x)(a+b-x)$
23i) $(3x-4)(x+3)$ ii) $(x+y)(y-2)$
iii) $3(x+2y)(x-2y)$
24) $(a^2+b^2)(a+b)(a-b)$ 25) $k=12$
26i) $x=\frac{4}{5}$; $y=-\frac{3}{5}$ ii) $t=\pm\frac{1}{2}$; $x=\pm\frac{4}{5}$
27) $\dfrac{V}{\pi h^2}+\dfrac{h}{3}$
28i) $(5x+9)(5x-9)$ ii) $(x+y+3)(x-y+1)$
iii) $(5x+1)(x+2)$ iv) $(x+a-2)(x+a-1)$
29) 53 30) $3(3x+4y)(3x-4y)$ 31) 37
32) $2x(x-2)(x+1)$ 33) $\dfrac{1}{(x+3)(x+2)}$
34) $\dfrac{3}{x(x+5)}$ 35) 73 36) $\dfrac{y}{p+q}-a$
37) $(a-3b)(c-2b)$ 38) $3(4p-3)(2p+1)$
39i) $3(2x+y)(2x-y)$ ii) $(x+y)(y+2)$
iii) $(3x+2)(x+4)$ 40) $-12\frac{5}{18}$
41i) $-3x+11y$ ii) $10x^2+10y^2$ 42) $-\dfrac{1}{3p}$
43i) -1 ii) 21 iii) 8
44i) $(2p-1)(p-2)$ ii) $(b+d)(a-2c)$
45) $\dfrac{7}{20}$ 46) $\dfrac{4}{x(x-4)(x+3)}$
47) $\dfrac{1}{x-3}$ 48) $c=\dfrac{2}{b-a}$ 49) 80
50i) $(3p-2)(2p+3)$ ii) $(5-a-7b)(5-a+7b)$
51) $x=-2$ 52) $x=8$, $y=-4$ 53) 1
54i) $\dfrac{V^2-U^2}{2a}$ ii) 13
55a) $\dfrac{100y}{x}$ b) $\dfrac{200y}{x(x-2)}$ 56) $\dfrac{3}{x^2(x^2-4)}$
57) $(f-c)$pence 58) 47p; 12km; $7+8n$; 7km
59) $x=2$, $y=-1$ 60) $x=3$ 61) $x=14$
62) $x=\dfrac{2}{7}$ 63) $p=4$, $q=-2$ 64; $(2x-17)$; 5
64) $(2x-17)$; 5 65) $x=1$ 66) $-\frac{3}{5}b$ 67) 9
68) $\dfrac{1-y}{3y+2}$ 69) $(3a+7)(4a-7)$ 70) $-48\frac{11}{36}$
71) $\dfrac{5c^2+4b^2}{4b}$ 72) $-\dfrac{10}{11}$ 73) $x=\dfrac{20}{3}$, $y=5$
74) $A=-1$, $B=4$ 75) $\dfrac{4a}{x+4}$ 76) $x=\dfrac{9}{4}$
77) $\dfrac{xz}{60y}$ 78) $x=\dfrac{10}{19}$, $y=-\dfrac{61}{38}$

ANSWERS TO CHAPTER 14

Exercise 41

1) 1·844 2) 2·862 3) 2·294 4) 3·039
5) 2·649 6) 1·735 7) 5·916 8) 9·445
9) 7·292 10) 9·110 11) 8·901 12) 7·072
13) 30 14) 26·94 15) 84·51 16) 298·3
17) 62·81 18) 29 890 19) 0·3921
20) 0·04121 21) 0·002665 22) 0·1987
23) 0·02798 24) 0·04447

Exercise 42

1) 2·25 2) 4·41 3) 73·96 4) 9·923
5) 58·98 6) 27·35 7) 18·18 8) 62·67
9) 64·27 10) 75·76 11) 529 12) 1 648
13) 9 565 000 14) 12 610 15) 9 628
16) 0·000361 17) 0·5317 18) 0·000017800
19) 0·08032 20) 0·0000003346 21) 5·864
22) 502·5 23) 10·99 24) 20 480
25) 6 526 26) 28·00 27) 26·67 28) 13·91

Exercise 43

1) 0·2941 2) 0·1221 3) 0·1899
4) 0·1082 5) 0·1426 6) 0·02857 7) 0·01121
8) 0·01881 9) 0·001111 10) 0·0001401
11) 6·509 12) 588·8 13) 25·34 14) 1277
15) 505·5 16) 0·004283 17) 53·10
18) 0·000016 19) 0·3499 20) 0·2311
21) 7·491 22) 0·03212 23) 0·1292
24) 0·2230 25) 2·383 26) 0·4440
27) 15·17 28) 0·06541

Self Test 14

1) c 2) a 3) b 4) b 5) a
6) c 7) b 8) c 9) a 10) b 11) b
12) b 13) b 14) c 15) b 16) a
17) c 18) b 19) b 20) c

ANSWERS TO CHAPTER 15

Exercise 44

1) a^{11} 2) z^{11} 3) y^{12} 4) 2^8 5) 3^8
6) $\frac{3}{32}a^6$ 7) a^3 8) m^7 9) 2^4 10) x^{15}
11) a^4 12) q^8 13) m 14) l^2 15) L^2
16) x^{12} 17) a^{15} 18) $9x^8$ 19) 2^6 20) 10^6
21) a^3b^6 22) $a^4b^8c^{12}$ 23) $2^5x^{10}y^{15}z^5$
24) $\frac{3^5m^{10}}{4^5n^{15}}$ 25) $\frac{1}{10}, \frac{1}{4}, \frac{1}{81}, \frac{1}{25}$ 26) 32, 625, 2, 3
27) $3^6, 3^{15}, 3^{12}, 3^{17}$ 28) $a^{1/5}, a^{2/3}, a^{4/7}, a^3$
29) 64, 1 000 000, 4, 8 30) 4, 8, $\frac{1}{3}$, $\frac{1}{3}$ 31) 10
32) $\frac{5}{6}$ 33) 1, $\frac{1}{5}$, 100 000 34) $\frac{1}{3x^2}, \frac{1}{3x^2}, \frac{1}{5a^4}$
35) $32a^5$ 36) x^4y^4 37) 12

Exercise 45

1) $1·96\times10$ 2) $3·85\times10^2$ 3) $5·6\times10$
4) $5·9876\times10^4$ 5) $1·897\times10^2$ 6) $1·5\times10^7$
7) $1·3\times10^{-2}$ 8) $6·98\times10^{-4}$ 9) $3·85\times10^{-3}$
10) $6·97\times10^{-1}$ 11) $1·44\times10^2$ 12) $3·2\times10^4$
13) $2·63\times10^{-2}$ 14) $4·5\times10^3$ 15) $2·6\times10^2$
16) $7·2\times10^{-4}$ 17) $2·1\times10$, 18) $1·5\times10^{-5}$
19) $1·342\times10^{-1}$ 20) $4·899\times10^{-1}$

Exercise 46

1) 1, 3, 0, 4, 2, 5, 1, 0, 2, 3, 3, 1, 2, 1
2a) 0·8451, 1·8451, 2·8451, 3·8451, 4·8451
b) 0·4914, 1·4914, 2·4914, 3·4914, 6·4914
c) 1·6839, 5·6839, 0·6839, 2·6839
d) 3·8974, 0·8974, 1·8974, 4·8974
e) 0·0013, 1·0013, 3·0013, 2·0013
3a) 2·089, 208·9, 20 890, 20·89
b) 1 884, 1·884, 18 840, 1 884 000
c) 3·969, 39·69, 39 690, 396·9
d) 78 500 000, 7·85, 785, 78 500
4) 362·1 5) 17 970 6) 148 900 7) 3 784
8) 1·941 9) 2·599 10) 9·906 11) 1·566
12) 71·93 13) 9·337 14) 6·634 15) 25·66
16) 393·2 17) 863·4 18) 1·596 19) 102·1
20) 49·82 21) 1·213 22) 1·647 23) 2·398
24) 2·487 25) 1·067 25) 1·888 27) 13·1
27) 56 130

Exercise 47

1a) 0·4498, $\bar{1}$·4498, $\bar{2}$·4498, $\bar{3}$·4498
b) 0·6625, $\bar{1}$·6625, $\bar{3}$·6625, $\bar{5}$·6625
c) $\bar{2}$·9898, $\bar{4}$·9898, $\bar{1}$·9898
d) $\bar{5}$·7690, $\bar{2}$·7690, $\bar{4}$·7690
2a) 0·2714 b) 0·006606 c) 0·0003537
d) 0·0307 e) 0·00000003403 f) 0·1052
3) 0, 1, $\bar{4}$, $\bar{4}$, $\bar{1}$, $\bar{2}$, $\bar{3}$, $\bar{9}$, $\bar{3}$, $\bar{2}$
4) 0·1, 2·3, $\bar{2}$·9, 0·1, 1·1, $\bar{2}$·7, $\bar{4}$·1, $\bar{2}$·0, 1·1, $\bar{1}$·2, $\bar{4}$·5, 4·4
5) $\bar{2}$·9172, 0·7389, $\bar{9}$·1650
6) $\bar{1}$, $\bar{3}$, $\bar{3}$, $\bar{3}$, 3, $\bar{1}$, 0, 3, $\bar{3}$, 3
7) 5·1, $\bar{3}$·2, 0·4, 2·3, $\bar{2}$·5, 4·2, $\bar{2}$·1, $\bar{2}$·9, $\bar{1}$·5, $\bar{4}$·5, $\bar{1}$·8, 1·8
8) 6·9094, $\bar{1}$·6204, $\bar{4}$·5424, 2·1238
9a) $\bar{2}$·8 b) $\bar{9}$·3 c) $\bar{1}$·4 d) $\bar{4}$·$\bar{4}$ e) $\bar{1}$·0
10a) $\bar{1}$·3 b) $\bar{1}$·3 c) $\bar{1}$·6 d) $\bar{1}$·5 e) $\bar{2}$·7
11) $4·747\times10^{-4}$ 12) $4·894\times10^{-1}$
13) $6·679\times10$ 14) $1·924\times10^{-5}$ 15) $2·900\times10^3$
16) $1·306\times10^{-3}$ 17) $4·839\times10^{-1}$
18) $5·480\times10^{-12}$ 19) $5·069\times10^{-1}$
20) $4·119\times10^{-1}$ 21) $9·210\times10^{-2}$
22) $9·305\times10^{-1}$

Exercise 48

1) 3·44 2) 3·61 3) 8·299 4) 225
5) 0·6955 6) 0·7688 7) 0·6437 8) 3·966
9) 25·89 10) $I=\frac{Wl^3}{48Ey}$; 37·55

Self Test 15

1) a 2) b 3) c 4) a 5) d 6) d
7) a 8) c 9) b 10) c 11) c 12) b
13) c 14) d 15) a 16) b 17) c
18) d 19) c 20) d 21) a 22) c 23) d
24) c 25) b

ANSWERS TO CHAPTER 16

Exercise 49

1) ±5 2) $\pm2·8284$ 3) ±4 4) ±4
5) ±4 6) $\pm1·732$ 7) 5 or 2 8) $\frac{4}{3}$ or -3
9) 0 or -7 10) 0 or 2·5 11) -8 or 4
12) -5 or -4 13) 3 14) -9 or 8
15) $\frac{1}{3}$ or 2 16) $\frac{4}{7}$ or $\frac{3}{2}$ 17) $-\frac{4}{3}$ or $\frac{7}{3}$
18) 0 or 3 19) 0 or -8 20) $-\frac{1}{2}$ or $\frac{3}{2}$

Exercise 50

1) 1·175 or $-0·425$ 2) 1·618 or $-0·618$
3) 0·573 or $-2·907$ 4) 0·211 or $-1·354$
5) 1 or $-0·2$ 6) 3·886 or $-0·386$
6) $-3·775$ or 0·442 8) $-9·18$ or 2·18
9) $-0·225$ or $-1·746$ 10) 11·14 or $-3·14$
11) $-6·275$ or 1·275 12) -11 or 6
13) 3·303 or $-0·303$ 14) 3·186 or 0·314

Exercise 51

1) $x=1$, $y=2$, or $x=2$, $y=1$
2) $x=-17$, $y=-20$ or $x=9$, $y=6$
3) $x=8{\cdot}5$, $y=0{\cdot}5$ or $x=6$, $y=3$
4) $x=9$, $y=3$ or $x=4$, $y=8$
5) $x=3{\cdot}2$, $y=-0{\cdot}4$ or $x=4$, $y=-2$
6) $x=2{\cdot}2$, $y=5{\cdot}4$ or $x=3$, $y=5$
7) $x=4$, $y=0{\cdot}5$ or $x=\frac{1}{3}$, $y=6$
8) $x=1{\cdot}5$, $y=-10{\cdot}5$ or $x=-5$, $y=-30$

Exercise 52

1) 6 2) 9 m and 8 m 3) 15 cm 4) 4 m × 1 m
5) 13·35 m 6) 5, 6 and 7 7) 14 cm 8) $\frac{2}{5}$
9) 8·83 m or 3·17 m
10) $d(d-2)+(d-4)^2=148$; $d=11$ cm
11a) $\dfrac{100}{x}$ b) $\dfrac{100}{x+1{\cdot}5}$; $x=2{\cdot}87$
12) $\dfrac{15\,600}{x}$; $x=12$
13) $\dfrac{240}{x}$; $\dfrac{240}{x+4}$; $x=16$; 2 m

Self Test 16

1) d 2) a 3) b 4) c 5) c 6) c
7) b 8) b 9) d 10) a 11) b 12) c, d.

ANSWERS TO MISCELLANEOUS EXERCISE

Exercise 53

1) $x=9$ or 1 2) $x=-1{\cdot}42$ or $0{\cdot}42$ 3) $\pm\frac{3}{2}$
4) 2 5) 0·3159 6) 1·58 or 0·42
7) $m=-\frac{1}{3}$; $n=-2$ 8) 2·02 9) 0 or $\frac{1}{9}$
10) 2·12 or −0·79
11a) 527·0 b) 0·4333 12) $a^{3/2}b^{11/8}$
13) 0·3359, 14) −11 or 6 15) $\dfrac{4a}{x+4}$; 16
16) 0·3760 17) $\frac{1}{2}$ or $2\frac{1}{2}$ 18) $\frac{5}{6}$ 19) 3·188
20) 3·19 or 0·31 21) 7 22a) 3·272 b) 56400
23) 39·81 24) 15; 532 25) −3·47 or 0·62
26a) 0·3044 b) 8·6605
27a) 0·8720 b) 16 28) $\frac{2}{7}$
29a) 5·971 b) 10·30 c) 0·001024
30a) 46·10 b) 7·813 c) 0·5718 31) 10^4
32) 32·13; 7·5 or 0·54 33) 40 34) 3·73 or 0·27
35a) x^6 b) $y^{3/2}$ c) z^8 36)a 27 b) $\frac{1}{4}$
37a) 1·38 b) 0·4391
38a) $x=\frac{4}{5}$; $y=-\frac{3}{5}$ b) $t=\pm\frac{1}{2}$; $x=\pm\frac{4}{5}$
39) $\dfrac{60d}{V}$, 6 km/h 40a) 23·40 b) 8·443
41a) $(3x+8)(x-4)$; $-\frac{8}{3}$ or 4 b) 0·46 or −0·86
42a) 1·854 b) 2·520 c) 1·380
43a) $x=3$ b) $x=4{\cdot}3$ or 0·70
44a) i) 0·4348 ii) 0·04317 b) $2{\cdot}356\times10^5$

ANSWERS TO CHAPTER 17

Exercise 54

1) 8 km 2) 15 Mg 3) 3·8 Mm
4) 1·891 Gg 5) 7 mm 6) 1·3 μm 7) 28 g
8) 360 mm 9) 64 mg 10) 3·6 mA
11) $5{\cdot}3\times10^4$m 12) $1{\cdot}8\times10^4$g 13) $3{\cdot}563\times10^6$g
14) $1{\cdot}876\times10^{10}$g 15) 7×10^{-2}m
16) $7{\cdot}8\times10^{-2}$g 17) $3{\cdot}58\times10^{-10}$m
18) $1{\cdot}82\times10^{-5}$m 19) $2{\cdot}706\times10^{14}$m
20) $2{\cdot}53\times10^{-4}$g

Exercise 55

1) 50 cm² 2) 0·0526 m² 3) 8·2 m²
4) 26 500 cm² 5) 123 800 cm² 6) 780 000 mm²

Exercise 56

1) 8·8 mm 2) 0·0128 m²
3a) 1200 mm² b) 275 mm² c) 259·5 mm²
d) 774 mm² e) 1050 mm² f) 1094 mm²
4) 22·1 cm² 5) 13·42 cm² 6) 9·62 cm²
7) 143 cm² 8) 53·7 m² 9) 28 cm²
10) 2·12 m 11) 3062 mm² 21) 15·7 cm
13a) 11 200 mm² b) 3·02 cm²
14a) 22·0 mm b) 86·8 m c) 26·4 cm
15a) 10·94 mm b) 5·900 cm c) 62·1 m
16) 6·2 cm² 17) 3·41 cm 18) 2 592 mm²
19) 909 20a) 1·047 cm b) 2·29 cm
21a) 119·6° b) 10·16° 22) 8·92 cm
23a) 4·71 m² b) 5·08 cm² c) 76·2 cm²
24) 866 mm²

Exercise 57

1) 5×10^6 cm³ 2) 8×10^7 mm³
3) $1{\cdot}8\times10^{10}$ mm³ 4) 0·83 m³
5) $8{\cdot}5\times10^{-4}$ m³ 6) 0·0785 m³ 7) 5 m³
8) 0·0025 l 9) 827 000 l 10) 8·275 l

Exercise 58

1) 76·19 m 2) 10·61 m 3) 0·00875 m³
4) 91·77 mm 5) 2475 m 6) 128 300 mm³
7) 6·543 cm 8) $1{\cdot}768\times10^6$
9) $V=t\left(\dfrac{\pi D^2}{4}-l^2\right)$; 2·46 cm³
10a) 37·14 m² b) 24·57 m² c) 38 320 l
11) 40 cm 12a) 47·19 cm b) 4 303 cm²
13) 2·56 cm 14) 55 800 kg
15) 55 mm; 13·5 mm
16a) 6877 l b) 15·82 m² c) 5·107 m²
17) 22·54 cm; $\frac{1}{2}$ 18) 52 kg
20) 3·832 cm; 4·69 cm; 214·1 cm²

Exercise 59

1) 5180 kg 2) 38·48 kg 3) 48·78 kg
4) 2·56 m²; 78·85 kg 5) 360·45 cm³; 8·04 g/cm³
6) 3·78 g/cm³ 7) 577·5 kg 8) 29 g
9) 18 mm 10) 22·68 kg; 10·96

Exercise 60

1) 0·053 m^3/s 2) 6·365 m^3/s 3) 5 h
4) 0·4 m; $266\frac{2}{3}$ s 5) 3·82 cm/min 6) 7977 s
7) 5·941 m; 127·3 m/s 8) 238 m^3; 2·11 m/s

Exercise 61

1) 113 cm^3; 524 cm^3 2) 27 cm^3 3) 9·35 cm
4) 66 900 mm^2; 10 500 mm^2 5) 46·1 cm^3
6) 22·7 kg; 13·10 kg 7) 560 kg; 16 cm
8) 3·5 cm; 1·59 cm 9) 41 900 mm^3; 41 570 mm^3
10) $\frac{1}{1000}$

Self Test 17

1) c 2) d 3) b 4) d 5) d 6) c
7) b 8) a, c 9) b, d 10) b 11) c
12) b, d 13) d 14) a 15) b 16) b
17) d 18) b 19) a, d 20) a 21) d
22) c, d 23) a 24) a, c 25) b 26) c
27) b 28) a, c 29) a, c 30) a 31) b, d
32) a 33) a, d 34) b 35) c

ANSWERS TO CHAPTER 18

Exercise 62

2) 7·5; 3·7 3) 254·3 cm^2 4) 45 5) 2·3 min

Exercise 63

5) 1, 3 6) -3, 4 7) -5, -2 8) 4, -3
9) $m=4$, $c=13$ 10) $m=2$, $c=-2$
11) $y=2x+1$ 12) $y=3x-2$ 13) $y=-2x-3$
14) $y=4-3x$ 15) $y=5x+7$ 16) $y=3x+4$
17) $P=4{\cdot}74\,Q+1{\cdot}0$ 18) $E=0{\cdot}51\,W+3$
19) 1·29; 20 20) $E=4I$

Exercise 64

6) $x=3$ or 4 7) $x=4$ 8) $x=\pm 3$
9) $x=-5{\cdot}4$ or 3·7
10a) $x=-6{\cdot}54$ or $-0{\cdot}46$ b) $x=-7{\cdot}28$ or 0·28
c) $x=-6$ or -1
11a) $x=-1$ or $\frac{1}{3}$ b) $x=-1{\cdot}39$ or 0·72
c) $x=-1{\cdot}21$ or 0·55
12a) $x=\pm 3$ b) $x=\pm 2{\cdot}24$ c) No solution

Exercise 65

1a) 1·15 b) $-0{\cdot}72$ or 1·39 c) 0 or 0·43
2) $-6{\cdot}16$ or 0·16 3) $x=4$, $y=1$
4) $x=7$, $y=3$ 5) $x=3$, $y=2$
6) $-0{\cdot}54$ or 1·40 7) 0, 3, 7·875, 0 and 2, 0·7
8) 4, 4·75, 5, 3·85 or 0·65 9) $x^2-x-11=0$, 3·85
10) 1, $-1{\cdot}75$, -2, $\frac{x^2}{4}+\frac{24}{x}+12=\frac{x}{3}-2$, 2·6, 5·4.

Self Test 18

1) True 2) True 3) False 4) True
5) True 6) False 7) True 8) True
9) False 10) False 11) True 12) False
13) False 14) True 15) True 16) True
17) False 18) False 19) True 20) True
21) b 22) b 23) a 24) c 25) b 26) a
27) c 28) b

ANSWERS TO CHAPTER 19

Exercise 66

1a) $y=kx^2$ b) $U=k\sqrt{V}$ c) $S=\frac{k}{T^3}$
d) $h=\frac{k}{\sqrt[3]{m}}$
2a) $\frac{81}{8}$ b) $\frac{4}{3}$ c) $\frac{8}{9}$ 3) 4·5 4) 36
5) $\sqrt{2}:1$

Exercise 67

1) 4 mm 2) $1\frac{7}{9}$ 3) 5·4 4) $p=12$ 5) 62
6) 4 7) -15 or 9 8) $v=84t-16t^2$; $5\frac{1}{4}$
9) 1930 10) $\frac{4}{25}$

Self Test 19

1) True 2) True 3) False 4) True
5) True 6) False 7) True 8) False
9) False 10) True 11) True 12) True
13) True 14) False 15) False 16) c
17) b 18) a 19) c 20) d 21) d
22) c 23) a 24) b

ANSWERS TO CHAPTER 20

Exercise 68

1) -5, 19 2) -4, 8 3) 3 4) 5·52 or $-6{\cdot}52$
5) -4, 6, 1·25 6) 8

Exercise 69

1) $2x$ 2) $7x^6$ 3) $12x^2$ 4) $30x^4$ 5) $1{\cdot}5t^2$
6) $2\pi R$ 7) $\frac{1}{2}x^{-1/2}$ 8) $6x^{1/2}$ 9) $x^{-1/2}$
10) $2x^{-1/3}$ 11) $-2x^{-3}$ 12) $-x^{-2}$
13) $-\frac{3}{5}x^{-2}$ 14) $-6x^{-4}$ 15) $-\frac{1}{2}x^{-3/2}$
16) $-\frac{2}{3}x^{-4/3}$ 17) $-7{\cdot}5x^{-5/2}$ 18) $\frac{3}{10}t^{-1/2}$
19) $-0{\cdot}01H^{-2}$ 20) $-5x^{-2}$ 21) $8x-3$
22) $9t^2-4t+5$ 23) $4u-1$ 24) $20x^3-21x^2+6x$
25) $21t^2-6t$ 26) $\frac{1}{2}x^{-1/2}+\frac{5}{2}x^{3/2}$ 27) $-3x^{-2}+1$
28) $\frac{1}{2}x^{-1/2}-\frac{1}{2}x^{-3/2}$ 29) $3x^2-1{\cdot}5x^{-3/2}$
30) $1{\cdot}3t^{0{\cdot}3}+0{\cdot}575t^{-3{\cdot}3}$ 31) $\frac{9}{5}x^2-\frac{4}{7}x-\frac{1}{2}x^{-1/2}$
32) $-0{\cdot}01x^{-2}$ 33) $46{\cdot}5x^{0{\cdot}5}-1{\cdot}44x^{-0{\cdot}4}$
34) $\frac{3}{2}x^2+5x^{-2}$ 35) $-6+14t-6t^2$

Exercise 70

1) -1 2) 6 3) $-\frac{1}{4}$ 4) $x=3$
5) $(\frac{1}{3}, -4\frac{5}{27})$; $(1, -3)$ 6) 16 7) 12

Exercise 71

1) 42 m/s 2) 6 m/s^2
3a) 9 b) 0 or 2 c) 6 d) 1 4) $-\frac{2}{27}, \frac{2}{27}$
5a) 5 b) 17 c) 4 6) 10·66 m 7) ± 6
8) 69 9) 10 10) 14, 36

Exercise 72

1) $-3\frac{1}{3}$ 2) $13\frac{1}{4}$ 3) 9, -23
4a) $-2{\cdot}25$ b) 3·3 or $-0{\cdot}3$ c) 2·4 or $-0{\cdot}4$
5) $2\frac{1}{4}$ 6) $\frac{7}{2}, -\frac{1}{2}, +\frac{1}{2}; -1, \frac{13}{24}$. 7) 2·43; 263
8) $7{\cdot}36 \times 7{\cdot}36 \times 3{\cdot}68$ 9) $x=\frac{4}{3}$, $k=\frac{2}{3}$
10) 625; $38{\cdot}23 \times 11{\cdot}77$

Exercise 73

1a) 11 (max), -16 (min) b) 4 (min), 0 (max)
c) 0 (max), -32 (min)
2a) 54 b) $x=5$ c) $x=-4$
3) $(3, -15)$, $(-1, 17)$ 4a) -2 b) 1 c) 9
5a) 12 b) 11·21 6) $x=1{\cdot}5$ 7) $x=10$
8) $r=4$ 9) 76·5 10) radius = height = 4·57 m

Self Test 20

1) True 2) False 3) True 4) True
5) True 6) True 7) False 8) True
9) True 10) False 11) True 12) True
13) False 14) True 15) True 16) True
17) False 18) True 19) True 20) True
21) True 22) True 23) False 24) True
25) True 26) False 27) True 28) True
29) True 30) False 31) a 32) b
33) c 34) c 35) c 36) c 37) d
38) b 39) d 40) b

ANSWERS TO CHAPTER 21

Exercise 74

1) $\frac{x^3}{3}+c$ 2) $\frac{x^9}{9}+c$ 3) $\frac{2x^{3/2}}{3}+c$

4) $-x^{-1}+c$ 5) $-\frac{x^{-3}}{3}+c$ 6) $2x^{1/2}+c$

7) $\frac{3x^5}{5}+c$ 8) $\frac{5x^9}{9}+c$ 9) $\frac{x^3}{3}+\frac{x^2}{2}+3x+c$

10) $\frac{x^4}{2}-\frac{7x^2}{2}-4x+c$ 11) $\frac{x^3}{3}-\frac{5x^2}{2}+2x^{1/2}-\frac{2}{x}+c$

12) $-\frac{4}{x^2}+\frac{2}{x}+\frac{2\sqrt{x^3}}{3}+c$ 13) $\frac{x^3}{3}-\frac{3x^2}{2}+2x+c$

14) $\frac{x^3}{3}+3x^2+9x+c$ 15) $\frac{4x^3}{3}-14x^2+49x+c$

Exercise 75

1) $y=\frac{x^2}{2}+1$ 2) $46\frac{2}{3}$

3) $y=10+3x-x^2$; 10; 12·25

4) $y=\frac{2x^3}{3}+\frac{3x^2}{2}+2x-\frac{1}{6}$ 5) $p=\frac{t^3}{3}-3t^2+9t-\frac{17}{3}$

7) $y=x^2-2x+4$ 8) $y=\frac{x^3}{3}+1$ 9) $\frac{10}{3}$

Exercise 76

1) $2\frac{1}{3}$ 2) 8 3) $8\frac{2}{3}$ 4) 4 5) $\frac{19}{6}$ 6) $\frac{2\sqrt{8}}{3}$

7) $\frac{2}{3}$ 8) $\frac{2}{3}$

Exercise 77

1) 136 2) 87 3) $5\frac{1}{6}$ 4) $\frac{2}{3}$ 5) 3·75
6) $1\frac{1}{3}$ 7) 60 8) $57\frac{1}{6}$ 9) 16 10) $\frac{1}{6}$

Exercise 78

1) 7·5 m 2) 20·5 m/s; 48 m
3) $s = 7{\cdot}5\,t^2 + 10t$; 160 m 4) 14·75 m
5) 60 m/s; 112 m 6) 136 m

Exercise 79

1) $25{\cdot}6\pi$ 2) 6π 3) $\frac{26\pi}{81}$ 4) $69\frac{1}{3}\pi$

5) $164\frac{2}{3}\pi$ 6) $30\frac{8}{15}\pi$ 7) 156π 8) 18π

9) $\frac{81\pi}{10}$ 10) $\frac{187\pi}{35}$

Self Test 21

1) b 2) b 3) d 4) c 5) a 6) b
7) b, c 8) d 9) a 10) b 11) c, d
12) a 13) c 14) a 15) a 16) c 17) c
18) d

ANSWERS TO MISCELLANEOUS EXERCISE

Exercise 80

1) $\sqrt{20}$ 2) $y=x^2-x+6$ 3) 4 4) -5
5) 31·25 6) 1·96 m; 3·10 m
7a) 4·15, 1·37 b) 1·18, 5·28 8) 7
9) $y=2x+1{\cdot}5x^2+2$ 10) 1·5
11) $x=1{\cdot}23$ or 2·91; $\frac{5}{6}$ 12) $\frac{1}{3}$ and $\frac{2}{3}$ seconds

13) $\frac{314}{3}\pi r^3$; 15·7 cm^3 14) $y=4$; $x=1$ or 4; 57π.

15) $3+18x^2$ 16) 2·5 17) 132 m
18) 24·9 cm 19) 1·35, 4·39, 34
20a) 15 m b) 20 c) 170 d) 240
21) 39 m 22) 15·88 cm
23) $x=0$ and $x=3$; $y=-2{\cdot}25$; 0; $8{\cdot}1\pi$

24) $x=0$, $y=0$ (max); $x=4$, $y=-32$ (min); $\frac{187\pi}{35}$

25) $17\frac{1}{3}$ 26) 67·42 m³ 27) ±6 28) $2\frac{1}{2}$
29) $x=1{\cdot}4$, $y=12{\cdot}8$ 30) 14; 15·8; 2479.

31) $\frac{x^4}{4}+x^2+\frac{1}{2x^2}+c$ 32) 9·042 cm

33) $3x^2-14x+12=0$; 3·53 34) 22, 2, 28, 16.
35) 375 36) $1\frac{1}{3}$ 37) 2·26, 4·21, 1·66, 4·64
38a) 0·34, 1·46 b) −3·1 39) −3
40) −1·45, 2·08

ANSWERS TO CHAPTER 22

Exercise 81

1) 28° 37′ 2) 69° 23′ 3) 14° 22′ 34″
4) 62° 48′ 11″ 5) 179° 11′ 25″ 6) 21° 3′
7) 22° 48′ 8) 7° 43′ 56″ 9) 5° 54′ 50″
10) 36° 58′ 11″ 11a) 3·206 b) 3·130
12a) 286° 26′ b) 99° 6′ c) 9° 6′
13a) 1·449 b) 3·299 c) 5·149 d) 0·909

Exercise 82

1) 20° 2) 100° 3) 35°
4) 70°, 110°, 110°, 70° 5) 54° 7) 65°
8) 230°, 32° 9) 65° 10) 80°

Self Test 22

1) d 2) a 3) b 4) a 5) b, d 6) b, d
7) b, d 8) c 9) c

ANSWERS TO CHAPTER 23

Exercise 83

1) $a=126°\ 41'$; $b=36°\ 41'$ 2) 140°
3) $p=80°$, $r=30°$, $q=70°$ 4) 70°
5) each 65° 48′

Exercise 85

1) 32 mm 2) $\frac{AB}{RQ}=\frac{AC}{RP}=\frac{BC}{QP}$

3) AE=3·2 cm; EH=2·4 cm 4) 1·6 cm 5) $\frac{1}{3}$
6) $1\frac{1}{2}$ cm 7) KB=2·6 cm; XY=1·5 cm
8) $\frac{1}{3}$

Exercise 86

1) 20 cm² 2) $6\frac{2}{3}$cm 3) 2·25:1; $3\frac{1}{3}$ cm; $4\frac{2}{3}$ cm
5) 20; 45 cm² 6i) $\frac{2}{5}$ ii) $\frac{9}{25}$ iii) $\frac{6}{25}$ 7i) $\frac{1}{4}$ ii) $\frac{1}{16}$
8) HK=4 cm; HB=1·67 cm

Exercise 87

1) $\frac{4}{5}$ 2) $3\frac{1}{3}$ 3) $\frac{3}{1}$ 5) 7·5 cm²
6a) 3 cm b) 24 cm 7) $\frac{5}{8}$; 10 cm²
8) BP=2·5 cm; PQ=7·5 cm

Exercise 88

1) 3·60 cm 2) 6·984 cm 3) 2·247 cm
4) 34·78 cm 5) $x=8{\cdot}602$ cm; $y=8{\cdot}485$ cm
6) 16·97 cm

Exercise 89

4) 74° 6) $\frac{3}{4}$ 8) $\frac{5}{3}$, $\frac{3}{8}$, $\frac{3}{8}$, $\frac{9}{64}$
9a) 5 cm; 4 cm; $\frac{4}{25}$ 11) 3·2 cm 12) 29°
13) 1·8 cm 15) $\frac{3}{4}$, 9 cm², 4 cm²

Self Test 23

1) b, c 2) c 3) d 4) a 5) b 6) b
7) a, b, c 8) d 9) b 10) a 11) c
12) c 13) b, c, d 14) a, c 15) b
16) a, c 17) b 18) b, d 19) d 20) a, d
21) c 22) a, b 23) b 24) a, b, c 25) b
26) d 27) b 28) c 29) a 30) c
31) c 32) a 33) b 34) b 35) d
36) b 37) d

ANSWERS TO CHAPTER 24

Exercise 90

1) 143° 2) 82° 3) 93° 4) $x=39°$; $y=105°$
5) 32° 6) 110° 7) 100°; 20° 9) 7·5 cm

Exercise 91

1a) 6 b) 12 c) 16 d) 20 right-angles
2a) 108° b) 135° c) 144° d) 150°
3) 131° 4) 12 5) 6° 6) 24 7) 7
8) 132°, 75° 9) 72°, 108° 10) 36°, 10
11) $n=30$ 12) 12

Exercise 92

1) 28 cm² 2) 84·84 cm² 3) $6\frac{2}{3}$ cm 4) 12 cm
4) 12 cm 5) 40 cm² 6) 240 cm²
7i) 14 cm² ii) $2\frac{2}{3}$ cm² 8) $5\frac{1}{3}$ cm

Self Test 24

1) d 2) a, b 3) a 4) b 5) b 6) d
7) a 8) b 9) c 10) b 11) a 12) c
13) a, d 14) c

ANSWERS TO CHAPTER 25

Exercise 93

1) 3 cm 2) 12 cm 3) 9·237 cm 4) 6·25 cm
5) 8·33 cm 7) 22·22 cm 8) 46·5 mm
9) 1·5 cm 10) 57·7 mm

Exercise 94

1) 30° 2) 100°, 60° 5) 55°, 62½°, 125°, 117½°
9) 32°, 148° 10) 62°, 59°

Exercise 95

1) 58°, 26° 3) 38°, 33° 5) 24 cm
6) 10 cm, 12 cm 9) 30° 12) $\sqrt{x(x-2r)}$

Self Test 25

1) b 2) d 3) a, c 4) a 5) b, d 6) d
7) c 8) a 9) d 10) b 11) c 12) b
13) d 14) b 15) a

ANSWERS TO CHAPTER 26

Exercise 96

4) 7 cm 6) 11·1 cm 7) 4·16 cm
8) 6·12 cm 12) 6·5 cm 18) 3·3 cm; 4·7 cm

ANSWERS TO CHAPTER 27

Exercise 97

4) 8 5) 4

Exercise 98

4) 4

ANSWERS TO MISCELLANEOUS EXERCISE

Exercise 99

1) 40° 2) — 3) 20° 4) — 5) 10·4 cm
6) 22·36 cm 7) a circle, centre 0, rad. 10 cm
8) 12 9) — 10) — 11) 31·5 cm 12) —
13) 108° 14) $3x°$ 15) 2·46 cm 16) —
17) 3·89 cm 18) — 19) 6 cm 20) 35°
21) — 22) — 23) 132° 24) 8·944 cm
25) — 26) — 27) — 28) 51°
29) 19·2 cm 30) 6·2 cm 31) — 32) 85°
33) 4 cm 34) 2·8 cm 35) 40° 36) 2·4 cm
37) 9·2 cm 38) — 39) 70° 40) 5 cm

ANSWERS TO CHAPTER 28

Exercise 100

1a) 0·5000 b) 0·7071 c) 0·9272
2a) 19° 28′ b) 48° 36′ c) 46° 3′
3a) 0·2079 b) 0·3123 c) 0·9646 d) 0·1285
e) 0·9991 f) 0·0032
4a) 9° b) 66° c) 81° 6′ d) 4° 36′
e) 78° 55′ f) 47° 41′ g) 2° 52′ h) 15° 40′
5a) 3·3808 b) 10·126 c) 25·94
6a) 41° 49′ b) 40° 47′ c) 22° 23′
7) 28·3 cm 8) 0·794 m 9) 21·6 cm
10) 7·47 m 11) 44° 44′, 44° 44′, 90° 32′

Exercise 101

1a) 0·9659 b) 0·9114 c) 0·2011 d) 1·0000
e) 0·2863 f) 0·7663
2a) 24° b) 70° c) 14° 42′ d) 64° 36′
e) 16° 32′ f) 89° 31′ g) 74° 52′ h) 61° 58′
3a) 9·33 cm b) 2·64 m c) 5·29 cm
4a) 60° 42′ b) 69° 20′ c) 53° 19′
5) 66° 7′, 66° 7′, 47° 46′, 3·84 cm 6) 2·876 cm
7) 1·969 cm 8) 91° 56′, 8·75 cm
9) 4·53 m, 2·11 m, 2·40 m, 5·65 m

Exercise 102

1a) 0·3249 b) 0·6346 c) 1·3613 d) 0·8229
e) 0·2004 f) 2·6583
2a) 24° b) 73° c) 4° 24′ d) 21° 41′
e) 19° 38′ f) 39° 34′ g) 62° 33′ h) 0° 56′
3a) 4·35 cm b) 9·289 cm c) 4·43 m
4a) 59° 2′ b) 15° 57′ c) 22° 42′
5) 7·70 cm 6) 2·78 cm 7) 33·31 cm
8) 2·86 m 9) 20·91 cm

Exercise 103

1a) $\bar{1}$·6794 b) $\bar{1}$·9837 c) $\bar{1}$·9958 d) $\bar{1}$·9599
e) $\bar{1}$·9927 f) $\bar{1}$·0940
2a) 57° 2′ b) 22° 25′ c) 75° 7′ d) 11° 42′
e) 80° 2′ f) 88° 24′
3a) 45° 43′ b) 18° 31′ c) 58° 1′
4) 9·285 5) 46° 14′ 6) 17° 42′ 7) 5·817 cm

Exercise 104

1) 0·9483, 0·3346 2) $\frac{3}{5}$, $\frac{4}{5}$ 3) 0·6561, 0·7547
4) $\frac{5}{13}$, $\frac{5}{12}$ 8a) $\frac{3}{4}$ b) $\frac{1}{3}$ c) $\frac{3}{4}$ 9a) ·5696
b) ·2061 c) 2·560 10) 41° 49′ 11) 48° 36′

Exercise 105

1) 11·548 m 2) 9·60 m 3) 50° 58′
4) 109·88 m 5) 189·2 m 6) 74·88 m, 20° 42′
7) 2·972 m 8) 11·77 m
9i) 28·47 m ii) 34·6 m iii) 29° 2′
10) 53·26 m

Exercise 106

1a) 92·62 km b) 151·5 km 2) 43·3 km; 25 km
3) 326° 19′ 4) 11·55 km; 10·774 km
5) 44·4 m, S 64° 16′ E 6) 24·72 km; 338° 51′
7) 24·41 km; 14·00 km 8) 127 km; 49° 34′
9) 11·5 km; N 61° 46′ E 10) 15·23 km; 66° 48′

Exercise 107

1) 3·624 cm 2) 10·84 cm 3) 1·04 cm; 3·46 cm
4) 14·34 cm 5) 15° 20′; 0·57 cm 6) 71·56 cm

Self Test 26

1) a 2) c 3) b 4) b 5) c 6) d
7) c 8) c 9) a 10) a 11) a 12) c
13) c 14) c 15) c 16) b

ANSWERS TO CHAPTER 29

Exercise 108

1a) 0·0854 b) −0·3123 c) −1·835
2) 14° 35′, 165° 25′ 3) 105° 42′, 254° 18′
4a) 54° 40′, 125° 20′ b) 151° 51′ 5) 3·0248
6) tan A $=\frac{12}{5}$ cos A $=\frac{5}{13}$ 7) $-\frac{7}{25}$ 8) $\frac{15}{17}$; $\frac{8}{15}$
9) 34° 40′ or 145° 20′; 28° 40′ or 331° 20′ 10) $\frac{11}{60}$

Exercise 109

1) ∠C=71° b=5·906 cm c=9·986 cm
2) ∠A=48° a=71·52 mm c=84·16 mm
3) ∠B=56° a=3·741 m b=9·528 m
4) ∠A=46° b=13·60 cm c=5·843 cm
5) ∠C=67° a=1·508 m c=2·361 m
6) ∠C=63° 32′ a=9·486 mm b=11·56 mm
7) ∠B=135° 38′ a=9·393 cm c=14·44 cm
8) ∠B=81° 54′ b=9·947 m c=3·609 m
9) ∠A=53° 39′ a=2·124 m b=2·390 m
10) ∠A=13° 51′ ∠B=144° 2′ b=17·16 m
11) ∠A=44° 46′ ∠C=49° 57′ a=10·69 cm
12) ∠B=93° 49′ ∠C=36° 52′ b=30·26 cm
13) ∠B=48° 31′ ∠C=26° 25′ c=4·247 cm
14) 21·75 cm 15) 117·4 mm 16) 11·25 m
17) 9·044 cm 18) 21·78 cm

Exercise 110

1) c=10·15 cm ∠A=50° 11′ ∠B=69° 49′
2) a=11·81 cm ∠B=44° 42′ c=79° 18′
3) b=4·989 m ∠A=82° 24′ ∠C=60° 18′
4) ∠A=38° 12′ ∠B=81° 38′ ∠C=60° 10′
5) ∠A=24° 42′ ∠B=44° 54′ ∠C=110° 24′
6) ∠A=34° 42′ ∠B=18° 6′ ∠C=127° 12′
7) 74° 24′ 8) 42° 53′ 9) 125° 52′ 10) 97° 42′

Self Test 27

1) a 2) b 3) c 4) c 5) c 6) a
7) c 8) b 9) d 10) c 11) a 12) a, c
13) b, d 14) c, d 15) a, c

ANSWERS TO CHAPTER 30

Exercise 111

1i) 70·41 m ii) 18·22 m iii) 11° 46′
2i) 34° 56′ ii) 33° iii) 31° 18′
3a) 17·32 cm b) 60° c) 61° 31′ d) 19·71 cm
4a) 61° 56′ b) 12 000 5i) 55° 9′ ii) 58° 54′
6a) 1524 m b) 43° 57′
7a) 45·18 m b) 927 m²
8a) 3·37 m b) 25° 31′ c) 7·83 m; 10 m

ANSWERS TO CHAPTER 31

Exercise 112

1) 1 km 2) 28 km 3) 500 m 4) 1:500 000
5) 1:10 000 000 6) 80 km 7) 75 cm
8) 1·6 cm 9) 2075 m 10) 5·6 km

Exercise 113

1) 120 000 m² 2) 80 000 m² 3) 1:3000
4) 1:25 000 5) 0·25 6) 700 km² 7) £14 400
8i) 13·41 cm² ii) 538·5 m iii) 134 100 m²
9) 10·4 km² 10) 35·50 hectares

Exercise 114

1) 1 in 25 2) 1 in 40 3) 100 m 4) 30 m
5) 7° 11′ 6) 1 in 4·80 7) 15 m 8) 1200 m

Exercise 115

1) 39° 48′ 2) 1 in 2·848 3) 17° 21′
4) 1103 m 5) 11° 19′ 6) 1 in 8·065

ANSWERS TO CHAPTER 32

Exercise 116

1) 345 km/h; 277° 2) 494 km/h; 096°
3) 395 km/h; 157° 4) 388 km/h; 308°
5) 269 km/h; 035° 6) 55 km/h; 307°
7) 120 km/h from 090° 8) 146 km/h from 290°
9) 154 km/h from 210° 10) 89 km/h from 024°
11) 255 km/h; 086° 12) 410 km/h; 145°
13) 370 km/h; 202° 14) 295 km/h; 300°
15) 250 km/h; 004° 16) 60 km/h; 259° or 341°
17) 156 km/h; 030°; 263° 18) 299°

Exercise 117

1) 44 km/h 2) 294°; 1¾ hours 3) 2 hours; 088°
4i) 69 km/h from 95° ii) 43°
5i) 277° ii) 169° iii) 242 km

Exercise 118

1) 30° upstream; 5·54 min 2) 23° 35′ upstream
3) 13·96 km/h; 086° 50′
4i) 23° 35′ upstream; ii) 32·88 m
5i) 33° 42′ downstream; 41° 44′ upstream
6i) 30 m ii) 3·873 m/s iii) 3·75 m/s
7) 19·7 km/h, 348°; 155°

ANSWERS TO CHAPTER 33

Exercise 119

1) 6115 km 2) 2000 km 3) 4447 km
4) 5516 km 5) 2224 km
6i) 3334 km ii) 94° E 7) 2·12 cm
8) Lat. 22· 42′ N; Long. 35° 13′ W
9) 1556 km; 28° 34′ W
10) 5356 km; 204 km longer
11) 3168 km; i) 3979 km ii) 53° 24′ N
12) 3663 km; 14° S 13) 2875 km; 2818 km
14) 9116 km; 15i) 47° 34′ ii) 4799 km

Self Test 28

1) c 2) d 3) b 4) c, d 5) a 6) d
7) c 8) a 9) c 10) d 11) b 12) b
13) d 14) b 15) c 16) c 17) d
18) c 19) a 20) b 21) c 22) b, d
23) d 24) a

ANSWERS TO MISCELLANEOUS EXERCISE

Exercise 120

1) 6·81 cm 2) 9·714 cm 3) 0·2816 4) 8 cm
5) $\frac{1}{5}$ 6) 49·08 cm 7) $\frac{1}{3}$ 8) $-\frac{24}{25}$
9) −2·9602 10) $50\sqrt{3}$ cm² 11) 50° 12′
12) 77° 10′ 13) 65 m; 30° 31′; 2156 m²; 40 m
14) $\frac{34}{25}$
15a) 10·21 cm; b) 6·610 cm c) 76·76
16) 4·33 cm 17) $\frac{15}{17}, \frac{8}{15}$ 18) 12° 46′
19) 6·889 cm
20i) $\frac{3}{7}$ ii) 6·708 iii) 48° 12′ iv) 45° 35′
21) 23·5 cm
22i) 3·852 ii) 0·3162 iii) 1·778 iv) $\bar{1}$·9174
23) 44·4 m 24) 14° 29′
25i) 5·292 cm; 54° 44′ 53° 16′
26) 34·64 m; 43° 51′; 60°
27a) 77·14 m; 59·1 m; b) 27° 24′
28) 2·052 cm; 24° 1′ 29a) 248 m b) 125° 45′
30) 27·7 m; 785 m²
31a) 815 m b) 2·01 cm to 1 km
32) 10° 7′; 5° 3′ 33) 2424 km 34) 77° 45′
35) 278 km 36) 16·67 hrs; Long. 124° 35′ E
37a) 23° 35′; 6·04 m

ANSWERS TO CHAPTER 34

Exercise 121

1a) {1, 3, 5, 7, 9, 11, 13, 15}
b) {1, 4, 9, 16, 25, 36, 49, 64}
c) {5, 10, 15, 20, 25, 30, 35, 40, 45, 50}
2a) No b) No c) Yes d) Yes e) No
3) No. $A \subset C \subset B$ 4) 16, 14 5) Yes
6a) True b) False c) False d) True
e) True f) True g) False h) True
7a) A′ b) $A \cap B$ c) $A \cup B$ d) $(A \cup B)'$
8) 20 9a) 3 b) 25 c) 7 10) 140
11a) {1, 3, 5, 7, 9, 10} b) {1, 3, 5,} c) {1, 3, 5}
12a) {5, 6} b) {5, 6} c) {3, 4, 5, 6, 7, 8}
d) {3, 4, 7, 8, 9, 10, 11, 12} e) same as (d)
f) ϕ 13) 40 14) 10 15) No, Not known

ANSWERS TO CHAPTER 35

Exercise 122

1a) $\begin{pmatrix} 5 & 3 \\ 7 & 8 \end{pmatrix}$ b) $\begin{pmatrix} 1 & 1 \\ 1 & 2 \end{pmatrix}$ c) $\begin{pmatrix} 12 & 9 \\ 23 & 19 \end{pmatrix}$
d) $\begin{pmatrix} 10 & 9 \\ 21 & 21 \end{pmatrix}$ e) $\begin{pmatrix} 3 & 4 \\ 2 & 5 \end{pmatrix}$ f) $\begin{pmatrix} 2 & 3 \\ 1 & 3 \end{pmatrix}$
g) $\begin{pmatrix} \frac{5}{7} & -\frac{2}{7} \\ -\frac{4}{7} & \frac{3}{7} \end{pmatrix}$ h) $\begin{pmatrix} 1 & -\frac{1}{3} \\ -1 & \frac{2}{3} \end{pmatrix}$
2) $k=0·5$ 3) $a=2, b=3$ 4) $\begin{pmatrix} 5 & 11 \\ 11 & 34 \end{pmatrix}$
5a) $x=1, y=2$ b) $x=3, y=4$ c) $x=2, y=1$

Exercise 123

1) $R=\begin{pmatrix} 4 \\ 6 \end{pmatrix}$; $\sqrt{52}$ 2) $R=\begin{pmatrix} 6 \\ 8 \end{pmatrix}$ 3) $\sqrt{20}$
4) (0, 1), (2, 5), (2, 7), (0, 3) 5) (12, 2)
6) (2, −1), (5, −1), (5, −4)
7) (−1, 1), (−3, 1), (−3, 4), (−1, 4)
8) (−1·598, 3·232) 9) $\begin{pmatrix} 9 & 4 \\ 2 & 1 \end{pmatrix}$ 10) $\begin{pmatrix} 1 & 4 \\ 0 & -1 \end{pmatrix}$

ANSWERS TO CHAPTER 36

Exercise 124

1a) 214_{10} b) 69_{10} c) 27_{10} 2) $101\,000_2$
3) $1\,010\,001_2$ 4) 1011_3; 31 5) 1151_8
6) 4422_8 7) 1111_2 8) $10\,101\,001_2$
9) 7130_8 10a) $101\,100\,110_2$ b) $111\,021_3$
c) 546_8 11) 1032_8 12) 10010_2
13) $11\,110_3$ 14) 120_8

Exercise 125

1a) 1	b) 2	c) 3	d) 4	e) 4	f) 3
g) 2	h) 4				
2a) 5	b) 1	c) 9	d) 7	e) 1	f) 5
g) 6	h) 7				
3a) 6	b) 6	c) 6	d) 0		

4) 6, 10, 12, 15, 20, 30, 60
5a) 4 b) 4 c) 4
6a) 2 b) 5 c) 4 d) 2 and 6 e) 2 and 8

ANSWERS TO CHAPTER 37

Exercise 126

3)

Mark	50–54	55–59	60–64	65–69	70–74
Frequency	1	2	11	10	12

Mark	75–79	80–84	85–89	90–94	95–99
Frequency	21	6	9	4	4

4)

Length	24·95	24·96	24·97	24·98	24·99
Frequency	1	1	4	4	5

Length	25·00	25·01	25·02	25·03	25·04
Frequency	5	7	6	4	2

Length	25·05
Frequency	1

5) 15 6) 29·89 cm 7) 723·3 hours
8a) 5 b) 8·5 c) 8 9) 47 10) £22·4
11) 5, 2·5 12) £22·5

ANSWERS TO MISCELLANEOUS EXERCISE

Exercise 127

1) $\begin{pmatrix} 0{\cdot}4 & 0{\cdot}1 \\ -0{\cdot}2 & 0{\cdot}2 \end{pmatrix}$ 2) 74_8 3) $x=4, y=-2$
4 1) {1, 2, 3, 5, 7, 8, 9, 11, 13} 2) {7, 9, 11}
5 1) 3 2) 1·8
6) 235 7) 63
8) $\begin{pmatrix} -1 & 6 \\ -1 & 1 \end{pmatrix}$ 9) 4
10) $x=2, y=5$
11a) $\begin{pmatrix} 5 & 0 \\ 0 & 5 \end{pmatrix}; \begin{pmatrix} 5 & 0 \\ 0 & 5 \end{pmatrix}$ b) $\frac{1}{5}\begin{pmatrix} 7 & 4 \\ -8 & -1 \end{pmatrix}$
12a) (0·7071, 2·1213) b) (0·7071, −2·1213)
13a) {1, 2, 5, 7, 9} b) ϕ c) {9} d) {2, 7, 8, 9}
14a) 5 b) 3 c) 8 d) 6
15) 1·000 kg
16) $x=6$ 17) {1}
18a) 10·4 b) 10·35
19) 9 20) 394_{10}

Self Test 29

1) b 2) c 3) d 4) d 5) a 6) d
7) c 8) b 9) a 10) a 11) c 12) c
13) c 14) b 15) a 16) c 17) a

ANSWERS TO CHAPTER 38

Exercise 128

1) 28 2) 47 3) −4 4) 117 5) 16
6) 12, 17, 22, 27, 32 and 37 7) 215 8) 702
9) 490 10) 10

Exercise 129

1) 640, 40 960 2) 12 288 3) 2·144
4) 5·091 5) 80·76 6) 65·98
7) 10·76, 12·59, 14·73, 17·24, 20·17 and 23·60
8) 30·78 9) 215·4 10) 495·8

ANSWERS TO CHAPTER 39

Exercise 130

1) £5287 2) £13 750 3) £40·96 4) £86·10
5) £400 6) 4 years 7) £800 8) 5 years

Exercise 131

1) 723 2) 523 3) −1 5) $p=q=1$

ANSWERS TO MISCELLANEOUS EXERCISE

Exercise 132

1) 196 608 2) 62·5 3) $r=2{\cdot}5$; $a=\frac{14}{33}$
4a) £105 b) £110.25 5) £500 6) 1
7) 4390 8) $r=1{\cdot}5$; 10·125

Self Test 30

1) b 2) c 3) a 5) d 6) c
7) a 8) a 9) c 10) c 11) b
12) b

ANSWERS TO CHAPTER 40

Exercise 133

1a) £1256 b) £351 c) £165·20
2a) £174·60 b) £304·50 c) £573·30
3a) £24 b) £81 c) £637·50
4a) £22·50 b) £125 c) £96
5a) 10·8% b) 2·35% c) 6·12%
6) £80; 10·67% 7) £738; £63; 8·54% 8) 32p
9) £1125 10) 4·4% 11) 46·4% 12) £11 800

Exercise 134

1a) £400 b) £362·31 c) £563·33
2a) £28 b) £60 c) £20
3) £681·81; £40·90; 6·81% 4) £240 5) £300
6) £218·18 7) £90 8) £4444·44; £66·67
9) £29·57 less 10a) £1515·38 b) £2561·57
c) £78

Exercise 135

1a) 40p in the £1 b) 9·12p in the £1
c) 33·57p in the £1
2a) £2100 b) £510·24 c) £2978·82
3) 44·67p in the £1 4) 64·24p in the £1
5) £9047

Self test 31

1) d 2) b 3) a 4) b 5) a 6) c
7) c 8) b

Index